Trends in Logic

Volume 64

TRENDS IN LOGIC
Studia Logica Library

VOLUME 64

The book series Trends in Logic covers essentially the same areas as the journal Studia Logica, that is, contemporary formal logic and its applications and relations to other disciplines. The series aims at publishing monographs and thematically coherent volumes dealing with important developments in logic and presenting significant contributions to logical research.

Volumes of Trends in Logic may range from highly focused studies to presentations that make a subject accessible to a broader scientific community or offer new perspectives for research. The series is open to contributions devoted to topics ranging from algebraic logic, model theory, proof theory, philosophical logic, non-classical logic, and logic in computer science to mathematical linguistics and formal epistemology. This thematic spectrum is also reflected in the editorial board of Trends in Logic. Volumes may be devoted to specific logical systems, particular methods and techniques, fundamental concepts, challenging open problems, different approaches to logical consequence, combinations of logics, classes of algebras or other structures, or interconnections between various logic-related domains.

This book series is indexed in SCOPUS.

Authors interested in proposing a completed book or a manuscript in progress or in conception can contact either chris.wilby@springer.com or one of the Editors of the Series.

Dorota Leszczyńska-Jasion

The Method of Socratic Proofs

From the Logic of Questions to Proof Theory

Springer

Dorota Leszczyńska-Jasion
Adam Mickiewicz University
Poznań, Poland

ISSN 1572-6126 ISSN 2212-7313 (electronic)
Trends in Logic
ISBN 978-3-031-82453-1 ISBN 978-3-031-82451-7 (eBook)
https://doi.org/10.1007/978-3-031-82451-7

This Springer imprint is published by the registered company Springer Nature Switzerland AG
The registered company address is: Gewerbestrasse 11, 6330 Cham, Switzerland

Preface

If it is true that questions are valuable
because they lead to judgments,
it may also be true that judgments are valuable
because they lead to inquiries.
—Felix S. Cohen [1, p. 351]

Conducting a formal proof is a particular example of scientific activity—a final act of justification of a truth found in the process of inquiry. A good proof is much like a good poem—to capture the essential nature of things is sometimes more important than formal correctness, and often the shorter the phrase, the better.

On the other hand, the beginning of an inquiry is a problem to be solved, and these are expressed by questions, hence in a conceptual order, questions must naturally precede proofs. Also in time order, inquiry precedes justification, and if the traditional proofs are formal reports of justifications, Socratic transformations and Socratic proofs—structures being the main subject of this monograph—are formal reports of the process of inquiry.

The considerations presented in this monograph are rooted in an assumption that questions can be, and should be, treated seriously as proof-theoretical objects not reducible to other objects. We seem to assume that the importance of questions lies in the fact that they lead to solutions. However, paraphrasing Cohen: it may also be true that the importance of proofs lies, at least partially, in the fact that they lead to further transformations of questions.

There are formal tools that allow for the modeling of both semantic and proof-theoretical relations between questions: the method of Socratic proofs and the relation of erotetic implication, both defined within the paradigm of Inferential Erotetic Logic created by Andrzej Wiśniewski. The main goal of this monograph is to develop a comprehensive understanding of these tools, which are applicable to classical logic—both propositional and first-order—as well as some normal modal propositional logics and intuitionistic propositional logic.

The Content of This Book

The content of this book evolved from another monograph published by the author in 2018 under the title *From Questions to Proofs: Between the Logic of Questions and Proof Theory* [2]. Its primary aim was to provide a systematic presentation of research at the junction of the logic of questions and proof theory, with special attention focused on proof-theoretical tools of Inferential Erotetic Logic. Many technical aspects, as the author thought at the time, of erotetic calculi were omitted, since they could be found elsewhere. It soon appeared, however, that there was a need for a more systematic account of the method of Socratic proofs together with the completeness part presented in a possibly uniform way for all erotetic calculi considered. This opened the issue of the structural analysis presented here.

Chapter 1 of this monograph is an extended and updated version of Chap. 1 of [2]. The content of the Chap. 2 of [2] was largely extended and finally divided into two chapters in the present monograph, treating separately classical logic and basic normal modal propositional logics. Numerous new results were added here—except from the adjustment of consistency properties to erotetic calculi, followed by structural analysis, we present erotetic calculi capturing *global* entailment in propositional modal logics. Chapter 4, treating intuitionistic propositional logic, has never been published before.

Chapters 5–7 correspond to Chaps. 3 and 4 of [2]. Again, their content was revised and supplemented with the analysis of intuitionistic logic in Chap. 7.

How to Read This Book?

The initial parts of Chap. 1, up to Sect. 1.2, were meant as a very general introduction to the logics of questions with a special focus on proof-theoretical findings within these logics. The remaning part of Chap. 1 introduces crumbs of formal notation, yet still no specific background is needed to read the whole chapter. Chapters 2–4 require some background knowledge in logic; nevertheless, they can be studied as a more formal introduction to the method of Socratic proofs. Further chapters are more technical as they are addressed to specialists in the field.

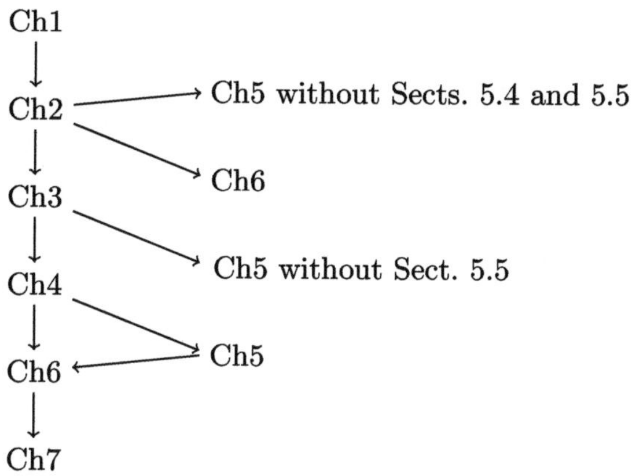

The above figure presents our recommendation for how to read this book. If one is interested only in classical logic, one can go from Chaps. 1 and 2 directly to the parts of Chap. 5 devoted to the erotetic issues in the classical setting, or to Chap. 6 concerning possible translations between proof methods. A path through Chaps. 3 and 5 concerns modal logics. One can also skip Chap. 5, concerning erotetic issues, if the proof-theoretic aspects are of greater importance.

Poznań, Poland — Dorota Leszczyńska-Jasion

References

1. Cohen, F. S. (1929). What is a question? *The Monist*, *39*(3), 350–364.
2. Leszczyńska-Jasion, D. (2018). *From questions to proofs. Between the logic of questions and proof theory*. Faculty of Social Science Publishers, Adam Mickiewicz University.

Acknowledgments

This monograph was prepared with the financial support of the National Science Center, Poland, grant no. 2017/26/E/HS1/00127.

I would like to thank the following people for helping me by reading and commenting on fragments of this monograph, or by kindly maintaining my enthusiasm to complete it: Szymon Chlebowski, Andrzej Indrzejczak, Paweł Łupkowski, Jerzy Pogonowski, Agata Tomczyk, and Andrzej Wiśniewski. My particular gratitude and respect go to Jerzy Pogonowski, who passed away before this book was published.

I am also indebted to the dean of the Faculty of Psychology and Cognitive Science of Adam Mickiewicz University, Mariusz Urbański, for granting me scientific leave, which allowed me to scatter and collect the necessary thoughts.

I would also like to thank two anonymous reviewers whose comments significantly contributed to the final revisions.

As always, my deepest gratitude goes to my children.

Contents

Symbols and Abbreviations

IEL	Inferential Erotetic Logic
IS	Inquisitive Semantics
CPL	Classical Propositional Logic
FOL	First-Order Logic
INT	Propositional Intuitionistic Logic
MiES	Minimal Erotetic Semantics

Languages

$\mathcal{L}_{\mathsf{CPL}}$	Language of CPL
$\mathcal{L}_{\mathsf{FOL}}$	Language of FOL
$\mathcal{L}^{?}_{\vdash\mathsf{L}}$	Language with sequents and questions built upon $\mathcal{L}_{\mathsf{L}}$, $\mathsf{L} \in \{\mathsf{CPL}, \mathsf{FOL}\}$
$\mathcal{M}$	Language of propositional modal logics
$\mathcal{M}^{?}_{\vdash}$	Language with sequents and questions built upon $\mathcal{M}$
$\mathcal{L}_{\mathsf{INT}}$	Language of INT
$\mathcal{L}^{?}_{\vdash\mathsf{INT}}$	Language with sequents and questions built upon $\mathcal{L}_{\mathsf{INT}}$
$\mathcal{L}^{?}_{\mathsf{CPL}}$	Language with questions built upon $\mathcal{L}_{\mathsf{CPL}}$
$\mathcal{L}^{?}_{\dashv\mathsf{L}}$	Language with reversed sequents and questions built upon $\mathcal{L}_{\mathsf{CPL}}$
$\mathcal{M}^{imp}_{\vdash}$	Language with sequents built upon $\mathcal{M}$
$\mathcal{L}^{imp}_{\vdash\mathsf{INT}}$	Language with sequents built upon $\mathcal{L}_{\mathsf{INT}}$

Types of Formulas

$-A$	Complement of formula A
α	Formulas of conjunctive types
β	Formulas of disjunctive type

γ Universal formulas
δ Existential formulas
κ Formulas with double negation or empty quantification
ν Necessity-type formulas
π Possibility-type formulas

Erotetic Calculi

$\mathbb{E}^{\mathbf{Q}}$ Erotetic calculus for FOL with both-sided sequents
$\mathbb{E}^{\mathbf{P}}$ Erotetic calculus for CPL with both-sided sequents
$\mathbb{E}^{\mathbf{RQ}}$ Erotetic calculus for FOL with right-sided sequents
$\mathbb{E}^{\mathbf{RP}}$ Erotetic calculus for CPL with right-sided sequents
$\mathbb{E}^{\mathbf{LQ}}$ Erotetic calculus for FOL with left-sided sequents
$\mathbb{E}^{\mathbf{LP}}$ Erotetic calculus for CPL with left-sided sequents
$\mathbb{E}^{\mathsf{L}}$ Erotetic calculi for propositional modal logics L $\in$ {K, D, T, KB, DB, B, K4, D4, S4, KB4, S5, K5, D5, K45, D45}
$\mathbb{E}^{\mathsf{INT}}$ Erotetic calculus for INT
$\mathbb{E}^{\mathsf{CPL}}_{res}$ Dual erotetic calculus for CPL
$\mathbb{E}^{\mathsf{K}\tau}$ Erotetic calculus for modal logic K corresponding to $\mathbb{G}^{\mathsf{K}\tau}$
$\mathbb{E}^{D\tau}$ Erotetic calculus for modal logic D corresponding to $\mathbb{G}^{\mathsf{D}\tau}$
$\mathbb{E}^{\mathsf{T}\tau}$ Erotetic calculus for modal logic T corresponding to $\mathbb{G}^{\mathsf{T}\tau}$
$\mathbb{E}^{\mathsf{K4}\tau}$ Erotetic calculus for modal logic K4 corresponding to $\mathbb{G}^{\mathsf{K4}\tau}$
$\mathbb{E}^{\mathsf{D4}\tau}$ Erotetic calculus for modal logic D4 corresponding to $\mathbb{G}^{\mathsf{D4}\tau}$
$\mathbb{E}^{\mathsf{S4}\tau}$ Erotetic calculus for modal logic S4 corresponding to $\mathbb{G}^{\mathsf{S4}\tau}$

Sequent Calculi

$\mathbb{G}^{\mathsf{Q}}$ Sequent calculus for FOL
$\mathbb{G}^{\mathsf{K}\tau}$ Sequent calculus for modal logic K
$\mathbb{G}^{\mathsf{D}\tau}$ Sequent calculus for modal logic D
$\mathbb{G}^{\mathsf{T}\tau}$ Sequent calculus for modal logic T
$\mathbb{G}^{\mathsf{K4}\tau}$ Sequent calculus for modal logic K4
$\mathbb{G}^{\mathsf{D4}\tau}$ Sequent calculus for modal logic D4
$\mathbb{G}^{\mathsf{S4}\tau}$ Sequent calculus for modal logic S4
$\mathbb{G}^{\mathsf{INT}\tau}$ Sequent calculus for INT

Some Entailment and Derivability Relations

$\Vert\!\models$	Multiple-conclusion entailment
$\mathsf{Im}(Q, X, Q^*)$	Erotetic implication
$\vdash_{\mathsf{L}}$	Consequence in $\mathsf{L} \in \{\mathsf{CPL}, \mathsf{FOL}\}$
$\vdash_{\mathsf{L}}^{bc}$	By-cases-consequence in $\mathsf{L} \in \{\mathsf{CPL}, \mathsf{FOL}\}$
$\Vert\!\vdash_{\mathsf{L}}$	Multiple-conclusion consequence in $\mathsf{L} \in \{\mathsf{CPL}, \mathsf{FOL}\}$

Algorithms

Procedures

Chapter 1
The Logics of Questions: An Overview

Our interest in the questioning process is a philosophical enterprise at least since Socrates. Today, the logic of questions is a branch of philosophical logic that aims at formal models of questioning and the way questions play their roles in natural languages. The discipline is often called 'erotetic logic' from the Greek word '$\grave{\varepsilon}\rho\omega\tau\varepsilon\mu\alpha$' (erotema), which means 'question'.[1]

On the other hand, proof theory is a well-established discipline studying mathematical proofs. In particular, proof theory models the process of argumentation, especially of proving, by means of formal systems in which proofs are constructed.

Finally, the relation between questions and proofs has not yet been thoroughly analyzed, although it is one of deep philosophical and cognitive significance. Despite the fact that both disciplines: that is, the logic of questions and proof theory, have a long-standing tradition and numerous widely recognized results, the idea of working strictly in the two areas seems relatively novel. The main task of this monograph is to explore this area. The main formal tool used is the method of Socratic proofs which will be briefly characterized later in this chapter, and systematically studied in the following ones. The aim of this chapter is introductory and almost no formalism is introduced here.

Proofs in the Logic of Questions, Questions in Proof Theory

A lot of time has passed since Belnap and Steel wrote about erotetic logic:

> Absolutely the wrong thing is to think it is a logic in the sense of a deductive system, since one would thus be driven to the pointless task of inventing an inferential scheme in which questions, or interrogatives, could serve as premises and conclusions. [7, p. 1]

---

[1] The first to use this term were the Priors [99].

D. Leszczyńska-Jasion, *The Method of Socratic Proofs*, Trends in Logic 64,
https://doi.org/10.1007/978-3-031-82451-7_1

The passage is nowadays recognized as evidence of a deeply missed attitude towards the analysis of questions.[2] Questions are now full citizens of the world of reasoning; many logicians examine and model reasoning in which questions occur as premises and/or conclusions. Following well-grounded achievements of proof theory we can build deductive systems including rules operating on questions, and we can examine structures containing questions, just as we examine proofs as such. Hence the logic of questions can gain significantly by borrowing proof-theoretical tools and techniques. But it also works the other way around, as witnessed by Felix S. Cohen cited in the Preface. The fact that we can examine proofs as final results of a cognitive process triggered by posing questions cannot be overestimated.

There are mutual influences between the two disciplines: the logic of questions and proof theory, and mutual benefits to be gained by working in cooperation. We believe that this point is illustrated by the content of this monograph. Before exploring the use of the method of Socratic proofs, in this chapter we draw a panorama of research conducted at the junction of the two disciplines and we place the issues undertaken in this book within that context.

To start with, we must say that there are very important issues in the logic of questions that will *not* be discussed in this book, or at least—not in detail; the most famous being, probably, the structure of questions and types of questions, the answerhood property, types of answers, and presuppositions of questions; but also operations on questions, effectiveness and completeness of question-answer systems, and many others. The reader can find many discussions of these topics elsewhere: [7, 36, 42, 47, 128, 136], there are also well-written articles available in online encyclopedias,[3] like [26, 46]. However, the issues listed at the beginning of this paragraph started long lasting discussions, creating the necessity to take a stance towards the solutions accepted by others. Some choices are inevitable, so, below, we explain those made when writing this book.

Interrogatives and Questions, and *What Is a Question?*

Most logical theories of questions distinguish between *interrogative sentences* and *questions*. Usually, the former are understood as expressions of a language and the latter as their meanings, just as *propositions* are assumed to be the meanings of *statements*. Moreover, since the influential paper by Hamblin [44] and the book by Belnap and Steel [7] the standard has been to explicate the meaning of an interrogative in terms of its set of possible answers.

---

[2] See [128, pp. 26–27], [15, p. 352], [16, p. 118]. Incidentally, the fact that numerous authors quote the passage to point out the authors' of [7] error highlights the great influence of the latter book.

[3] The reader should be careful here—the focus of the entry [26] on the inquisitive semantics approach seems disproportionate with respect to the other frameworks. Similarly, we found the article [101] completely silent about the framework of Inferential Erotetic Logic, although it explicitly describes 'proposition-set theory' as an alternative with respect to inquisitive-semantic theories of questions. See also the rest of this chapter.

The question 'what is a question' was raised by Cohen in 1929 in [21], probably the first paper tackling the topic of questions by means of formal tools of modern logic.[4] The paper does not concern the syntactic issue—the identity of the interrogative—but the meaning, hence it undertakes an erotetic counterpart of the philosophical question 'What is a proposition?'.

Some logicians and philosophers claim that questions cannot, or should not, be treated as separate entities, and that they are totally reducible to the meanings of other sentences. We may call this standpoint an **anti-realistic** one, by an analogy to the great philosophical debates. This standpoint is represented by, *i.a.*, the epistemic-imperative trend, which will be described later in this chapter. At the moment, let us consider, by way of an example, Jeffreys' standpoint expressed in his *Theory of Probability*.[5] Jeffreys says that the question 'Is Mr. Smith at home?' can be reformulated as the following three statements expressing the entire content of the question:

1. 'I do not know whether Mr. Smith is at home.'
2. 'I want to know whether Mr. Smith is at home.'
3. 'I believe you know whether Mr. Smith is at home.'

Some other logicians and philosophers disagree with this claim and tend to construct a new logical apparatus capable of describing the *real* nature of questions. The early representatives of this **realistic** view were, for example, F. Cohen, already mentioned, and Charles Hamblin. We will cite their main pro-realism arguments, as the arguments correspond with and complement the general assumption taken in this monograph, namely, that questions can be, and should be, treated seriously as proof-theoretical objects not reducible to other objects.

The main argument by Cohen [21] is that an interrogative is a propositional function, grammatically resembling a sentence; however, treating it as a sentence leads to serious confusions. Normally, Cohen says, we understand '$\sin^2 x + \cos^2 x = 1$' as a generally quantified proposition, and '$\sin^2 x = 1$' as a question about the range of values of x. We can, Cohen admits, interpret the former as an indeterminate question about the value of x, and the latter as a false universal proposition, but we would not take it as a natural interpretation. The example clearly shows that sentences of similar forms can express propositions or questions. Severe problems can emerge when logicians confuse the two different meanings of sentences.[6]

Later, the idea that an interrogative is a sentential function with an interrogative operator became common in the literature of the subject. Kubiński's works constitute a good example in this aspect. In the last decades, however, semantic ideas (e.g. inquisitive semantics) became more powerful.

The main argument against reducing questions to propositions, as put forth by Hamblin [44], is straightforward. Hamblin asserts that describing the meaning of a

---

[4] But see also [1, 2]: short notes referring research by Kazimierz Ajdukiewicz in the twenties of XX century, where the use of formal tools in the analysis of interrogative sentences is explicitly mentioned.

[5] We refer to [59] after [44] where Hamblin argues against the standpoint expressed by Jeffreys.

[6] The argument appears in [21] in the context of a critical discussion with the authors of *Principia Mathematica*, a topic vivid in the time.

question in terms of desires or demands confuses the *content* of a question with the *context* in which the question was posed. Sentences 1–3 proposed by Jeffreys describe the context of asking question 'Is Mr. Smith at home?' instead of the content of the question itself. Hamblin points out that Jeffreys would not make a similar mistake in the case of statements—he would not confuse p with '*I believe p*'—and he concludes that the reason of this conceptual confusion is that questions are denied a logic.

In the literature on the subject, there have long been other types of arguments directed against the idea of reducing interrogative sentences to expressions of other syntactic categories. For instance, Koj [61] argued that interrogative sentences do not have adequate paraphrases when formulated exclusively using expressions of other syntactic categories. It follows that erotetic logic needs to employ terms that are not definable elsewhere. This argument serves as a strong counter to the notion of reducing questions (the meanings of interrogative sentences) to the meanings of other expressions.

Additional support for our realistic approach to questions comes from the emerging field of 'zetetic epistemology'. The term 'zetetic', introduced by Friedman [30], comes from the Greek word '$\zeta\eta\tau\grave{\epsilon}\omega$' (zeteo), meaning to 'seek for' or 'inquire after'. In her earlier works [27–29], Friedman argues for the need to analyze *interrogative attitudes*, which have questions as their content. These interrogative attitudes, also called *zetetic attitudes*, are attitudes of an agent *towards questions*, and are expressed by words like *asks*, *inquires*, *investigates*, or *wonders*.

In her recent publication [31], Friedman formulates the 'zetetic entanglement thesis', which asserts that traditional doxastic[7] attitudes (such as knowing, believing, conceiving, doubting) are best understood in reference to their role in the process of inquiry.

That said, let us add that one can be less assertive about the nature of questions, but still more apt to use the tools that almost obtrude themselves when adopting the anti-realistic position. The apparatus that has already proved useful in the analysis of other types of expressions can be used to analyze the properties of interrogatives, their relations to answers, relations between interrogatives, and so on. The tools of syntax analysis can be used successfully, even if we do not decide whether the meanings of interrogatives can be reduced to the meanings of other sentences or not. This can be the starting point for the perspective developed in Tadeusz Kubiński's works [65–67], and it may lead to what Andrzej Wiśniewski calls **semi-reductionistic** approach (see [135, Chap. 2], especially Sect. 2.3).

Like any philosophical standpoint, those listed above must be eventually evaluated with respect to their consistency and productiveness. This book follows the semi-reductionistic approach, though, as the reader probably can see, in this battle the heart of the writer is on the realists' side. Since the rest of this book is on formal logic rather than philosophy, and since the main standpoints have been settled, from now on, in most cases we will ignore the distinction between interrogatives and questions and speak of questions as expressions of language.

---

[7] Another term derived from the Greek word: '$\delta o\xi\alpha$', meaning 'belief'.

Types of Questions and Answers

Types of questions constitute another important issue, addressed by logicians, linguists and philosophers on the grounds of many diverse theories. To give at least some examples, wh-questions and polar questions belong to frequently used and analyzed categories. *Polar questions*, also called *yes-no questions*, are questions that have two mutually exclusive direct answers exhausting all possibilities: the affirmative one, and the negative one, usually summarized as simple '*Yes.*' and '*No.*', where the particles are sentence equivalents. *Wh-questions* are questions built using *variables* of everyday speech such as *who*, *which*, *what*, *why*, and so on. As a matter of fact, this book will not reach for this wealth of different forms, since polar questions will be sufficient for our considerations.

We will use the notion of *direct answer to a question*. As put by Harrah, a direct answer 'gives exactly what the question calls for', it is logically sufficient and immediate (see [47, p. 1]). This notion is distinguished from that of a *partial answer* 'which may give some (but perhaps not all) of what the question calls for' (op.cit.). In the formal framework presented here, direct answers are statements; no other kinds of possible replies to questions are modeled in this book.[8] This attitude is fully compatible with the first of the so-called Hamblin's postulates [44, pp. 162–163]:

Hamblin's Postulates

1. An answer to a question is a statement.
2. Knowing what counts as an answer is equivalent to knowing the question.
3. The possible answers to a question are an exhaustive set of mutually exclusive possibilities.

As with many other writers, we follow the second postulate in adopting the standard that—no matter reducible or not—the meaning of an interrogative (question) should be explicated in terms of the set of its possible answers. It is also worth mentioning another formulation of Hamblin's third postulate [44, p. 166]:

3.' A question is equivalent to a decomposition (or section, or division) of the possible universes.

The postulate gave rise to, *i.a.*, the idea of partition semantics. In partition semantics, the meaning of a question is a function assigning, to each possible world, a true and exhaustive answer to that question in that world, and the set of all such propositions must form a *partition* of the logical space. Partition semantics was a predecessor of the rich inquisitive semantics framework. It was proposed in [41] and further developed in a dynamic setting *i.a.* in [38], see also [18, Chap. 9].

Contrary to some popular and influential frameworks, such as those established in [7] or the inquisitive semantics framework, in this book we do not assign a logical

---

[8] Although it is perfectly possible to do so. As far as the paradigm of Inferential Erotetic Logic is concerned, the reader will find a report on empirical research on answering a query with a query in [37, 49, 85, 86, 89] present extensions of this work.

value to a question. Answers to questions can be true or false, whereas questions are *truly answerable*, or not. This descriptive term was also used by Belnap [7]. In contemporary literature questions are usually described as *sound* or *unsound*. A sound question is one which has a true direct answer, while an unsound question is one which has no true direct answer, such as:

- *Is Ann in the bedroom or in the living room?*

asked in a situation where Ann is in the kitchen. Usually, an unsound question is one with a false *presupposition*, as in the case of the following one—question children might ask while trying to grasp The Whole:

- *What is the largest number?*

Presuppositions

The very notion of presupposition comes from Strawson [107], but it was mainly Belnap who adjusted it to the realm of erotetic logic. The Strawsonian notion applies to statements as follows: that a statement *A presupposes B* (equivalently, *B is a presupposition of A*) means that the truth of B is necessary for A to have a truth value, that is, to be true or false at all. A statement with a false presupposition in the Strawsonian sense lacks a truth value. According to Belnap, that a question Q *presupposes B* means that the truth of B is necessary for Q to be sound.[9] A question with a false presupposition in the Belnapian sense has only false direct answers. This is the case with the two above questions. The first question falsely presupposes that Ann is in one of the two mentioned rooms, hence both of its direct answers: '*Ann is in the bedroom.*' and '*Ann is in the living room.*', are false. So is the sentence '*n is the largest natural number*', no matter what we put for '*n*' (we have made here the bold assumption that children have *natural* numbers in mind).

In a formal account, the notion of soundness is relativized to some sort of a model, just like the notion of truth is. Moreover, a question which is sound in every possible model is called *safe*; otherwise, it is *risky*. To give an example, the following question, asked with an indication of a colorful surface:

- Is it blue?

is safe because the surface must either be blue or not blue. In contrast, another question is risky:

- Who won the presidential election held on May 10, 2020 in Poland?

The false presupposition in this case is that someone won the election.[10]

---

[9] Technically, a presupposition of a question is usually defined as a declarative sentence that follows from *every* direct answer to the question. See also Sect. 5.2 in this monograph for a formal account.

[10] The election had been scheduled on May 10, 2020 before the spread of coronavirus in Europe. Officially, the election was never canceled but in fact—no election took place that day in Poland.

Set-of-Answers Methodology

As explained above, we follow the *semi-reductionistic* approach to questions. It is in line with this approach to construct questions as expressions falling under the scheme '?Θ', where 'Θ' is an expression, by and large, indicating the set of direct answers to the question.[11] In the contexts developed in this book it means that questions are identified with expressions of a formal language of the following form:

$$?\{A_1, \ldots, A_n\}$$

where $A_1, \ldots, A_n$, $n > 1$, are all the direct answers to the question and are pairwise syntactically distinct. In the literature, the identification of questions with expressions *displaying* their sets of direct answers is called 'set-of-answers methodology'. Expressing this in the language of the 'realist *vs* anti-realist' dispute, the set-of-answers methodology identifies an interrogative with an expression of a certain form and claims nothing about the nature of its meaning, thus postponing the dispute itself.

The described approach is often viewed as a methodological explication of the second postulate by Hamblin. Its main advantage is that it opens the possibility of using the semantic apparatus defined for the statements in defining relations between questions, and the tools allow for a giant step towards formalization.

Finally, the set-of-answers methodology does not preclude infinite sets of answers. Questions with infinitely many answers are considered, e.g., in [128]. However, the limitation to questions with finitely many answers is justified by the aims and range of considerations of this book.

1.1 Inferential Erotetic Logic

The method of Socratic proofs emerged within the framework of Inferential Erotetic Logic (let us use 'IEL' from now on), a logic created and developed by Andrzej Wiśniewski in the late 1980s and 1990s. From the very beginning, IEL was developed as an alternative to the 'received view', that is, mainly the one developed by Belnap and Steel [7]. On the other hand, the ideas of IEL were developed on the grounds of the Polish tradition of the logic of questions whose main representative at that time was Tadeusz Kubiński. His paper [65], published in 1960, already contains an exposition of important ideas that influenced the development of the logic of questions in Poland.

This chapter was not meant as a historical survey, but we cannot overlook the contributions of the Polish logicians who built this tradition. These include Ajdukiewicz [1, 2], Hiż [58], Kubiński [65–67], Koj [62], Leszko [79], and Andrzej Wiśniewski.

[11] See [135, p. 17] for a more precise account.

Readers can find an exposition of this part of logic-in-Poland history, for example, in: Sect. 1.6 of [128] or in [145]; the book [128] also provides an exhaustive reference section.

It is widely acknowledged that the book 'The Logic of Questions and Answers' [7], written by Belnap and Steel and published in 1976, was the first monograph to develop a logical account of questions. However, Tadeusz Kubiński had already developed such an account in his book 'Wstęp do logicznej teorii pytań' [66], which was published earlier, in 1971. Since the latter was written in Polish, and its English-language version, 'An Outline of the Logical Theory of Questions' [67], was not published until 1980, Belnap's account had already become a standard by then.

The book by Belnap and Steel represents a paradigm in the logic of questions which was vital in the eighties. As we have indicated above, according to this paradigm *it is not* the aim of the logic of questions to work on inferential systems. There is no deductive system to be searched for, since questions do not occur in our everyday reasoning as premises and conclusions—this is the standpoint of Belnap and Steel which was definitely rejected by Wiśniewski. According to Wiśniewski, there are important cognitive processes with questions involved, and the primary goal of IEL is their logical analysis. More precisely, IEL points out different types of erotetic reasoning, that is, reasoning in which questions play the role of premises and/or conclusion, and explicates criteria of validity of such reasoning. The main notion introduced for this purpose, which will be defined and used also in this monograph, is that of *erotetic implication*.[12]

There are two conditions defining the notion of erotetic implication. First, if question Q_1 erotetically implies question Q_2, then the soundness of the first question, Q_1, warrants the soundness of the second question, Q_2. This *transmission of soundness* captures the intuition that if Q_1 is *well posed* in a given situation, then so should be Q_2. Or in other words, if in a given situation it is reasonable to ask Q_1, then it is also reasonable to ask Q_2. For example, *if* question:

- *Is Ann in the bedroom or in the living room?*

is sound in a given situation where we can hear Ann's voice, then so is:

- *Does her voice come from the bedroom or from the living room?*

in the same situation. The second condition defining the notion of erotetic implication amounts to the fact that Q_2 is *asked for a purpose*: every direct answer to Q_2 must bring one closer to answering Q_1.

IEL adopts the set-of-answers methodology. Let us recall that it is in line with the first and second postulate of Hamblin. On the other hand, the only restriction put on the direct answers to a question $Q = ?\{A_1, \ldots, A_n\}$ is that they differ syntactically,

[12] This monograph is self-contained as far as IEL is concerned. However, for a concise introduction written by the author of the IEL paradigm, the reader can consult [130]. The book [128] presents the results obtained until the early nineties; it has been republished in 2010 and 2013. Another book by Wiśniewski [135], presents IEL's state of the art in 2013. The latest account of IEL can be found in [140]. The very notion of erotetic implication was introduced by Wiśniewski [125], the reader can also consult any of [127, 130, 135].

that is, any two direct answers among $A_1, \ldots, A_n$ are *different expressions*. If *different expressions* can be viewed as representing *mutually exclusive possibilities*, then IEL can be claimed to adopt a very week version of the third postulate by Hamblin, however, IEL does not follow the postulate in any stronger sense. The postulate is often conceived as the most controversial among the three—see [47, p. 5]. It is also worth to notice that the inquisitive semantics paradigm evolved around discussions concerning, *i.a.*, the third postulate, accepting it in its stronger versions (see also the previous comment concerning partition semantics).

IEL provides one of the predominant contemporary frameworks for systematic logical reflection on questions. The others are the *Interrogative Model of Inquiry* created by Jaakko Hintikka (IMI, for short), the paradigm based on *inquisitive semantics* started by Jeroen Groenendijk, and an epistemic account of the logic of questions, which was initiated by Lennart Åqvist, and is nowadays represented by many researchers.[13]

The different paradigms certainly influence each other. For example, IEL was in many respects inspired by Hintikka's account of the logic of questions as a general theory of reasoning. However, even though questions are present in IMI, they do not occur as conclusions, and, eventually, they are not taken seriously as constitutive parts of reasoning. In this respect, IEL was developed as an alternative to IMI (see, for example [130] for this view). The epistemic approach to questions developed by Czech logicians is clearly inspired by IEL (see [96]), and there are also less transparent, but still significant, interconnections between the inquisitive and other paradigms.[14]

Below we describe two kinds of deductive systems that emerged within the IEL paradigm: the method of Socratic proofs and the method of synthetic tableaux.

The Method of Socratic Proofs

Here comes a brief introduction of the main character of this book. The method of Socratic proofs offers a formal explication of the idea of answering questions by pure questioning. In the realm of the very logic of questions it is a formal method of transforming questions by the so-called 'erotetic rules', which analyze the logical structure of a question and, in a way, simplify this structure, while still preserving certain important semantic properties of the transformed question. Here is an example of such a transformation; question

- Q_1 = *Is number 252 divisible by 18?*

could be transformed into

- Q_2 = *Is number 252 divisible by 9, and is it divisible by 2 as well?*

To consider a less domain-dependent example, let us take question

[13] The ideas of IMI are best represented in papers collected in [53]. For the inquisitive semantics paradigm see [14, 16–19] and the next section of this chapter. The main representatives of various versions of the epistemic approach are, for example [3, 50, 96].

[14] See [43, 82, 138, 141].

- Q_1 = *Does* $q \to p$ *follow from* $p \to q$?

If 'follow' is understood as 'is logically entailed in Classical Logic', then this question can be transformed into

- *Does* p *follow from* $p \to q$ *and* q?

and further

- Q_2 = *Does* p *follow from* $\neg p$ *and* q, *and does it follow also from* q *and* q?

which is, obviously, not the case. Let us observe that the two conditions defining the relation of erotetic implication are satisfied by the above examples—clearly, we have the transmission of soundness, and in both cases the answer to question Q_2 leads directly to the answer to question Q_1.

The erotetic rules governing the transformations of questions form erotetic calculi. In the realm of proof-theory, each erotetic calculus constitutes a proof system for a certain specific logic. Thus the method of Socratic proofs is a proof method. More specifically, it is a proof method which enables answering questions concerning the validity of a formula or the entailment, like in the above example, by transforming the questions into a form which may be called 'rhetoric'; an answer to such a rhetoric question may be considered obvious, because the question concerns some fundamental, well-known properties of validity or entailment in a given logic—such as the fact that q *does not* entail p. It is also worth highlighting that the cognitive process modeled by such transformations of questions is the process of inquiry into a formal problem the solution of which may later be presented as a formal proof.[15]

The method has been introduced by Wiśniewski [132], developed for the first-order level in [133, 143], and then further adjusted to the following non-classical logics:

- propositional paraconsistent logics, mainly **CLuN** and **CLuNs** in [144],
- propositional modal logics: a first account, in [68],
- propositional intuitionistic logic, in the form of Socratic procedures presented in [102]; a completely new account is given in Chap. 4 of this monograph,
- more propositional modal logics in [69–71], this account was used also in [72]; a refreshed account of this approach, together with a proof-theoretic account of global entailment relation and a new completeness proof, is presented in Chap. 3 of this monograph,
- the minimal LFI (Logic of Formal Inconsistency) logic, **mbC**, in [13],
- FDE (First Degree Entailment) in [109].

In [9] a fresh account of classical first-order logic was presented, together with the so-called dual erotetic calculi. The rules of dual erotetic calculi transform questions by analyzing complex formulas and simulating, at the same time, the resolution method inside the erotetic machinery. On the level of interpretation of Socratic transformation one can summarize the difference between the *canonical* (not dual) and the dual calculi by the following metaphor: if a Socratic transformation constructed

[15] This idea is explored further in [73].

in a canonical erotetic calculus reports the pursuit of truth, then a Socratic transformation constructed in a dual erotetic calculus reports the hunting for falsehood. Nevertheless, the meaning of the term *duality* can be read on numerous levels, and we shall come back to this topic in Sect. 6.3 of Chap. 6.

Some applications of the method of Socratic proofs are analyzed in [10, 11, 76–78, 119]. An exposition together with intuitive background of the method may be found in [135]. The next chapter describes the method in detail, and Chaps. 6, 7 are focused on applications of the method in the area of translations of derivations between various proof systems.

It is worth emphasizing that the translations *via* erotetic rules are possible due to their *semantic invertibility*. Generally speaking, semantic invertibility is a property complementing *semantic correctness* (the latter is also called *adequacy* or *soundness* of a rule). If a rule is semantically correct, then one cannot go from a true (or tautological) premise to a false (not tautological) conclusion.[16] If a rule is semantically invertible, then one cannot go from a false premise to a true conclusion. Thus correct rules transform truths only into truths, and invertible rules derive truths only from truths. In the realm of erotetic calculi, semantic correctness of a rule amounts to transmission of soundness of a question from a premise to a conclusion and semantic invertibility of a rule amounts to transmission of soundness of a question from a conclusion to a premise.

Due to the properties described above the method of Socratic proofs enables transforming questions in a way which warrants that all semantic information is transmitted from question to question. This property makes backtracking redundant, hence the source of possible merits for implementation. Last but not least, the semantic property of invertibility may be viewed as just another consequence of the fact that erotetic calculi *are calculi of questions* designed to maintain the property of erotetic implication between questions.

The Method of Synthetic Tableaux

From the proof-theoretical perspective, the method of synthetic tableaux belongs to a different family of methods than the method of Socratic proofs. While the method of Socratic proofs is to some extent based on sequent calculi, the method of synthetic tableaux creates a class on its own, staying in close proximity with the standard method of truth-tables and with resolution. It was introduced to the literature by Mariusz Urbański, the reader will find the original account of the method in [111, 114].

By and large, a synthetic tableau is created by taking complementary, mutually exclusive assumptions, like 'p' and 'not p', and deriving from them conclusions by the so-called synthetic rules, that is, rules that build complex sentences from simpler ones. The aim is to derive the same conclusion from each set of assumptions

[16] There is an important difference between transmitting validity and transmitting truth—the latter, transmitting truth, is a stronger property. We deliberately simplify this issue for the sake of this presentation.

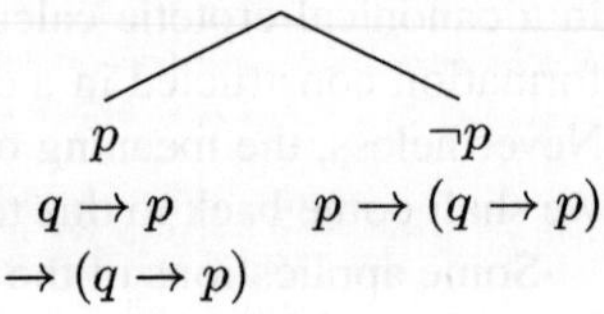

Fig. 1.1 A synthetic tableau proof of the law of antecedent

taken at the start. If a sentence can be derived from each such set of complementary, mutually exclusive assumptions, then it must be a law of logic. Let us consider, as an example, the law of antecedent $F = p \to (q \to p)$. The diagram presented on Fig. 1.1 is a visualization of the synthetic tableau proof of F and it represents the following reasoning. On the one hand, if p is the case, then the material implication $q \to p$ trivially holds for arbitrary q. By the same token F holds in this situation. On the other hand, if p is not the case, then F is an implication with false antecedent and for this reason it holds in this situation. Either way, F is the case.

Early developments of the method were described in [112, 113, 115, 116], extension to the first-order level was developed by Leszczyńska-Jasion and Chlebowski in [74]. The method found some interesting applications in the domain of abduction: [63, 64, 117].[17]

Contrary to appearances, the method of synthetic tableaux is not a tableau method in the widely accepted meaning of the term, but we shall not go into detail here, as that would be going beyond the scope of this book. But it is worth explaining that the method has been invented and developed partially to serve in analysis of *erotetic search scenarios*, structures which may be viewed as erotetic decision trees. Roughly speaking, an erotetic search scenario shows which questions should be asked, and when they should be asked, in order to solve some initial problem. It starts with a question expressing the initial problem, and each subsequent question is erotetically implied by some previous one.[18] An erotetic search scenario contains questions and declarative sentences, and a synthetic tableau may be viewed as a declarative part of such a scenario. Thus, in a way, the method of synthetic tableaux may be considered a tool for generating part of an erotetic search scenario. Nevertheless, the method constitutes a proof-theoretical category on its own.

1.2 Key Paradigms in the Logic of Questions

In this section, we briefly describe other significant paradigms in the contemporary logic of questions. Since the focus is on *logical* approaches, we will not delve into the extensive linguistic literature on this topic; for those interested, we suggest resources like [36, 42, 136].

[17] Recently, heuristics of synthetic tableau construction were examined in computational [106] and graph-theoretical setting [105].

[18] For more information on erotetic search scenarios see [83, 112, 131].

Interrogative Model of Inquiry

In the early eighties[19] Hintikka introduced what is known as the Interrogative Model of Inquiry, a model that has inspired ongoing exploration in the logic of questions for decades. To put it in Hintikka's own words:

> Its basic idea is the same as that of the oldest explicit form of reasoning in philosophy, the Socratic method of questioning or *elenchus*. In it, all new information enters into an argument or a line of reasoning in the form of answers to questions that the inquirer addresses to a suitable source of information. [54, p. 18]

Hintikka models scientific inquiry as a question-driven process, which may also be conceived as a game between the Inquirer, who asks questions, and the Nature providing answers. The logic of questions and tools from mathematical game theory play essential roles in this framework.

IMI gave rise to the idea of interrogative tableaux described in Sect. 1.3.1 of this chapter. However, before we turn to this subject, let us also note that IMI adopts the imperative-epistemic approach to questions.

Epistemic Approaches to Questions

The logic of questions developed by Hintikka leans heavily on epistemic logic. As we have already explained, according to the epistemic (or imperative-epistemic) approach, questions are essentially *epistemic requests*, and the act of answering fulfils this request. For instance, the question:

- *Is Ann in the bedroom or in the living room?*

expresses a request (imperative) for knowledge interpreted as:

- *Bring it about that I know that Ann is in the bedroom or I know that Ann is in the living room.*
- *Make it the case that I know that Ann is in the bedroom or that I know that Ann is in the living room.*

In this account, the epistemic state described by the sentence:

- *I know that Ann is in the bedroom or I know that Ann is in the living room.*

is called the *desideratum* of the question. The expression 'I know that' is modeled by modal epistemic logic, that is, a modal logic in which the necessity operator '$\Box$' is interpreted as knowledge.

We have discussed shortly the anti-realistic commitment of this approach; however, the approach can be understood methodologically, without subscribing to those philosophical commitments. The power of modal logics entering the stage is definitely a virtue.

[19] The idea was described in [57], the name 'Interrogative Model' was introduced later in [51] where also the model was described more formally.

The epistemic trend traces back to Lennart Åqvist and his [3] (first edition published in 1965), and was later expanded by Hintikka in, *i.a.*, [50] (see also the summary and bibliography in [47]). The approach to questions described above is often called the 'Make Me Know' approach, as it takes questions to be synonymous with an imperative of the form 'Let it be the case that I know...' or 'Make it the case that I know...' (see [47, p. 24]).

The model has evolved further in epistemic extensions of IEL developed, *i.a.*, in [88, 96–98, 108]. The representatives of this approach borrow certain concepts from IEL and redefine them in an epistemic setting, thus opening the possibility to model the questioning-answering process taking into account the questioner's knowledge and ignorance (see especially [96]). In [97, 98, 108] the approach is further extended with the apparatus of public announcement logic, which makes the whole approach a dynamic one.

The epistemic framework develops the idea of *askability*, a much more refined version of question soundness. The criteria for askability specify the conditions under which an agent is justified in asking a question. While the idea is rooted in pragmatic considerations, its formalization is expressed in semantics. Intuitively, a question is *askable* if the following three conditions are met [96, p. 66]:

1. **Non-triviality**: It is not reasonable to ask a question if the answer is known.
2. **Admissibility**: Each direct answer is considered as possible.
3. **Context**: At least one of the direct answers must be the right one.

The three conditions are further clarified using modal logics, leading to the definition of *epistemic erotetic implication* in terms of *transmission of askability*. In this context, if question Q_1 implies question Q_2, it means that whenever the conditions of askability for Q_1 are met in a given situation, the askability conditions for Q_2 are also met in that same situation (*cf.* [96, p. 71]).

The epistemic analysis initiated and developed by Åqvist, Hintikka and others has influenced also the inquisitive semantics paradigm which we now turn to.

Inquisitive Semantics

The framework of inquisitive semantics (IS, for short) develops a general theory of linguistic meaning whose central notions are those of *information state*, *information exchange*, or *issues raised in a conversation*—instead of the linguistically superficial, traditional *truth conditions*. As far as questions are concerned, IS builds on the notion of information state instead of that of a possible answer.

One of the main aims of this framework is to unify the analysis of declarative and interrogative sentences, as a theory of meaning cannot treat the two separately. For this purpose IS develops the support-conditions semantics, where *support* is a certain generalization of *truth*, equally applicable to declarative formulas and interrogatives.

Here are the main assumptions underlying IS. The fundamental difference between a statement and a question is in their *content*, not in their form. Every sentence has an *informative content* and an *inquisitive content*. If the first content, that is, the information carried by a sentence, is sufficient to resolve the issue it raises, that is, its inquisitive content, then the sentence is a statement, otherwise it is a question. For example, the sentence:

(S) *Ann is in the living room.*

raises no issue that is unsettled by its informative content, whereas:

(I) *Is Ann in the bedroom or in the living room?*

raises an issue it cannot resolve. For this reason the former sentence is a statement, whereas the latter is a question. According to this approach, the sentence:

- *Ann is in the bedroom or in the living room.*

has the same informative and inquisitive content as the aforementioned question, and hence it is also a question.[20] As we can see, in the IS approach, the categories of statements and questions are primarily defined semantically, not syntactically.

As explained above, the inquisitive theory of meaning is contrasted with a truth-functional account of meaning. The central notion of the latter is that of truth and truth conditions, whereas the basic notion of the former is that of information and *support* of a statement by an information state. Support generalizes the traditional concept of truth. Technically, it is defined on the grounds of semantics of modal logics, where an information state is a set of possible worlds. For example, statement (S) given above is supported by each information state, which consists solely of possible worlds where Ann is actually in the living room.

Although the concepts of support and truth are contrasted, truth can be reconstructed through the idea of support. In this reconstruction, truth is extended to include questions. In this way the representatives of IS follow Belnap, and take questions to be true or false in a given information state. Simply put, the earlier question (I) is considered true in a given state if the state resolves the issue it raises, that is, supports one of its direct answers: either in each world of the state Ann is in the bedroom or in each world of the state Ann is in the living room.

Assigning truth values to questions contrasts sharply with the position taken in this monograph. However, this does not undermine the fact that extending the notion of truth through support creates a framework, where the semantic properties of both declarative statements and questions can be analyzed using the same tools.

The power of this approach is revealed in the unification of concepts. For instance, the relation of entailment defined in this framework unifies the following logical notions: (i) a statement being a logical consequence of another statement, (ii) a statement logically resolving a question, (iii) a question logically presupposing a statement, (iv) a question logically determining another. The last relation is also called *dependency* between questions.

Inquisitive semantics evolves into a general theory of meaning, developing formal tools to model such diverse phenomena as semantic content, conversational context and propositional attitudes, *i.e.* the attitudes of an agent towards propositions expressed in words such as *knows*, *believes* or *doubts*. This concept is further

---

[20] There are numerous differing accounts of the matters among the inquisitive-semantics researchers, we simplify this presentation out of necessity. Our presentation is based mainly on [14, 16, 18].

generalized to cover issue-directed attitudes like *wondering*, *inquiring* and *asking*.[21] The aim to unify the analysis of declarative and interrogative sentences has been successfully achieved by IS, although taking the perspective adopted in this monograph yields that certain features of the IS-analysis of meaning constitute a price to be paid for the success rather than an advantage—such as some intuitively awkward semantic consequences like assigning questions a truth value. However, from the IS perspective the two are justified, as we saw. Among other things, one should realize that the truth value assigned to questions is not the traditionally understood one, but one defined by the notion of support.

The particular differences between the frameworks of IEL and IS, as observed in this section, can be shown to have a common root. They all follow from one fundamental difference between the two frameworks: IEL's primary interest is in *cognition*, while IS's is in *conversation*. And so, IEL starts with defining questions purely syntactically, because reasoning, the skeleton of cognition, is traditionally analyzed by formal logic in terms of the *form* of premises and conclusion. IS starts with the semantic and pragmatic considerations because 'reasoning is just one particular language game', [38, p. 109], far from being central in our daily conversations. Further differences can be described around the third postulate by Hamblin, as we have already observed. Having said that, let us note that there is also a clear trend in the development of the IS paradigm that puts inferences with questions in the very centre of its attention [14, 16].[22]

1.3 Between the Logic of Questions and Proof Theory: An Overview of Research Goals

In the preceding parts of this chapter, we have sketched the erotetic landscape by introducing the main conceptual tools of the logic of questions and its most important contemporary paradigms. In this section, we provide a systematic presentation of formal results that follow from combining questions and proofs in a single framework. The results are organized along an axis determined by the following research goals:

G_1 Designing problem-solving strategies with reference to the questioning-answering process.

G_2 Defining relations between questions that may be called *inferential* or *deductive*, as well as conducting purely semantic analysis of these relations.

---

21 Let us recall that the 'issue-directed' attitudes have been called 'zetetic' attitudes by Friedman [30]. As a side note, in [32], Friedman advocates for a paradigm shift in epistemology—from a doxastic model, focused on belief, to an inquiry-first model. In fact, it seems clear that we are witnessing such a shift; see [122, 123]. What is more, in [122] the authors set up a highly promising research program of connecting IEL to interrogative epistemology.

22 For further description of the IS framework the reader can consult [18] or any of: [14–17, 19, 40]. We recommend [18] which is a comprehensive introduction for a newcomer.

G_3 Formal study of structures containing questions. Just as proofs, which are structures containing declarative sentences, are objects of formal study, so too are structures containing questions.

1.3.1 G_1*: Questions and Problem-Solving Strategies*

The common feature of the developments described in this section is that they analyze how to use questions to solve a problem or to obtain a proof that something is true. Often, a formal model of an information-seeking or knowledge-seeking process is involved, and its description includes both declarative and interrogative steps.

Interrogative Tableaux

As we have explained above, IMI (*Interrogative Model of Inquiry*) developed by Hintikka, aims to model rational inquiry in science, and it does so by, *i.a.*, formulating rules for question-and-answer procedures in the style of tableau systems.

Interrogative tableaux were proposed, *i.a.*, in [52, 56], developed in [48] in terms of game-theoretical semantics, then proposed in a more expanded form in [55]. Proof-theoretically, the system is based on Beth's semantic tableaux or on a sequent calculus. In the case of tableaux, a formula is analyzed in a tableau divided into columns with true formulas and false formulas. A formula to be considered is placed in the 'false' column, and then one tries to derive a contradiction. The formalism is extended with the possibility of considering different answers to questions asked during inquiry. No syntactical representation of questions is required. It is just assumed that some of the questions of the Inquirer are answered by the Nature and hence the answers enrich the set of premises available in the deductive moves.

Generally speaking, an interrogative tableau indicates the moves of Nature and the Inquirer. The declarative steps are guided by standard tableau rules; for example, if the Inquirer finds a true statement of the form '$A \& B$' in their part of the tableau, then they may add both A and B to this part of the tableau. Interrogative moves are guided by the interrogative rules, and they amount to, first, asking questions the answers to which may be included in the appropriate part of a tableau, and, second, considering the results of adding the different answers.

Unfortunately, in this model there is no place to look for relations between questions. Similarly, the very process of *arising* of a question is not an issue here.

On the other hand, since [48] it has been known that interrogative tableaux are subject to game-theoretical interpretations. The interpretations have been further developed, e.g., in [34, 35].

Erotetic Search Scenarios

In a way, erotetic search scenarios (*e-scenarios*, for short) explicate closely related ideas. As roughly explained above, an e-scenario for a given problem is a tree with a

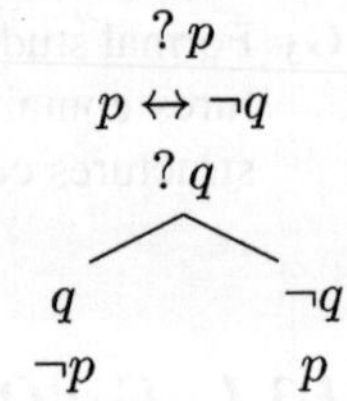

Fig. 1.2 An erotetic search scenario for question $?p$

question expressing the initial problem at the root, a set of declarative premises under the root, and further questions, each of which is erotetically implied by a previous one. Some of these questions, called *queries*, are answered, and the e-scenario shows what to do next if such-and-such an answer to a query is arrived at. Each branch of the tree ends with a direct answer to the initial problem. An example is presented in Fig. 1.2, where $?p$ is shorthand for the polar question $?\{p, \neg p\}$, whose direct answers are p and $\neg p$; similarly for $?q$. The latter question is a query—it is followed by its direct answers, which constitute different branches of the e-scenario. Then, on each branch, the declarative premise $p \leftrightarrow \neg q$ is utilized to derive the answer to the initial question.

The idea of e-scenarios was introduced by Wiśniewski [131], and later applied in proof-theory [112], dialogue modeling [83, 84, 118], and automatic query generation [87]; e-scenarios were also extended to non-classical settings [75, 88], and their pragmatic account was applied in the Turing Test adequacy debate [81, 90] (see also [135, Part III]).

As far as our goal $\mathbf{G_1}$ is concerned, the main improvement with respect to interrogative tableaux is that asking queries in an e-scenario is, in a way, guided by the relation of erotetic implication, whereas in the case of interrogative tableaux, no such guidance is provided. Many e-scenarios and schemes of e-scenarios are presented in [131, 135]. The schemes can be conceived as ready-made scenarios for solving problems of a given logical structure; moreover, depending on the logical structure, the construction of an e-scenario can be automated.[23] These ready-made schemes were also used in [12] in the field of automated e-scenario application.

E-scenarios are typically presented as ready-made structures, and in the general case, there are no rules guiding the user in creating their own e-scenario. In order to provide a description of e-scenarios in terms of rules of inference, one needs a corresponding description of erotetic implication, as the latter underpins those moves in an e-scenario that generate questions. However, prior to papers [22, 93, 94] (to be mentioned later in this chapter), there was no proof-theoretical description of erotetic implication of this kind. We believe that it can now be claimed that, with the rules capturing erotetic implication, the task of providing a satisfactory proof-theoretical description of e-scenarios is attainable, though still—not to the best of our knowledge—entirely completed.

[23] A Prolog-implementation of a straightforward algorithm for the CPL case is available at https://intquestpro.wordpress.com/software-data/.

Natural Deduction Systems for Inquisitive Logics

A natural deduction (ND, for short) system for inquisitive propositional logic InqB and a possibly incomplete ND-system for first-order logic InqBQ were presented by Ciardelli [14]; see also [15, 16]. Interestingly, the interrogative tableaux described above, e-scenarios, and natural deduction systems for inquisitive logics may be viewed as formalizing the same idea. In Ciardelli's words:

> [Q]uestions have a very important role to play in inferences: they make it possible to formalize arguments involving generic information of a certain type, such as *where Alice lives* [...] questions may be used as placeholders for arbitrary information of the corresponding type. By manipulating such placeholders, we may then provide formal proofs [...]. [14, p. 77]

Roughly speaking, in the context of interrogative tableaux, 'manipulating placeholders' is tantamount to analyzing the result of adding different direct answers to a suitable part of a tableau. In the e-scenarios setting, it is tantamount to choosing a specific set of direct answers to the queries asked in an e-scenario. In fact, 'choosing a specific set of answers' has been formally described as an operation of *contraction* performed on an e-scenario.[24]

Both in the propositional and first-order cases, the inquisitive system is based on natural deduction, with special rules for the inquisitive connectives characterizing questions in this framework.[25] More specifically, in the propositional case, inquisitive content is generated by the connective of inquisitive disjunction $\mathbin{\vee\!\!\!\vee}$, so the natural deduction rules for this connective can be viewed as rules for introducing and eliminating questions. An example is presented in Fig. 1.3, where '$?p$' is short for '$p \mathbin{\vee\!\!\!\vee} \neg p$', and similarly for '$?q$'. Let us observe that the derivation presented in Fig. 1.3 *proves* the dependency between questions $?p$ and $?q$ under the assumption expressed by the formula '$p \leftrightarrow \neg q$'. This fact should be compared with the e-scenario presented in Fig. 1.2, where the dependency is indeed used to derive the direct answers, but it can hardly be seen as a *proof* of it.

The rules of this ND system are expressive enough to generate proofs in which a question is derived as the final conclusion from assumptions that may include both

$$\dfrac{\dfrac{[?p]_2 \qquad \dfrac{\dfrac{[p]_1 \quad p \leftrightarrow \neg q}{\neg q}\,(\leftrightarrow \mathsf{e})}{?q}\,(\mathbin{\vee\!\!\!\vee}\mathsf{i}) \qquad \dfrac{\dfrac{[\neg p]_1 \quad p \leftrightarrow \neg q}{q}\,(\leftrightarrow \mathsf{e})}{?q}\,(\mathbin{\vee\!\!\!\vee}\mathsf{i})}{?q}\,(\mathbin{\vee\!\!\!\vee}\mathsf{e}, 1)}{?p \rightarrow ?q}\,(\rightarrow \mathsf{i}, 2)$$

Fig. 1.3 An ND-style proof of dependency between questions $?p$ and $?q$ under the assumption that $p \leftrightarrow \neg q$. The example comes from [16, pp. 106–107]

---

[24] See [135, Part III], especially the last chapter, 'E-scenarios and Problem Solving'.

[25] To some extent, the inquisitive connectives behave just like intuitionistic connectives. The relation between the two, however, is more subtle. See [19]. It is also worth noting that there is a display calculus for InqB studied in [33].

statements and questions. The main result concerning the system is the inductive definition of a procedure that, for each such proof deriving a question, and for each choice of answers (called resolutions) to questions occurring in the proof as premises, transforms it into a proof of a specific answer (resolution) to the final question, with statements as premises. This whole enterprise leads to an interesting computational interpretation of proofs containing questions, 'reminiscent of the proofs-as-programs interpretation of intuitionistic logic' [16, p. 3].

The act of choosing resolutions to questions occurring in a proof and calculating the answer to the final (derived) question corresponds very accurately to what is defined as contraction in an e-scenario. The main difference between e-scenarios and the ND system described lies in the direction of building the respective structures (e-scenarios, ND derivations/proofs) as defined by the rules. An e-scenario starts with the main question to be solved, whereas an ND derivation leads to a question, obtained as the final conclusion. In the next section, we also highlight a similar difference between IS and IEL frameworks when discussing semantic relations between questions.

1.3.2 G_2: Questions and Relations Between Them

Back in the fifties, Hamblin had already suggested defining a relation of *containment of questions* (see [44, 47] for a discussion). The idea is that one question contains another when one can deduce an answer to the second question from every answer to the first. This general idea of resolving one question by answering another is vital and apparent in many other definitions of relations between questions.

Another early idea was that of Bromberger, put forward in [8], where the author describes and provides examples of the relation of *giving rise to an interrogative by a declarative sentence* and *declarative sentence entailing an interrogative*. Similar ideas are to be found later in Kubiński's writings under the concept of *evocation of a question by a set of declarative sentences*.

Belnap defines the relation of *erotetical equivalence* between interrogatives (see [7, Sect. 3.4] or, again, [47] for a discussion). Two interrogatives are in this relation if and only if, for every direct answer to one of them, there is an equivalent direct answer to the second interrogative, and vice versa: for every direct answer to the second interrogative, there is a direct answer to the first that is equivalent to it. Belnap, in [7, Chap. 3], also defines a *propositional implication*, which is a generalization of entailment between statements. He calls an interrogative *true in an interpretation* if some direct answer to the interrogative is true in the interpretation. One can then define *propositional implication* between a set X of *quasi-wffs* (that is, expressions which may be propositional formulas or interrogatives) and a quasi-wff A as the transition of truth: X implies A in this sense if and only if A is true in every interpretation in which every element of X is true.

Tadeusz Kubiński was probably the first logician who considered *systems of questions* with the aim of describing diverse relations between questions and possibly defining deductive moves between them (see [66, 67]).[26] Perhaps the most important relation defined by Kubiński is that of *equipollence of questions*. Two questions are *equipollent* if and only if there exists a bijection (that is, a function which establishes a one-to-one correspondence) that assigns to each direct answer to one question an equivalent direct answer to the second question. The existence of a bijection makes this relation *stronger* than the relation of erotetical equivalence defined by Belnap, since in Belnap's account it is possible for erotetically equivalent questions to have sets of direct answers with different cardinalities, which is, in turn, excluded by the existence of the bijection. Kubiński also considered, among many others, relations of being *weaker than* and being *stronger than* between questions, and formulated systems of questions to capture these relations.

Starting in the eighties, Wiśniewski, a former student of Kubiński, defined the relations of *evocation of a question by a set of declarative premises* and *erotetic implication*.[27] We have already described the latter, so let us turn to the former. The relation of evocation explicates the criteria of validity for inferences in which a question *is posed* on the basis of some declarative premises (statements), such as:

- *The Twin Prime Conjecture is one of the still unsettled problems of number theory. It's fascinating because of the simplicity of its expression. Hm. Number n is a prime. Is* $n - 2$ *or* $n + 2$ *prime as well?*[28]

Evocation is defined by two conditions: (i) if, in a given situation, the premises are true, then the question *evoked* by them is sound; (ii) no direct answer to the evoked question follows from the premises. This is the case in the above example. First, the question of whether $n - 2$ or $n + 2$ is prime is sound, and second, no answer to it follows from the premises. Obviously, no answer to the question leads to a solution of the famous number theory problem; nevertheless, the key point is that the problem triggers the question.

As in the case of erotetic implication, the notion of evocation captures a frequently occurring and important cognitive phenomenon. Evocation is defined to explicate the arising of a question from some premises.[29] A recent attempt to capture the same kind of phenomenon can be found in [23], where Moritz Cordes introduces *performators*, a category of illocutionary expressions. With performators such as WHETHER, WHICH, and WHAT, the act of asking can be regulated by rules. Here is an example of such a rule (simplified for the sake of this presentation): if one has inferred an existential

---

[26] The reader can find more information in a book devoted to the memory of Tadeusz Kubiński: [145]. For more information on his systems of questions, the reader may consult [129] in [145]; we also recommend [137].

[27] Strictly speaking, the first notion introduced was that of *generation*. See [62, 124–126] and further footnotes 29 and 36 in this chapter.

[28] The Twin Prime Conjecture asserts that there are infinitely many twin primes, that is, primes differing by 2, such as 3 and 5, 5 and 7, 11 and 13…

[29] There are also special cases of evocation considered by Wiśniewski, such as that of generation, though we will not delve deeper into this topic. The reader can consult [126], where the concept of erotetic argument is introduced and the relations of weak generation and generation are defined. See also the first chapter of [128].

formula stating that there is an object with property P, then one can ask which object is such that P. In a language with performators this can be presented as:

$$\frac{\exists x P(x)}{\text{WHAT } P(a)}$$

Further behavior of expressions with performators can be modeled in a speech act calculus as that presented in [24].

The framework of inquisitive semantics defines the notion of *dependency of questions*. The relation holds between questions Q_1 and Q_2 whenever settling the first question, Q_1, entails settling the second. In this situation, we say that Q_2 *depends on* Q_1. For example, the question:

- $Q_2 =$ *Is Ann in the bedroom or in the living room?*

depends, in this sense, on the question:

- $Q_1 =$ *Does her voice come from the bedroom or from the living room?*

Let us observe that the relation has, in a sense, the *opposite direction* to erotetic implication. Namely, dependency holds between Q_1 and Q_2 in our example, whereas the direction of erotetic implication, as presented in its definition, is rather from Q_2 to Q_1.[30] Although this is ostensibly but a shallow observation, it reveals something more significant. In the IS framework, the more specific information must come before the more general information. In this sense, it reflects a natural order of entailment, and the relation of dependency is a special case of the generalized notion of entailment in inquisitive semantics. That is to say, in the IS framework, the direction of entailment (dependency) between questions is driven by the issue:

- ***inquisitive dependency direction:*** *What other question can be settled by information settling Q?*

On the other hand, in the erotetic framework, the direction of erotetic implication between questions is driven by the issue:

- ***erotetic implication direction:*** *What other question should I ask in order to gain information that will settle Q?*

Further refinements of the notion of dependency are possible. For example, [100], extending the IS framework to relevant logic, allows for capturing relevant relations between questions. Another refinement of dependency is the notion of *compliance*, which 'judges whether a certain dialogue move is coherent with respect to previous moves' [40]. However, delving deeper into this concept would take us far afield from the topic of this book.

Finally, in [138], Wiśniewski demonstrates the interconnections between the relation of inquisitive entailment and a relation he coined *constructive generalized*

30 Erotetic implication holds between Q_2 and Q_1 on the basis of the additional declarative premise that Ann's voice can be heard coming from either the bedroom or the living room.

entailment. The latter is a semantic relation between a family of sets of formulas and a set of formulas, and it can be viewed as a generalization of the former. Unfortunately, [138] does not provide a proof-theoretical description of this semantic relation.

1.3.3 $\mathbf{G_3}$: *Questions and Systems of Questions*

The main research objective in $\mathbf{G_3}$ is to investigate systems (or calculi) of questions, with a particular focus on sequences (or structures) of questions generated within these systems, much like formal proofs are investigated. The remainder of this monograph is largely dedicated to the exploration of $\mathbf{G_3}$, with extensions to the other two areas as needed. We shall use concepts, tools and methods specific to proof theory, but with a focus on question transformations.

When we delve deeper into proof-theoretical considerations, it turns out that the method of Socratic proofs has some properties that make it particularly suitable for proof translations between different proof systems. Specifically, it can be used to define functions that transform a Socratic proof of sequent '$\vdash A$' into, e.g., a Gentzen-style proof of the same sequent, or a proof of formula A in another proof system. As previously mentioned, the main feature which makes the method of Socratic proofs suitable for this kind of translation is the invertibility of the rules, which applies to the *entire structures* contained within the questions. Typically, these structures consist of finite sequences of sequents from a formal language, with each sequence preserving all the semantic information relevant to the problem at hand.

Let us now list some other results that clearly fall within the scope of $\mathbf{G_3}$. The very problem of defining systems of questions capturing specific inferential relations between them, as explored by Kubiński, is a prime example of a $\mathbf{G_3}$-type endeavor. Until the end of this chapter, we shall briefly discuss the following:

- Proof systems capturing evocation and/or erotetic implication (here the method of Socratic proofs is excluded),
- Axiomatic-style accounts of other logics of questions, particularly inquisitive ones,
- Issues related to effectiveness and completeness in the logic of questions.

Proof Systems Capturing Evocation and/or Erotetic Implication

Under this heading, we list proof systems designed specifically to capture the relations of evocation and/or erotetic implication as defined by Wiśniewski, but *other than* the erotetic calculi that are the subject of the rest of this monograph.

In [103], Skura and Wiśniewski develop an axiomatic account of proper multiple-conclusion entailment. Later, in [137], Wiśniewski extends the ideas from [103] into an axiomatic account of classical propositional evocation in a sequent calculus style. The language described in [137] involves *erotetic sequents*, which are sequents containing both declarative formulas and questions. The calculus is sound and complete with respect to the specified relation of evocation. However, the proof-theoretic construction has one element of nonprocedural character, which is the following rule, R_2:

$$\frac{X \vdash ?\{A_1, \ldots, A_n, B\} \quad \text{where } B \leftrightarrow C \text{ is a thesis of } \mathsf{CPL}}{X \vdash ?\{A_1, \ldots, A_n, C\} \quad \text{and } C \neq A_i \text{ for } i \in \{1, \ldots, n\}}$$

The rule states that from the fact that X evokes a question, one can derive the conclusion that the same X also evokes a transformed version of the question, where the transformation consists of reformulating one of the direct answers to the question, B, into C. It is assumed that the system is built upon some proof-theoretic description of CPL (classical propositional logic), which justifies the lack of explicit description of the rules guiding the process of reformulating B into C.

In [139], Wiśniewski introduces a calculus that captures the property of a set of formulas of CPL being minimally inconsistent. The proof-theoretic description of this property can be utilized to formulate another calculus for classical propositional evocation, although this possibility is not explicitly presented in [139].

In [93, 94], the author, Jared Millson, develops sequent systems that can be claimed to successfully capture specific types of reasoning involving evocation and erotetic implication, both within classical propositional logic. The sequents used by Millson are similar to erotetic sequents. They also incorporate both declarative formulas and questions, and can be viewed as describing special cases of the claimed relations. For example, the sequent

$$p \vee q \vdash ?\{p, q\}$$

represents the fact that the declarative formula '$p \vee q$' evokes the question asking which one, p or q, is the case. Similarly, the sequent

$$?\{p, q\}, r \vee s, r \to p, s \to q \vdash ?\{r, s\} \tag{1.1}$$

corresponds to the fact that question $?\{p, q\}$—asking whether it is the case that p or whether it is the case that q—together with the declarative premises $r \vee s$, $r \to p$, and $s \to q$, erotetically implies question $?\{r, s\}$. The sequent-calculus rules are then designed to build complex sequents of this kind from axioms. The entire system is constructed very much in the spirit of structural proof theory.

There is an additional aspect of *defeasibility* captured by the system. Technically, an additional component is added to the sequents used in the two papers. These sequents are termed *defeasible* and, in addition to the usual antecedent and succedent, include a third set, called *defeater set*, which is written beneath the horizontal bar of the turnstile symbol. This set is used for the bookkeeping of the information that a particular reasoning will no longer be legitimate once certain information is obtained. For example,

$$p \vee q \vdash_{\{\{p\},\{q\}\}} ?\{p, q\}$$

can be interpreted to mean that an agent entitled to believe that $p \vee q$ is entitled to inquire whether p or q, so long as she is neither entitled to believe that p nor entitled to believe that q.

The sequent system presented in [93] is sound and complete with respect to *regular* erotetic implication, which is a special case of erotetic implication that ensures each answer to the second, implied question entails an answer to the first, implying question. In [94], the erotetic implication relation aimed at is not only regular but also *strong*, which means that no direct answer to the implying question is entailed by the set of declarative premises. Strong erotetic implication was chosen by Millson due to its defeasible character. Both papers include a fruitful epistemological background that justifies the proof-theoretical concepts introduced by the author.

The paper [22] by Moritz Cordes also develops a calculus for classical propositional erotetic implication and contains interesting observations and ideas concerning the first-order level.

The calculus for erotetic implication is based on the fact that the relation is semantically defined by two entailment conditions. Since we have sequent calculi that are sound and complete with respect to classical entailment, these calculi can be used to decide each of the entailment conditions separately. For example, to establish that erotetic implication holds in the case captured by sequent (1.1) above, one must demonstrate that the following entailment holds[31]:

$$p \vee q, r \vee s, r \rightarrow p, s \rightarrow q \models r \vee s$$

for the first condition, and that both:

$$r, r \vee s, r \rightarrow p, s \rightarrow q \models p$$

$$s, r \vee s, r \rightarrow p, s \rightarrow q \models q$$

hold for the second condition. This can be accomplished by three proof-trees. The proof trees, constructed within Gentzen's sequent calculus LK, can be grouped together into a graph called by the author *PEI-development*, where 'PEI' stands for 'propositional erotetic implication'.

The calculus provides a successful framework both for practical applications and metalogical considerations. It can be utilized to implement a decision procedure for erotetic implication in the classical propositional case. As far as metalogic is concerned, in [22, p. 17], the author presents a proof of the following result concerning evocation in classical first-order logic (we simplify the formulation here for the sake of this presentation).

[31] We simplify the conditions; see Chap. 5 of this book for a precise account.

Theorem 1.1 (Cordes' theorem) *There is no calculus defining derivability relation* $\vdash_{\mathsf{FOE}}$ *such that for each set X of propositional formulas (in classical first-order logic) and each question Q, X evokes Q if and only if* $X \vdash_{\mathsf{FOE}} Q$.

That is, no calculus that is both sound and complete with respect to classical first-order evocation *can exist*, at least as long as we understand 'calculus' in a familiar way: as a finite set of schemes of rules governing the derivability process. The theorem can be viewed as a consequence of the first-order undecidability connected with the negative condition defining evocation—the condition stating that no direct answer to the evoked question is entailed by the evoking set. It is worth stressing that the very undecidability of first-order logic accounts only for the lack of a negative test for first-order evocation, and that the incompleteness result demonstrated by Cordes is stronger than that, as it proves that there is no *positive test*[32] for the relation either. This means that for each evocation-calculus (understood in a familiar sense mentioned above), we can find questions satisfying the definition of evocation at the first-order level for which no finite construction obtainable within the calculus establishes that the relation holds.

That there is no positive test for first-order evocation was also observed by Meheus [92]. The paper concerns adaptive logics for question evocation and develops dynamic proof systems for them. In [20], the authors present a similarly fashioned proof system for question evocation in CPL, and in [92], Meheus also provides a question evocation system for modal logic S5. However, since the three systems are underlain by the same formal and technical ideas, we focus on classical first-order evocation.

Describing a complex research framework in a paragraph is a thankless task, yet we must give it a try. The first adaptive logic was developed by Diderik Batens around 1980, with the result published in [4]. The subject of the paper is an inconsistency adaptive logic, that is, a logic that localizes inconsistent consequences of a theory and, in a way, *adapts* itself to the undesirable behavior of inconsistent premises. This adaptation process is expressed *via* the dynamic proof theory developed for these logics. Roughly speaking, the idea is to use the full strength of, for example, first-order logic to derive consequences from a given theory, and to restrict the logic's derivative power once inconsistency arises. Derivations and proofs in adaptive logic systems are written in a linear natural deduction style. Parts of a derivation are conditional: they hinge upon the assumption that the premises of the derivation do not exhibit undesirable behavior. The assumptions are followed as much as possible, 'that is, *unless and until* they are explicitly violated'.[33] In the case of *inconsistency* adaptive logics, once an assumption is violated, 'the rules of inference are restricted in order to avoid triviality. However, where this is not the case, the rules can be applied in their full strength.'[34]

[32] However, the reader should be aware that neither Cordes nor Meheus in [92] explicates this lack-of-positive-test result on the grounds of a computation model—such as recursive functions, Turing machines, etc. Certainly, it would be an interesting task to conduct such an analysis.

[33] See [91, p. 59].

[34] *Ibid.* See also an introduction to adaptive logics in [5].

In the case of the adaptive logics for question evocation, there is one rule that allows for the derivation of a question and a mechanism, called *marking*, used to *withdraw* a question once it is established that a direct answer to the question follows from the premises. The former rule, which derives a question, allows for the introduction of a question at a line based on the derivation of a specific sort of *presupposition* to the question, one that ensures the question is sound (see the introduction to this chapter for presuppositions). Questions are derived under the assumption that the premises of the derivation do not exhibit undesirable behavior—that is, no direct answer to the question is derivable from the premises. The question is considered derived, *unless and until* the assumption is explicitly violated. Once a direct answer is derived, the line containing the question is marked, and the question is no longer considered evoked by the premises. Thus, in Meheus' system for question evocation, the logic adapts itself to the possibility of withdrawing a question that was previously considered evoked.

The declarative part of the system under consideration is described in a nonprocedural manner: no rules guiding the derivations of direct answers, presuppositions, or other declarative formulas are provided in [92]. It is assumed that the mechanism for deriving declarative formulas is somehow given by the derivability relation of some proof-theoretical description of the underlying logic, as there are various sound and complete formalizations of it. Setting this aside, the notion of derivability in these systems has some interesting peculiarities. Following the content of Definitions 11 and 12, [92, p. 147], we say that a line of a derivation[35] containing a question is *marked* when a direct answer to the question is derived at some other line of the derivation, and it is *unmarked* otherwise. A question is said to be *derived at a given stage of a derivation* if it is derived on a line that is unmarked at that stage. A question is *finally derived* if it is derived on a line that is unmarked and *will not be marked in any further extension of the proof.*

Here we come to the intriguing concept of final derivability; the reader can sense the unconstructive challenge this presents. In fact, there is no 'positive test' for final derivability in adaptive logics in general (see [121, p. 744]), which makes the development of proof-procedures, heuristics, and at least partial criteria for final derivability an engaging pursuit. See, for example, [6, 121] for further discussion.

To sum up, we began with Cordes' Theorem, which expresses the limits of developing a proof-theoretic description of first-order evocation. These limits were noted in [92], although no formal proof of this fact is provided there. However, the adaptive logics program introduces its own notion with very similar limits. In fact, by referring to Cordes' Theorem and the results described in [92], one can establish that there is no calculus (i.e., a finite set of rule schemes) that defines the final derivability relation for the system of adaptive logic for first-order question evocation.

[35] The term *proof,* not *derivation,* is used in [92], but in view of the conditional character of the status 'derived,' we find it more appropriate to use the latter term in this description.

Axiomatic-Style Accounts of the Logics of Questions

Axiomatic systems were the first to be formally presented and thoroughly examined in the history of proof theory. In the development of philosophical logics, axiomatic systems were usually introduced not to search for proofs or examine them as such, but for reasons going beyond proof theory: to explicate new ideas, to explore the meaning and consequences of new concepts, to capture specific types of reasoning, and so on. Although, in general, these are not the goals adopted in this book, we present below axiomatic descriptions of a few erotetic systems that (except for IEL) were not previously mentioned in this chapter.

One of the early contributions in this area is [25], where the author, Maxwell J. Cresswell, analyzes the phrase 'p is the answer to d', symbolized as 'Spd', on axiomatic grounds using modal first-order logic. The following can be expressed in this language:

$$\Box(p \equiv q) \wedge Spd \rightarrow Sqd \tag{1.2}$$

$$(d = e) =^{df} \forall p \Box(Spd \equiv Spe) \tag{1.3}$$

Equation (1.2) states that if p is the answer to question d, then any statement logically equivalent to p also answers d. Equation (1.3) defines the identity of two questions.

Early axiomatic developments within IEL can be found in [124], where Wiśniewski proposes an axiomatic system for the classical propositional generation of questions.[36] The presented system is threefold, as it includes rules for assertion (standard axiomatic rules), rules for rejection (both sets of rules based on the ideas of Łukasiewicz, [80]), and proper erotetic rules. Here is an example of a thesis of this system, which corresponds to sequent (1.1) discussed above; the sign '$\leadsto$' is used for the indicated erotetic relation.

$$p \vee q, r \vee s, r \rightarrow p, s \rightarrow q \ \leadsto \ ?\{r, s\} \tag{1.4}$$

In [110], the authors develop an axiomatic presentation of the Logic of Interrogation, abbreviated as 'LoI', as presented by Groenendijk [38]. LoI was established before the emergence of inquisitive semantics and serves as a logic for reasoning about the relevance or *aboutness* of linguistic utterances. To provide at least a glimpse of this work, expressions (1.5) and (1.6) present the rules for the connectives '$\neg$' and '$\wedge$', respectively, while expression (1.7) is an axiom for equality:

[36] To be precise, in [124], the system is described as capturing the semantic relation of *arising of a question on the basis of a finite set of declarative sentences*. A decade later, in [128], the term *arising of a question* is used rather as a descriptive term referring to a cognitive phenomenon, for which the normative criteria can be explicated by the notion of *evocation of a question*. *Generation* is then a special case of evocation. See also footnote no 29 in this chapter.

$$?\phi \vdash ?\neg\phi, \tag{1.5}$$

$$?\phi, ?\psi \vdash ?(\phi \wedge \psi), \tag{1.6}$$

$$\vdash ?(x = y). \tag{1.7}$$

Rule (1.5) for negation states that it is licensed to ask whether $\neg\phi$ holds after the question whether ϕ was posed. Rule (1.6) for conjunction indicates that after the questions $?\phi$ and $?\psi$ have been raised in a conversation, it is relevant to ask whether $\phi \wedge \psi$ holds. Finally, axiom (1.7) for equality states that it is always licensed to ask whether x equals y.

The axiomatic systems can be quite complex, as illustrated by the proof system introduced in [120], which is an axiomatic-style system built above **S5**, featuring numerous additional axioms and rules of inference that characterize the additional modalities: K for knowledge, U for universal modality, Q for the issue-equivalence relation, and R as an intersection modality. The expression '$Q\phi$' signifies that 'locally in a given world, the current structure of the issue relation has ϕ true,' while '$R\phi$' indicates that 'ϕ holds in all epistemically indistinguishable and issue-equivalent worlds' [120, p. 636]. Using this language, one can express the following (cited from [120, p. 637]): the fact that the structure of the current issue settles ϕ:

$$U(Q\phi \vee Q\neg\phi),$$

the fact that an agent considers it possible that ϕ is not settled by the structure of the current issue:

$$\neg K\neg(\phi \wedge \neg Q\phi),$$

that ϕ is neither known by the agent nor settled by the structure of the issue, but is true upon resolution:

$$\neg Q\phi \wedge \neg K\phi \wedge R\phi.$$

[120] represents another influential trend in the dynamic epistemic logic of questions (see also: [95, 104]). To some extent, the topic of [120] can be viewed as a continuation of research conducted in the field of inquisitive semantics, with the problem of information and issue management at the center of the paper's focus. The authors of [120] make a distinction between *interrogative logics*, which are 'mainly about *general inquiry* and learning about the world' (op.cit. p. 634), and *logics of questions*, which represent a 'stream of work on *questions* in natural language' (p. 634 op.cit.). A prominent representative of the former is Hintikka, while the latter is represented by Groenendijk and Stokhof. The aim of [120] is to conduct the 'logic of questions' by the tools of the 'interrogative logic.' Following this distinction, the book presented to the reader addresses a specified topic within the realm of 'interrogative logics.'

Effectiveness and Completeness in the Logic of Questions

Last but not least, the intriguing problems of effectiveness and completeness—analogues (more or less) of similar problems in the realm of declaratives—serve as good examples of **G₃**-problems that one may encounter when studying systems of questions in a formal manner. For instance, Harrah's incompleteness theorem states that if the set of questions of a formal language is effectively enumerable and the set of direct answers to a question of the language is also effectively enumerable, then *not every* set of sentences of the language can serve as the set of direct answers to a question.[37]

Clearly, the notion of completeness can be understood in many ways. One may be tempted, for example, to provide a *complete* axiomatic account of a system that captures a certain relation involving questions, and one can achieve this successfully: see, for example, [103, 137], where axiomatic accounts for evocation in classical propositional logic are developed. Conversely, incompleteness results in this field can also be demonstrated, as shown in [22].

In this chapter, we presented the main conceptual tools of the logic of questions and sketched the key contemporary paradigms that focus on the proof-theoretic findings developed within these frameworks. In the next chapter, we will examine one such development: erotetic calculi for first-order logic.

References

1. Ajdukiewicz, K. (1923). O intencyi pytania "co to jest P" [*On the intention of question "what is P"*, the issue contains a short note summarizing Ajdukiewicz' lecture under this title given for the Polish Philosophical Society in February 1923. https://www.wbc.poznan.pl/dlibra]. *Ruch Filozoficzny*, *7*(9–10), 152b–153a.
2. Ajdukiewicz, K. (1926). Analiza semantyczna zdania pytajnego [*Semantic analysis of an interrogative sentence*, the issue contains a short note summarizing Ajdukiewicz' lecture under this title given for the Polish Psychological Society in November 1926. https://www.wbc.poznan.pl/dlibra]. *Ruch Filozoficzny*, *10*(7–10), 194b–195b.
3. Åqvist, L. (1975). *A new approach to the logical theory of interrogatives: Analysis and formalization*. Verlag G. Narr.
4. Batens, D. (1989). Dynamic dialectical logics. In G. Priest, R. Routley, & J. Norman (Eds.), *Paraconsistent logic. Essays on the inconsistent* (pp. 187–217). Philosophia Verlag.
5. Batens, D. (2001). A general characterization of adaptive logics. *Logique et Analyse, 173–175*, 45–68.
6. Batens, D. (2005). A procedural criterion for final derivability in inconsistency-adaptive logics. *Journal of Applied Logic, 3*(1), 221–250.
7. Belnap, N. D., & Steel, T. P. (1976). *The logic of questions and answers*. Yale University Press.
8. Bromberger, S. (1966). Questions. *The Journal of Philosophy, 63*(20), 597–606.
9. Chlebowski, S. (2018). *Canonical and dual erotetic calculi for first-order logic* (Doctoral dissertation). Adam Mickiewicz University.

[37] For a more precise account, see [45]; [47, Sect. 3] and [128, p. 98]. See also [134, 142] for a strengthening of Harrah's incompleteness theorem.

10. Chlebowski, S., & Gajda, A. (2017). Abductive question-answer system (AQAS) for classical propositional logic. In H. Christiansen, H. Jaudoin, P. Chountas, T. Andreasen, & H. L. Larsen (Eds.), *Flexible query answering systems, FQAS 2017, Lecture Notes in Computer Science* (Vol. 10333, pp. 3–14). Springer. https://doi.org/10.1007/978-3-319-59692-1_1
11. Chlebowski, S., Gajda, A., & Urbański, M. (2022). An abductive question-answer system for the minimal logic of formal inconsistency mbC. *Studia Logica, 110*, 479–509.
12. Chlebowski, S., Komosinski, M., & Kups, A. (2017). Automated generation of erotetic search scenarios: Classification, optimization, and knowledge extraction. *ACM Transactions on Computational Logic, 18*(2), 8:1–8:27. https://doi.org/10.1145/3056537
13. Chlebowski, S., & Leszczyńska-Jasion, D. (2015). Dual erotetic calculi and the minimal LFI. *Studia Logica, 103*, 1245–1278. https://doi.org/10.1007/s11225-015-9617-0
14. Ciardelli, I. A. (2016). *Questions in logic* (Doctoral dissertation). University of Amsterdam.
15. Ciardelli, I. A. (2018). Questions as information types. *Synthese, 195*, 321–365. https://doi.org/10.1007/s11229-019-02321-7
16. Ciardelli, I. A. (2022). *Inquisitive Logic. Consequence and inference in the realm of questions*. Springer.
17. Ciardelli, I. A., Groenendijk, J., & Roelofsen, F. (2015). On the semantics and logic of declaratives and interrogatives. *Synthese, 192*(6), 1689–1728. https://doi.org/10.1007/s11229-013-0352-7
18. Ciardelli, I. A., Groenendijk, J., & Roelofsen, F. (2019). *Inquisitive semantics*. Oxford University Press.
19. Ciardelli, I. A., & Roelofsen, F. (2011). Inquisitive logic. *Journal of Philosophical Logic, 40*, 55–94. https://doi.org/10.1007/s10992-010-9142-6
20. Clercq, K. D., & Verhoeven, L. (2004). Sieving out relevant and efficient questions. *Logique et Analyse, 185–188*, 189–216. https://www.jstor.org/stable/44084772
21. Cohen, F. S. (1929). What is a question? *The Monist, 39*(3), 350–364.
22. Cordes, M. (2020). Calculizing classical inferential erotetic logic. *The Review of Symbolic Logic, 1–22*. https://doi.org/10.1017/S1755020320000131
23. Cordes, M. (2021). How to arrive at questions. In M. Cordes (Ed.), *Asking and answering. Rivalling approaches to interrogative methods* (pp. 165–175). Narr Francke Attempto Verlag.
24. Cordes, M., & Reinmuth, F. (2017). Commentary and illocutionary expressions in linear calculi of natural deduction. *Logic and Logical Philosophy, 26*(2), 163–196. https://doi.org/10.12775/LLP.2017.002
25. Cresswell, M. J. (1965). The logic of interrogatives. In J. N. Crossley & M. A. E. Dummet (Eds.), *Formal systems and recursive functions. Proceedings of the 8th Logic Colloquium, Oxford, July 1963* (pp. 8–11).
26. Cross, C., & Roelofsen, F. (2022). Questions. In E. N. Zalta (Ed.), *The Stanford Encyclopedia of Philosophy* (Summer 2022 edition). https://plato.stanford.edu/archives/sum2022/entries/questions/
27. Friedman, J. (2013). Question-directed attitudes. *Philosophical Perspectives, 27*(1), 145–174. https://doi.org/10.1111/phpe.12026
28. Friedman, J. (2017). Why suspend judging? *Noûs, 51*(2), 302–326.
29. Friedman, J. (2019). Inquiry and belief. *Noûs, 53*(2), 296–315. https://doi.org/https://doi.org/10.1111/nous.12222
30. Friedman, J. (2020). The epistemic and the zetetic. *Philosophical Review, 129*(4), 501–536. https://doi.org/10.1215/00318108-8540918
31. Friedman, J. (forth.[a]). *Inquiry in epistemology: Doxastic attitudes as zetetic attitudes* [The Blackwell Companion to Epistemology, 3rd ed.].
32. Friedman, J. (forth.[b]). *Zetetic epistemology* [Towards an Expansive Epistemology: Norms, Action, and the Social Sphere. Baron Reed & A. K. Flowerree (eds.)].
33. Frittella, S., Greco, G., Palmigiano, A., & Yang, F. (2016). A multi-type calculus for inquisitive logic. In J. Väänänen, Å. Hirvonen, & R. de Queiroz (Eds.), *Logic, language, information, and computation* (pp. 215–233). Springer.

34. Genot, E. J. (2009). The game of inquiry: The interrogative approach to inquiry and belief revision theory. *Synthese, 171*(2), 271–289. https://doi.org/10.1007/s11229-009-9639-0
35. Genot, E. J., & Jacot, J. (2010). How can questions be informative before they are answered? *Strategic information in interrogative games. Episteme, 9*(2), 189–204. https://doi.org/10.1017/epi.2012.8
36. Ginzburg, J. (2011). Questions: Logic and interactions. In J. van Benthem & A. ter Meulen (Eds.), *Handbook of logic and language* (2nd ed., pp. 1133–1146). Elsevier.
37. Ginzburg, J., Yusupujiang, Z., Li, C., Ren, K., & Łupkowski, P. (2019). Characterizing the response space of questions: A corpus study for English and Polish [Stockholm, Sweden, 11–13 of September 2019, Association for Computational Linguistics]. *Proceedings of the SIGDial 2019 Conference*, 320–330.
38. Groenendijk, J. (1999). The logic of interrogation [reprinted as (39)]. In T. Matthews & D. Strolovitch (Eds.), *Semantics and linguistic theory* (pp. 109–126). Cornell University Press.
39. Groenendijk, J. (2007). The logic of interrogation. In M. Aloni, A. Butler, & P. Dekker (Eds.), *Questions in Dynamic Semantics* (pp. 43–62). Elsevier.
40. Groenendijk, J., & Roelofsen, F. (2009). Inquisitive semantics and pragmatics. In J. M. Larrazabal & L. Zubeldia (Eds.), *Meaning, content, and argument: Proceedings of the ILCLI International Workshop on Semantics, Pragmatics, and Rhetoric* (pp. 41–72). University of the Basque Country Publication Service.
41. Groenendijk, J., & Stokhof, M. (1984). *Studies on the semantics of questions and the pragmatics of answers* (Doctoral dissertation). University of Amsterdam.
42. Groenendijk, J., & Stokhof, M. (2011). Questions. In J. van Benthem & A. ter Meulen (Eds.), *Handbook of logic and language* (2nd ed., pp. 1059–1131). Elsevier.
43. Hamami, Y. (2015). The interrogative model of inquiry meets dynamic epistemic logics. *Synthese, 192*(6), 1609–1642. https://doi.org/10.1007/s11229-014-0460-z
44. Hamblin, C. L. (1958). Questions. *Australasian Journal of Philosophy, 36*(3), 159–168.
45. Harrah, D. (1969). On completeness in the logic of questions. *American Philosophical Quarterly, 6*, 158–164.
46. Harrah, D. (1998). Questions. In *Routledge encyclopedia of philosophy*. Taylor and Francis. https://doi.org/10.4324/9780415249126-U031-1
47. Harrah, D. (2002). The logic of questions. In D. M. Gabbay & F. Guenthner (Eds.), *Handbook of philosophical logic* (2nd ed., Vol. 8, pp. 1–60). Springer.
48. Harris, S. (1994). Gts and interrogative tableaux. *Synthese, 99*(3), 329–343. https://doi.org/10.1007/BF01063992
49. Hautli-Janisz, A., Budzynska, K., McKillop, C., Plüss, B., Gold, V., & Reed, C. (2022). Questions in argumentative dialogue. *Journal of Pragmatics, 188*, 56–79. https://doi.org/10.1016/j.pragma.2021.10.029
50. Hintikka, J. (1976). *The semantics of questions and the questions of semantics: Case studies in the interrelations of logic, semantics and syntax* (Vol. 28 of Acta Philosophica Fennica). North-Holland Publishing Company.
51. Hintikka, J. (1984). The logic of science as a model-oriented logic. In *PSA: Proceedings of the Biennial Meeting of the Philosophy of Science Association* (pp. 177–185).
52. Hintikka, J. (1992). The interrogative model of inquiry as a general theory of argumentation. *Communication and Cognition: An Interdisciplinary Quarterly Journal, 25*(2–3), 221–242.
53. Hintikka, J. (1999). *Inquiry as inquiry: A logic of scientific discovery*. Kluwer Academic Publishers.
54. Hintikka, J. (2007). *Socratic epistemology. Explorations of knowledge-seeking by questioning*. Cambridge University Press.
55. Hintikka, J., Halonen, I., & Mutanen, A. (1999). Interrogative logic as a general theory of reasoning. In *Inquiry as inquiry: A logic of scientific discovery* (pp. 47–90). Kluwer Academic Publishers.
56. Hintikka, J., & Harris, S. (1988). On the logic of interrogative inquiry. In *PSA: Proceedings of the Biennal Meeting of the Philosophy of Science Association* (pp. 233–240).

57. Hintikka, J., & Hintikka, M. B. (1982). Sherlock Holmes confronts modern logic. Toward a theory of information-seeking through questioning. In E. Barth & J. Martens (Eds.), *Argumentation: Approaches to theory formation* (pp. 55–76). John Benjamins Publishing Company.
58. Hiż, H. (1962). Questions and answers. *The Journal of Philosophy, 59*, 253–265.
59. Jeffreys, H. (1948). *Theory of probability* (2nd ed.). The Clarendon Press.
60. Koj, L. (1971). Analiza pytań I. Problem terminów pierwotnych logiki pytań. *Studia Semiotyczne, 2*, 99–113.
61. Koj, L. (1989). The problem of the primitive terms of the logic of questions [A translation of (60)]. In *Inquiries into the generating and proper use of questions* (pp. 11–32). The Publishing House of Maria Curie-Skłodowska University in Lublin.
62. Koj, L., & Wiśniewski, A. (1989). *Inquiries into the generating and proper use of questions*. The Publishing House of Maria Curie-Skłodowska University in Lublin.
63. Komosinski, M., Kups, A., Leszczyńska-Jasion, D., & Urbański, M. (2014). Identifying efficient abductive hypotheses using multi-criteria dominance relation. *ACM Transactions on Computational Logic*, *15*(4). https://doi.org/10.1145/2629669
64. Komosinski, M., Kups, A., & Urbański, M. (2012). Multi-criteria evaluation of abductive hypotheses: Towards efficient optimization in proof theory. In *Proceedings of the 18th International Conference on Soft Computing* (pp. 320–325).
65. Kubiński, T. (1960). An essay in the logic of questions. In *Atti del XII Congresso Internazionale di Filosofia (Venezia 1958)* (pp. 315–322).
66. Kubiński, T. (1971). *Wstęp do logicznej teorii pytań*. PWN.
67. Kubiński, T. (1980). *An outline of the logical theory of questions*. Akademie-Verlag.
68. Leszczyńska, D. (2004). Socratic proofs for some normal modal propositional logics. *Logique et Analyse, 47*(185–188), 259–285.
69. Leszczyńska, D. (2007). *The method of Socratic proofs for normal modal propositional logics*. Adam Mickiewicz University Press.
70. Leszczyńska-Jasion, D. (2008). The method of Socratic proofs for modal propositional logics: K5, S4.2, S4.3, S4M, S4F, S4R and G. *Studia Logica, 89*(3), 371–405.
71. Leszczyńska-Jasion, D. (2009). A loop-free decision procedure for modal propositional logics K4, S4 and S5. *Journal of Philosophical Logic, 38*(2), 151–177.
72. Leszczyńska-Jasion, D. (2018). *From questions to proofs. Between the logic of questions and proof theory*. Faculty of Social Science Publishers, Adam Mickiewicz University.
73. Leszczyńska-Jasion, D. (2021). The method of Socratic proofs. From the logic of questions to proof theory. In M. Cordes (Ed.), *Asking and answering. Rivalling approaches to interrogative methods* (pp. 183–198). Narr Francke Attempto Verlag.
74. Leszczyńska-Jasion, D., & Chlebowski, S. (2019). Synthetic tableaux with unrestricted cut for first-order theories. *Axioms, 8*(4).
75. Leszczyńska-Jasion, D., & Łupkowski, P. (2016). Erotetic search scenarios and three-valued logic. *Journal of Logic, Language and Information, 25*(1), 51–76.
76. Leszczyńska-Jasion, D., Petrukhin, Y., & Shangin, V. (2019). The method of Socratic proofs meets correspondence analysis. *Bulletin of the Section of Logic, 48*(2), 99–116.
77. Leszczyńska-Jasion, D., Petrukhin, Y., Shangin, V., & Jukiewicz, M. (2019). Functional completeness in CPL via correspondence analysis. *Bulletin of the Section of Logic, 48*(1), 45–76.
78. Leszczyńska-Jasion, D., Urbański, M., & Wiśniewski, A. (2013). Socratic trees. *Studia Logica, 101*(5), 959–986. https://doi.org/10.1007/s11225-012-9404-0
79. Leszko, R. (1997). Graphs and matrices of compound numerical questions. In A. Wiśniewski & J. Zygmunt (Eds.), *Erotetic logic, deontic logic and other logical matters: Essays in memory of Tadeusz Kubiński*. Wroclaw University Press.
80. Łukasiewicz, J. (1951). *Aristotle's syllogistic from the standpoint of modern formal logic*. Oxford University Press.
81. Łupkowski, P. (2011). A formal approach to exploring the interrogator's perspective in the Turing test. *Logic and Logical Philosophy*, *20*(1–2), 139–158. https://doi.org/10.12775/LLP.2011.007

82. Łupkowski, P. (2015). Question dependency in terms of compliance and erotetic implication. *Logic and Logical Philosophy, 24*, 357–376.
83. Łupkowski, P. (2016). *Logic of questions in the wild. Inferential Erotetic Logic in information seeking dialogue modelling*. College Publications.
84. Łupkowski, P. (2017). IEL-based formal dialogue system for tutorials. *Logic and Logical Philosophy, 26*(3), 287–320.
85. Łupkowski, P., & Ginzburg, J. (2013). A corpus-based taxonomy of question responses. In *Proceedings of the 10th International Conference on Computational Semantics (IWCS 2013), March 2013* (pp. 354–361).
86. Łupkowski, P., & Ginzburg, J. (2016). Query responses. *Journal of Language Modelling, 4*(2), 245–292.
87. Łupkowski, P., & Leszczyńska-Jasion, D. (2015). Generating cooperative question-responses by means of erotetic search scenarios. *Logic and Logical Philosophy, 24*(1), 61–78.
88. Łupkowski, P., Majer, O., Peliš, M., & Urbański, M. (2018). Epistemic erotetic search scenarios. *Logic and Logical Philosophy, 27*(3), 301–328.
89. Łupkowski, P., Urbański, M., Wiśniewski, A., Błądek, W., Juska, A., Kostrzewa, A., Pankow, D., Paluszkiewicz, K., Ignaszak, O., Urbańska, J., Żyluk, N., Gajda, A., & Marciniak, B. (2017). Erotetic Reasoning Corpus. A data set for research on natural question processing. *Journal of Language Modelling, 5*(3), 607–631.
90. Łupkowski, P., & Wiśniewski, A. (2011). Turing interrogative games. *Minds and Machines, 21*, 435–448.
91. Meheus, J. (1999). Erotetic arguments from inconsistent premises. *Logique et Analyse, 165–166*, 49–80.
92. Meheus, J. (2001). Adaptive logics for question evocation. *Logique et Analyse, 173–175*, 135–164.
93. Millson, J. A. (2019). A cut-free sequent calculus for defeasible erotetic inferences. *Studia Logica, 107*, 1279–1312.
94. Millson, J. A. (2021). A defeasible calculus for zetetic agents. *Logic and Logical Philosophy, 30*, 3–37. https://doi.org/10.12775/LLP.2020.019
95. Minică, Ş. (2011). *Dynamic logic of questions (Doctoral dissertation)*. Institute for Logic: Language and Computation, University of Amsterdam.
96. Peliš, M. (2016). *Inferences with ignorance: Logics of questions*. Karolinum Press.
97. Peliš, M., & Majer, O. (2010). Logic of questions from the viewpoint of dynamic epistemic logic. In M. Peliš (Ed.), *The Logica Yearbook 2009* (pp. 157–172). College Publications.
98. Peliš, M., & Majer, O. (2011). Logic of questions and public announcements. In N. Bezhanishvili, S. Löbner, K. Schwabe, & L. Spada (Eds.), *Logic, language, and computation* (pp. 145–157). Springer.
99. Prior, M., & Prior, A. (1955). Erotetic logic. *Philosophical Review, 64*(1), 43–59.
100. Punčochář, V. (2020). A relevant logic of questions. *Journal of Philosophical Logic, 49*, 905–939. https://doi.org/10.1007/s10992-019-09541-9
101. Roelofsen, F. (2019). Semantic theories of questions. In *Oxford research encyclopedia of linguistics*. Oxford University Press. https://doi.org/10.1093/acrefore/9780199384655.013.504
102. Skura, T. F. (2005). Intuitionistic Socratic procedures. *Journal of Applied Non-Classical Logics, 15*(4), 453–464.
103. Skura, T. F., & Wiśniewski, A. (2015). A system for proper multiple-conclusion entailment. *Logic and Logical Philosophy, 24*, 241–253.
104. Smets, S., Boddy, R., & Baltag, A. (2018). Group knowledge in interrogative epistemology. In H. Ditmarschvan Ditmarsch & G. Sandu (Eds.), *Jaakko Hintikka on knowledge and game theoretical semantics* (pp. 131–164). Springer.
105. Sochański, M., & Leszczyńska-Jasion, D. (2025). *Methods of construction of synthetic tableaux based on cographs and permutations* [unpublished]. Submitted to *Journal of Logic and Computation*.

106. Sochański, M., Leszczyńska-Jasion, D., Chlebowski, S., Tomczyk, A., & Jukiewicz, M. (2022). Synthetic tableaux: Minimal tableau search heuristics. In J. Blanchette, L. Kovács, & D. Pattinson (Eds.), *Automated Reasoning. IJCAR 2022* (pp. 407–425). Springer, Cham. https://doi.org/10.1007/978-3-031-10769-6_25
107. Strawson, P. F. (1950). On referring. *Mind, 59*(235), 320–344.
108. Švarný, P., Majer, O., & Peliš, M. (2014). Erotetic epistemic logic in private communication protocol. In M. Dančák & V. Punčochář (Eds.), *The Logica Yearbook 2013* (pp. 223–238). College Publications.
109. Szczepiński, R. (2018). *The method of Socratic proofs for first degree entailment* [Unpublished manuscript received from the author in March 2018].
110. ten Cate, B., & Shan, C.-C. (2007). Axiomatizing Groenendijk's Logic of Interrogation. In M. Aloni, A. Butler, & P. Dekker (Eds.), *Questions in dynamic semantics* (pp. 63–82). Elsevier.
111. Urbański, M. (2001). Remarks on synthetic tableaux for classical propositional calculus. *Bulletin of the Section of Logic, 30*(4), 194–204.
112. Urbański, M. (2001). Synthetic tableaux and erotetic search scenarios: Extension and extraction. *Logique et Analyse, 173-174-175*, 69–91.
113. Urbański, M. (2002). Synthetic tableaux for Łukasiewicz's calculus Ł3. *Logique et Analyse, 177–178*, 155–173.
114. Urbański, M. (2002). *Tabele syntetyczne a logika pyta ń*. Maria Curie-Skłodowska University Press.
115. Urbański, M. (2004). How to synthesize a paraconsistent negation. The case of CLuN. *Logique et Analyse, 185–188*, 319–333.
116. Urbański, M. (2005). *Tableaux, abduction and truthlikeness (Research Report)*. Poznań: Adam Mickiewicz University.
117. Urbański, M. (2010). *Rozumowania abducyjne*. Adam Mickiewicz University Press.
118. Urbański, M., & Łupkowski, P. (2010). Erotetic search scenarios: Revealing interrogator's hidden agenda. In P. Łupkowski & M. Purver (Eds.), *Semantics and pragmatics of dialogue* (pp. 67–74). Polish Society for Cognitive Science.
119. Urbański, M., & Wiśniewski, A. (2015). On search for law-like statements as abductive hypotheses by Socratic transformations. In C. Baskent (Ed.), *Perspectives on interrogative models of inquiry* (pp. 111–127). Springer.
120. van Benthem, J., & Minică, Ş. (2012). Toward a dynamic logic of questions. *Journal of Philosophical Logic, 41*, 633–669.
121. Verdee, P. (2013). A proof procedure for adaptive logics. *Logic Journal of the IGPL, 21*(5), 743–766.
122. Whitcomb, D., & Millson, J. (2023). Inquiring attitudes and erotetic logic: Norms of restriction and expansion. *Journal of the American Philosophical Association, 1–23*. https://doi.org/10.1017/apa.2023.13
123. Willard-Kyle, C., Millson, J., & Whitcomb, D. (2024). Evoked questions and inquiring attitudes. *The Philosophical Quarterly*. https://doi.org/10.1093/pq/pqae083
124. Wiśniewski, A. (1985). Propositional logic and erotetic inferences [with erratum published in the next number of the journal]. *Bulletin of the Section of Logic, 14*(2), 72–78.
125. Wiśniewski, A. (1990). Implied questions. *Manuscrito, 13*(2), 23–38.
126. Wiśniewski, A. (1991). Erotetic arguments: A preliminary analysis. *Studia Logica, 50*(2), 261–274.
127. Wiśniewski, A. (1994). Erotetic implications. *Journal of Philosophical Logic, 23*, 173–195.
128. Wiśniewski, A. (1995). *The posing of questions: Logical foundations of erotetic inferences*. Kluwer Academic Publishers.
129. Wiśniewski, A. (1997). Kubiński's theory of questions. In A. Wiśniewski & J. Zygmunt (Eds.), *Erotetic logic, deontic logic and other logical matters: Essays in memory of Tadeusz Kubiński* (pp. 29–50). Wroclaw University Press.
130. Wiśniewski, A. (2001). Questions and inferences. *Logique et Analyse, 173–175*, 5–43.
131. Wiśniewski, A. (2003). Erotetic search scenarios. *Synthese, 134*(3), 389–427.
132. Wiśniewski, A. (2004). Socratic proofs. *Journal of Philosophical Logic, 33*(3), 299–326.

133. Wiśniewski, A. (2006). *A right-sided Socratic calculus for classical logic (Research Report)*. Poznań: Adam Mickiewicz University.
134. Wiśniewski, A. (2013). Effectiveness of question-answer systems. In *Essays in logical philosophy* (pp. 53–62). LIT-Verlag.
135. Wiśniewski, A. (2013). *Questions, inferences, and scenarios*. College Publications.
136. Wiśniewski, A. (2015). Semantics of questions. In S. Lappin & C. Fox (Eds.), *The handbook of contemporary semantic theory* (2nd ed., pp. 273–313). Wiley-Blackwell.
137. Wiśniewski, A. (2016). An axiomatic account of question evocation: The propositional case. *Axioms, 5*.
138. Wiśniewski, A. (2017). Generalized entailments. *Logic and Logical Philosophy, 26*, 321–356. https://doi.org/10.12775/LLP.2017.014
139. Wiśniewski, A. (2021). Entailment, Transmission of Truth, and Minimality. *Journal of Applied Logics, 8*(10), 2471–2522.
140. Wiśniewski, A. (2021). An essay on Inferential Erotetic Logic. In M. Cordes (Ed.), *Asking and answering. rivalling approaches to interrogative methods* (pp. 105–138). Narr Francke Attempto Verlag.
141. Wiśniewski, A., & Leszczyńska-Jasion, D. (2015). Inferential erotetic logic meets inquisitive semantics. *Synthese, 192*, 1585–1608.
142. Wiśniewski, A., & Pogonowski, J. (2010). Interrogatives, recursion, and incompleteness. *Journal of Logic and Computation, 20*(6), 1187–1199.
143. Wiśniewski, A., & Shangin, V. (2006). Socratic proofs for quantifiers. *Journal of Philosophical Logic, 35*(2), 147–178. https://doi.org/10.1007/s10992-005-9000-0
144. Wiśniewski, A., Vanackere, G., & Leszczyńska, D. (2005). Socratic proofs and paraconsistency: A case study. *Studia Logica, 80*(2–3), 433–468.
145. Wiśniewski, A., & Zygmunt, J. (Eds.). (1997). *Erotetic logic, deontic logic and other logical matters: Essays in memory of Tadeusz Kubiński*. Wroclaw University Press.

Chapter 2
Erotetic Calculi for the Classical Logic

This chapter presents erotetic calculi for classical logic—both propositional and first-order. As explained in the first chapter, erotetic calculi are systems that transform questions of certain formal languages. The intuitive interpretation underlying this formalism is that the aim of such transformations is to solve the problem expressed by the initial question. Thus, the entire process of transforming one question into further questions can be viewed as a model of *question processing*, understood as a cognitive, problem-solving activity. In the previous chapter, we also referred to this process as *inquiring*, and we completed the picture by comparing this process against that of *justifying*, where the latter can be seen as the intuitive counterpart of a proof in more traditional proof systems, such as axiomatic systems, sequent systems, or natural deduction systems.

Our intuitive understanding of transformations of questions is supported by the fact that erotetic rules constitute an inferential relation between questions, whose semantic counterpart is a special case of erotetic implication. We will strengthen our interpretation of the method of Socratic proofs in Chap. 5, where the syntactic counterpart of this inferential relation will be examined.

In the meantime, let us note that questions acted upon by erotetic rules can concern, i.a., the validity of a formula. In this way, an erotetic calculus constitutes a proof system for the underlying logic. In this chapter, we will focus on the proof-theoretical aspects of erotetic calculi for classical logic. We will start with technical preliminaries, then add sequents and questions to the language of classical logic (Sect. 2.2). In Sect. 2.3, we present the erotetic rules. In Sect. 2.4, their soundness is demonstrated, and Sect. 2.5 contains a general discussion on structural, derivable, and admissible erotetic rules. Finally, Sect. 2.6 demonstrates the completeness of the calculi; consistency properties are used for that purpose (see [8]).

D. Leszczyńska-Jasion, *The Method of Socratic Proofs*, Trends in Logic 64,
https://doi.org/10.1007/978-3-031-82451-7_2

In this chapter, we present three erotetic calculi for classical first-order logic, followed by three erotetic calculi for the propositional fragment, which are easily derived from the former. Let us briefly justify the development of this plethora of systems. The basic syntactic structure used in erotetic calculi is the sequent, as we shall see, and the three calculi differ in the types of sequents they employ. A sequent is a rich structure; in particular, we can define various versions of sequents, which makes them suitable for simulating diverse proof methods. We will focus on this aspect of the method of Socratic proofs in subsequent chapters. In the next two chapters, we will also see how different types of sequents are helpful in formalizing non-classical logics.

2.1 Preliminaries

This section contains basic definitions and notation conventions. The material can be skipped until needed.

Quotation marks (i.e., ‘ ’) are used in various roles throughout this book. We believe, however, that this will not cause confusion. In addition to their usual functions, quotation marks are used to indicate that an expression is mentioned (not used), and as the so-called Quinean corners. The primary goal in writing this book is readability, followed by preciseness; therefore quotation marks in metalogical functions are sometimes omitted.

We will use ‘$\varnothing$’ for the empty set, ‘$\mathbb{N}_0$’ for the set of natural numbers including 0, and ‘$\mathbb{N}$’ for the set of natural numbers without 0. The symbol ‘$\subseteq$’ is for set inclusion, and ‘$\subset$’ for proper set inclusion. If X is a set, the symbol ‘2^X’ will be used for the power set of X, that is, the set of all subsets of X. Since parentheses are used in many contexts throughout this book, we will use angled brackets ‘$\langle$’ and ‘$\rangle$’ for finite sequences.

The abbreviation ‘iff’ is used for ‘if and only if’. In inductive proofs, ‘IH’ stands for ‘inductive hypothesis’. We will also use ‘CPL’ for classical propositional logic and ‘FOL’ for classical first-order logic. The symbol ‘$\mathcal{L}_{\mathsf{CPL}}$’ stands for the language of CPL, which contains the following symbols: countably infinitely many propositional variables $p_1, p_2, p_3, \ldots$, propositional connectives $\neg, \wedge, \vee, \rightarrow$, and parentheses: (,). We will use p, q, r for propositional variables of $\mathcal{L}_{\mathsf{CPL}}$, and p_i as a metavariable. ‘*VAR*’ denotes the set of all propositional variables of $\mathcal{L}_{\mathsf{CPL}}$. The notion of a formula of the language $\mathcal{L}_{\mathsf{CPL}}$ is given by the following definition.

Definition 2.1 (*Formula of Language $\mathcal{L}_{\mathsf{CPL}}$*)

1. Each element of VAR is a formula of language $\mathcal{L}_{\mathsf{CPL}}$.
2. If A is a formula of language $\mathcal{L}_{\mathsf{CPL}}$, then the expression of the form ‘$\neg A$’ is also a formula of language $\mathcal{L}_{\mathsf{CPL}}$.

3. If A and B are formulas of language $\mathcal{L}_{\mathsf{CPL}}$, then the expressions of the forms: '$(A \wedge B)$', '$(A \vee B)$', '$(A \rightarrow B)$' are also formulas of language $\mathcal{L}_{\mathsf{CPL}}$.
4. Nothing else is a formula of language $\mathcal{L}_{\mathsf{CPL}}$.

The set of all formulas of language $\mathcal{L}_{\mathsf{CPL}}$ will be denoted by '$FORM_{\mathsf{CPL}}$'.

As usual, it is assumed that $\neg$ (negation) binds more strongly than binary connectives, and that $\wedge$ (conjunction) and $\vee$ (disjunction) bind more strongly than $\rightarrow$ (implication). Parentheses are omitted when there is no risk of confusion. If needed, equivalence ($\leftrightarrow$) is introduced by definition:

$$A \leftrightarrow B =^{df} (A \rightarrow B) \wedge (B \rightarrow A). \tag{2.1}$$

The notion of a formula of a language is sometimes defined using Backus-Naur form, commonly referred to as BNF. The BNF notation involves specifying a syntactic category (e.g., that of a formula) and then providing recursive rules to generate the members of this category. For example, the following:

$$A ::= p \mid \neg A \mid A \wedge A \mid A \vee A \mid A \rightarrow A$$

produces the set of all formulas of the language $\mathcal{L}_{\mathsf{CPL}}$.[1] From now on, we will describe formal languages using BNF notation.

$\mathcal{L}_{\mathsf{FOL}}$ stands for the language of FOL with the following symbols:

- the propositional connectives of $\mathcal{L}_{\mathsf{CPL}}$, that is, $\neg, \wedge, \vee, \rightarrow$,
- quantifiers: $\forall$ (the universal quantifier, 'for all') and $\exists$ (the existential quantifier, 'there is'),
- countably infinitely many individual variables $x_1, x_2, x_3, \ldots$; for simplicity, we will use x, y, z,
- countably infinitely many predicates of arbitrary arity $n \in \mathbb{N}$: $P_1^n, P_2^n, P_3^n, \ldots$; for simplicity, we will use P, R when the arity of a predicate is clear from the context,
- countably infinitely many function symbols of arbitrary arity $n \in \mathbb{N}$: F_1^n, F_2^n, $F_3^n, \ldots$; as above, we will use F, G for simplicity,
- countably infinitely many individual constants: $a_1, a_2, a_3, \ldots$; similarly, a, b, c will be used for simplicity,
- comma: , and parentheses: (,).

It is assumed that the sets of individual variables of $\mathcal{L}_{\mathsf{FOL}}$ and that of individual constants of $\mathcal{L}_{\mathsf{FOL}}$ are disjoint.

The greater expressiveness of the language $\mathcal{L}_{\mathsf{FOL}}$ manifests itself, i.a., in two categories of 'meaningful' expressions. Namely, $\mathcal{L}_{\mathsf{FOL}}$ includes *terms*, corresponding to natural-language names, and *formulas*, corresponding to natural-language sentences. Terms of $\mathcal{L}_{\mathsf{FOL}}$ are individual variables, individual constants, and complex terms built from simpler ones using function symbols. Thus, if $t_1, \ldots, t_n$ are terms

---

[1] Strictly speaking, this need not be the same set of expressions as that defined in Definition 2.1, with a possible difference involving the use of parentheses.

of $\mathcal{L}_{\mathsf{FOL}}$, and n is the arity of the function symbol F_i^n, then '$F_i^n(t_1, \ldots, t_n)$' is a term of the language as well.

Predicates are used to build *atomic formulas of the language* $\mathcal{L}_{\mathsf{FOL}}$. If $t_1, \ldots, t_n$ are terms of the language $\mathcal{L}_{\mathsf{FOL}}$ and n is the arity of the predicate P_i^n, then '$P_i^n(t_1, \ldots, t_n)$' is an atomic formula of $\mathcal{L}_{\mathsf{FOL}}$. Finally, *formulas of the language* $\mathcal{L}_{\mathsf{FOL}}$ are defined by:

$$A ::= P_i^n(t_1, \ldots, t_n) \mid \forall x_i A \mid \exists x_i A \mid \neg A \mid A \wedge A \mid A \vee A \mid A \to A$$

The set of all formulas of the language $\mathcal{L}_{\mathsf{FOL}}$ will be denoted by '$FORM_{\mathsf{FOL}}$'.

As in $\mathcal{L}_{\mathsf{CPL}}$, equivalence can be introduced by (2.1).

Symbols: x_i, a_i, F_i^n, and P_i^n will be used as the respective metavariables, as in the above definitions. The letters t, t_i will be used as metavariables for terms of the language $\mathcal{L}_{\mathsf{FOL}}$.

Individual variables can occur *free* or *bound* in a formula of $\mathcal{L}_{\mathsf{FOL}}$, as shown in the following examples:

$$\forall x P(x)$$

where 'x' is bound by '$\forall$',

$$P(x) \to \neg Q(x)$$

where 'x' occurs twice, and is free in both cases, and

$$P(x) \to \exists x \neg Q(x)$$

where 'x' is free in its first occurrence (from the left), and bound by '$\exists$' in its second occurrence. Intuitively, whether an individual variable is free or bound in a given position within a formula affects the *meaning* of the formula. We have already seen this semantic and pragmatic aspect in the first chapter, when discussing the idea of treating questions as expressed by sentential functions. For this reason, we need a handful of definitions to be precise on the matter.

The *set of free variables of a term t of* $\mathcal{L}_{\mathsf{FOL}}$, symbolically denoted as $FV(t)$, is defined as follows: If t is an individual variable, x_i, then: $FV(t) = \{x_i\}$. If t is an individual constant, then $FV(t) = \varnothing$. Finally, for complex terms, we define:

$$FV(F_i^n(t_1, \ldots, t_n)) = \bigcup_{i=1}^{n} FV(t_i).$$

Now we can define the *set of free variables of a formula A of the language* $\mathcal{L}_{\mathsf{FOL}}$, in symbols, $FV(A)$. We say that an individual variable x_k belongs to the set of free variables of an atomic formula $P_j^n(t_1, \ldots, t_n)$ iff x_k occurs in one of the terms of the formula. Symbolically:

$$x_k \in FV(P_j^n(t_1, \ldots, t_n)) \text{ iff } x_k \in \bigcup_{i=1}^{n} FV(t_i).$$

Further, if A is of the form '$\neg B$', then $FV(A) = FV(B)$. If A is of the form '$B \otimes C$', where $\otimes \in \{\wedge, \vee, \rightarrow\}$, then $FV(A) = FV(B) \cup FV(C)$. Finally, if A is of the form '$Qx_i B$', where $Q \in \{\forall, \exists\}$, then $FV(A) = FV(B) \setminus \{x_i\}$. Last but not least, if a variable x_i occurs in a formula A, but $x_i \notin FV(A)$, then the variable is *bound in* A.

Now we can state that if the set of free variables of a formula A of $\mathcal{L}_{\mathsf{FOL}}$ is non-empty, then A is called a *sentential function of the language* $\mathcal{L}_{\mathsf{FOL}}$; otherwise, A is called a *sentence of* $\mathcal{L}_{\mathsf{FOL}}$.

In the case of $\mathcal{L}_{\mathsf{CPL}}$, we will use the notions 'formula' and 'sentence' interchangeably. We will also drop the phrase 'of language...' when this does not lead to confusion. The letters A, B, C, D, E, F, possibly with subscripts, will be used as metavariables for formulas, while the letters X, Y, Z, also possibly with subscripts, will serve as metavariables for sets of formulas. This notation applies to both the languages of CPL and FOL. In the case of $\mathcal{L}_{\mathsf{FOL}}$, the letter F may be used as a metavariable for both function symbols and formulas, but context should clarify any potential confusion.

By a *closed term of the language* $\mathcal{L}_{\mathsf{FOL}}$, we mean a term of the language that does not contain any individual variables. Closed terms of $\mathcal{L}_{\mathsf{FOL}}$ can be thought of as 'genuine' names of the language.

For substitutions, we use the notation: $A[x_i/t]$, which refers to the result of substituting a term t for an individual variable x_i in a formula A. First, we define a substitution of a term for an individual variable in another term. This is set recursively as follows.

$$t[x_i/x_j] = \begin{cases} t & \text{if } i = j; \\ x_j & \text{if } i \neq j; \end{cases} \qquad \text{and} \qquad t[x_i/a_j] = a_j,$$

for the base case. For complex terms we set:

$$t[x_i/F_j^n(t_1, \ldots, t_n)] = F_j^n(t[x_i/t_1], \ldots, t[x_i/t_n]).$$

Substitutions in formulas are also defined recursively. In the base case we put:

$$P_j^n(t_1, \ldots, t_n)[x_i/t] = P_j^n(t[x_i/t_1], \ldots, t[x_i/t_n]);$$

and further, for $\otimes \in \{\wedge, \vee, \rightarrow\}$:

$$(A \otimes B)[x_i/t] = (A[x_i/t] \otimes B[x_i/t]);$$

$$(\neg A)[x_i/t] = \neg(A[x_i/t]).$$

For the cases with quantifiers, $Q \in \{\forall, \exists\}$:

$$(Qx_i A)[x_i/t] = (Qx_i A); \tag{2.2}$$

and finally, if $i \neq j$:

$$(Qx_jA)[x_i/t] = Qx_j(A[x_i/t]) \tag{2.3}$$

One more technical notion is in order here. We say that a *term t is free for x_i in A* provided that whenever condition (2.3) is met, $x_j \notin FV(t)$. If $x_j \in FV(t)$, then, as a result of substitution, one obtains a bound variable where previously there was a free variable. In this case, we say that there is a *clash of variables*, or that a *free variable of t collides with the quantifier bounding x_j*. Generally, substitutions in a clash-of-variables situation should be avoided. Why? Because we want to have a version of substitution that ensures a free variable can be thought of as 'whatever'. If a formula describing 'whatever' truly describes *the whatever*, then it should remain true after substituting any other name for 'whatever'. In other words, whatever is true of the extension of 'whatever' must also be true of the extension of any other name. This somewhat enigmatic statement expresses a condition that must be satisfied by a substitution operation if the operation is *truth-preserving*.

The following formula provides a prime example illustrating why an operation of substitution that does not avoid the clash of variables is not truth-preserving and, hence, is not appropriate. Interpreting R as referring to the greater-than relation among the natural numbers, the formula

$$\exists x R(x, y) \tag{2.4}$$

says that there is always an x greater than y, which is obviously true under the indicated interpretation. However, after substituting $F(x)$ for y we arrive at:

$$\exists x R(x, F(x))$$

which asserts that there is an x greater than $F(x)$. Here 'F' can stand, for instance, for the successor operation, leading to the sentence 'There is an x greater than $x + 1$', which is plainly false under the intended interpretation of '$+1$' and 'greater than'.

One should also remember that the definition of substitution is recursive, therefore the clash-situation can happen at a lower level of formula's structure. For example, term $F(x)$ is not free for y in the formula:

$$\forall z(P(y, z) \to \exists x R(x, y))$$

since condition (2.3) is met in the substitution $\exists x R(x, y)[y/F(x)]$, but $x \in FV(F(x))$.

It is often assumed that the very use of the notation '$A[x_i/t]$' implies that t is free for x_i in A (see [26, Sect. 1.1.2]). The reason for this assumption is that the clash-of-variables situation can always be avoided by applying the following law, known as the 'renaming of bound variables':

$$Qx_iA \leftrightarrow Qx_jA[x_i/x_j] \tag{2.5}$$

where $Q \in \{\forall, \exists\}$, x_j is not free in A, and x_j is free for x_i in A. For example, we can apply this law to formula (2.4) as follows: first, we choose an individual variable for x_j that does not occur in the formula, say $x_j = z$, so we are guaranteed that x_j is free for x_i in A. We then obtain '$\exists z R(z, y)$' on the right side of (2.5), and we can perform the substitution $\exists z R(z, y)[y/F(x)]$ without encountering a clash of variables. Henceforth in this work, whenever substitution is performed, we assume that there is no clash of variables.

Finally, we will use S, T to denote finite, possibly empty, sequences of formulas of the language $\mathcal{L}_{\mathsf{CPL}}/\mathcal{L}_{\mathsf{FOL}}$. If $S = \langle A_1, \ldots, A_{n-1}, A_n \rangle$ is such a sequence, then by '$\bigvee S$' and '$\bigwedge S$' we refer to the following disjunction, respectively, conjunction, of the terms of S:

$$A_1 \vee (\ldots \vee (A_{n-1} \vee A_n) \ldots),$$

$$A_1 \wedge (\ldots \wedge (A_{n-1} \wedge A_n) \ldots).$$

A Note Concerning the Use of Parameters

Parameters are additional name symbols commonly employed in the construction of proof systems for FOL. If we wish to consider the language $\mathcal{L}_{\mathsf{FOL}}$ as an adequate model of languages of mathematical theories, then parameters are appropriate proof-theoretical tools. Our constants introduced as symbols of $\mathcal{L}_{\mathsf{FOL}}$ represent *names* such as '1' or '∅'. Conversely, in mathematical reasoning, *some* objects are represented by additional meta-level symbols. For instance, if in the course of a proof we arrive at the conclusion that 'it is the case that $\exists x P(x)$', and we then wish to reason about the object having the property P, we would choose a generic symbol as a name for that object rather than a specific name like '1' or '∅', since these already carry meaning in the underlying model.

Thus, it seems reasonable to formulate the rules of proof systems with a supply of parameters. Moreover, for adequate modeling of a mathematical theory, parameters should belong to the meta-language rather than the object-level language of the theory.

However, it should be understood that parameters are not strictly necessary to formulate an adequate proof system for FOL, nor are they essential for conducting a meta-level proof of completeness using consistency properties. Although the lack of additional supply of uncommitted names presents a challenge in demonstrating the saturation property (see Lemma 2.10), many other aspects of the completeness proof simplify significantly when we drop the 'layered' language construction that includes both individual constants and parameters. It is one good reason to formulate a proof system without them. In addition, all other versions of the Socratic proofs method for FOL, ranging from [29] to [2], have been formulated using parameters (and various constructions for completeness proofs), so this version is already well-explored in the erotetic setting. For these reasons, a non-parameter version is developed in this book.

2.2 Supplementing Languages with Sequents and Questions

Erotetic calculi are worded in languages that include questions. However, the primary declarative expressions of these languages are sequents. As explained in the previous chapter, the initial motivation for constructing erotetic calculi was to create calculi that transform questions concerning significant (semantic or proof-theoretical) properties or relations defined within the underlying logic—such as theoremhood, consequence relation, validity, and entailment. Consequently, the use of sequents was a natural choice, as a sequent represents a crucial unit of proof-theoretical information.

Roughly speaking, a sequent is an expression of a formal language falling under the following scheme:

$$A_1, \ldots, A_n \Rightarrow B_1, \ldots, B_m.$$

It is assumed that this expression concerns *inferences* or *deductions* from the premises $A_1, \ldots, A_n$ to a conclusion, which is a single formula among $B_1, \ldots, B_m$. This type of expressions was introduced and studied by Paul Hertz[2] to examine the formal properties of proofs in axiomatic systems. The first sequent calculi (for classical and intuitionistic logic) was introduced by Gentzen in [10] as technical tools used to prove certain properties of the calculus of natural deduction, mainly its completeness and normalization. Gentzen used the arrow '$\Rightarrow$' as the sequent sign; the use of the turnstile symbol '$\vdash$' in this context comes from Haskell B. Curry (see [4]). In this book, we follow Curry's account.

Actually, there are more types of sequents and of erotetic calculi than those discussed here. The other possibilities, specifically the so-called *dual* erotetic calculi, are examined in [2]. While the systematic study of dual erotetic calculi is beyond the scope of this book, we will briefly refer to one of the dual calculi for CPL in Sect. 6.3 of Chap. 6.

The description of the syntax of languages with sequents and questions follows [30, pp. 22–24]. Thus, let $\mathcal{L}_{\mathsf{L}}$ stand for the language of L, where L is either CPL or FOL. For each L, a richer language, labeled '$\mathcal{L}^?_{\vdash\mathsf{L}}$', is construed. The vocabulary of $\mathcal{L}^?_{\vdash\mathsf{L}}$ includes the vocabulary of $\mathcal{L}_{\mathsf{L}}$, as well as the comma ',' (if this is not already present in the language), the semicolon ';' and the following signs: ?, $\vdash$, *ng* ($\mathcal{L}^?_{\vdash\mathsf{L}}$-negation), & ($\mathcal{L}^?_{\vdash\mathsf{L}}$-conjunction). The language $\mathcal{L}^?_{\vdash\mathsf{L}}$ has two categories of well-formed expressions: *declarative well-formed formulas* (d-wffs, for short) and *erotetic well-formed formulas*, simply referred to as *questions*. Among the d-wffs of $\mathcal{L}^?_{\vdash\mathsf{L}}$, we distinguish *atomic d-wffs of* $\mathcal{L}^?_{\vdash\mathsf{L}}$, which are sequents.

A *both-sided sequent of language* $\mathcal{L}^?_{\vdash\mathsf{L}}$ is an expression of the form:

$$S \vdash A \tag{2.6}$$

[2] Hertz studied sentences (*sätze*) of the form '$A_1, \ldots, A_n \longrightarrow B$,' that is, *single succedent* sequents, as we shall call them, already in the 1920s. For more on the history of the topic, see [12, 17].

where S is a finite, possibly empty, sequence of formulas of $\mathcal{L}_{\mathsf{L}}$ and A is a formula of $\mathcal{L}_{\mathsf{L}}$. The turnstile symbol '$\vdash$', which is an object-level sign in this context, refers to the derivability relation in the underlying logic L. This interpretation justifies the use of the turnstile symbol instead of the arrow '$\Rightarrow$'. Thus, the expression (2.6) can be understood as representing a metalevel statement 'formula A is derivable from sequence S in logic L,' where derivability from a sequence is understood as derivability from the set of terms of the sequence.

A *right-sided sequent of language* $\mathcal{L}^{?}_{\vdash\mathsf{L}}$ is an expression of the form:

$$\vdash S \tag{2.7}$$

where S is a finite and non-empty sequence of formulas of $\mathcal{L}_{\mathsf{L}}$. In this context, the turnstile symbol may be interpreted as referring to the validity of the disjunction $\bigvee S$ in L.

A *left-sided sequent of* $\mathcal{L}^{?}_{\vdash\mathsf{L}}$ is an expression of the form:

$$S \vdash \tag{2.8}$$

where, again, S is a finite and non-empty sequence of formulas of $\mathcal{L}_{\mathsf{L}}$. Sequents of this form may be interpreted as stating that the set of terms of sequence S is inconsistent in L.

The presentation of the three types of sequents is now generalized as follows. A *sequent of language* $\mathcal{L}^{?}_{\vdash\mathsf{L}}$ is an expression of the form:

$$S \vdash T \tag{2.9}$$

where S and T are finite sequences of formulas of $\mathcal{L}_{\mathsf{L}}$, and either T is a one-term sequence, in which case we write (2.6) instead of $S \vdash \langle A \rangle$, or S is empty but T is not, then we have (2.7), or, finally, T is empty and S is not, and then we have (2.8).[3]

Traditionally, S is referred to as the *antecedent* of sequent (2.9), and T as its *succedent*. Sequents that allow exactly one formula in the succedent are called *single succedent* or *single-conclusioned*, while those that allow more than one formula are referred to as *multisuccedent* or *multi-conclusioned*.

The standard interpretation of a multisuccedent sequent of the form (2.9), often called *denotational* or *material*, is in terms of the validity of the formula '$\bigwedge S \rightarrow \bigvee T$'; see [19, p. 47]; [24, p. 33]. We will return to the issue of sequent interpretation in Sect. 5.6 of Chap. 5, where we develop a purely syntactic interpretation of multisuccedent sequents.

---

[3] We do not consider the case where both T and S are empty, though it makes sense to allow the empty sequent when considering both canonical and dual sequents. This syntactic option is explored in [2].

In structural proof theory, it is more common to use multisets[4] instead of sequences as antecedents and succedents of sequents.[5] In fact, for many logics (including those examined in this book) the order of occurrence of formulas in a sequent is irrelevant for provability. Therefore using a structure like a multiset, which disregards order, is convenient.

However, it seems natural for people to write things in a particular order, and it becomes necessary when implementing a theorem-prover to display things in some ordered manner. Finite sequences naturally align with common data structures such as lists, arrays, and tables. When we use multisets, we implicitly declare that order is unimportant, whereas using sequences requires us either to prove that the order is irrelevant or to introduce explicit rules allowing changes in order. For this reason, while it may be more laborious, it seems more appropriate to use sequences, especially in contexts where implementation is a concern.

Returning to the main topic, after the generalized presentation, we can state that an *atomic d-wff* of the language $\mathcal{L}^{?}_{\vdash\mathsf{L}}$ is an expression of the form (2.9), where S and T are appropriately restricted. We will typically use the term 'sequent' instead of 'atomic d-wff of $\mathcal{L}^{?}_{\vdash\mathsf{L}}$.' Lower-case Greek letters ϕ, ψ, possibly with subscripts, will be used for sequents. The notion of a *d-wff of* $\mathcal{L}^{?}_{\vdash\mathsf{L}}$ is defined as follows.

$$\mathsf{t} ::= \phi \mid ng\mathsf{t} \mid \mathsf{t}\&\mathsf{t}$$

Each d-wff of $\mathcal{L}^{?}_{\vdash\mathsf{L}}$ that is not a sequent is called *compound*. However, we will not make use of compound d-wffs of language $\mathcal{L}^{?}_{\vdash\mathsf{L}}$ until Chap. 5, where we introduce the notion of an answer to a question of $\mathcal{L}^{?}_{\vdash\mathsf{L}}$. Until then, we will deal with single *sequents* and/or their *sequences*. To denote *sequences of sequents* of $\mathcal{L}^{?}_{\vdash\mathsf{L}}$ we will use upper-case Greek letters Φ, Ψ, with subscripts as needed.

Having addressed the declaratives, we now turn to terminology for the erotetic part. *Questions of the language* $\mathcal{L}^{?}_{\vdash\mathsf{L}}$ are expressions of the form '$?(\Phi)$', where Φ is a finite and non-empty sequence of sequents. We will omit the angled brackets indicating a sequence and write:

$$?(S_1 \vdash T_1\ ;\ \ldots\ ;\ S_n \vdash T_n) \tag{2.10}$$

instead of '$?(\langle S_1 \vdash T_1\ ;\ \ldots\ ;\ S_n \vdash T_n\rangle)$'. As shown, the semicolon separates the sequents within the sequence. The sequents '$S_1 \vdash T_1$',..., '$S_n \vdash T_n$' are called *constituents of the question* (2.10). We will also refer to question (2.10) as being *based on the sequents* '$S_1 \vdash T_1$',..., '$S_n \vdash T_n$'. The letter Q, possibly with subscripts, will be used as a metavariable for questions of $\mathcal{L}^{?}_{\vdash\mathsf{L}}$.

[4] Formally, by a multiset we understand a set X together with a function $f : X \longrightarrow \mathbb{N}$, that is, a pair $\langle X, f\rangle$. The function assigns to each member of X the number of its occurrences. To borrow the expressions from [26, p. 5], a multiset is a 'set with multiplicity' or a 'sequence *modulo* the ordering.'

[5] See [19, 26]. Other variants, such as sets, are also considered. Poggiolesi [22] provides a clear classification of all these variants in the context of sequent calculi for modal logics.

Let us point out that the semicolon used to separate sequents behaves like a metalevel conjunction—a question of $\mathcal{L}^{?}_{\vdash\mathsf{L}}$ of the form (2.10) asks whether, for each i, the meta-level statement expressed by the d-wff '$S_i \vdash T_i$' holds true. Thus, erotetic calculi constructed in languages with both-sided sequents of the form (2.6) are calculi that transform questions concerning statements about derivability or, semantically, entailment. Erotetic calculi using sequents of the form (2.8) transform questions about inconsistency—whether syntactically or semantically defined. Naturally, the following issue arises: if the validity of a disjunction is the semantically defined subject of questions based on right-sided sequents of the form (2.7), how should this be interpreted in syntactic terms? This issue is addressed in Chap. 5, particularly in Sect. 5.6.

Let us finish this section with a brief note about hypersequent systems. The structures involving sequences of sequents may have already brought hypersequents to the reader's mind.[6] However, this resemblance is misleading. First of all, the separator between sequents in a hypersequent represents a metalevel disjunction. Derivations in a hypersequent system are typically organized as trees, with hypersequents at the nodes, and a hypersequent is considered an axiom if *at least one* of its sequents is axiomatic. In contrast, as we will see in the next section, the structure of Socratic proofs is different: for a Socratic transformation to count as a proof, *each* constituent of the last question must be of the appropriate form.

In [31], the term 'seqsequent' was introduced to describe finite sequences of sequents, deliberately avoiding the associations tied to the term 'hypersequent.' The word 'seqsequent' is derived from the Latin *sequentia*, meaning 'sequence' (see [31, p. 2499]). We will use this term interchangeably with 'finite sequence of sequents.'

There are at least two ways to compare sequences of sequents in erotetic calculi with hypersequents. The former was designed to model cognitive phenomena that lay in the scope of interest of the logic of questions, whereas the latter serves as a powerful framework for analyzing non-classical logics. As such, they are simply different objects created for distinct purposes. However, a more intriguing perspective arises when considering the duality between conjunctive and disjunctive readings. One might consider higher-order structures that incorporate both: conjunctive seqsequents in the antecedent and hypersequents in the succedent. Such structures have been explored by Došen in [5] (see also [14, 15] for a broader discussion), where sequents of arbitrary levels are introduced—a sequent of level n has collections of sequents of level $n - 1$ on both sides of the turnstile.

[6] Hypersequents were introduced by Pottinger [23] and independently by A. Avron in the seminal paper [1] where the term 'hypersequent' was first used.

2.3 Erotetic Rules for the Classical Case

Below we present the rules of three erotetic calculi for FOL and three erotetic calculi for CPL. Each calculus is *syntactically uniform* in the following sense: its rules operate on questions whose constituents are of only one of the three forms: (2.6), (2.7), or (2.8). Thus, both-sided sequents (2.6) are syntactic units for calculus $\mathbb{E}^{\mathbf{Q}}$, whereas right-sided sequents (2.7) and left-sided sequents (2.8) are syntactic units for, respectively, calculi $\mathbb{E}^{\mathbf{RQ}}$ and $\mathbb{E}^{\mathbf{LQ}}$.[7]

In presenting the calculi, we use a specific version of uniform notation, originally introduced by Raymond Smullyan (see [25]). As the reader will see in the next chapter, we also use the uniform notation with respect to modal logics. The uniform notation for classical logic is displayed in Table 2.1. The difference from the original approach is that we define the notation using the complement operation instead of negation. The notion of complement, $-A$, of a formula A, is defined as follows:

$$-A = \begin{cases} B & \text{if } A \text{ is of the form } \neg B \\ \neg A & \text{otherwise} \end{cases}$$

Typically, the complement is symbolically written as '$\overline{A}$', but this is not sufficiently readable.

Formula $\gamma(t)$ is called *an instance of* γ and formula $\delta(t)$ is called *an instance of* δ.

The following corollary, derived from the assumed definitions, examines the interplay between negation and complement in our uniform notation. The terms below will prove useful. A formula F is called *positive* when negation is not its main func-

Table 2.1 α-, β-, γ-, δ-formulas

α	α_0	α_1	β	β_0	β_1
$A \wedge B$	A	B	$-(A \wedge B)$	$-A$	$-B$
$-(A \vee B)$	$-A$	$-B$	$A \vee B$	A	B
$-(A \to B)$	A	$-B$	$A \to B$	$-A$	B
γ	$\gamma(t)$		δ	$\delta(t)$	
$\forall x_i A$	$A[x_i/t]$		$-\forall x_i A$	$-A[x_i/t]$	
$-\exists x_i A$	$-A[x_i/t]$		$\exists x_i A$	$A[x_i/t]$	

[7] The erotetic calculus for classical propositional logic was first presented in [28], and the first-order version in [32]. The right-sided first-order version was considered in [29], while the left-sided propositional erotetic calculus was discussed in [28, pp. 316–318].

Calculus $\mathbb{E}^{\mathbf{Q}}$ is a notational variant of calculus $\mathbb{E}^{\mathsf{FOL}}$ examined in [2], which, in turn, was an extension of the original $\mathbf{E}^{\mathbf{PQ}}$ introduced in [32]. $\mathbb{E}^{\mathsf{FOL}}$ is an extension of $\mathbf{E}^{\mathbf{PQ}}$ to the full language of the first-order logic. The main differences between $\mathbb{E}^{\mathbf{Q}}$, $\mathbb{E}^{\mathsf{FOL}}$ and $\mathbf{E}^{\mathbf{PQ}}$ lie in the definition of the uniform notation. The same applies to the other two erotetic calculi, $\mathbb{E}^{\mathbf{RQ}}$ and $\mathbb{E}^{\mathbf{LQ}}$, which are notational variants of the previously existing calculi.

tor, i.e. F is not of the form '$\neg A$'. When it is, F is called *negative*. Moreover, if F is of the form '$\neg\neg B$', we call F *doubly negated.*

Corollary 2.1 *For each formula F of language $\mathcal{L}_\mathsf{L}$ the following holds.*

1. *If F is positive, then $\neg F = -F$. If F is negative, then $F = \neg - F$, and hence also $\neg F = \neg\neg - F$.*
2. *If F is of the form '$\neg\neg A$' (doubly negated), then $- - F = A$; otherwise $- - F = F$.*
3. *If $-F = -A$, then*
   (a) *either $F = A$, or*
   (b) *$F = \neg\neg A$, where A is positive, or*
   (c) *$A = \neg\neg F$, where F is positive.*
4. *If F is an α-formula, then '$-F$' is a β-formula and conversely: if F is a β-formula, then '$-F$' is an α-formula. Moreover, it then holds for each $i \in \{0, 1\}$ that:*
$$\alpha_i = \begin{cases} -\beta_i \text{ and } -\beta_i = \neg\beta_i; \ \beta_i \text{ is then positive; or} \\ -\beta_i \text{ and } -\beta_i \neq \neg\beta_i; \ \beta_i \text{ is then negative; or} \\ \neg\beta_i \text{ and } \neg\beta_i \neq -\beta_i, \text{ then both } \beta_i, \alpha_i \text{ are negative;} \end{cases}$$
and similarly
$$\beta_i = \begin{cases} -\alpha_i \text{ and } -\alpha_i = \neg\alpha_i; \ \alpha_i \text{ is then positive; or} \\ -\alpha_i \text{ and } -\alpha_i \neq \neg\alpha_i; \ \alpha_i \text{ is then negative; or} \\ \neg\alpha_i \text{ and } \neg\alpha_i \neq -\alpha_i; \text{ then both } \alpha_i, \beta_i \text{ are negative.} \end{cases}$$
5. *If F is a γ-formula, then '$-F$' is a δ-formula and conversely: if F is a δ-formula, then '$-F$' is a γ-formula. Moreover, for each term t:*
$$\gamma(t) = \begin{cases} -\delta(t) \text{ and } -\delta(t) = \neg\delta(t); \ \delta(t) \text{ is positive; or} \\ -\delta(t) \text{ and } -\delta(t) \neq \neg\delta(t); \ \delta(t) \text{ is negative; or} \\ \neg\delta(t) \text{ and } \neg\delta(t) \neq -\delta(t); \text{ both instances are negative;} \end{cases}$$
and similarly
$$\delta(t) = \begin{cases} -\gamma(t) \text{ and } -\gamma(t) = \neg\gamma(t); \ \gamma(t) \text{ is positive; or} \\ -\gamma(t) \text{ and } -\gamma(t) \neq \neg\gamma(t); \ \gamma(t) \text{ is negative; or} \\ \neg\gamma(t) \text{ and } \neg\gamma(t) \neq -\gamma(t); \text{ both instances are negative.} \end{cases}$$

Proof Clauses 1–3 follow from the definition of complement. The other clauses can be proved directly by inspection of Table 2.1. Below, we provide another table that clearly displays all the cases in question for α- and β-formulas (clause 4). One case will be analyzed in detail.

α	α_0	$-\alpha_0$	$\neg\alpha_0$	α_1	$-\alpha_1$	$\neg\alpha_1$
$A \wedge B$	A	$-A$	$\neg A$	B	$-B$	$\neg B$
$\neg(A \vee B)$	$-A$	$-A$	$\neg - A$	$-B$	$-B$	$\neg - B$
$\neg(A \to B)$	A	$-A$	$\neg A$	$-B$	$-B$	$\neg - B$
$A \to B$	$-A$	$-A$	$\neg - A$	B	$-B$	$\neg B$
$A \vee B$	A	$-A$	$\neg A$	B	$-B$	$\neg B$
$\neg(A \wedge B)$	$-A$	$-A$	$\neg - A$	$-B$	$-B$	$\neg - B$
β	β_0	$-\beta_0$	$\neg\beta_0$	β_1	$-\beta_1$	$\neg\beta_1$

Let $\alpha = \neg(A \to B)$. Clearly, '$A \to B$' is the complement of this formula and it is a β-formula. We use the fact that every formula of $\mathcal{L}_\mathsf{L}$ is negative or not doubly negated (where 'or' is not exclusive here). By clause 1, if A is negative, then $\alpha_0 = \neg\beta_0$. By clause 2, if A is not doubly negated, then $\alpha_0 = -\beta_0$. We can also see that for this α-formula, $\alpha_1 = -\beta_1$. Conversely, for $\beta = A \to B$, $\beta_0 = -\alpha_0$. For $\beta_1 = B$, if B is negative, then $\beta_1 = \neg\alpha_1$, and if B is not doubly negated, then $\beta_1 = -\alpha_1$.

For clause 5, we consider a γ-formula of the form '$\forall x_i A$'. Its complement is a δ-formula '$\neg\forall x_i A$'. Then, for a term t, we have: $\gamma(t) = A[x_i/t]$, $-\delta(t) = - - A[x_i/t]$, and $\neg\delta(t) = \neg - A[x_i/t]$. If the formula '$A[x_i/t]$' is negative, then $\gamma(t) = \neg\delta(t)$, and if '$A[x_i/t]$' is not doubly negated, then $\gamma(t) = -\delta(t)$. The reasoning is analogous in all the remaining cases. □

Clauses 4 and 5 of the above corollary indicate that the complement operation is the correct way to express the duality between α- and β-formulas, as well as between γ- and δ-formulas.

In the presentation of the rules for quantifiers, we will also use the 'κ-notation' defined in Table 2.2. This notation was introduced in the paper [32] to capture the cases of double negation, quantifiers in the scope of negation, and empty quantification within a single schema. Since we are using the γ- and δ-notation here, 'κ' will refer only to double negation and empty quantification. To prevent trivial applications of the rules, we assume that cases of empty quantification and genuine quantification are split into κ-cases and γ-, δ-cases. In other words, from now on, when we use, e.g., the notation 'γ', '$\gamma(t)$' in one context in order to refer to '$\forall x A$' and '$A[x/t]$', we assume that x has at least one free occurrence in A.

Erotetic calculi (of appropriate 'sidedness') for propositional logic are obtained from the first-order version by omitting the rules for quantifiers and restricting κ to refer only to formulas of the form '$\neg\neg A$'. In this way, the rules L_κ and R_κ in the propositional case become the rules for double negation. For the propositional parts of $\mathbb{E}^{\mathbf{Q}}$, $\mathbb{E}^{\mathbf{RQ}}$, and $\mathbb{E}^{\mathbf{LQ}}$ we will use the notations $\mathbb{E}^{\mathbf{P}}$, $\mathbb{E}^{\mathbf{RP}}$, and $\mathbb{E}^{\mathbf{LP}}$, respectively.

The impatient reader can already preview the rules in Tables 2.3 and 2.4. In the meantime, we introduce some further explanations. Let $\mathbb{E} \in \{\mathbb{E}^{\mathbf{Q}}, \mathbb{E}^{\mathbf{RQ}}, \mathbb{E}^{\mathbf{LQ}}, \mathbb{E}^{\mathbf{P}}, \mathbb{E}^{\mathbf{RP}}, \mathbb{E}^{\mathbf{LP}}\}$. The notion of *proof in* erotetic calculus $\mathbb{E}$ is introduced as a finite sequence,

Table 2.2 κ–notation for $\mathcal{L}_{\mathsf{CPL}}$ and for $\mathcal{L}_{\mathsf{FOL}}$

	κ	κ_0
$\mathcal{L}_{\mathsf{CPL}}$	$\neg\neg A$	A
$\mathcal{L}_{\mathsf{FOL}}$	$\neg\neg A$	A
	$\left.\begin{array}{r}\forall x_i\, A\\ \exists x_i\, A\end{array}\right\}$ where x_i is not free in A	A
	$\left.\begin{array}{r}\neg\forall x_i\, A\\ \neg\exists x_i\, A\end{array}\right\}$ where x_i is not free in A	$-A$

contrary to the derivations-as-trees tradition in structural proof theory. The counterpart of a *derivation* is a *Socratic transformation*, which is defined as follows (see [28]):

Definition 2.2 (*Socratic Transformation*) Let Q be a question of the language $\mathcal{L}^{?}_{\vdash\mathsf{L}}$. A *Socratic transformation of question* Q *in calculus* $\mathbb{E}$ is a sequence $\langle Q_1, Q_2, \ldots\rangle$ of questions of the language $\mathcal{L}^{?}_{\vdash\mathsf{L}}$ such that $Q = Q_1$, and each Q_i $(i > 1)$ results from Q_{i-1} by applying one of the rules of $\mathbb{E}$.

In the propositional case, Socratic transformations are always finite, but in the first-order case, they may be infinite.

In the presentation of erotetic rules, we will use a comma to separate formulas of $\mathcal{L}_{\mathsf{L}}$ (where $\mathsf{L} \in \{\mathsf{CPL}, \mathsf{FOL}\}$) in a sequence of formulas, and a semicolon to separate sequents in a seqsequent. Reading the rules correctly can be a tiring task, especially when the schemes are dense with symbols, and both the erotetic and declarative contexts appear in the same scheme. For this reason, the following conventions are proposed.

Let us recall that letter S refers to finite sequences of formulas of the language $\mathcal{L}_{\mathsf{L}}$. The expression ‘$S(A)$’ refers to a finite sequence of formulas of $\mathcal{L}_{\mathsf{L}}$ where A is a term of that sequence. In other words, we can say that a sequence is of the form ‘$S(A)$’, or that, for some i ($1 \leqslant i \leqslant n$, n is the number of terms of S), it takes the form:

$$\langle F_1, \ldots, F_{i-1}, A, F_{i+1}, \ldots, F_n\rangle. \tag{2.11}$$

Alternatively, we can say that the sequence is of the form:

$$S_1\ '\ \langle A\rangle\ '\ S_2 \tag{2.12}$$

where the sign ‘′’ refers to the operation of concatenation of finite sequences (with undefined arity). In each case, the expression refers to the same class of sequences.

To indicate that both A and B are terms of S, we write ‘$S(A)(B)$’. The order of A and B in S is arbitrary; thus, ‘$S(A)(B)$’ and ‘$S(B)(A)$’ have the same meaning (i.e., they refer to the same class of sequences). If we use ‘$S(A)$’ and then ‘$S(B)$’ in the same context (for example, in the premise and conclusion of a rule), then ‘$S(B)$’

refers to the result of *replacing* the distinguished term of the sequence '$S(A)$', which is A, with B. More specifically, if '$S(A)$' is written as (2.11), then '$S(B)$' takes the form:

$$\langle F_1, \ldots, F_{i-1}, B, F_{i+1}, \ldots, F_n \rangle.$$

It's important to stress that this replacement operation applies only to a specific *occurrence* of a formula; other occurrences of A—if there are any—are unaffected.

Furthermore, in the context of '$S(A_1)(A_2)$' the expression '$S(B_1)(B_2)$' refers to the result of the simultaneous replacement of term A_1 with B_1, and term A_2 with B_2, and so on. If '$S(\_)$' occurs in the same context as '$S(A)$', understood as (2.11), it indicates the removal of term A, resulting in:

$$\langle F_1, \ldots, F_{i-1}, F_{i+1}, \ldots, F_n \rangle.$$

Thus, '$S(\_)$' is used to indicate *some* position in sequence S. Finally, the inscription '$S(B_1, B_2)$' in the context of '$S(A)$' refers to the result of replacing A with *two* terms: B_1, B_2. More generally, in the context of '$S(A)$', understood as in (2.11), '$S(B_1, \ldots, B_k)$' refers to a sequence of the form:

$$\langle F_1, \ldots, F_{i-1}, B_1, \ldots, B_k, F_{i+1}, \ldots, F_n \rangle.$$

The same conventions apply to finite sequences of sequents (denoted by Φ).

2.3.1 Calculi $\mathbb{E}^{\mathbf{Q}}$, $\mathbb{E}^{\mathbf{P}}$

The rules of erotetic calculus $\mathbb{E}^{\mathbf{P}}$ are presented in Table 2.3, while the quantifier rules are outlined in Table 2.4. Erotetic calculus $\mathbb{E}^{\mathbf{Q}}$ encompasses all the rules displayed in these two tables. In the presentation, instead of 'closed term of language $\mathcal{L}_{\mathsf{FOL}}$' and 'individual constant of language $\mathcal{L}_{\mathsf{FOL}}$' we write simply 'closed term' and 'constant,' respectively. The calculi are articulated in the language $\mathcal{L}^{?}_{\vdash\mathsf{L}}$, which employs both-sided sequents. We will introduce the concept of Socratic proof, illustrated by an example.

In rule L_α, according to the conventions just introduced, the sequence '$S(\alpha_0, \alpha_1)$' displayed in the conclusion refers to the sequence of formulas obtained from '$S(\alpha)$' by replacing a distinguished occurrence of α with its components. Similarly, in rule L_γ (Table 2.4), the sequence '$S(\gamma)$' in the premise transforms into '$S(\gamma, \gamma(t))$' in the conclusion, which means that the occurrence of a γ-formula on the left has been rewritten and an occurrence of its instance added. The same conventions apply to seqsequents, as previously explained. For instance, in rule R_α, the seqsequent '$\Phi\ (S \vdash \alpha_0\ ;\ S \vdash \alpha_1)$' in the conclusion results from '$\Phi\ (S \vdash \alpha)$' by replacing the occurrence of sequent '$S \vdash \alpha$' with the two displayed occurrences of sequents. Throughout this text, we will typically overlook the distinction between an

Table 2.3 Rules of $\mathbb{E}^{\mathbf{Q}}$

$$\frac{?(\ \Phi\ (\ S(\alpha)\ \vdash\ C\ )\ )}{?(\ \Phi\ (\ S(\alpha_0,\alpha_1)\ \vdash\ C\ )\ )}\ L_\alpha \qquad \frac{?(\ \Phi\ (\ S\ \vdash\ \alpha\ )\ )}{?(\ \Phi\ (\ S\ \vdash\ \alpha_0\ ;\ S\ \vdash\ \alpha_1\ )\ )}\ R_\alpha$$

$$\frac{?(\ \Phi\ (\ S(\beta)\ \vdash\ C\ )\ )}{?(\ \Phi\ (\ S(\beta_0)\vdash;\ S(\beta_1)\vdash C\ )\ )}\ L_\beta$$

$$\frac{?(\ \Phi\ (\ S(\_)\ \vdash\ \beta\ )\ )}{?(\ \Phi\ (\ S(-\beta_i)\ \vdash\ \beta_{1-i}\ )\ )}\ R_\beta \qquad \text{where } i \in \{0, 1\}$$

$$\frac{?(\ \Phi\ (\ S(\kappa)\ \vdash\ C\ )\ )}{?(\ \Phi\ (\ S(\kappa_0)\ \vdash\ C\ )\ )}\ L_\kappa \qquad \frac{?(\ \Phi\ (\ S\ \vdash\ \kappa\ )\ )}{?(\ \Phi\ (\ S\ \vdash\ \kappa_0\ )\ )}\ R_\kappa$$

Table 2.4 Rules of $\mathbb{E}^{\mathbf{Q}}$ for γ-,δ-formulas

$$\frac{?(\ \Phi\ (\ S(\gamma)\ \vdash\ C\ )\ )}{?(\ \Phi\ (\ S(\gamma,\gamma(t)\ )\ \vdash\ C\ )\ )}\ L_\gamma \qquad \frac{?(\ \Phi\ (\ S\ \vdash\ \gamma\ )\ )}{?(\ \Phi\ (\ S\ \vdash\ \gamma(a_i)\ )\ )}\ R_\gamma$$

t is any closed term — a_i is a constant which does not occur in the premise

$$\frac{?(\ \Phi\ (\ S(\delta)\ \vdash\ C\ )\ )}{?(\ \Phi\ (\ S(\delta(a_i))\ \vdash\ C\ )\ )}\ L_\delta \qquad \frac{?(\ \Phi\ (\ S(\_)\ \vdash\ \delta\ )\ )}{?(\ \Phi\ (\ S(-\ \delta)\ \vdash\ \delta(t)\ )\ )}\ R_\delta$$

a_i is a constant which does not occur in the premise — t is any closed term

expression and its occurrence; we have highlighted this distinction here to emphasize that the replacement operation is distinct from substitution.

Rule R_β encompasses two cases where either the first or second component of β is transitioned to the left.[8] This switching effectively complements the formula. Additionally, the use of '$S(\_)$' and '$S(-\beta_i)$' indicates our freedom to select any position in S for '$-\beta_i$'. This remark similarly applies to rule R_δ, as shown in Table 2.4.

Drawing an analogy to tableau systems, we classify erotetic rules that introduce two sequents in place of a single one as *branching rules*, while those that do not are termed *non-branching rules*. In this framework, R_α and L_β are branching rules, whereas L_α, R_β, L_κ, and R_κ are non-branching. All the quantifier rules, presented in Table 2.4, are non-branching. This terminology will be consistently applied to other erotetic calculi discussed in this book.

Let us briefly discuss the provisos for the applicability of R_γ and L_δ. In Sect. 2.4 we will see that, as far as semantic correctness and invertibility of rules are concerned, the proviso can be formulated in a *local* version, that is, with a restriction that the constant does not occur in the *sequent-premise*. However, as we will see in the completeness section, sometimes it is useful to require that the proviso is *global*, that

[8] In the original erotetic calculi in [28, 29] the rule captures one case with $i = 0$ which is sufficient for completeness.

is, that it concerns the entire *question-premise*. In summary, for the soundness of the rules, a local proviso suffices, but the completeness proofs within the methodology of this book, necessitate a global proviso.

A sequent '$S \vdash A$' of language $\mathcal{L}^{?}_{\vdash\mathsf{L}}$ is called *closed* if it contains no sentential functions. It follows that for $\mathsf{L} = \mathsf{CPL}$ all sequents are closed.

Definition 2.3 (*Socratic Proof in* $\mathbb{E}^{\mathbf{Q}}$) Let $\mathbb{E} \in \{\mathbb{E}^{\mathbf{Q}}, \mathbb{E}^{\mathbf{P}}\}$, and let '$S \vdash A$' stand for a closed sequent of $\mathcal{L}^{?}_{\vdash\mathsf{L}}$. A *Socratic proof of '$S \vdash A$' in* $\mathbb{E}$ is a finite Socratic transformation of question '$?(S \vdash A)$' in $\mathbb{E}$ such that each constituent of the last question of the transformation is of one of the following forms:

(*a*) $T(B) \vdash B$, or
(*b*) $T(B)(-B) \vdash C$.

If there is a Socratic proof of a sequent in $\mathbb{E}$, then we say that the sequent is *provable in* $\mathbb{E}$.

The following Socratic proof, conducted in $\mathbb{E}^{\mathbf{P}}$, illustrates the application of R_β alongside the reductions of negation signs through the complement operation. To the right of the proof, we indicate the specific rule applied to each question.

In the first step, the entire implication is recognized as a β-formula; β_0 is thus '$\neg(p \vee (q \wedge r))$', and hence $-\beta_0$ is the antecedent of the implication. The first component, β_0, of formula '$\neg(\neg\neg\neg p \wedge \neg q)$', simplifies to '$\neg\neg p$'. The next application of R_β moves the complement of '$\neg\neg p$' to the left side of the sequent. Similar reductions immediately produce q on the right side.

Example 2.1

$$\begin{array}{cl}
?(\vdash (p \vee (q \wedge r)) \to \neg(\neg\neg\neg p \wedge \neg q)) & \\ \hline
?(p \vee (q \wedge r) \vdash \neg(\neg\neg\neg p \wedge \neg q)) & R_\beta \\ \hline
?(p \vee (q \wedge r), \neg p \vdash q) & R_\beta \\ \hline
?(p, \neg p \vdash q \,;\, q \wedge r, \neg p \vdash q) & L_\beta \\ \hline
?(p, \neg p \vdash q \,;\, q, r, \neg p \vdash q) & L_\alpha
\end{array}$$

The next example presents a Socratic proof in $\mathbb{E}^{\mathbf{Q}}$. The initial sequent is $\forall x \forall y (P(x, y) \to \neg P(y, x)) \vdash \neg \exists x P(x, x)$.

Example 2.2 Below A stands for formula '$\forall x \forall y (P(x, y) \to \neg P(y, x))$', and B is for '$\forall y (P(a, y) \to \neg P(y, a))$'.

$$\begin{array}{cl}
?(A \vdash \neg\exists x P(x, x)) & \\ \hline
?(A \vdash \neg P(a, a)) & R_\gamma \\ \hline
?(A, B \vdash \neg P(a, a)) & L_\gamma \\ \hline
?(A, B, P(a, a) \to \neg P(a, a) \vdash \neg P(a, a)) & L_\gamma \\ \hline
?(A, B, \neg P(a, a) \vdash \neg P(a, a) \,;\, A, B, \neg P(a, a) \vdash \neg P(a, a)) & L_\beta
\end{array}$$

To summarize, calculi $\mathbb{E}^{\mathbf{P}}$ and $\mathbb{E}^{\mathbf{Q}}$ transform questions based on both-sided sequents. As previously mentioned, calculus $\mathbb{E}^{\mathbf{Q}}$ is a notation variant of $\mathbb{E}^{\mathsf{FOL}}$, as examined in [2]. Simultaneously, it serves as a linguistic extension and a notation variant of $\mathbf{E^{PQ}}$, which was presented in [32]. The calculi $\mathbf{E^{PQ}}$, $\mathbf{E^{RPQ}}$, and $\mathbf{E^{LPQ}}$, originally proposed by Wiśniewski, were formulated for the language of FOL without function symbols, using uniform notation in the Smullyan's version. However, this notation applied to propositional formulas, but not to quantified ones. Thus, calculus $\mathbb{E}^{\mathbf{Q}}$ can be seen as both a notation variant and an extension of $\mathbf{E^{PQ}}$. In [2], the author considers the full language of FOL—as we do here—employing uniform notation consistently at the quantifier level. However, our notation differs by utilizing complement instead of negation in defining uniform notation.

As mentioned earlier, the operation of complement appears to be the most suitable choice for expressing the duality between α- and β-formulas, as well as between γs and δs. It is reasonable to say that, while standard Gentzen-sequent systems characterize logical operators—including negation—the uniform-notation version characterizes the α- and β-truth-functions. Consequently, the negation sign is not treated as a standalone operator in this context.

Terminology: Question-Premise, Question-Conclusion, and Others

Consider a Socratic transformation **s**, where question s_{n+1} results from question s_n through the application of erotetic rule **r** in either $\mathbb{E}^{\mathbf{Q}}$ or $\mathbb{E}^{\mathbf{P}}$. We call s_n the *question-premise* and s_{n+1} the *question-conclusion* of this application of **r**. The constituent of s_n with a schema distinguished in the question-premise of **r** is termed the *sequent-premise* of question s_n.

Furthermore, the constituent (or both constituents, if applicable) of question s_{n+1}, whose schema is distinguished in the question-conclusion of **r**, *results from* the sequent-premise of s_n and is called the *sequent-conclusion of* s_{n+1}. Each question of **s** has exactly one sequent-premise (or none, if it is the final question) and at most two sequents-conclusions, which are 'new' relative to the previous question. We also say that sequent ψ *results in* **s** *from* sequent ϕ *via erotetic rule* **r** when, for some term s_n of **s**, s_{n+1} results from s_n by **r**, ϕ is the sequent-premise of s_n, and ψ is the sequent-conclusion (possibly, one of the two sequents-conclusions) of s_{n+1}.

This terminology will be consistently applied across other erotetic calculi discussed in this work. Additionally, we employ the standard concept of an *active formula* of a rule, referring to the formula-occurrence whose schema is distinguished in the sequent-premise of the rule.

Before we continue, let's clarify a key point. As per Definition 2.3, a Socratic proof in calculus $\mathbb{E}^{\mathbf{Q}}$ begins with a question that has only a closed sequent as its constituent. The calculus rules prohibit the introduction of a sentential function into a question-conclusion. Therefore, every formula in a Socratic proof is a sentence of $\mathcal{L}_{\mathsf{FOL}}$. The same restriction applies to the other calculi presented in this section.

By omitting sentential functions in the proof-theoretical apparatus, we avoid complications related to α-conversion, which usually arise in sequent calculi that handle quantifier introduction through individual variables but lack the rule of cut. Such calculi are incomplete without considering α-conversion, which renames bound variables in a formula. Without the cut rule, these sequent calculi can only prove theorems *modulo* renaming bound variables (see [26, pp. 57–58]).[9] On the other hand, a sentential function is valid in FOL if and only if its universal generalization is.

2.3.2 Calculi $\mathbb{E}^{\mathbf{RQ}}$, $\mathbb{E}^{\mathbf{RP}}$

In the following sections, we turn our attention to the quantifier level. The right-sided calculus $\mathbb{E}^{\mathbf{RQ}}$ is formulated in the language $\mathcal{L}^{?}_{\vdash\mathsf{FOL}}$, but it operates exclusively with right-sided sequents. Table 2.5 presents the rules of the calculus.

As before, we provide the definition of a Socratic proof in $\mathbb{E}^{\mathbf{RQ}}$, followed by an example. Similar to both-sided sequents, a sequent '$\vdash S$' of the language $\mathcal{L}^{?}_{\vdash\mathsf{FOL}}$ is termed *closed* if it contains no sentential functions.

Definition 2.4 (*Socratic Proof in* $\mathbb{E}^{\mathbf{RQ}}$) Let '$\vdash S$' be a closed sequent of $\mathcal{L}^{?}_{\vdash\mathsf{FOL}}$. A *Socratic proof of* '$\vdash S$' *in* $\mathbb{E}^{\mathbf{RQ}}$ is a finite Socratic transformation of the question '$?(\vdash S)$' in $\mathbb{E}^{\mathbf{RQ}}$, such that each constituent of the final question in the transformation has the following form:

(c) $\vdash T(B)(-B)$.

If a sequent has a Socratic proof in $\mathbb{E}^{\mathbf{RQ}}$, we say that it is *provable in* $\mathbb{E}^{\mathbf{RQ}}$.

Table 2.5 Rules of $\mathbb{E}^{\mathbf{RQ}}$

$$\frac{?(\,\Phi\,(\vdash S(\alpha)\,)\,)}{?(\,\Phi\,(\vdash S(\alpha_0)\,;\,\vdash S(\alpha_1)\,)\,)}\;R_\alpha$$

$$\frac{?(\,\Phi\,(\vdash S(\beta)\,)\,)}{?(\,\Phi\,(\vdash S(\beta_0,\beta_1)\,)\,)}\;R_\beta \qquad \frac{?(\,\Phi\,(\vdash S(\kappa)\,)\,)}{?(\,\Phi\,(\vdash S(\kappa_0)\,)\,)}\;R_\kappa$$

$$\frac{?(\,\Phi\,(\vdash S(\gamma)\,)\,)}{?(\,\Phi\,(\vdash S(\gamma(a_i))\,)\,)}\;R_\gamma \qquad \frac{?(\,\Phi\,(\vdash S\,(\delta)\,)\,)}{?(\,\Phi\,(\vdash S(\delta,\delta(t))\,)\,)}\;R_\delta$$

a_i is a constant which does not occur in the premise — t is any closed term

[9] That this issue is subtle is demonstrated by the fact that calculi **G3i** and **G3c** discussed in [19] are not complete without assuming α-conversion, contrary to the authors' claims [19, p. 62]. It is easy to check that sequent $\forall x \forall y(Ry \wedge Qx) \Rightarrow Qy$ is valid, but not provable in **G3c** without the use of cut, contradicting the Substitution Lemma (Lemma 4.1.2, [19, p. 68]), which in fact is incorrect. The said sequent is used in [26] to illustrate the issue of α-conversion and cut.

Example 2.3 The sequent $\vdash \forall x(P(x) \to R(x)) \to (\forall x P(x) \to \forall x R(x))$ is provable in $\mathbb{E}^{\mathbf{RQ}}$. Below, A stands for '$\neg\forall x(P(x) \to R(x))$', and B stands for '$\neg\forall x P(x)$'.

$$\cfrac{\cfrac{\cfrac{\cfrac{\cfrac{\cfrac{?(\vdash \forall x(P(x) \to R(x)) \to (\forall x P(x) \to \forall x R(x)))}{?(\vdash A, \forall x P(x) \to \forall x R(x))}\,R_\beta}{?(\vdash A, B, \forall x R(x))}\,R_\beta}{?(\vdash A, B, R(a))}\,R_\gamma}{?(\vdash A, \neg(P(a) \to R(a)), B, R(a))}\,R_\delta}{?(\vdash A, \neg(P(a) \to R(a)), B, \neg P(a), R(a))}\,R_\delta}{?(\vdash A, P(a), B, \neg P(a), R(a)\,;\, \vdash A, \neg R(a), B, \neg P(a), R(a))}\,R_\alpha$$

2.3.3 *Calculi* $\mathbb{E}^{\mathbf{LQ}}$, $\mathbb{E}^{\mathbf{LP}}$

As before, calculus $\mathbb{E}^{\mathbf{LQ}}$ is expressed in the language $\mathcal{L}^?_{\vdash\mathsf{FOL}}$, but this time with the left-sided sequents. Table 2.6 presents the rules governing this calculus.

A sequent '$S \vdash$' is *closed*, if it contains no sentential functions.

Definition 2.5 (*Socratic Proof in* $\mathbb{E}^{\mathbf{LQ}}$) Let '$S \vdash$' be a closed sequent of $\mathcal{L}^?_{\vdash\mathsf{FOL}}$. A *Socratic proof of* '$S \vdash$' *in* $\mathbb{E}^{\mathbf{LQ}}$ is a finite Socratic transformation of '$?(S \vdash)$' in $\mathbb{E}^{\mathbf{LQ}}$, such that each constituent of the last question in the transformation has the form:

(*d*) $T(B)(-B) \vdash$

If such a proof exists for a sequent, we say the sequent is *provable in* $\mathbb{E}^{\mathbf{LQ}}$.

Example 2.4 The sequent $\neg\exists x(P(x) \to \forall x P(x)) \vdash$ is provable in $\mathbb{E}^{\mathbf{LQ}}$. Here, C stands for '$\neg\exists x(P(x) \to \forall x P(x))$'.

Table 2.6 Rules of $\mathbb{E}^{\mathbf{LQ}}$

$$\frac{?(\,\Phi\,(\,S(\alpha) \vdash))}{?(\,\Phi\,(\,S(\alpha_0, \alpha_1) \vdash))}\,L_\alpha \qquad \frac{?(\,\Phi\,(\,S(\kappa) \vdash))}{?(\,\Phi\,(\,S(\kappa_0) \vdash))}\,L_\kappa$$

$$\frac{?(\,\Phi\,(\,S(\beta) \vdash))}{?(\,\Phi\,(\,S(\beta_0) \vdash\,;\, S(\beta_1) \vdash))}\,L_\beta$$

$$\frac{?(\,\Phi\,(\,S(\gamma) \vdash))}{?(\,\Phi\,(\,S(\gamma, \gamma(t)) \vdash))}\,L_\gamma \qquad \frac{?(\,\Phi\,(\,S(\delta) \vdash))}{?(\,\Phi\,(\,S(\delta(a_i)) \vdash))}\,L_\delta$$

t is any closed term — a_i is a constant which does not occur in the premise

$$
\dfrac{\dfrac{\dfrac{\dfrac{\dfrac{?(\,C\vdash)}{?(\,C,\neg(P(a)\to\forall xP(x))\vdash)}\,L_\gamma}{?(\,C,P(a),\neg\forall xP(x)\vdash)}\,L_\alpha}{?(\,C,P(a),\neg\forall xP(x),\neg P(b)\vdash)}\,L_\delta}{?(\,C,\neg(P(b)\to\forall xP(x)),P(a),\neg\forall xP(x),\neg P(b)\vdash)}\,L_\gamma}{?(\,C,P(b),\neg\forall xP(x),P(a),\neg\forall xP(x),\neg P(b)\vdash)}\,L_\alpha
$$

As in analytic tableau systems, we demonstrate validity of A in erotetic calculus $\mathbb{E}^{\mathbf{LQ}}$ by showing *inconsistency* of '$\neg A$'.

2.4 Soundness of Erotetic Calculi

We begin by reviewing the familiar semantics of classical propositional logic (CPL) and first-order logic (FOL), and then extend these to provide semantics for the declarative components of $\mathcal{L}^?_{\vdash\mathsf{CPL}}$ and $\mathcal{L}^?_{\vdash\mathsf{FOL}}$.

Semantics of CPL and FOL

For CPL, we adopt the standard notion of a *Boolean valuation* (*valuation*, for short). A valuation, denoted v, is a function from the set of all formulas of $\mathcal{L}_{\mathsf{CPL}}$ to the set $\{0, 1\}$ of truth values. This function satisfies the following well-known conditions:

(i) $v(\neg A) = 1$ iff $v(A) = 0$,
(ii) $v(A \wedge B) = 1$ iff $v(A) = v(B) = 1$,
(iii) $v(A \vee B) = 0$ iff $v(A) = v(B) = 0$,
(iv) $v(A \to B) = 0$ iff $v(A) = v(\neg B) = 1$.

If $v(A) = 1$, we say that A is *true under* v; otherwise it is *false under* v. A formula of $\mathcal{L}_{\mathsf{CPL}}$ that is true under every valuation is called CPL-*valid*. Similarly, if a set X of formulas is such that each formula in X is true under a valuation v, we say that X is *satisfied by* v, or that v is a CPL-*model of* X.

For $\mathcal{L}_{\mathsf{FOL}}$, we employ *model-theoretic semantics*. The key concept here is an *interpretation of language*. An interpretation $\mathcal{I}$ is a structure $\langle U, f\rangle$, where U is a non-empty set known as the *universe* or *domain of the interpretation*, and f is a function assigning 'meaning' to the non-logical symbols of the language, such as individual constants, function symbols, and predicates. Specifically:

(i) For each constant a_i, $f(a_i)$ is an element of U (i.e., $f(a_i) \in U$).
(ii) For each predicate P_i of arity n, $f(P_i)$ is an n-argument relation on the universe U (i.e., $f(P_i) \subseteq U^n$).
(iii) For each function symbol F_i of arity n, $f(F_i)$ is an n-argument function that takes elements from U and returns values in U (i.e., $f(F_i) \subseteq U^n \times U$).

From this concept of interpretation, we define *model-theoretic truth*. For an atomic formula $P_i(t_1, \dots, t_n)$, the formula is true under interpretation $\mathcal{I}$ if the relation $f(P_i)$ holds between the objects assigned to terms $t_1, \dots, t_n$ in U. To interpret formulas with individual variables, we need an *object assignment in $\mathcal{I}$*, which is a function σ that maps individual variables to elements in U. With this in hand, we define the *interpretation of a term t in $\mathcal{I} = \langle U, f \rangle$ under assignment σ*, denoted $t^{\mathcal{I}}[\sigma]$, as follows:

$$t^{\mathcal{I}}[\sigma] = \begin{cases} \sigma(x_i) & \text{if } t \text{ is an individual variable } x_i, \\ f(a_i) & \text{if } t \text{ is an individual constant } a_i, \\ f(F_i^n)(t_1^{\mathcal{I}}[\sigma], \ldots, t_n^{\mathcal{I}}[\sigma]) & \text{if } t \text{ is of the form } F_i^n(t_1, \ldots, t_n). \end{cases}$$

As we can see, the definition is recursive. For terms of the form '$F_i^n(t_1, \ldots, t_n)$', the interpretation first evaluates the inner terms $t_1, \ldots, t_n$, and then applies the function $f(F_i)$ to these results.

An atomic formula $P_i(t_1, \ldots, t_n)$ is *satisfied by assignment σ in $\mathcal{I}$*, symbolically:

$$\mathcal{I} \models P_i(t_1, \ldots, t_n)\ [\sigma]$$

if the relation $f(P_i)$ holds between the objects $t_1^{\mathcal{I}}[\sigma], \ldots, t_n^{\mathcal{I}}[\sigma]$.

If σ is an object assignment and $u \in U$ is an object, let $\sigma(x_i/u)$ stand for the object assignment that is identical to σ except that it maps x_i to u. Now we may define inductively:

1. $\mathcal{I} \models \neg A\ [\sigma]$ iff $\mathcal{I} \not\models A\ [\sigma]$.
2. $\mathcal{I} \models A \wedge B\ [\sigma]$ iff $\mathcal{I} \models A\ [\sigma]$ and $\mathcal{I} \models B\ [\sigma]$.
3. $\mathcal{I} \models A \vee B\ [\sigma]$ iff $\mathcal{I} \models A\ [\sigma]$ or $\mathcal{I} \models B\ [\sigma]$.
4. $\mathcal{I} \models A \rightarrow B\ [\sigma]$ iff $\mathcal{I} \not\models A\ [\sigma]$ or $\mathcal{I} \models B\ [\sigma]$.
5. $\mathcal{I} \models \forall x_i A\ [\sigma]$ iff $\mathcal{I} \models A\ [\sigma(x_i/u)]$ for each $u \in U$.
6. $\mathcal{I} \models \exists x_i A\ [\sigma]$ iff $\mathcal{I} \models A\ [\sigma(x_i/u)]$ for some $u \in U$.

A formula A is *true under $\mathcal{I}$*, denoted '$\mathcal{I} \models A$,' if $\mathcal{I} \models A\ [\sigma]$ holds for every object assignment σ in $\mathcal{I}$. A formula that is true under every interpretation of $\mathcal{L}_{\mathsf{FOL}}$ is said to be FOL-*valid*.

If a set X of formulas of $\mathcal{L}_{\mathsf{FOL}}$ is such that each formula in X is true under an interpretation $\mathcal{I}$, we say that $\mathcal{I}$ is a FOL-*model of* X. To unify our terminology, we refer to Boolean valuations as CPL-*interpretations* (of the language $\mathcal{L}_{\mathsf{CPL}}$) and model-theoretic interpretations as FOL-*interpretations* (of $\mathcal{L}_{\mathsf{FOL}}$), denoting both by $\mathcal{I}$. If $\mathcal{I}$ is an L-model of a singleton set $X = \{A\}$, we call $\mathcal{I}$ an L-*model of formula* A. Conversely, if $\mathcal{I}$ is not an L-model of a sentence A, then we say that the sentence A *is false under $\mathcal{I}$*, or that *$\mathcal{I}$ falsifies A*.

We write

$$X \models_{\mathsf{L}} A \tag{2.13}$$

for the *entailment in* L, which is a semantic relation understood as the *transmission of truth*. Formally, a set X of formulas of the language $\mathcal{L}_{\mathsf{L}}$ *entails in* L a formula A of the same language, symbolically (2.13), iff every L-model of X is also an L-model of A. Furthermore, a set X of formulas of $\mathcal{L}_{\mathsf{L}}$ is called L-*satisfiable* or *semantically consistent in* L if there exists at least one L-model of X. Otherwise, it is termed L-*unsatisfiable* or *semantically inconsistent in* L.

We conclude this section with a few straightforward observations. The first one concerns the uniform notation and expresses the semantic characterization of formulas of the respective categories. We skip the proof.

Corollary 2.2 *Let* $\mathsf{L} \in \{\mathsf{CPL}, \mathsf{FOL}\}$, *and let* $\mathcal{I}$ *stand for an* L-*interpretation. The following is true.*

1. *For each* κ*-formula of* $\mathcal{L}_{\mathsf{L}}$, $\mathcal{I}$ *is an* L-*model of* κ *iff* $\mathcal{I}$ *is an* L-*model of* κ_0.
2. *For each sentence of* $\mathcal{L}_{\mathsf{L}}$ *which is an* α*-formula,* $\mathcal{I}$ *is an* L-*model of* α *iff* $\mathcal{I}$ *is an* L-*model of both* α_0 *and* α_1.
3. *For each sentence of* $\mathcal{L}_{\mathsf{L}}$ *which is a* β*-formula,* $\mathcal{I}$ *is an* L-*model of* β *iff* $\mathcal{I}$ *is an* L-*model of at least one of* β_0 *and* β_1.

The above corollary does not cover γ- and δ-formulas, which will be addressed separately in Corollary 2.18.

It is also straightforward to observe the following.

Corollary 2.3 *If* $\mathcal{I}$ *is an* L-*model of a set* X *of formulas of* $\mathcal{L}_{\mathsf{L}}$, *then* $\mathcal{I}$ *is also an* L-*model of each subset of* X.

Semantics of the Declarative Part of $\mathcal{L}^?_{\vdash\mathsf{L}}$

The semantics of sequents is typically described through a natural and straightforward generalization of the semantics of the underlying logic. For instance, in the case of CPL, a sequent $S \vdash T$ is considered *correct under a valuation* v if the corresponding formula '$\bigwedge S \to \bigvee T$' is true under v. This approach can be easily extended to FOL as well. (Recall the *denotational* or *material* interpretation of sequents mentioned in the previous chapter.)

Now, let '$S \vdash T$' be a closed sequent of $\mathcal{L}^?_{\vdash\mathsf{L}}$. As discussed above, in proof-theoretical contexts, the sequents of $\mathcal{L}^?_{\vdash\mathsf{L}}$ consist solely of sentences of $\mathcal{L}_{\mathsf{L}}$. Let $\mathcal{I}$ be an L-interpretation of $\mathcal{L}_{\mathsf{L}}$. We say that '$S \vdash T$' is *correct under* $\mathcal{I}$ if either a sentence in S is false under $\mathcal{I}$ or a sentence in T is true under $\mathcal{I}$. This leads to the following conclusions:

Corollary 2.4 *Let* S *and* T *stand for finite sequences of sentences of* $\mathcal{L}_{\mathsf{L}}$, A *for a single sentence of* $\mathcal{L}_{\mathsf{L}}$, *and let* $\mathcal{I}$ *be an* L-*interpretation of* $\mathcal{L}_{\mathsf{L}}$.

1. *A both-sided sequent '*$S \vdash A$*' of language* $\mathcal{L}^?_{\vdash\mathsf{L}}$ *is correct under* $\mathcal{I}$ *iff the formula of the form '*$\bigwedge S \to A$*' is true under* $\mathcal{I}$.
2. *A left-sided sequent '*$S \vdash$ *' of language* $\mathcal{L}^?_{\vdash\mathsf{L}}$ *is correct under* $\mathcal{I}$ *iff the formula of the form '*$\bigwedge S$*' is false under '*$\mathcal{I}$*'.*
3. *A right-sided sequent '*$\vdash T$*' of language* $\mathcal{L}^?_{\vdash\mathsf{L}}$ *is correct under* $\mathcal{I}$ *iff the formula of the form '*$\bigvee T$*' is true under* $\mathcal{I}$.

Soundness and Invertibility of Erotetic Rules. Inquiring and Justifying Revisited

The next lemma asserts that all rules of erotetic calculi for classical logic, with the exception of R_γ and L_δ of their respective $\mathbb{E}$s, preserve the joint correctness of all constituents of a question in both directions: from the question-premise of a rule to its question-conclusion, and conversely, from the question-conclusion to the question-premise. It is important to note that the first direction (premise to conclusion) corresponds to what is typically referred to as the *soundness*, or *correctness*, of a rule. The converse (conclusion to premise), on the other hand, aligns with the *invertibility* of a rule.

Notably, while in sequent calculus the proof of soundness typically depends on the soundness of the rules, in erotetic calculi, it is the invertibility of the rules that is crucial for proving their soundness. This difference arises from the fact that sequent calculi and erotetic calculi have opposite directions of *proving* as defined by their rules. It is well-known that in sequent calculi, the direction of proving as dictated by the rules is reversed compared to the actual process of proof-search. In erotetic calculi, however, these two relations—the proving relation defined by the rules and the actual process of proof-search—are unified.

Whether or not this unification appears natural from a proof-theoretic perspective, it is justified by our view that questions play a significant role in reasoning.[10] The reversal of these directions in declarative reasoning reflects the idea that the transition from premise to conclusion models a step in the justification process. In contrast, the search for a solution of an unsettled problem may follow an entirely different, and as yet unknown, path of discovery.

On the other hand, the rules of erotetic calculi model a cognitive situation where an agent asks themselves a question and proceeds to formulate a subsequent question, typically in context of searching for a solution. We have termed this process *inquiring*. The second question is both *derived* from the first and serves as a means to answer it.

As mentioned in the previous chapter, in certain cases, the processes of inquiring and justifying complement one another, with the former typically leading to the latter. Let us illustrate this with an example.

Suppose our task is to *prove* (i.e., formally *justify*) the following identity:

$$\log_b x = \log_b a \cdot \log_a x. \tag{2.14}$$

Having the definition of a logarithm in mind ($\log_c y = d$ iff $c^d = y$) we can reformulate (2.14) into:

$$b^{\log_b x} = b^{\log_b a \cdot \log_a x}.$$

Given this equation and the definition, we can restate the problem as follows:

$$x = b^{\log_b a \cdot \log_a x}.$$

[10] See also the discussion concerning IEL in the first chapter.

Applying further mathematical transformations yields:

$$x = (b^{\log_b a})^{\log_a x},$$

which reduces to $x = a^{\log_a x}$, and, ultimately, to $x = x$.

This reasoning, however, *does not constitute a formal proof* of the original statement. Rather, it reflects an *inquiry into the grounds* for the identity. In contrast, a formal proof can be laid out as follows:

$$\dfrac{\dfrac{\dfrac{\dfrac{\Rightarrow x = x \quad \Rightarrow \text{def. of log}}{\Rightarrow x = a^{\log_a x}}\,(rep) \quad \Rightarrow \text{def. of log}}{\Rightarrow x = (b^{\log_b a})^{\log_a x}}\,(rep)}{\Rightarrow x = b^{\log_b a \cdot \log_a x}}\,(math) \quad \Rightarrow \text{def. of log}}{\dfrac{\Rightarrow b^{\log_b x} = b^{\log_b a \cdot \log_a x}}{\Rightarrow \log_b x = \log_b a \cdot \log_a x}\,(math)}\,(rep)$$

We used (rep) for a replacement operation and $(math)$ to indicate some basic transformations. Both are possible to capture in a sequent calculus.

In the above proof, the steps move systematically from the premise to the conclusion, adhering to the conventions of formal justification. This contrasts with the process of *inquiry*, which can be reconstructed as follows:

$$\dfrac{\text{What is the ground for } \log_b x = \log_b a \cdot \log_a x \text{ ?}}{\dfrac{\text{What is the ground for } b^{\log_b x} = b^{\log_b a \cdot \log_a x} \text{ ?}}{\dfrac{\text{What is the ground for } x = b^{\log_b a \cdot \log_a x} \text{ ?}}{\dfrac{\text{What is the ground for } x = (b^{\log_b a})^{\log_a x} \text{ ?}}{\dfrac{\text{What is the ground for } x = a^{\log_a x} \text{ ?}}{\text{What is the ground for } x = x \text{ ?}}}}}}$$

In this form, each question reflects a step of inquiry that investigates the underlying grounds for (2.14). While this process mirrors aspects of justification, its nature is one of exploration rather than proving, with each new question serving as a means to clarify or resolve the original inquiry. Here, the roles of *inquiring* and *justifying* are clearly distinguished: *inquiring* initiates the search for the grounds of the identity, while *justifying* follows the steps required to formally establish it. We first inquire to construct a justification of what was established.

Finally, within the framework of erotetic calculi, the inquiry process can be captured as presented in Fig. 2.1. In this formalization, the process of inquiry and the derived questions are integral to understanding the logical structure of the problem, reflecting the dynamic nature of cognitive reasoning when searching for solutions.

$$\cfrac{\cfrac{\cfrac{\cfrac{\cfrac{?\,\{\,\log_b x = \log_b a \cdot \log_a x\,\}}{?\,\{\,b^{\log_b x} = b^{\log_b a \cdot \log_a x}\,\}}\,(math)}{?\,\{\,x = b^{\log_b a \cdot \log_a x}\,;\,x = b^{\log_b x}\,\}}\,(rep)}{?\,\{\,x = (b^{\log_b a})^{\log_a x}\,;\,x = b^{\log_b x}\,\}}\,(math)}{?\,\{\,x = a^{\log_a x}\,;\,a = b^{\log_b a}\,;\,x = b^{\log_b x}\,\}}\,(rep)}{?\,\{\,x = x\,;\,x = a^{\log_a x}\,;\,a = b^{\log_b a}\,;\,x = b^{\log_b x}\,\}}\,(rep)$$

Fig. 2.1 Inquiring into the grounds for (2.14)—a formal reconstruction

Back to Correctness and Invertibility of Erotetic Rules

After presenting the philosophical motivations behind the adopted form of erotetic rules—specifically, the direction of the transition from premise to conclusion—we must finally acknowledge that our choice of this direction introduces certain peculiarities in the metalogical analysis. We will see this in the next section; for now, however, we will return to the correctness of erotetic calculi.

Let $\mathbb{E} \in \{\mathbb{E}^{\mathbf{Q}}, \mathbb{E}^{\mathbf{RQ}}, \mathbb{E}^{\mathbf{LQ}}, \mathbb{E}^{\mathbf{P}}, \mathbb{E}^{\mathbf{RP}}, \mathbb{E}^{\mathbf{LP}}\}$. Since we aim at the transmission of correctness in *Socratic proofs*, we assume that the constituents of the analyzed question do not contain sentential functions.

Lemma 2.1 (Correctness and Invertibility of Erotetic Rules) *Suppose that Q^* results from Q by a rule of erotetic calculus $\mathbb{E}$, excluding R_γ and L_δ. Let $\mathcal{I}$ be an* L-*interpretation of $\mathcal{L}_{\mathsf{L}}$. Then, each constituent of Q is correct under $\mathcal{I}$ iff each constituent of Q^* is correct under $\mathcal{I}$.*

Proof Let $\mathcal{I} = \langle U, f \rangle$ be an arbitrary but fixed L-interpretation of $\mathcal{L}_{\mathsf{L}}$. The proof proceeds by cases, beginning with the propositional rules.

Case 1: Assume $Q^* = ?(\Phi(\, S(\alpha_0, \alpha_1) \vdash C\,))$ results from $Q = ?(\Phi(\, S(\alpha) \vdash C\,))$ by rule L_α of calculus $\mathbb{E}^{\mathbf{Q}}$ (or $\mathbb{E}^{\mathbf{P}}$). Suppose that one constituent of Q^* is not correct under $\mathcal{I}$. If the incorrect constituent is part of Φ, it is also present in Q. Now, suppose that the sequent '$S(\alpha_0, \alpha_1) \vdash C$' is not correct under $\mathcal{I}$. Then each sentence of S, together with α_0 and α_1, is true under $\mathcal{I}$, while C is false. By clause 2 of Corollary 2.2, this implies that the sequent $S(\alpha) \vdash C$ is also incorrect under $\mathcal{I}$. Hence, correctness of '$S(\alpha) \vdash C$' under $\mathcal{I}$ implies correctness of '$S(\alpha_0, \alpha_1) \vdash C$' under $\mathcal{I}$. The converse is equally straightforward.

The same reasoning applies to: R_β in $\mathbb{E}^{\mathbf{RQ}}, \mathbb{E}^{\mathbf{RP}}$ and L_α in $\mathbb{E}^{\mathbf{LQ}}, \mathbb{E}^{\mathbf{LP}}$ (and the same pertains to the propositional parts of the three calculi).

Case 2: Assume $Q^* = ?(\Phi(\, S(-\beta_i) \vdash \beta_{1-i}\,))$ results from $Q = ?(\Phi(\, S \vdash \beta\,))$ by the rule R_β of $\mathbb{E}^{\mathbf{Q}}$ (or $\mathbb{E}^{\mathbf{P}}$). Here, a formula is shifted from the succedent to the antecedent. By clause 3 of Corollary 2.2, for any sentence of type β, the formula is false under $\mathcal{I}$ iff formula '$-\beta_i$' is true under $\mathcal{I}$ and β_{1-i} is false under $\mathcal{I}$. Therefore, sequent '$S \vdash \beta$' is correct under $\mathcal{I}$ iff sequent '$S(-\beta_i) \vdash \beta_{1-i}$' is correct under $\mathcal{I}$, which is what we need.

Case 3: This case will be represented by rule L_β of $\mathbb{E}^{\mathbf{Q}}$. Assume $Q^* = ?(\Phi(S(\beta_0) \vdash C\,;\, S(\beta_1) \vdash C))$ results from $Q = ?(\Phi(S(\beta) \vdash C))$. Similarly as before, we skip the context Φ. If at least one of the two sequents '$S(\beta_0) \vdash C$' and '$S(\beta_1) \vdash C$', is not correct under $\mathcal{I}$, then for some $i \in \{0, 1\}$, β_i is true under $\mathcal{I}$, and thus so is

β (by clause 3 of Corollary 2.2). If, conversely, the sequent-premise is not correct under $\mathcal{I}$, then β must be true, and hence one of $\beta_i, i \in \{0, 1\}$ is true as well. Again, this is enough to conclude that the sequent-premise is correct under $\mathcal{I}$ iff the two sequent-conclusions are correct under $\mathcal{I}$. The same reasoning applies to R_α in $\mathbb{E}^{\mathbf{Q}}$, $\mathbb{E}^{\mathbf{RQ}}$ and L_β in $\mathbb{E}^{\mathbf{LQ}}$.

We omit the κ-cases.

Case 4: Finally, assume $Q^* = ?(\Phi(S(-\delta) \vdash \delta(t)))$ results from $Q = ?(\Phi(S(\_) \vdash \delta))$ by rule R_δ of $\mathbb{E}^{\mathbf{Q}}$. Suppose the sequent-conclusion is incorrect under $\mathcal{I} = \langle U, f \rangle$. Then sentence δ must be false under $\mathcal{I}$, and thus '$S(\_) \vdash \delta$' is incorrect under $\mathcal{I}$. Let us observe that the presence of '$-\delta$' on the left side of the sequent is necessary to conduct this argument, since the falsity of $\delta(t)$ under $\mathcal{I}$, for one term t, would not guarantee that δ itself is false.

Conversely, if '$S(\_) \vdash \delta$' is not correct under $\mathcal{I}$, then all sentences in '$S(-\delta)$' are true under $\mathcal{I}$, and $\mathcal{I} \nvDash \delta$. Hence, there is at least one object assignment σ under which:

$$\mathcal{I} \nvDash \delta\, [\sigma]. \tag{2.15}$$

If δ is of the form '$\exists x_j A$', then $\delta(t)$ takes the form '$A[x_j/t]$', where t is a closed term. Then (2.15) holds iff for each object $u \in U$:

$$\mathcal{I} \nvDash A\, [\sigma(x_j/u)]. \tag{2.16}$$

In particular, for the object $t^{\mathcal{I}}[\sigma]$,[11] (2.16) yields $\mathcal{I} \nvDash A\, [\sigma(x_j/t^{\mathcal{I}}[\sigma])]$, which shows that also:

$$\mathcal{I} \nvDash A[x_j/t]\, [\sigma(x_j/t^{\mathcal{I}}[\sigma])], \tag{2.17}$$

and this entails that $\mathcal{I} \nvDash A[x_j/t]$. If δ is of the form '$\neg\forall x_j A$', then the reasoning goes along the same lines, but in (2.16) we have for all $u \in U$:

$$\mathcal{I} \nvDash -A\, [\sigma(x_j/u)],$$

in (2.17):

$$\mathcal{I} \nvDash -A[x_j/t]\, [\sigma(x_j/t^{\mathcal{I}}[\sigma])]$$

which yields:

$$\mathcal{I} \nvDash -A[x_j/t].$$

In both cases we arrive at the conclusion that sequent '$S(-\delta) \vdash \delta(t)$' is not correct under $\mathcal{I}$.

The reasoning is analogous for: L_γ of $\mathbb{E}^{\mathbf{Q}}$, R_δ of $\mathbb{E}^{\mathbf{RQ}}$, and L_γ of $\mathbb{E}^{\mathbf{LQ}}$. □

As to rules R_γ and L_δ, the correctness of each constituent of a question-premise under a fixed FOL-interpretation warrants the correctness of each constituent of the

[11] The object is independent of σ, since t is a closed term. This does not influence the reasoning, however, which is correct for sentential functions as well.

respective question-conclusion under the same FOL-interpretation, but not the other way around. For the bottom-up direction we can only prove the transmission of the stronger property: being correct under *every* FOL-interpretation.

Lemma 2.2 (Correctness and Invertibility of R_γ, L_δ) *Suppose that Q^* results from Q by rule R_γ or L_δ of $\mathbb{E}$. Then:*

1. *If each constituent of Q is correct under a FOL-interpretation $\mathcal{I}$, then each constituent of Q^* is correct under $\mathcal{I}$ as well.*
   Moreover,
2. *If each constituent of Q^* is correct under **every** FOL-interpretation, then each constituent of Q is correct under **every** FOL-interpretation.*

Proof We will consider only the case of rule R_γ in $\mathbb{E}^{\mathbf{Q}}$ with $\gamma = \forall x_j A$, the remaining cases are analogous. Both clauses *1* and *2* are proved by transposition.

Thus suppose that Q and Q^* are of the following forms:

$$Q = ?(\ \Phi\ (\ S \vdash \forall x_j A\ )\ ), \qquad Q^* = ?(\ \Phi\ (\ S \vdash A[x_j/a_i]\ )\ ),$$

where a_i does not occur in the question-premise. For the first clause, suppose that '$S \vdash A[x_j/a_i]$' is not correct under $\mathcal{I}$. Then, every term of sequence S is true under $\mathcal{I}$, but $\mathcal{I} \nvDash A[x_j/a_i]$, thereby establishing that $\mathcal{I} \nvDash \forall x_j A$. Thus, the sequent-premise of Q is not correct under $\mathcal{I}$ either.

For the second clause assume that sequent '$S \vdash \forall x_j A$' is not correct under an interpretation $\mathcal{I} = \langle U, f \rangle$. Then, each term of S is true under $\mathcal{I}$, but '$\forall x_j A$' is not. Hence, there is an object assignment σ, and an object $u \in U$, such that $\mathcal{I} \nvDash A\ [\sigma(x_j/u)]$. We indicate an interpretation that falsifies the sequent-conclusion. Let $\mathcal{I}^* = \langle U, f^* \rangle$ be a FOL-interpretation that differs from $\mathcal{I}$ at most with respect to the assignment $f^*(a_i) = u$. Since a_i is new with respect to '$S \vdash \forall x_j A$', interpretation $\mathcal{I}^*$ makes true each term of S just as $\mathcal{I}$ does, but at the same time $\mathcal{I}^* \nvDash A[x_j/a_i]$, as required. □

It is also easy to see that:

Corollary 2.5 *Every sequent of language $\mathcal{L}^?_{\vdash \mathsf{L}}$ in one of the forms: (a)–(d) specified in Definitions 2.3, 2.4, and 2.5 is correct under every L-interpretation of $\mathcal{L}_{\mathsf{L}}$.*

The above corollary, together with Lemmas 2.1 and 2.2, yields soundness of $\mathbb{E}$ with respect to the correctness of sequents under L-interpretations, and also with respect to the semantics of the underlying logic L.

Theorem 2.2 (Soundness of $\mathbb{E}$) *If a closed sequent '$S \vdash T$' of $\mathcal{L}^?_{\vdash \mathsf{L}}$ has a Socratic proof in $\mathbb{E}$, then the sequent is correct under every L-interpretation of $\mathcal{L}_{\mathsf{L}}$. Moreover,*

1. *the both-sided case and the right-sided case: if there is a Socratic proof of sequent '$\vdash A$' in $\mathbb{E}$, then A is L-valid.*
2. *the left-sided case: if there is a Socratic proof of sequent '$-A \vdash$' in $\mathbb{E}$, then A is L-valid.*

Proof Suppose that **p** is a Socratic proof of sequent '$S \vdash T$' in $\mathbb{E}$ and assume that **p** has n terms (questions). According to the definition of a Socratic proof and the above corollary, each constituent of the last (n-th) term of **p** is correct under every L-interpretation. By the invertibility of the rules of $\mathbb{E}$ (Lemmas 2.1, 2.2) also each constituent of the $(n-1)$st term of **p** is correct under every L-interpretation. The argument is analogous in the inductive step, therefore, by decreasing induction with respect to $i : 1 \leqslant i \leqslant n$, sequent '$S \vdash T$' is correct under every L-interpretation. We skip the remaining details. □

A Socratic proof *proves* the correctness of a sequent under every L-interpretation. However, as we have established, sequents represent meta-level statements concerning properties or relations defined in the underlying logic. Therefore, we may say that a Socratic proof proves the truth of the corresponding meta-level statement. We will go back to this issue in Chap. 5, where we also present the semantics for the erotetic parts of languages with sequents.

2.5 Admissible, Derivable, and Structural Rules

This section is devoted to introducing two types of the so-called structural rules into the erotetic realm—namely, the rules of weakening and contraction—and demonstrating that these rules are *admissible* in a certain sense of the term. The primary aim of this analysis is to use of the rules of weakening and contraction in the proofs of completeness presented in Sect. 2.6. The analysis of structural rules lies at the heart of structural proof theory, and thus the behavior of these rules in an erotetic context may be of interest in itself.

We were also interested in studying how the frequently used strategy for proving admissibility results, developed by Negri and von Plato [19, 21], works in the context of erotetic calculi. Certain modifications to this strategy were necessary due to, first, the 'sequence-nature' of the seqsequent-based rules, as opposed to derivations defined as trees, and second, the decision to allow non-atomic formulas in the definition of constituents ending a Socratic proof. In the aforementioned strategy, axioms of sequent calculi are restricted to sequents with an atomic formula occurring in both the antecedent and succedent of the sequent.

We begin this section with a number of technical results to be used later. The proper analysis of the notion of admissibility starts later in this section. Corollaries 2.6–2.9 concern renaming individual constants in Socratic transformations. A reader familiar with admissibility proofs for FOL will recognize a parallel between these corollaries and results concerning substitutions and the so-called α-conversion.

As before, until the end of this section, unless explicitly stated otherwise, $\mathbb{E}$ stands for any of $\mathbb{E}^{\mathbf{Q}}$, $\mathbb{E}^{\mathbf{RQ}}$, $\mathbb{E}^{\mathbf{LQ}}$, $\mathbb{E}^{\mathbf{P}}$, $\mathbb{E}^{\mathbf{RP}}$, or $\mathbb{E}^{\mathbf{LP}}$. In the sequel, it will be useful to generalize the notion of a Socratic proof in the following way:

Definition 2.6 (*Successful Socratic Transformation*) Let Q be a question of the language $\mathcal{L}^{?}_{\vdash\mathsf{FOL}}$. A Socratic transformation of question Q in $\mathbb{E}$ is *successful* if it is finite and each constituent of its last question is of one of the forms (a)–(d) specified in Definitions 2.3, 2.4, and 2.5.

A Socratic proof leads to the resolution of a problem expressed by a specific question of the language $\mathcal{L}^?_{\vdash\mathsf{L}}$—a question with only one constituent. On the other hand, a successful Socratic transformation leads to the resolution of a question of $\mathcal{L}^?_{\vdash\mathsf{L}}$ of an arbitrary form.

Here are some additional useful definitions and conventions. By *length of a finite sequence*, we mean the number of terms in the sequence. For simplicity, any sequent of the language $\mathcal{L}^?_{\vdash\mathsf{L}}$ that takes one of the forms (a)–(d) specified in Definitions 2.3, 2.4, or 2.5 will be called *basic*.

A rule is understood in this monograph set-theoretically, as a set of ordered pairs, each pair containing a premise and a conclusion as the first and the second element, respectively. If $\mathbf{r}$ is a rule, then each element of $\mathbf{r}$ is called *an instance of* $\mathbf{r}$. Additionally, from now on, instead of:

$$\frac{Q}{Q^*}\ \mathbf{r}$$

we will often simply write: $\mathbf{r} = Q/Q^*$. Whether this notation refers to the general scheme that defines the rule syntactically or to a particular instance of the rule will be clear from the context.

Renaming Individual Constants

We begin with a minor result that will be used shortly.

Corollary 2.6 (Renaming Individual Constants: New for New) *Let* $\mathbf{s}$ *be a finite Socratic transformation of a question in* $\mathbb{E} \in \{\mathbb{E}^{\mathbf{Q}}, \mathbb{E}^{\mathbf{RQ}}, \mathbb{E}^{\mathbf{LQ}}\}$*. Suppose rule* $\mathbf{r} \in \{R_\gamma, L_\delta\}$ *is applied in* $\mathbf{s}$ *and introduces a new individual constant .*a_i *in the sequent-conclusion of* $\mathbf{r}$*. If* a_k*, where* $k \neq i$*, is another constant not occurring in* $\mathbf{s}$*, then replacing every occurrence of* a_i *in* $\mathbf{s}$ *with* a_k *results in a finite Socratic transformation of the same question in* $\mathbb{E}$ *and of the same length. Moreover, if* $\mathbf{s}$ *is successful, so is the resulting transformation.*

Proof The result is obvious, so we will not delve into formal details, but we will note the important points.

First, it is easy to see that if questions Q and Q^* occur in $\mathbf{s}$, and Q/Q^* is an instance of an erotetic rule $\mathbf{r}$, then the pair remains an instance of the same rule $\mathbf{r}$ after the replacement operation is performed on Q and Q^*. This holds for $\mathbf{r} \in R_\gamma, L_\delta$, given the assumption that a_k does not occur in $\mathbf{s}$.

Second, if a basic sequent is of the form (a) (specified in Definition 2.3) due to the presence of, say, $P(a_i)$ on both sides of the turnstile, where a_i is the constant to be replaced, then after the replacement we obtain a sequent that still falls under the same pattern (a). The same applies to the other schemes (b)–(d) specified in Definition 2.3–2.5.

Third, individual constants cannot be eliminated from questions in the course of a Socratic transformation. Therefore, if an individual constant occurs in the *first* question of our Socratic transformation, it cannot be reused as the new constant a_i by $\mathbf{r} \in \{R_\gamma, L_\delta\}$, and hence it will not be replaced. Hence, the result is still a Socratic transformation of *the same question*.

Finally, the replacement operation does not affect the length of the Socratic transformation. □

Here is an opportunity to explain why we need the global version of the proviso for the applicability of $\mathbf{r} \in \{R_\gamma, L_\delta\}$ instead of the local one. Suppose a Socratic transformation of a question of the form:

$$?(\ \exists x A(x) \vdash P(a) \wedge Q(b)\ )$$

is being considered, and the next step is an application of R_α, resulting in:

$$?(\ \exists x A(x) \vdash P(a)\ ;\ \exists x A(x) \vdash Q(b)\ ).$$

In this case, it can happen that b no longer appears in the first sequent, and if the proviso were local, an application of L_δ could use this constant in eliminating the existential quantifier. However, this would result in the replacement operation described above altering the *initial question*, and thus Corollary 2.6 would no longer hold.

This corollary, however, is crucial for the proof of the Separation Lemma (which will be used multiple times later) and for Lemma 2.4, which concerns the admissibility of the rule of weakening. These lemmas, in turn, are essential for conducting the completeness proof of our erotetic calculi within the assumed methodology of consistency properties.

First, however, let us notice that the following holds true.

Corollary 2.7 *If Q/Q^* is an instance of a rule $\mathbf{r}$ of $\mathbb{E}$, and there is a successful Socratic transformation of question Q^* in $\mathbb{E}$ of length n, then there is also a successful Socratic transformation of question Q in $\mathbb{E}$ of length $n+1$.*

Proof It suffices to observe that there is no restriction on the applicability of erotetic rules that concerns the entire Socratic transformation; only the question-premise and the question-conclusion are relevant. Thus, if $\mathbf{s} = \langle s_1, \ldots, s_n \rangle$ is a successful Socratic transformation of Q^*, and Q/Q^* is an instance of $\mathbf{r}$, then $\langle Q, s_1, \ldots, s_n \rangle$ is a successful Socratic transformation of Q of the required length. □

Further, we shall practice some more renaming. The following result uses Corollary 2.7.

Corollary 2.8 *Let Q be a question of $\mathcal{L}^?_{\vdash \mathsf{FOL}}$, and let ϕ stand for a constituent of Q. Moreover, let a_k be an arbitrary individual constant of $\mathcal{L}_{\mathsf{FOL}}$.*

If Q has a successful Socratic transformation in $\mathbb{E} \in \{\mathbb{E}^{\mathbf{Q}}, \mathbb{E}^{\mathbf{LQ}}, \mathbb{E}^{\mathbf{RQ}}\}$ of length n, then for each individual constant a_i that occurs in ϕ, the question resulting from Q by replacing every occurrence of a_k in ϕ with an occurrence of a_i also has a successful Socratic transformation in $\mathbb{E}$ of length n.

Proof If the individual constant a_k does not occur in ϕ at all, then the replacement operation changes nothing. Thus, we assume that a_k occurs in ϕ, and that $a_i \neq a_k$. The proof proceeds by induction on the length of a Socratic transformation of Q.

Let **s** be a successful Socratic transformation of Q in the calculus $\mathbb{E} \in \{\mathbb{E}^{\mathbf{Q}}, \mathbb{E}^{\mathbf{LQ}}, \mathbb{E}^{\mathbf{RQ}}\}$. Assume also that the individual constants a_i and a_k occur in ϕ, with $a_i \neq a_k$. In the base case **s** is of length 1, ϕ is a basic sequent, and it is clear that this property is not altered by replacing a_k with a_i.

Inductive step. Assume that $\mathbf{s} = \langle Q_0, Q_1, \ldots, Q_n \rangle$ is of length $n + 1$, and that Q_0/Q_1 is an instance of an erotetic rule **r**. Suppose that **r** acts upon ϕ, and let ψ be the sequent resulting from ϕ by replacing every occurrence of a_k with an occurrence of a_i.

First, assume that **r** is a non-branching rule. Let $Q_1 = ?(\Phi(\phi^*))$, where ϕ^* is the sequent-conclusion of **r**. The sequence $\langle Q_1, \ldots, Q_n \rangle$ is a successful Socratic transformation of Q_1 in $\mathbb{E}$ of length n, therefore, by IH, question resulting from Q_1 by replacing every occurrence of a_k in ϕ^* with an occurrence of a_i has a successful Socratic transformation in $\mathbb{E}$ of length n. Let ψ^* be the result of applying the replacement operation to ϕ^*. The described replacement operation does not affect the applicability of rules requiring a fresh individual constant, since the fresh constant is neither a_k nor a_i.

Next, since Q_0/Q_1, i.e., $?(\Phi(\phi))/?(\Phi(\phi^*))$, is an instance of **r**, so is $?(\Phi(\psi))/?(\Phi(\psi^*))$. This is straightforward; for example, consider $\mathbf{r} = L_\gamma$, $a_i = a_1$, and $a_k = a_2$. Since the following is an instance of L_γ:

$$\frac{?(\ P_1(a_1) \to P_2(a_2), \neg\exists x\, Q_1(x, a_1) \vdash P_2(a_1)\ )}{?(\ P_1(a_1) \to P_2(a_2), \neg\exists x\, Q_1(x, a_1), \neg Q_1(f(a_2), a_1) \vdash P_2(a_1)\ )}\ L_\gamma$$

so is:

$$\frac{?(\ P_1(a_1) \to P_2(a_1), \neg\exists x\, Q_1(x, a_1) \vdash P_2(a_1)\ )}{?(\ P_1(a_1) \to P_2(a_1), \neg\exists x\, Q_1(x, a_1), \neg Q_1(f(a_1), a_1) \vdash P_2(a_1)\ )}\ L_\gamma$$

Finally, since question $?(\Phi(\psi^*))$ has a successful Socratic transformation in $\mathbb{E}$ of length n, and $?(\Phi(\psi))/?(\Phi(\psi^*))$ is an instance of **r**, by Corollary 2.7 there is a successful Socratic transformation of question $?(\Phi(\psi))$ in $\mathbb{E}$ of length $n + 1$.

Now, assume **r** is a branching rule, and let ϕ_0 and ϕ_1 be the sequents-conclusions of **r**. The sequence $\langle Q_1, \ldots, Q_n \rangle$ is then an n-term successful Socratic transformation of $Q_1 = ?(\Phi(\phi_0, \phi_1))$ in $\mathbb{E}$. Let ψ_0 and ψ_1 represent the results of replacing every occurrence of a_k by a_i in ϕ_0 and ϕ_1, respectively. By IH, the question $Q_1^* = ?(\Phi(\psi_0, \phi_1))$ has a successful Socratic transformation in $\mathbb{E}$ of length n, and by IH again, $?(\Phi(\psi_0, \psi_1))$ has a successful Socratic transformation in $\mathbb{E}$ of length n. Finally, since the pair $?(\Phi(\psi))/?(\Phi(\psi_0; \psi_1))$ is an instance of **r**, $?(\Phi(\psi))$ has a successful Socratic transformation in $\mathbb{E}$ of length $n + 1$. □

The following result is a direct consequence of the previous corollary.

Corollary 2.9 (Renaming Individual Constants: Old for New)

1. *If a sequent of the form '$S(\delta(a_k))(\delta(a_i)) \vdash T$', where a_k occurs exactly once in the sequent, has a Socratic proof in $\mathbb{E} \in \{\mathbb{E}^{\mathbf{Q}}, \mathbb{E}^{\mathbf{LQ}}\}$, then there is also a Socratic proof in $\mathbb{E}$ of the sequent '$S(\delta(a_i))(\delta(a_i)) \vdash T$'.*
2. *If a sequent of the form '$S(-\gamma(a_k)) \vdash \gamma(a_i)$', where a_k occurs exactly once in the sequent, is provable in $\mathbb{E}^{\mathbf{Q}}$, then there is also a Socratic proof in $\mathbb{E}^{\mathbf{Q}}$ of the sequent '$S(-\gamma(a_i)) \vdash \gamma(a_i)$'.*
3. *If a sequent of the form '$\vdash T(\gamma(a_k))(\gamma(a_i))$', where a_k occurs exactly once in the sequent, is provable in $\mathbb{E}^{\mathbf{RQ}}$, then there is also a Socratic proof in $\mathbb{E}^{\mathbf{RQ}}$ of the sequent '$\vdash T(\gamma(a_i))(\gamma(a_i))$'.*

Corollary 2.9 is crucial for the proof of the admissibility of contraction presented later in this section.

Separation Lemma

Let us now reference the following result by Chlebowski [2, pp. 37–38] for further use (the formulation has been slightly modified from the original). This result is presented in [2] among several similar findings concerning the structural properties of erotetic calculi. Although somewhat technical, we provide a detailed proof of the lemma here, using this opportunity to explain the use of algorithmic tools employed throughout this book. Readers who find the proof either too lengthy or too elementary, may skip to this result.

Lemma 2.3 (Separation lemma) *Let $\phi_1, \ldots, \phi_n$ stand for sequents of $\mathcal{L}^?_{\vdash\mathsf{L}}$. The following conditions are equivalent for any erotetic calculus $\mathbb{E}$:*

1. *There is a successful Socratic transformation of the question '$?(\phi_1, \ldots, \phi_n)$' in $\mathbb{E}$.*
2. *For each i, where $1 \leqslant i \leqslant n$, there is a successful Socratic transformation of the question '$?(\phi_i)$' in $\mathbb{E}$.*

Proof We follow the core idea of the proof by Chlebowski, but present it in a new format (cf. [2, pp. 38–39]).

Moving from condition *1* to *2*, we must 'extract' the relevant elements from the Socratic transformation of '$?(\phi_1, \ldots, \phi_n)$' that pertain specifically to a given ϕ_i. This process involves removing a considerable amount of external context. Conversely, in moving from *2* to *1*, we need to combine the transformations of the individual $?(\phi_i)$, which requires adding a significant amount of external context. Below, we provide an algorithmic description of the 'extraction' and 'combination' processes. Many results in this monograph will follow a similar proof structure.[12]

Suppose we have a Socratic transformation of the question '$?(\phi_1, \ldots, \phi_n)$', say $\mathbf{s} = \langle s_1, \ldots, s_m \rangle$ is the transformation, and let $i \in \{1, \ldots, n\}$ be chosen. We will examine each question in $\mathbf{s}$ and either select some of its constituents—if a rule was

[12] We use pseudocode as, for example, in [3].

applied relative to a sequent relevant to ϕ_i—or select none, if a rule was applied to a non-relevant constituent. The relevant constituents will then be added to the resulting Socratic transformation, which will be stored under the sequence '**t**'. Thus, the algorithm starts as follows.

t $\leftarrow \langle ?(\phi_i) \rangle$
for $l = 1$ **to** $m - 1$ **do**
⋮
end

We must take into account that applying a branching rule to a relevant sequent increases the number of relevant sequents, as one relevant sequent-premise has two relevant sequent-conclusions. To track this, we will store the number of relevant sequents that need to be selected from the current question as 'w' (for *width*). Thus, before entering the for-loop, we add:

$w \leftarrow 1$

What happens inside the for-loop depends on the character of the rule applied to the current question, therefore inside the for-loop we begin with these assignments:

r $\leftarrow$ the rule applied to question s_l
$k \leftarrow$ the ordinal number of the active constituent in s_l

Index i will be used to indicate the position of the first relevant sequent in the currently analyzed question. Initially, the value of i is provided. If **r** is a non-branching rule, three scenarios can arise, as described below. In the description of case (2), we use the predicate '**append**', which is a built-in function in many programming languages. It works like the concatenation of two finite sequences. We assume **append** operates on two lists, passed as the first and second arguments, and produces a third list containing the consecutive elements of the first list followed by those of the second. The resultant list is provided as the third argument of **append**.

---

1 **if** $k < i$ **then**
  do nothing, just go to the next question by increasing the value of l:
  $l \leftarrow l + 1$
end
2 **if** $i \leqslant k < i + w$ **then**
  u $\leftarrow$ the sequence of constituents of s_{l+1} from i-th to $(i + w - 1)$-st
  append(**t**, $\langle ?(\mathbf{u}) \rangle$, $Temp$)
  t $\leftarrow Temp$
  $l \leftarrow l + 1$
end
3 **if** $i + w \leqslant k$ **then**
  do nothing, just put $l \leftarrow l + 1$
end

The three cases specified above capture the following situations: (1) **r** acts left of the sequence of our relevant sequents, (2) **r** acts upon one of the relevant sequents, (3)

r acts right of the sequence. The three cases are distinguished in the same way for branching **r**. However, we have to take into account that when a branching rule is applied, the first case increases the value of i. Thus for branching **r** we put:

```
if k < i then
  | i ← i + 1
  | l ← l + 1
end
```

In the second case, the value of w is increased, whereas the third case changes nothing.

```
if i ⩽ k < i + w then
  | u ← the sequence of constituents of s_{l+1} from i-th to (i + w)-th
  | append(t, ⟨?(u)⟩, Temp)
  | t ← Temp
  | w ← w + 1
  | l ← l + 1
end
```

Finally, once the if-then conditions are embedded within the for-loop, the instruction '$l \leftarrow l + 1$' becomes redundant. This is because the loop's structure inherently ensures that the control index, in this case l, is incremented after each iteration. The complete algorithm is presented as Algorithm 2.1.

We now need to address the reverse direction. The structure of the corresponding algorithm is relatively straightforward and is presented as Algorithm 2.2.

However, a new problem emerges. We begin with a successful Socratic transformation $\mathbf{s}^i$ for each $?(\phi_i)$, where $1 \leqslant i \leqslant n$. These Socratic transformations are independent of one another, hence it is perfectly possible that one of them contains occurrences of individual constants that are introduced by R_γ or L_δ as new constants in another transformation. Since the proviso of applicability of these rules is global, the rules' applicability could be lost after running the algorithm.

To ensure the rules remain applicable, we must first use Corollary 2.6. This is done as follows: Let IC represent the set of all individual constants that occur in any of $\mathbf{s}^1, \ldots, \mathbf{s}^n$. Clearly, this set is finite. Starting with the first application of $\mathbf{r} \in \{R_\gamma, L_\delta\}$ (if any) in $\mathbf{s}^1$, we replace each occurrence of the new individual constant introduced by **r** with one that does not belong to IC. The replacement operation is performed on the entire $\mathbf{s}^1$, as described in Corollary 2.6, and by virtue of this Corollary, the result is still a successful Socratic transformation of the same question. After performing this replacement, the new individual constant is added to IC, ensuring it cannot be reused in the same manner. This process is repeated for each application of R_γ and L_δ, and subsequently for each $\mathbf{s}^i$, where $1 \leqslant i \leqslant n$.

Algorithm 2.1: Algorithm producing a Socratic transformation of $?(\phi_i)$ from that of $?(\phi_1, \ldots, \phi_n)$

```
Data: Socratic transformation s = ⟨s_1, ..., s_m⟩ of ?(φ_1, ..., φ_n), i: 1 ⩽ i ⩽ n
Result: Socratic transformation t of ?(φ_i)
w ← 1
t ← ⟨?(φ_i)⟩
for l = 1 to m − 1 do
    if r is a non-branching rule then
        r ← the rule applied to question s_l
        k ← the number of the active constituent in s_l
        if i ⩽ k < i + w then
            u ← the sequence of constituents of s_{l+1} from i-th to (i + w − 1)-st
            append(t, ⟨?(u)⟩, Temp)
            t ← Temp
        end
    end
    if r is branching then
        if k < i then
            i ← i + 1
        end
        if i ⩽ k < i + w then
            u ← the sequence of constituents of s_{l+1} from i-th to (i + w)-th
            append(t, ⟨?(u)⟩, Temp)
            t ← Temp
            w ← w + 1
        end
    end
end
```

Let $\mathbf{t}^1, \ldots, \mathbf{t}^n$ stand for the new Socratic transformations derived from $\mathbf{s}^1, \ldots, \mathbf{s}^n$ through the replacement process described in the previous paragraph. These Socratic transformations $\mathbf{t}^i$ serve as the input data for Algorithm 2.2. In line 4 of the algorithm, it is assumed that during the final iteration of the outer for-loop, the structure '$\langle \phi_{i+1}, \ldots, \phi_n \rangle$' is simply the empty sequence (denoted as '$\varnothing$'). □

The following is a direct consequence of the Separation Lemma.

Corollary 2.10 (Length-Preserving Separation) *If there is a successful Socratic transformation of the question '$?(\phi_1, \ldots, \phi_n)$' in $\mathbb{E}$ of length l, then for each $i : 1 \leqslant i \leqslant n$, there is a successful Socratic transformation of the question '$?(\phi_i)$' of length no greater than l.*

It is worth noting that length preservation is not guaranteed in the opposite direction: constructing a Socratic transformation of the question '$?(\phi_1, \ldots, \phi_n)$' from the transformations of '$?(\phi_i)$' may result in a much longer sequence than the maximal

Algorithm 2.2: Algorithm producing a Socratic transformation of $?(\phi_1, \ldots, \phi_n)$ from those of $?(\phi_1), \ldots, ?(\phi_n)$

Data: Socratic transformation $\mathbf{t}^i = \langle t^i_1, \ldots, t^i_{m_i} \rangle$ of $?(\phi_i)$ for each $i : 1 \leqslant i \leqslant n$
Result: Socratic transformation **s** of question $?(\phi_1, \ldots, \phi_n)$

$\mathbf{s} \leftarrow \langle ?(\phi_1, \ldots, \phi_n) \rangle$
$L \leftarrow \varnothing$ `/* left context empty at the start */`
for $i = 1$ **to** n **do**
  $R \leftarrow \langle \phi_{i+1}, \ldots, \phi_n \rangle$ `/* right context */`
  for $k = 2$ **to** m_i **do**
    find question $t^i_k = ?(\Phi)$
    append$(L, \langle \Phi \rangle, Temp1)$
    append$(Temp1, R, Temp2)$
    append$(\mathbf{s}, \langle ?(Temp2) \rangle, Temp3)$
    $\mathbf{s} \leftarrow Temp3$
  end
  $?(\Phi) \leftarrow$ the last question of **s**
  append$(L, \Phi, Temp)$
  $L \leftarrow Temp$ `/* left context updated */`
end

length of any Socratic transformation of '$?(\phi_i)$'. In this respect, the situation differs from standard sequent calculi, where proofs are defined as trees.[13]

Admissibility and Derivability of the Erotetic Rules

In [33], several rules admissible in the erotetic calculus $\mathbf{E^{PQ}}$ are introduced and analyzed.[14] Some, but not all, of these rules are derivable in the calculus. The authors also introduce a set of admissible structural rules; however, the precise notions of derivability and admissibility within the erotetic context are not examined in detail in [33].

Sara Negri and Jan von Plato define the notion of admissibility as follows [19, p. 20]:

> 'Given a system of rules **G**, we say that a rule with premisses $S_1, \ldots, S_n$ and conclusion S is *admissible* in **G** if, whenever an instance of $S_1, \ldots, S_n$ is derivable in **G**, the corresponding instance of S is derivable in **G**.'

Here, derivability of a sequent S_i in **G** is defined as the existence of the relevant derivation. In other words, if a rule is admissible, its addition to a proof system (or system of rules, to follow the quotation) does not alter the derivability relation generated by the system. Any conclusion derivable from a set of premises after the addition must have already been derivable beforehand. It means, among other things, that adding an admissible rule to a sound calculus must result in a sound calculus.

[13] This is one reason why the strategy from [19] requires modifications to apply it in the erotetic setting. See also footnotes 19 and 16 in this chapter for references to structural proof theory.
[14] $\mathbf{E^{PQ}}$ is the original version of the erotetic system for FOL introduced by Wiśniewski. See also footnote no 7 of this chapter.

In the simplest case, admissibility is met in the form of derivability of a rule.[15] Referring back to the earlier quotation, we would say that a rule with premises $S_1, \ldots, S_n$ and conclusion S is *derivable* in **G** if every instance of S is derivable in **G** from the respective instances of $S_1, \ldots, S_n$. Typically, this means that we can develop a derivation schema leading from premises $S_1, \ldots, S_n$ to conclusion S, using only the rules of **G**. Consequently, a derived rule is often seen as a tool for creating useful shortcuts in a proof. Adjusting the notion of derivability to erotetic rules seems straightforward:

Definition 2.7 (*Derivability of a Rule in* $\mathbb{E}$) Let $\mathbf{r} = Q/Q^*$ be an erotetic rule. We will say that $\mathbf{r}$ is *derivable in* erotetic calculus $\mathbb{E}$ iff there exists a finite Socratic transformation $\mathbf{s}$ of question Q in $\mathbb{E}$ such that Q^* is its last term.

This interpretation aligns with the notion as understood in [33].

The notion of admissibility is more nuanced. Recall the fact that the direction of proving defined by erotetic rules is converse to the direction of proving defined by the rules of a standard sequent calculus. We have noticed one consequence of this fact: the soundness proof of an erotetic calculus depends on the invertibility of the rules, whereas in sequent calculi, soundness hinges on the correctness of the rules. The following definition of admissibility highlights another consequence of the above mentioned fact.

Definition 2.8 (*Admissibility of a Rule in* $\mathbb{E}$) Let $\mathbf{r} = Q/Q^*$ be an erotetic rule. We say that $\mathbf{r}$ is *semi-admissible in* erotetic calculus $\mathbb{E}$ iff the following condition holds:

1. if there is a successful Socratic transformation of question Q^* in $\mathbb{E}$, then there is a successful Socratic transformation of question Q in $\mathbb{E}$.

If, in addition to condition 1, the rule $\mathbf{r}$ also satisfies condition 2:

2. if there is a successful Socratic transformation of question Q in $\mathbb{E}$, then there is a successful Socratic transformation of question Q^* in $\mathbb{E}$,

then we say that $\mathbf{r}$ is *admissible in* erotetic calculus $\mathbb{E}$.

A comment is definitely needed here. On the face of it, it may seem that clause *2* of Definition 2.8 captures the basic intuition behind admissibility, which is expressed in the quotation cited above. However, the point is that the 'final conclusion' of a successful Socratic transformation is represented by its *first* element, not the last one, as it is in standard proofs. For example, suppose we are able to prove the following dependency:

(d) if there exists a successful Socratic transformation in $\mathbb{E} \in \{\mathbb{E}^{\mathbf{RP}}, \mathbb{E}^{\mathbf{RQ}}\}$ of the question '$?(\vdash T(A)\,;\, \vdash T(\neg A)\,)$', then there exists a successful Socratic transformation in $\mathbb{E}$ of the question '$?(\vdash T(\_)\,)$'.

[15] See also the account of admissibility and derivability by Indrzejczak in [13, p. 22].

Given dependency (d), when investigating the provability of the sequent '$\vdash T(\_)$', one can pass from the question '$?(\vdash T(\_))$' to the question '$?(\vdash T(A) \,;\, \vdash T(\neg A))$', since finishing a Socratic transformation of the later question successfully shows that there exists a successful Socratic transformation of the former question. In other words, proving dependency (d) demonstrates the admissibility of the following rule in the calculus $\mathbb{E}^{\mathbf{RQ}}/\mathbb{E}^{\mathbf{RP}}$:

$$\frac{?(\vdash T(\_))}{?(\vdash T(A) \,;\, \vdash T(\neg A))}$$

It should be stressed that the dependency expressed by (d) *does not* indicate that there is a Socratic transformation *leading* from the question '$?(\vdash T(\_))$' to the question '$?(\vdash T(A) \,;\, \vdash T(\neg A))$'; in fact, no such Socratic transformation exists, as the rule is not derivable in any $\mathbb{E}$.

The primary aim of transitioning from question Q to Q^* by means of an erotetic rule is to simplify the process of resolving question Q. Therefore, an admissible rule Q/Q^* should be introduced with the purpose of leveraging the fact that resolving Q^* is easier than resolving Q. In other words, the existence of a solution to Q^* must guarantee the existence of a solution to Q, as we saw in the above example. In standard sequent calculi, it is the existence of a *resolution* of premises $S_1, \ldots, S_n$, i.e. their derivation from axioms, that guarantees the existence of a *resolution* (derivation) of the conclusion S. Thus, we can clearly see how the different direction of proving influences the notion of admissibility.

All of this leads to the conclusion that it is condition 1 in Definition 2.8 that expresses the idea of admissibility known from other proof systems. Now, we can ask what the purpose of calling the property *semi-admissibility* is, and why condition 2 is involved. The point is that condition 1 alone does not guarantee the correctness of a rule. Let us consider, for example, the following:

$$\frac{?(\vdash A, \neg A)}{?(\vdash A \wedge \neg A)}$$

where A is a sentence of $\mathcal{L}_{\mathsf{FOL}}$. Unfortunately, for each calculus $\mathbb{E}$ considered in this book (except for the left-sided ones), the rule satisfies condition 1 of Definition 2.8, as *there is no* successful Socratic transformation of the question '$?(\vdash A \wedge \neg A)$' in $\mathbb{E}$. But the rule is clearly incorrect. Hence, we need condition 2 in Definition 2.8 to ensure correctness. This is another consequence of the fact that a Socratic proof begins with a question about the *final conclusion* and proceeds to axioms. In the usual setting, the admissibility of a rule in a sequent calculus ensures its correctness, but—as we can see—this is not the case with erotetic calculi.

Finally, to sum up the discussion of admissibility and derivability, we expect the latter to be a special case of the former. However, we will wait until the end of Sect. 2.6 to prove this result.

Structural Rules

Structural rules and their properties are central to structural proof theory, where a distinction is commonly made between *logical rules*—those with a particular connective displayed in their schemes—and *structural rules*, which modify the structure of a sequent independently of the form of the formulas it contains. While erotetic calculi do not require structural rules for soundness and completeness, these rules prove extremely useful in the metatheory, much like they do in more conventional calculi of sequents.[16]

Below, we consider some particular semi-admissible rules, notably the structural rules of weakening and contraction.[17] To proceed, we will introduce the concept of a subsequence of a sequence, which can be roughly understood as the result of deleting certain elements from a sequence while preserving the order of the remaining elements. More formally, a subsequence of a sequence can be viewed as the composition $g \circ f$ of two functions, f and g. The function f selects the positions of the elements in sequence S that are retained, ensuring that their original order is preserved. The function g then maps these selected positions to the corresponding values in S. Here is the definition:

Definition 2.9 (*Subsequence*) Let S stand for an n-term sequence of objects from a set X, that is, a function $S : \{1, \ldots, n\} \longrightarrow X$. Suppose that $0 \leqslant k \leqslant n$. Let g be the restriction of S to a k-element domain $\{i_1, \ldots, i_k\} \subseteq \{1, \ldots, n\}$, where $i_1 < i_2 < \cdots < i_k$. Finally, let f be a function from $\{1, \ldots, k\}$ to $\{i_1, \ldots, i_k\}$, such that $f(j) = i_j$. Then the composition $g \circ f$ of functions f and g is called a *subsequence of* S. Moreover,

- If $0 < k$, then $g \circ f$ is called a *non-empty subsequence of* S.
- If $k < n$, then $g \circ f$ is called a *proper subsequence of* S.

Finally,

- If S^* is a (non-empty, proper) subsequence of S, then S is called a *(non-empty, proper) supersequence of* S^*.

In the marginal cases of $k = 0$ and $k = n$ we arrive at, respectively, the empty sequence and sequence S itself. In cases where the subsequence is proper, we observe that $\{i_1, \ldots, i_k\} \subset \{1, \ldots, n\}$, that is, the range of f is a proper subset of the domain of S.

With the notion of a subsequence established, the rules of weakening can be expressed in the following general form:

$$\frac{?(\,\Phi\,(\,S \vdash T\,)\,)}{?(\,\Phi\,(\,S^* \vdash T\,)\,)}\; L_{weak} \qquad \frac{?(\,\Phi\,(\,S \vdash T\,)\,)}{?(\,\Phi\,(\,S \vdash T^*\,)\,)}\; R_{weak}$$

where S^* is a non-empty proper subsequence of S, and similarly, T^* is a non-empty proper subsequence of T. The requirement that S^* and T^* be non-empty ensures

[16] For more on structural proof theory, see [19, 20, 26, 27].

[17] Weakening is also referred to as *thinning* in [9, 16] or *addition* in [6].

that the empty sequent is avoided, as it is not included in the language. In the both-sided case, the exclusion of the empty sequent is guaranteed by the presence of at least one formula in the succedent. Therefore, when L_{weak} is added to $\mathbb{E}^{\mathbf{Q}}$ or $\mathbb{E}^{\mathbf{P}}$, S^* is permitted to be empty, but it must still be a *proper* subsequence of S. The 'properness' condition serves to prevent trivial repetitions of the same question.

To summarize, considering the 'sidedness' of erotetic calculi: rule L_{weak} applies to calculi $\mathbb{E}^{\mathbf{Q}}$ and $\mathbb{E}^{\mathbf{P}}$, where S^* is a proper subsequence of S; for calculi $\mathbb{E}^{\mathbf{LQ}}$ and $\mathbb{E}^{\mathbf{LP}}$, S^* in L_{weak} must be a non-empty proper subsequence of S; and rule R_{weak} is used in calculi $\mathbb{E}^{\mathbf{RQ}}$ and $\mathbb{E}^{\mathbf{RP}}$, where T^* is a non-empty proper subsequence of T.

At first glance, the structure of our rules of weakening may seem surprising; however, this perception changes when one recalls the issue of converted direction. In the erotetic setting, the rules of weakening add formulas to what constitutes a premise of the rule. Consequently, when read top-down, they effectively remove information. Here are some specific instances of these rules, which can be interpreted as processes for removing irrelevant information:

$$\frac{?(\,p \wedge q, r \vdash p\,)}{?(\,p \wedge q \vdash p\,)}\, L_{weak} \qquad \frac{?(\vdash p, q, \neg p\,)}{?(\vdash p, \neg p\,)}\, R_{weak}$$

In fact, the above examples are quite typical applications of the rules of weakening, which, in standard sequent calculi, are employed when the axiomatic basis of the calculus is restricted to axiomatic sequents of, more or less, the form '$P \vdash P$'. Weakening is necessary to add context to these skeleton axioms.

Thus, it can be said that the application of weakening in a Socratic transformation corresponds to the removal of informational noise that needs to be eliminated in order to focus on what is relevant to the ongoing inquiry.[18]

However, irrespective of this erotetic interpretation, the primary reason for employing these rules in this book is that weakening is necessary to adapt the technique of consistency properties, which works so smoothly for set-based structures like tableaux.

The following lemma is an analogue of height-preserving admissibility of weakening (cf. [19, Sect. 3.2]):

Lemma 2.4 (Semi-Admissibility of L_{weak}, R_{weak} in $\mathbb{E}$) *Let* $\mathbb{E} \in \{\mathbb{E}^{\mathbf{Q}}, \mathbb{E}^{\mathbf{LQ}}, \mathbb{E}^{\mathbf{RQ}}, \mathbb{E}^{\mathbf{P}}, \mathbb{E}^{\mathbf{LP}}, \mathbb{E}^{\mathbf{RP}}\}$. *If there is a successful Socratic transformation* **s** *of the question-conclusion of one of the rules* L_{weak}, R_{weak} *in* $\mathbb{E}$, *and* **s** *is of length* n, *then there is a successful Socratic transformation of the question-premise of the rule in* $\mathbb{E}$, *which is also of length* n.

Proof Suppose there is a successful Socratic transformation **s** of the question '$?(\Phi(S^* \vdash T))$' in calculus $\mathbb{E}$. Assume **s** is of length n, and that sequence S^* is a non-empty, proper subsequence of some S.

---

[18] The topic of the erotetic interpretation of the rules of weakening and contraction is explored more broadly in [18].

The general idea of this kind of proof is that, since **s** is successful, it remains so even when we add additional formulas to the constituents of its consecutive questions. The only issue arises when the added formulas introduce new individual constants in cases involving applications of R_γ or L_δ. Importantly, the addition of these formulas does not change the length of the transformation. The proof proceeds by induction on n, which is a common approach to admissibility proofs.[19]

Base case: Suppose that **s** has exactly one element, which must be the question '$?(\Phi(S^* \vdash T))$.' Since the one-term Socratic transformation is successful, each of its constituents must be basic, i.e., it falls under one of the schemes (a)–(d) specified in Definitions 2.3–2.5. If we add elements to S^*, the sequent still falls under the appropriate scheme. Thus, the one-term sequence containing only the question '$?(\Phi\,(S \vdash T\,))$' is also successful.

Inductive step: Suppose that $\mathbf{s} = \langle Q_0, Q_1, \ldots, Q_n \rangle$ has $n + 1$ terms, and assume that Q_1 results from $Q_0 = ?(\Phi(S^* \vdash T))$ by a rule **r** of $\mathbb{E}$. We aim to show that there is a successful Socratic transformation of '$?(\Phi(S \vdash T))$' of length $n + 1$.

We begin by addressing the problematic cases where $\mathbf{r} \in \{R_\gamma, L_\delta\}$. Here, extending S^* may introduce an individual constant that was used in the application of **r** as a 'fresh' one. For example, the following instance of L_δ (on the left) becomes invalid when we weaken the antecedent by adding $P(b)$:

$$\frac{?(\,P(a), \exists x Q(x) \vdash\ Q(a) \vee P(a)\,)}{?(\,P(a), Q(b) \vdash\ Q(a) \vee P(a)\,)}\ L_\delta\ \frac{?(\,P(a), P(b), \exists x Q(x) \vdash\ Q(a) \vee P(a)\,)}{?(\,P(a), P(b), Q(b) \vdash\ Q(a) \vee P(a)\,)}$$

To resolve this, we carefully *rename* the individual constants to preserve the applicability of L_δ. Since sequents are finite structures, only a finite number of individual constants can appear in a sequent, while we have an infinite supply of constants. Thus, when $\mathbf{r} \in \{R_\gamma, L_\delta\}$ is applied, we list all constants added by weakening and all constants already in the question-premise. Then, we replace each occurrence of the new constant introduced by **r** with one that does not appear in the list. By Corollary 2.6 (i.e., renaming individual constants, the 'new for new' case), this renaming preserves the property of being a successful Socratic transformation, and the length of the transformation remains n. For the remainder of the proof, we assume that this problem is solved.

Returning to the general case of rule **r**: If **r** is applied with respect to a constituent of Φ other than '$S^* \vdash T$,' then the resulting question Q_1 is of the form '$?(\Psi(S^* \vdash T))$.' The n-term sequence '$\langle Q_1, \ldots, Q_n \rangle$' is a successful Socratic transformation of Q_1, so by the induction hypothesis (IH), there is a successful Socratic transformation of the question '$?(\Psi(S \vdash T))$,' and the transformation is of length n. As Q_0/Q_1 is an instance of **r**, so is $?(\Phi(S \vdash T))/?(\Psi(S \vdash T))$. Hence, by Corollary 2.7, there is a successful Socratic transformation of $?(\Phi(S \vdash T))$ in $\mathbb{E}$ of length $n + 1$, as required.

---

[19] More specifically, derivations in sequent calculi are defined as trees, so inductive proofs are typically organized with respect to the height of a derivation, which corresponds to the length (the number of elements) of the maximal branch.

Next, assume **r** is applied with respect to '$S^* \vdash T$'. We consider the following cases:

(a) $\mathbf{r} \in \{R_\kappa, R_\gamma\}$, the rule is non-branching, and the antecedent remains unchanged.
(b) $\mathbf{r} \in \{R_\beta, R_\delta\}$, non-branching, with formula $-\beta / -\delta$ added to S^*.
(c) $\mathbf{r} = R_\alpha$, Q_1 is of the form '$?(\Phi(\, S^* \vdash T_0\, ;\ S^* \vdash T_1\,))$.'
(d) $\mathbf{r} \in \{L_\kappa, L_\alpha, L_\gamma, L_\delta\}$, the rule is non-branching.
(e) **r** is L_β.

In (a), the antecedent remains unchanged, so the reasoning follows as in the general case.

In (b), there is one additional formula switched to the left. Assume that $\mathbf{r} = R_\delta$ (the reasoning is analogous for R_β):

$$\frac{?(\,\Phi\,(\,S^*(\_) \,\vdash\, \delta\,)\,)}{?(\,\Phi\,(\,S^*(-\delta) \,\vdash\, \delta(t)\,)\,)}\, R_\delta \tag{2.18}$$

The n-term sequence '$\langle Q_1, \ldots, Q_n\rangle$' is a successful Socratic transformation of the question-conclusion of (2.18), so by IH, there is a successful transformation of length n of the question '$?(\Phi(S(-\delta) \vdash \delta(t)))$,' where '$S(-\delta)$' differs from S only in the presence of the additional $-\delta$. The following pair is also an instance of R_δ, since (2.18) is:

$$\frac{?(\,\Phi\,(\,S(\_) \,\vdash\, \delta\,)\,)}{?(\,\Phi\,(\,S(-\delta) \,\vdash\, \delta(t)\,)\,),}$$

therefore, and by Corollary 2.7, there is a successful Socratic transformation of question '$?(\Phi(S \vdash \delta))$,' which is of length $n+1$, as required.

In case (c), we need to apply IH twice, weakening each of the sequents-conclusions separately. Hence, first, by IH applied with respect to Q_1, we conclude that there is a successful Socratic transformation of length n of the question '$?(\Phi(S \vdash T_0\,;\ S^* \vdash T_1))$.' Second, by IH applied with respect to this question, we conclude that there is a successful Socratic transformation of '$?(\Phi(S \vdash T_0\,;\ S \vdash T_1))$' of length n. Let us observe that we need the 'height-preserving,' or rather 'length-preserving,' property here, otherwise we would be stuck after the first application of IH.

For case (d), assume that $\mathbf{r} = L_\alpha$. The antecedent of the sequent-premise of Q_0 is of the form '$S^*(\alpha)$,' whereas the antecedent of the sequent-conclusion in Q_1 is '$S^*(\alpha_0, \alpha_1)$.' Let '$S(\alpha_0, \alpha_1)$' be a sequence such that '$S^*(\alpha_0, \alpha_1)$' is its subsequence, differing from S exactly in that it has two formulas, α_0 and α_1, where S had one formula, α. Since the question $Q_1 = ?(\Phi(S^*(\alpha_0, \alpha_1) \vdash T))$ has a successful Socratic transformation of length n, by IH, there is also a successful Socratic transformation of length n of the question '$?(\Phi(S(\alpha_0, \alpha_1) \vdash T))$.' Then, the appropriate instance of L_α shows that, by Corollary 2.7, question '$?(\Phi(S(\alpha) \vdash T))$' also has a successful Socratic transformation. The cases of other rules are analogous.

For case (e), we reason as in case (d), but we need to apply IH twice, as in case (c).

The reasoning is analogous for rule R_{weak}. □

The following is a straightforward corollary from the above lemma.

Corollary 2.11 (Eliminability of Weakening)

1. *Let* $\mathbb{E} \in \{\mathbb{E}^{\mathbf{Q}}, \mathbb{E}^{\mathbf{P}}\}$, *and let* Q *stand for a question of* $\mathcal{L}^{?}_{\vdash \mathsf{L}}$ *with both-sided constituents. If there is a successful Socratic transformation of* Q *in* $\mathbb{E} \cup \{L_{weak}\}$, *then there is also one in* $\mathbb{E}$.
2. *Let* $\mathbb{E} \in \{\mathbb{E}^{\mathbf{RQ}}, \mathbb{E}^{\mathbf{RP}}\}$, *and let* Q *stand for a question of* $\mathcal{L}^{?}_{\vdash \mathsf{L}}$ *with right-sided constituents. If there is a successful Socratic transformation of* Q *in* $\mathbb{E} \cup \{R_{weak}\}$, *then there is also one in* $\mathbb{E}$.
3. *Let* $\mathbb{E} \in \{\mathbb{E}^{\mathbf{LQ}}, \mathbb{E}^{\mathbf{LP}}\}$, *and let* Q *stand for a question of* $\mathcal{L}^{?}_{\vdash \mathsf{L}}$ *with left-sided constituents. If there is a successful Socratic transformation of* Q *in* $\mathbb{E} \cup \{L_{weak}\}$, *then there is also one in* $\mathbb{E}$.

Proof The proof boils down to eliminating the consecutive applications of weakening by applying Lemma 2.4. If the reader feels the need for further rigor, we can carry out this proof by induction.

Assume that Q has a successful Socratic transformation, $\mathbf{s}$, in $\mathbb{E} \cup \{\mathbf{r}\}$, where $\mathbf{r}$ is the appropriate rule of weakening, and suppose that $\mathbf{r}$ was applied n times in $\mathbf{s}$. In the base case we take $n = 0$; there is nothing further to prove.

In the inductive step, we consider the first instance of $\mathbf{r}$ in $\mathbf{s}$—say it is Q_1/Q_2. By IH, there is a successful Socratic transformation of Q_2 in $\mathbb{E}$, and by Lemma 2.4, there is a successful Socratic transformation of Q_1 in $\mathbb{E}$, as required. □

The rules of weakening clearly do not satisfy clause 2 of Definition 2.8; thus, they are semi-admissible but not admissible in the respective $\mathbb{E}$. It is sufficient to consider the following instances of the said rules:

$$\frac{?(p, q \vdash p)}{?(q \vdash p)}\; L_{weak} \qquad \frac{?(\vdash \neg p, p)}{?(\vdash p)}\; R_{weak} \qquad \frac{?(p, \neg p \vdash)}{?(p \vdash)}\; L_{weak}$$

The following lemmas and corollaries address special cases of the invertibility of rules. There are two important notions of invertibility that should not be confused. What we have analyzed in the previous section is a semantic notion of invertibility, whereas the one discussed here is defined by reference to derivability; hence, we call it 'proof-theoretic'.[20]

Typically, in proofs of the admissibility of contraction, a stronger property of rules is required, known as *height-preserving invertibility*. In our case, an analogue to the preservation of height would be the preservation of length, though this will not be required. In general, proving admissibility without preserving the height of a derivation is more difficult than proving the height-preserving version. However, the challenge in our case is that length preservation is not equivalent to height preservation, and simply replicating the strategy presented and applied in [19] would not work for us.

---

[20] Again, for background, see [19, Chap. 2].

In what follows, we show that the rules of $\mathbb{E}$ are invertible. We start with a lemma concerning doubly negated formulas.

Lemma 2.5 (Transparency of Double Negation)

1. *Let* $\mathbb{E} \in \{\mathbb{E}^{\mathbf{Q}}, \mathbb{E}^{\mathbf{P}}, \mathbb{E}^{\mathbf{LQ}}, \mathbb{E}^{\mathbf{LP}}\}$. *A sequent of the language* $\mathcal{L}^{?}_{\vdash\mathsf{L}}$ *of the form '*$S(\neg\neg A) \vdash T$*' has a Socratic proof in* $\mathbb{E}$ *iff a sequent of* $\mathcal{L}^{?}_{\vdash\mathsf{L}}$ *of the form '*$S(A) \vdash T$*' has a Socratic proof in* $\mathbb{E}$.
2. *Let* $\mathbb{E} \in \{\mathbb{E}^{\mathbf{Q}}, \mathbb{E}^{\mathbf{P}}, \mathbb{E}^{\mathbf{RQ}}, \mathbb{E}^{\mathbf{RP}}\}$. *A sequent '*$S \vdash T(\neg\neg A)$*' has a Socratic proof in* $\mathbb{E}$ *iff a sequent '*$S \vdash T(A)$*' has a Socratic proof in* $\mathbb{E}$.

Proof We focus on clause 1, as the second is analogous. Let $\mathbb{E} \in \{\mathbb{E}^{\mathbf{Q}}, \mathbb{E}^{\mathbf{P}}, \mathbb{E}^{\mathbf{LQ}}, \mathbb{E}^{\mathbf{LP}}\}$.

If the question '$?(S(A) \vdash T)$' has a successful Socratic transformation in $\mathbb{E}$, then, by Corollary 2.7, '$?(S(\neg\neg A) \vdash T)$' also has a successful Socratic transformation, as $?(S(\neg\neg A) \vdash T)/?(S(A) \vdash T)$ is an instance of L_{κ}.

For the if-then direction, we use induction on the length of the Socratic transformation. More specifically, we will prove by induction the following statement ($\mathbf{S^N}$). This will allow us to handle the branching case smoothly.

($\mathbf{S^N}$) If a question Q has a successful Socratic transformation in $\mathbb{E}$, then for each constituent of Q that is of the form '$S(\neg\neg A) \vdash T$', the sequent '$S(A) \vdash T$' has a Socratic proof in $\mathbb{E}$.

Let us emphasize that the preservation of the length of the Socratic transformation is not required here. Once we prove ($\mathbf{S^N}$), it will easily follow that if a sequent '$S(\neg\neg A) \vdash T$' has a Socratic proof, meaning there is a successful Socratic transformation of '$?(S(\neg\neg A) \vdash T)$', then '$S(A) \vdash T$' also has a Socratic proof.

Hence, assume that Q has a successful Socratic transformation in $\mathbb{E}$. In the base case, the transformation is of length 1, which means that each constituent of Q is basic. Assume that a constituent of Q is of the form '$S(\neg\neg A) \vdash T$'. There are several possible cases:

(a) There is a formula B in both the antecedent and succedent, but $B \neq \neg\neg A$.
(b) The sequent has the form '$S(\neg\neg A) \vdash \neg\neg A$'.
(c) There is a formula B such that both B and '$-B$' occur in '$S(\neg\neg A)$', but neither of them is '$\neg\neg A$'.
(d) There is B and '$-B$' present in S, and one of them is '$\neg\neg A$'.

In cases (a) and (c), sequent '$S(A) \vdash T$' is basic. If (d) holds and $B = \neg\neg A$, then the sequent '$S(A) \vdash T$' is also basic, as it contains both A and $\neg A = -B$ in the antecedent. Needless to say, every basic sequent has a Socratic proof.

In case (b) and the remaining situation in (d), we need to refer to an instance of R_{κ}, or respectively L_{κ}. For example, the following:

$$\frac{?(\ S(A) \vdash \neg\neg A\ )}{?(\ S(A) \vdash A\ )}\ R_{\kappa}$$

is a Socratic proof of '$?(S(A) \vdash \neg\neg A)$' in $\mathbb{E}$, which proves (b). Let us emphasize that while it is not proven that the sequent is basic, it is proven that it has a Socratic proof. Clearly, we lose the preservation of length at this point.

In the inductive part, we assume that ($\mathbf{S^N}$) holds for questions whose successful Socratic transformation is of length n. Suppose that '$?(\Phi(S(\neg\neg A) \vdash T))$' has a successful Socratic transformation $\langle Q_0, Q_1, \ldots, Q_n \rangle$ of length $n+1$. If the distinguished sequent is the sequent-premise and the distinguished formula '$\neg\neg A$' is active in Q_0—that is, $Q_1 = ?(\Phi(S(A) \vdash T))$—then the sequence $\langle Q_1, \ldots, Q_n \rangle$ is a successful Socratic transformation of Q_1 in $\mathbb{E}$, and by the Separation Lemma (Lemma 2.3), the sequent '$S(A) \vdash T$' has a Socratic proof in $\mathbb{E}$.

If a rule $\mathbf{r}$ is applied with respect to a sequent other than the distinguished one, then by IH there is a Socratic proof of '$S(A) \vdash T$'.

Now, assume that the sequent is the sequent-premise, but the chosen occurrence of the formula '$\neg\neg A$' is not active in Q_0. For ease of reference, let us restate:

$$Q_0 = ?(\ \Phi\ (\ S(\neg\neg A) \vdash T\ )\ )$$

Then Q_1 is of one of the following forms:

1. $?(\Phi(\ S(\neg\neg A) \vdash T^*\ ))$, where $\mathbf{r}$ is a non-branching rule acting in the succedent, but $\mathbf{r} \notin \{R_\beta, R_\delta\}$ (no formulas are switched from the right to the left).
2. $?(\Phi(\ S(\neg\neg A)(-B) \vdash C\ ))$, where $\mathbf{r} \in \{R_\beta, R_\delta\}$, $B \in \{\beta_i, \delta(t)\}$, and $C \in \{\beta, \delta\}$.
3. $?(\Phi(\ S(\neg\neg A) \vdash T_0\ ;\ S(\neg\neg A) \vdash T_1\ ))$, where $\mathbf{r} = R_\alpha$ (branching rule acting in the succedent).
4. $?(\Phi(\ S^*(\neg\neg A) \vdash T\ ))$, where $\mathbf{r}$ is a non-branching rule acting in the antecedent.
5. $?(\Phi(\ S_0^*(\neg\neg A) \vdash T\ ;\ S_1^*(\neg\neg A) \vdash T\ ))$, where $\mathbf{r}$ is a branching rule acting in the antecedent.

In case 1, we use IH to establish that there is a Socratic proof of '$S(A) \vdash T^*$'. Since Q_0/Q_1 is an instance of $\mathbf{r}$, the pair $?(S(A) \vdash T)/?(S(A) \vdash T^*)$ is also an instance of $\mathbf{r}$, hence by Corollary 2.7, the sequent '$S(A) \vdash T$' is provable. The reasoning is analogous in cases 2 and 4.

In case 3, by IH applied twice, both '$S(A) \vdash T_0$' and '$S(A) \vdash T_1$' have Socratic proofs. By Lemma 2.3, there is a successful Socratic transformation of the question '$?(S(A) \vdash T_0; S(A) \vdash T_1)$'. Since Q_0/Q_1 is an instance of R_α, so is:

$$\frac{?(\ S(A)\ \vdash\ T\ )}{?(\ S(A)\ \vdash\ T_0\ ;\ S(A)\ \vdash\ T_1\ )}$$

By Corollary 2.7, '$S(A) \vdash T$' is provable in $\mathbb{E}$, as required. The reasoning is similar in case 5. □

The following corollary generalizes Lemma 2.5.

Corollary 2.12 (Proof-Theoretic Invertibility of κ-rules) *If '$S(\kappa) \vdash T$' has a Socratic proof in $\mathbb{E} \in \{\mathbb{E}^{\mathbf{Q}}, \mathbb{E}^{\mathbf{P}}, \mathbb{E}^{\mathbf{LQ}}, \mathbb{E}^{\mathbf{LP}}\}$, then '$S(\kappa_0) \vdash T$' also has a Socratic* *proof in $\mathbb{E}$. Similarly, if '$S \vdash T(\kappa)$' has a Socratic proof in $\mathbb{E}$, then '$S \vdash T(\kappa_0)$' also has a Socratic proof in $\mathbb{E}$.*

Proof If the scheme of the κ-rule covers the case of doubly negated formulas, then the result is already established in Lemma 2.5. Therefore, we are left with the cases of empty quantification.

The proof proceeds by induction on the length of the Socratic transformation. The statement to be proven is formulated similarly to the one in the proof of Lemma 2.5.

($\mathbf{S}^{\kappa}$) If a question Q has a successful Socratic transformation in $\mathbb{E}$, then for each constituent of Q that is of the form '$S(\kappa) \vdash T$', the sequent '$S(\kappa_0) \vdash T$' also has a Socratic proof in $\mathbb{E}$. Analogously for the right side.

Now, let **s** stand for a Socratic proof of '$S(\kappa) \vdash T$' in $\mathbb{E} \in \{\mathbb{E}^{\mathbf{Q}}, \mathbb{E}^{\mathbf{P}}, \mathbb{E}^{\mathbf{LQ}}, \mathbb{E}^{\mathbf{LP}}\}$. In the base case, **s** is a one-term sequence, so each constituent of Q is basic, including '$S(\kappa) \vdash T$'. As in the proof of the previous lemma, we show that '$S(\kappa_0) \vdash T$' is either basic or has a two-term Socratic proof.

The inductive step follows the pattern of the proof of Lemma 2.5. If a rule is applied to a constituent other than the displayed one, the result follows by IH. If '$S(\kappa) \vdash T$' is the sequent-premise and L_κ is applied, the result follows directly from the Separation Lemma (Lemma 2.3). For other rules, we continue by following the same pattern; in the branching cases, we again refer to Lemma 2.3. □

The same pattern is followed in the proof of the next corollary.

Corollary 2.13 (Proof-Theoretic Invertibility of Rules L_α, R_α, L_β, R_β) *The following holds:*

1. *If '$S(\alpha) \vdash T$' has a Socratic proof in $\mathbb{E} \in \{\mathbb{E}^{\mathbf{Q}}, \mathbb{E}^{\mathbf{P}}, \mathbb{E}^{\mathbf{LQ}}, \mathbb{E}^{\mathbf{LP}}\}$, then '$S(\alpha_0, \alpha_1) \vdash T$' has a Socratic proof in $\mathbb{E}$.*
2. *If '$S \vdash T(\alpha)$' has a Socratic proof in $\mathbb{E} \in \{\mathbb{E}^{\mathbf{Q}}, \mathbb{E}^{\mathbf{P}}, \mathbb{E}^{\mathbf{RQ}}, \mathbb{E}^{\mathbf{RP}}\}$, then '$S \vdash T(\alpha_0)$' and '$S \vdash T(\alpha_1)$' both have Socratic proofs in $\mathbb{E}$.*
3. *If '$S(\beta) \vdash T$' has a Socratic proof in $\mathbb{E} \in \{\mathbb{E}^{\mathbf{Q}}, \mathbb{E}^{\mathbf{P}}, \mathbb{E}^{\mathbf{LQ}}, \mathbb{E}^{\mathbf{LP}}\}$, then '$S(\beta_0) \vdash T$' and '$S(\beta_1) \vdash T$' both have Socratic proofs in $\mathbb{E}$.*
4. *If '$\vdash T(\beta)$' has a Socratic proof in $\mathbb{E} \in \{\mathbb{E}^{\mathbf{RQ}}, \mathbb{E}^{\mathbf{RP}}\}$, then '$\vdash T(\beta_0, \beta_1)$' has a Socratic proof in $\mathbb{E}$.*
5. *If '$S(\_) \vdash \beta$' has a Socratic proof in $\mathbb{E} \in \{\mathbb{E}^{\mathbf{Q}}, \mathbb{E}^{\mathbf{P}}\}$, then '$S(-\beta_i) \vdash \beta_{1-i}$' has a Socratic proof in $\mathbb{E}$.*

Proof In the base step for clause 1, we consider the case where '$S(\alpha) \vdash T$' is basic, given that '$-\alpha$' is present in '$S(\alpha)$'. We recall Corollary 2.1, clause 4, which entails that '$-\alpha$' is a β-formula, and $\alpha_i \in \{-\beta_i, \neg\beta_i\}$. This shows that α_i and β_i form a pair of complementary formulas.

The following shows that there is a Socratic proof of '$?(S(\alpha_0, \alpha_1) \vdash T)$' in $\mathbb{E}$. The sequence '$S(\alpha_0, \alpha_1)$' is written as '$S(\alpha_0, \alpha_1)(\beta)$' to indicate that there is an occurrence of β in S.

$$\frac{?(\ S(\alpha_0,\alpha_1)(\beta) \vdash T\ )}{?(\ S(\alpha_0,\alpha_1)(\beta_0) \vdash T\ ;\ S(\alpha_0,\alpha_1)(\beta_1) \vdash T\ )}\ L_\beta$$

Another possible case occurs when '$S(\alpha) \vdash T$' is basic because both α and $\neg\alpha$ are present in the antecedent, but $\neg\alpha$ is a κ-formula rather than a β-formula. In this case, we apply the previous corollary first, and then follow the above scheme. The rest of the proof, including clauses 2–5, can be easily verified. □

For the next lemma, we will need the notion of a permutation of a finite non-empty sequence. Let us recall that a finite, n-term ($n > 0$) sequence is a function from $\{1, \ldots, n\}$ to some set X. If $S : \{1, \ldots, n\} \longrightarrow X$, then by a *permutation of* S, we mean any bijective function from $\{1, \ldots, n\}$ to the range of S.

Further, let '$S \vdash T$' be a sequent of $\mathcal{L}^?_{\vdash\mathsf{L}}$. If the sequent is not right-sided, then by *left-permutation of the sequent* we mean the sequent '$S^* \vdash T$', where S^* is a permutation of S. Similarly, by *right-permutation of* '$S \vdash T$', where the sequent is not left-sided, we mean the sequent '$S \vdash T^*$', where T^* is a permutation of T.

Finally, if sequent ϕ is a left-permutation or a right-permutation of sequent ψ, we may simply say that ϕ is a *permutation of* ψ.

Lemma 2.6 (Permutation) *Each permutation of a sequent provable in* $\mathbb{E}$ *is provable in* $\mathbb{E}$.

Proof The proof is organized similarly to the proof of the if-then part of the previous lemma—we obtain the result by inductively proving a statement linking questions of an arbitrary form with a sequent of a specific form.

($\mathbf{S^P}$) If $Q = ?(\phi_1, \ldots, \phi_m)$ has a successful Socratic transformation in $\mathbb{E}$, then each permutation ψ_j of ϕ_j, $1 \leqslant j \leqslant m$, has a Socratic proof in $\mathbb{E}$.

We proceed by induction on the length of the Socratic transformation. Assume that Q has a successful Socratic transformation $\mathbf{s} = \langle Q_0, \ldots, Q_n \rangle$ in $\mathbb{E}$.

Base case: each constituent of Q is a basic sequent, and it is easy to see that permuting the antecedent or succedent does not alter the property of being basic. Therefore, by the Separation Lemma (Lemma 2.3), each permutation of a basic constituent of Q is a provable sequent.

Inductive step: We focus on left-permutations; the reasoning for the right side is analogous. There are as many subcases as rules of $\mathbb{E}$. Assume that $\mathbb{E} \in \{\mathbb{E}^{\mathbf{Q}}, \mathbb{E}^{\mathbf{P}}\}$, and that Q_0 and Q_1 take the following forms:

$$\frac{Q_0 = ?(\ \Phi\ (\ S(\_) \ \vdash\ \beta\ )\ )}{Q_1 = ?(\ \Phi\ (\ S(\beta_i)\ \vdash\ \beta_{1-i}\ )\ )}\ R_\beta$$

We need to show that a permutation of any constituent of Q_0, particularly the one displayed, is provable in $\mathbb{E}$. By assumption, $\mathbf{s}$ is a successful Socratic transformation of Q_0 in $\mathbb{E}$, so the sequence '$\langle Q_1, \ldots, Q_n \rangle$' is a successful Socratic transformation of Q_1 in $\mathbb{E}$. For the constituents of Q_0 other than the sequent-premise, the result follows by IH.

Let '$S^* \vdash \beta$' be an arbitrary left-permutation of the sequent-premise of Q_0. By IH, ($\mathbf{S^P}$) applies to question Q_1, thus it holds for any left-permutation of the sequent '$S(\beta_i) \vdash \beta_{1-i}$'. In particular, it applies to '$S^*(\beta_i) \vdash \beta_{1-i}$', where the position of β_i within S^* is arbitrary. By IH, the sequent '$S^*(\beta_i) \vdash \beta_{1-i}$' has a Socratic proof in $\mathbb{E}$. The pair $?(S^*(\_) \vdash \beta)/?(S^*(\beta_i) \vdash \beta_{1-i})$ is an instance of the rule R_β, so by Corollary 2.7, the sequent '$S^* \vdash \beta$' has a Socratic proof in $\mathbb{E}$. The reasoning is similar for R_δ and even simpler for R_κ and R_γ, since they don't involve switching to the left side.

Now assume that R_α was applied to Q_0:

$$\frac{Q_0 = ?(\ \Phi\ (\ S \vdash \alpha\ )\ )}{Q_1 = ?(\ \Phi\ (\ S \vdash \alpha_0\ ;\ S \vdash \alpha_1\ )\ )}\ R_\alpha$$

Let '$S^* \vdash \alpha$' be an arbitrary left-permutation of the sequent-premise. By IH, there are Socratic proofs of both '$S^* \vdash \alpha_0$' and '$S^* \vdash \alpha_1$' in $\mathbb{E}$, where these are left-permutations of the sequent-conclusions. By Lemma 2.3, there is a successful Socratic transformation of '$?(S^* \vdash \alpha_0\ ;\ S^* \vdash \alpha_1)$', and hence, there is also a Socratic proof of '$S^* \vdash \alpha$'.

For the left side, we consider only L_α. In the scheme displayed below, for $\mathbb{E}^{\mathbf{LQ}}$ and $\mathbb{E}^{\mathbf{LP}}$, T is empty, whereas for $\mathbb{E}^{\mathbf{Q}}$ and $\mathbb{E}^{\mathbf{P}}$, T is a one-term sequence:

$$\frac{Q_0 = ?(\ \Phi\ (\ S(\alpha)\ \vdash\ T\ )\ )}{Q_1 = ?(\ \Phi\ (\ S(\alpha_0, \alpha_1)\ \vdash\ T\ )\ )}\ L_\alpha$$

Let '$S^*(\alpha) \vdash T$' be a left-permutation of the sequent-premise. Here we must be cautious, as the positions of α in $S(\alpha)$ and $S^*(\alpha)$ are independent. Assume the active α is the kth term of S^*, while its position in S is arbitrary. We define a permutation of sequence '$S(\alpha_0, \alpha_1)$' as follows: from the 1st to the $(k-1)$st position, we place the respective terms of S^*; then we place α_0 in the kth position and α_1 in the $(k+1)$st position. Afterward, we insert the remaining terms of S^* from the $(k+1)$st position onward. Let '$S^*(\alpha_0, \alpha_1)$' stand for the resulting sequence. IH applies to Q_1, so for each left-permutation of the sequent-conclusion, including '$S^*(\alpha_0, \alpha_1) \vdash T$', there is a Socratic proof in $\mathbb{E}$, which is what we need: we refer to an instance of L_α and Corollary 2.7.

For L_β, Lemma 2.3 is applied as in the previous cases.

Finally, if a sequent ϕ is provable in $\mathbb{E}$, then there is a successful Socratic transformation of the question $?(\phi)$, and by ($\mathbf{S^P}$), each permutation of ϕ has a Socratic proof in $\mathbb{E}$. □

The next lemma deals with cases specific for the both-sided sequents only.

Lemma 2.7 (Switching) *Let* $\mathbb{E} \in \{\mathbb{E}^{\mathbf{Q}}, \mathbb{E}^{\mathbf{P}}\}$. *If the sequent* '$S(-A) \vdash B$' *has a Socratic proof in* $\mathbb{E}$, *then* '$S(-B) \vdash A$' *also has a Socratic proof in* $\mathbb{E}$.

Proof The proof is organized as the previous one. We show that the following holds:

($\mathbf{S}^{\mathbf{S}}$) If a question Q has a successful Socratic transformation in $\mathbb{E}$, then for each constituent of Q that takes the form '$S(-A) \vdash B$', the sequent '$S(-B) \vdash A$' also has a Socratic proof in $\mathbb{E}$.

The proof is by induction on the length of the Socratic transformation of Q.

In the base case, each constituent of Q is basic. If Q contains a constituent of the form '$S(-A) \vdash B$', one of the following situations arises:

1. Some formulas F and $-F$ are present in the antecedent, with neither being $-A$.
2. Formulas F and $-F$ are present in the antecedent, and one of them is $-A$.
3. B is present in the antecedent but is not identical to $-A$.
4. $B = -A$.

In cases 1 and 3, the sequent '$S(-B) \vdash A$' is also basic. Now suppose case 2 holds, where $F = -A$. The initial sequent $\phi = S(-A) \vdash B$ and the corresponding sequent $\psi = S(-B) \vdash A$ can be displayed as follows:

$$\phi = S(-A)(- - A) \vdash B \qquad \psi = S(-B)(- - A) \vdash A$$

If $A = - - A$, then ψ is already basic. Otherwise, by Corollary 2.1, clause 2, we observe that A is a doubly negated formula. Thus, by Lemma 2.5 (Transparency of double negation), the sequent '$S(-B)(- - A) \vdash A$' has a Socratic proof in $\mathbb{E}$ iff the sequent '$S(-B)(A) \vdash A$' does. All other cases follow analogously to those already discussed.

Inductive part: Suppose the question '$?(\Phi(S(-A) \vdash B))$' has a Socratic proof $\mathbf{s} = \langle Q_0, Q_1 \ldots, Q_n \rangle$ of length $n + 1$. Also assume that Q_0/Q_1 is an instance of an erotetic rule $\mathbf{r}$. If the distinguished sequent $\phi = S(-A) \vdash B$ is not the sequent-premise, then by IH, the sequent $\psi = S(-B) \vdash A$ is provable.

Now assume that ϕ is the sequent-premise. Here is the list of possible cases:

1. $\mathbf{r}$ is a non-branching rule acting in the antecedent, and '$-A$' is not active.
2. $\mathbf{r}$ is a non-branching rule acting in the antecedent, and '$-A$' is active.
3. $\mathbf{r}$ is a branching rule acting on the left, where $-A$ may be active or not (two subcases).
4. $\mathbf{r}$ is a non-branching rule acting in the succedent, with $\mathbf{r} \in \{R_\kappa, R_\gamma\}$.
5. $\mathbf{r}$ is a non-branching rule acting in the succedent, with $\mathbf{r} \in \{R_\beta, R_\delta\}$.
6. $\mathbf{r}$ is a branching rule acting in the succedent, meaning $\mathbf{r} = R_\alpha$.

We skip case 1.

In case 2, $\mathbf{r} \in \{L_\kappa, L_\alpha, L_\gamma, L_\delta\}$. Let $\mathbf{r} = L_\kappa$ be the case of double negation. The active formula '$-A$' has the form '$\neg\neg F$', which yields that $A = \neg\neg\neg F$. This gives us the following instance of L_κ:

$$\frac{Q_0 = ?(\ \Phi\ (\ S(\neg\neg F) \vdash B\ )\ )}{Q_1 = ?(\ \Phi\ (\ S(F) \vdash B\ )\ )}\ L_\kappa$$

By IH, the sequent '$S(-B) \vdash \neg F$' has a Socratic proof in $\mathbb{E}$. Applying Corollary 2.7 with the following instance of R_κ (recall that $A = \neg\neg\neg F$):

$$\frac{?(\,S(-B) \vdash A\,)}{?(\,S(-B) \vdash \neg F\,)}\,R_\kappa$$

we conclude that '$S(-B) \vdash A$' is provable in $\mathbb{E}$, as required. A similar situation arises when κ refers to empty quantification. For example, if '$-A$' takes the form '$\neg\forall x_i F$' where x_i is not free in F, we have:

$$\frac{Q_0 = ?(\,\Phi\,(\,S(\neg\forall x_i F) \vdash B\,))}{Q_1 = ?(\,\Phi\,(\,S(-F) \vdash B\,))}\,L_\kappa$$

By IH, the sequent '$S(-B) \vdash F$' is provable in $\mathbb{E}$. Depending on whether $A = \forall x_i F$ or $A = \neg\neg\forall x_i F$, we apply one of the following:

$$\frac{?(\,\Phi\,(\,S(-B) \vdash \forall x_i F\,))}{?(\,\Phi\,(\,S(-B) \vdash F\,))}\,R_\kappa \qquad \cfrac{\cfrac{?(\,\Phi\,(\,S(-B) \vdash \neg\neg\forall x_i F\,))}{?(\,\Phi\,(\,S(-B) \vdash \forall x_i F\,))}\,R_\kappa}{?(\,\Phi\,(\,S(-B) \vdash F\,))}\,R_\kappa$$

and show that the sequent '$S(-B) \vdash A$' has a Socratic proof in $\mathbb{E}$. All other subcases of L_κ are analogous.

Next, we consider $\mathbf{r} \in \{L_\alpha, L_\gamma, L_\delta\}$. For $\mathbf{r} = L_\alpha$, we make extensive use of Corollary 2.1 and Lemma 2.5. We will present the most complex subcase for clarity.

Assume that the formula '$-A$' is an α-formula and $A = \neg\alpha$. The question Q_1 then takes the form '$?(S(\alpha_0, \alpha_1) \vdash B)$'. By IH, sequent '$S(-B, \alpha_1) \vdash \neg\alpha_0$', has a Socratic proof in $\mathbb{E}$. Now, we observe from Corollary 2.1, clause 4, that for $\alpha = -A$, the formula '$- - A$' is a β-formula. Assume further that $\neg\alpha_0$ is equal to β_0, but $\alpha_1 = \neg\beta_1 \neq -\beta_1$. An example of this situation occurs when $\alpha = p \wedge \neg\neg q$. Here, the corresponding β has components $\beta_0 = \neg p$ and $\beta_1 = \neg q$, and hence $-\beta_1 = q$. Consequently, $\alpha_1 = \neg\neg - \beta_1$. Since we aim to apply R_β, we need to eliminate the double negation in front of '$-\beta_1$'. Using Lemma 2.5 (Transparency of double negation), we conclude that the sequent '$S(-B, -\beta_1) \vdash \beta_0$' is provable in $\mathbb{E}$. Therefore, '$S(-B) \vdash \beta$' has a Socratic proof. Finally, with the help of an instance of R_κ we demonstrate that '$S(-B) \vdash A$' has a Socratic proof in $\mathbb{E}$.

We summarize the reasoning schematically as follows:

$$\frac{?(\,\Phi\,(\,S(-A) \vdash B\,))}{?(\,\Phi\,(\,S(\alpha_0, \alpha_1) \vdash B\,))}\,L_\alpha \quad \xrightarrow{\text{by IH}} \quad \underset{\text{provable}}{?(\,S(-B, \alpha_1) \vdash \neg\alpha_0\,)}$$

$$\Downarrow \text{ by Lemma 5}$$

$$\cfrac{\cfrac{?(\,S(-B) \vdash A\,)}{?(\,S(-B) \vdash \beta\,)}\,R_\kappa}{?(\,S(-B, -\beta_1) \vdash \neg\alpha_0\,)}\,R_\beta \quad \xleftarrow{\text{final}} \quad \underset{\text{provable}}{?(\,S(-B, -\beta_1) \vdash \neg\alpha_0\,)}$$

The reasoning for $\mathbf{r} \in \{L_\gamma, L_\delta\}$ is similar and is omitted here for brevity.

Case 3. We consider only the case where '$-A$' is the active β-formula. We can express the situation as follows:

$$\frac{Q_0 = ?(\ \Phi\ (\ S(-A)\ \vdash\ B\ )\ )}{Q_1 = ?(\ \Phi\ (\ S(\beta_0)\ \vdash B\ ;\ S(\beta_1) \vdash\ B\ )\ )}\ L_\beta$$

By IH, we conclude that both '$S(-B) \vdash \neg\beta_0$' and '$S(-B) \vdash \neg\beta_1$' have Socratic proofs in $\mathbb{E}$. Furthermore, by Corollary 2.1, A must be either α or $\neg\neg\alpha$, where α satisfies $\neg\beta_i = \alpha_i$ or $\neg\beta_i = \neg\neg\alpha_i$ for $i \in \{0, 1\}$. In the double-negation case, we apply Lemma 2.5. In each subcase, we utilize Lemma 2.3 (Separation) to conclude that the question '$?(S(-B) \vdash \alpha_0\ ;\ S(-B) \vdash \alpha_1)$' admits a successful Socratic transformation in $\mathbb{E}$. Therefore, by an instance of R_α and Corollary 2.7, the sequent '$S(-B) \vdash A$' has a Socratic proof in $\mathbb{E}$.

For the remaining cases, we only consider R_β. Let:

$$\frac{?(\ \Phi\ (\ S(-A)\ \vdash\ \beta\ )\ )}{?(\ \Phi\ (\ S(-A)(-\beta_i)\ \vdash\ \beta_{1-i}\ )\ )}\ R_\beta$$

Here, the challenge lies in the uncertainty regarding the position of '$-\beta_i$' in the sequent conclusion. Our goal is to secure an instance of rule L_α, which requires that the components occupy specific positions. Thus, R_β (and R_δ) will be the only cases in this proof necessitating Lemma 2.6 (Permutation).

According to IH, there is a Socratic proof of '$S(-\beta_{1-i})(-\beta_i) \vdash A$'. As in all analogous cases discussed in this and the preceding lemmas, we can assert that there is a Socratic proof of '$S(\alpha_{1-i})(\alpha_i) \vdash A$', where $\alpha = -\beta$. The justification for this conclusion is as follows: If an occurrence of double negation is absent, we invoke an instance of L_κ along with Corollary 2.7; if there is a redundant occurrence of double negation, we apply Lemma 2.5. Next, we use Lemma 2.6 (Permutation) to conclude that there is also a Socratic proof of '$S(\alpha_0, \alpha_1) \vdash A$', where the left-permutation is structured such that α_0 occupies the position previously held by $-\beta_{1-i}$, followed by the second component of α. The remaining parts of the reasoning follows as previously. □

The next goal is semi-admissible rules of contraction. The rules are defined in a manner akin to weakening. First, an auxiliary notion: if S is a sequence, then $\mathsf{set}(S)$ is the set of terms of S.

$$\frac{?(\ \Phi\ (\ S^* \vdash T\ )\ )}{?(\ \Phi\ (\ S \vdash T\ )\ )}\ L_{contr} \qquad \frac{?(\ \Phi\ (\ S \vdash T^*\ )\ )}{?(\ \Phi\ (\ S \vdash T\ )\ )}\ R_{contr}$$

where S^* is a non-empty proper subsequence of S and $\mathsf{set}(S) = \mathsf{set}(S^*)$. The same applies to T^* and T in R_{contr}. Similarly to the case of weakening, when added to $\mathbb{E}^{\mathbf{Q}}$, $\mathbb{E}^{\mathbf{P}}$, rule L_{contr} can be used with a slightly generalized proviso. Namely, in the both-sided case, T contains one formula; suppose it is A. Then S^* is a proper subsequence of S such that $\mathsf{set}(S^*) = \mathsf{set}(S)$ *or* $\mathsf{set}(S^*) \cup \{-A\} = \mathsf{set}(S)$. This means that the following is an instance of the rule:

$$\frac{?(\vdash A\,)}{?(\,-A \vdash A\,)}$$

Applying contraction rules amounts to adding occurrences of formulas that already occur in the antecedent or succedent of the sequent-premise, respectively. As seen in the example above, one can also repeat a formula in the succedent by placing its complement in the antecedent. Again, compared to the usual contraction rules of sequent calculi, the above rules are inverted. A reader unfamiliar with structural proof theory may wonder about the purpose of applying such rules.

First of all, when different variants of rules are considered, the rule is sometimes necessary to secure information. For example, in the case of the so-called *additive* rules, also known as rules *with independent contexts*, such as the following:

$$\frac{A, \Gamma \vdash C \quad B, \Delta \vdash C}{A \vee B, \Gamma, \Delta \vdash C} \quad \text{where } \Gamma, \Delta \text{ are finite multisets of formulas}$$

it can happen that repeating some part of Γ or Δ in the premises is necessary to complete the proof. This aspect is invisible in erotetic calculi, as all erotetic rules are designed to *share contexts* (in other terminology: erotetic rules are *multiplicative*), precisely to secure all the information contained in the question-premise, and thus ensure erotetic implication. Nevertheless, if other variants of erotetic rules were considered, the contraction rules in the erotetic setting would clearly be interpreted as rules for securing information.

Since the proofs of semi-admissibility of contraction are generally more complex than those for weakening, we present below a more standard form of the rules where exactly one 'contraction-formula' is added in the conclusion. It is easy to see that every application of the general rules L_{contr} and R_{contr} can be simulated by multiple applications of these more specific rules. We retain the name R_{contr} for the rules, as this should not lead to misunderstandings. Focusing on one instance of the contraction-formula will make the subsequent proofs much more transparent.

$$\frac{?(\,\Phi\,(\,S(A)(\_) \vdash T\,)\,)}{?(\,\Phi\,(\,S(A)(A) \vdash T\,)\,)}\; L^{ls}_{contr} \qquad \frac{?(\,\Phi\,(\,S(\_) \vdash A\,)\,)}{?(\,\Phi\,(\,S(-A) \vdash A\,)\,)}\; L^{sw}_{contr}$$

$$\frac{?(\,\Phi\,(\vdash T(A)(\_)\,)\,)}{?(\,\Phi\,(\vdash T(A)(A)\,)\,)}\; R_{contr}$$

Let us recall that (_) indicates an arbitrary position in the relevant sequent. Rule L^{ls}_{contr}, with T empty, can be added to calculus $\mathbb{E}^{\mathbf{LQ}}$. For calculus $\mathbb{E}^{\mathbf{Q}}$, T is a one-term sequence. 'ls' stands for 'left side'. Rule L^{sw}_{contr}, where sw stands for 'switching,' fits calculus $\mathbb{E}^{\mathbf{Q}}$, and rule R_{contr} can be added to calculus $\mathbb{E}^{\mathbf{RQ}}$.

Before proceeding to the proof of the semi-admissibility of contraction rules in erotetic calculi, we need to establish several technical results. To start, here is an extension of Corollary 2.13.

Corollary 2.14 (Proof-Theoretic Invertibility of Rules L_γ, R_γ, L_δ, R_δ) *The following holds:*

1. *If '$S(\gamma) \vdash T$' has a Socratic proof in $\mathbb{E} \in \{\mathbb{E}^{\mathbf{Q}}, \mathbb{E}^{\mathbf{LQ}}\}$, then for each closed term t of $\mathcal{L}_{\mathsf{FOL}}$, '$S(\gamma, \gamma(t)) \vdash T$' has a Socratic proof in $\mathbb{E}$.*
2. *If '$\vdash T(\delta)$' has a Socratic proof in $\mathbb{E}^{\mathbf{RQ}}$, then for each closed term t of $\mathcal{L}_{\mathsf{FOL}}$, '$\vdash T(\delta, \delta(t))$' has a Socratic proof in $\mathbb{E}^{\mathbf{RQ}}$.*
3. *If '$S(\_) \vdash \delta$' has a Socratic proof in $\mathbb{E}^{\mathbf{Q}}$, then for each closed term t of $\mathcal{L}_{\mathsf{FOL}}$, '$S(-\delta) \vdash \delta(t)$' has a Socratic proof in $\mathbb{E}^{\mathbf{Q}}$.*
4. *If '$S(\delta) \vdash T$' has a Socratic proof in $\mathbb{E} \in \{\mathbb{E}^{\mathbf{Q}}, \mathbb{E}^{\mathbf{LQ}}\}$, then for each individual constant a_i that does not occur in '$S(\delta) \vdash T$', the sequent '$S(\delta(a_i)) \vdash T$' has a Socratic proof in $\mathbb{E}$.*
5. *If '$S \vdash T(\gamma)$' has a Socratic proof in $\mathbb{E} \in \{\mathbb{E}^{\mathbf{Q}}, \mathbb{E}^{\mathbf{RQ}}\}$, then for each individual constant a_i that does not occur in '$S \vdash T(\gamma)$', the sequent '$S \vdash T(\gamma(a_i))$' has a Socratic proof in $\mathbb{E}$.*

Proof Clause 1 follows directly from the semi-admissibility of weakening: since $?(S(\gamma, \gamma(t)) \vdash T)/?(S(\gamma) \vdash T)$ is an instance of L_{weak}, Lemma 2.4 ensures that if '$S(\gamma) \vdash T$' has a Socratic proof in $\mathbb{E}$, then so does '$S(\gamma, \gamma(t)) \vdash T$'. The same holds for clause 2.

Clause 3 is more demanding, but, after all, it also relies on weakening. Let t be an arbitrary closed term of $\mathcal{L}_{\mathsf{FOL}}$. By Corollary 2.1, clause 5, '$-\delta$' is a γ-formula with $\delta(t) \in \{-\gamma(t), \neg\gamma(t)\}$. By Lemma 2.4, the sequent '$S(\gamma(t)) \vdash \delta$' is provable in $\mathbb{E}^{\mathbf{Q}}$. By Lemma 2.7 (Switching), '$S(-\delta) \vdash \neg\gamma(t)$' is provable in $\mathbb{E}^{\mathbf{Q}}$. If $\delta(t) \neq \neg\gamma(t)$, we apply Lemma 2.5 (double negation), concluding that '$S(-\delta) \vdash \delta(t)$' has a Socratic proof in $\mathbb{E}^{\mathbf{Q}}$.

The proofs of clauses 4 and 5 follow similar pattern to Corollaries 2.12 and 2.13, with slight modification due to the existential quantifier. We consider only one possible situation for the base case. Assume that the considered sequent is of the form '$S(\delta) \vdash \delta$'. Then the following:

$$\frac{?(S(\,\delta(a_i)\,) \vdash \delta\,)}{?(\,S(\,\delta(a_i), -\delta\,) \vdash \delta(a_i)\,)}\ R_\delta$$

is a Socratic proof of '$\delta(a_i) \vdash \delta$' in $\mathbb{E}^{\mathbf{Q}}$.

In the inductive step, we apply IH as in Corollaries 2.12 and 2.13. This is enough to finish the reasoning, except for the case where L_δ is used:

$$\frac{?(\,\Phi\,(\,S(\,\delta\,) \vdash T\,)\,)}{?(\,\Phi\,(\,S(\,\delta(a_k)\,) \vdash T\,)\,)}\ L_\delta$$

where a_k does not occur in the question-premise. By Lemma 2.3 (Separation), '$S(\delta(a_k)) \vdash T$' has a Socratic proof in $\mathbb{E}^{\mathbf{Q}}$. If $i = k$, we are done. Assume that $i \neq k$. Since a_i does not occur in '$S(\delta) \vdash T$', it is new also to '$S(\delta(a_k)) \vdash T$'. In this case Corollary 2.6 applies (renaming of individual constants), therefore '$S(\delta(a_i)) \vdash T$' is provable in $\mathbb{E}^{\mathbf{Q}}$. The proof is analogous for clause 5. □

Let us now return to contraction. We refer to the sequent '$S(A)(\_) \vdash T$', defined as in the question-premise of rule L^{ls}_{contr}, as the *left-contraction of sequent* '$S(A)(A) \vdash T$'. The sequent '$S(\_) \vdash A$' will be called the *sw-left-contraction of sequent* '$S(-A) \vdash A$'. Similarly, '$\vdash T(A)(\_)$' will be termed the *right-contraction of sequent* '$\vdash T(A)(A)$'. Finally, every left-contraction, sw-left-contraction, and right-contraction of a sequent will be collectively referred to as a *contraction of this sequent*.

We now prove that every contraction of a provable sequent is itself provable. The lemma is analogous to the admissibility of contraction but lacks the height-preserving property (cf. [19, Sect. 3.2]). The general idea is straightforward: applying one of the contraction rules removes (reading the rule bottom-up) an occurrence of a formula. However, due to the set-of-terms condition, another occurrence of the removed formula (or of its complement, but on the opposite side of the sequent) remains, therefore provability is not lost.

Though this may appear simple, the proofs of admissibility of contraction can be complex, and not every variant of sequent calculus for classical logic admits this property.[21] As previously mentioned, the proof presented here draws upon the elegant pattern for admissibility results described in [19, 21], though it does not adhere to that pattern entirely. That approach depends on height-preserving invertibility of rules, used to reduce the complexity of contraction-formulas, thus making induction on complexity viable. However, in the context of Socratic transformations, length-preservation does not always align with height-preservation in analogous cases. The non-atomic nature of basic sequents is not the sole factor contributing to this issue.

We have explored alternative methods to address this problem, such as introducing a more refined measure of the length of transformations to handle branching cases. Unfortunately, the final outcome is somewhat unsatisfactory when compared to the original pattern. For this reason, we opted to develop a proof that does not rely on length preservation.

As in [19], we rely on the proof-theoretic invertibility of rules and employ induction on the *complexity* of contraction-formulas. For this reasons, we adopt the following measure of complexity for formulas of the language $\mathcal{L}_{\mathsf{L}}$ (cf. [8, pp. 22, 184]):

Definition 2.10 (*Rank of a Formula*) The *rank of a formula of the language* $\mathcal{L}_{\mathsf{L}}$ is defined as a function $r : FORM_{\mathsf{L}} \longrightarrow \mathbb{N}_0$, where:

---

[21] For example, sequent calculi for FOL that do not repeat the formula with the quantifier in the counterparts of our rules L_γ, R_δ require contraction to repeat the formulas before certain applications of these rules. Similarly, the left-implication rule in intuitionistic logic necessitates either the presence of the implication formula in the left premise of the rule or the inclusion of the contraction rule in the system.

1. In $\mathsf{L} = \mathsf{CPL}$, for each propositional variable p_i, $r(p_i) = r(-p_i) = 0$;
2. In $\mathsf{L} = \mathsf{FOL}$, for each atomic formula A, $r(A) = r(-A) = 0$;
3. For each formula κ of $\mathcal{L}_\mathsf{L}$, $r(\kappa) = r(\kappa_0) + 1$;
4. For each formula α of $\mathcal{L}_\mathsf{L}$, $r(\alpha) = r(\alpha_0) + r(\alpha_1) + 1$;
5. For each formula β of $\mathcal{L}_\mathsf{L}$, $r(\beta) = r(\beta_0) + r(\beta_1) + 1$;
6. In $\mathsf{L} = \mathsf{FOL}$, for each γ, $r(\gamma) = r(\gamma(t)) + 1$, where t is an arbitrary term;
7. In $\mathsf{L} = \mathsf{FOL}$, for each δ, $r(\delta) = r(\delta(t)) + 1$, where t is an arbitrary term.

We now state the following result:

Corollary 2.15 *A contraction of a basic sequent is provable in* $\mathbb{E}$.

Proof In most cases, the contracted sequent remains basic, given the set-of-terms condition. Interesting cases arise when the sequent '$S(-A) \vdash A$' is basic due to the presence of '$-A$' in S, and we consider its sw-left-contraction '$S(\_) \vdash A$', where '$-A$' no longer appears in the sequent. If the original sequent is basic because both A and '$-A$' are in the antecedent, then A is a term of the sequence $S(\_)$, making the sw-left-contraction '$S(\_) \vdash A$' basic as well. However, if '$S(-A) \vdash A$' is basic because '$- - A$' occurs in S, whit $A \neq - - A$, then by Corollary 2.1, clause 2, A is a doubly negated formula, specifically of the form '$\neg\neg(- - A)$'. By Lemma 2.5 (double negation), '$S(\_) \vdash A$' has a Socratic proof in $\mathbb{E}$ iff '$S(\_) \vdash - - A$' does, and the latter sequent is basic. □

In the main result on the semi-admissibility of contraction, we will use induction with respect to the rank of the contraction-formula. However, γ-formulas typically introduce certain complications in this type of reasoning, due to the additional occurrence of γ that remains in the sequent-conclusion. Therefore, in the next lemma, we handle γ-formulas separately, using induction on the length of a Socratic transformation.

Lemma 2.8 (Contraction of Generally Quantified Formulas)

1. *If '$S(\gamma)(\gamma) \vdash T$' has a Socratic proof in $\mathbb{E} \in \{\mathbb{E}^{\mathbf{LQ}}, \mathbb{E}^{\mathbf{Q}}\}$, then '$S(\gamma)(\_) \vdash T$' also has a Socratic proof in $\mathbb{E}$.*
2. *If '$S(-\delta) \vdash \delta$' has a Socratic proof in $\mathbb{E}^{\mathbf{Q}}$, then '$S(\_) \vdash \delta$' also has a Socratic proof in $\mathbb{E}^{\mathbf{Q}}$.*
3. *If '$\vdash T(\delta)(\delta)$' has a Socratic proof in $\mathbb{E}^{\mathbf{RQ}}$, then '$\vdash T(\delta)(\_)$' also has a Socratic proof in $\mathbb{E}^{\mathbf{RQ}}$.*

Proof The proof proceeds by induction on the length of the Socratic proof of the considered sequent. If the length is 1, the sequent is basic, and by Corollary 2.15, its contraction is provable in $\mathbb{E}$.

For the inductive step, we consider only selected cases relevant to the calculus $\mathbb{E}^{\mathbf{Q}}$. Assume we are dealing with a sequent of the form (1) $S(\gamma)(\gamma) \vdash C$ or (2) $S(-\delta) \vdash \delta$, and that it has a Socratic proof $\mathbf{s} = \langle Q_0, Q_1, \ldots, Q_n \rangle$ in $\mathbb{E}$ of length $n + 1$. The inductive hypothesis (IH) follows the familiar formulation: if a constituent of Q_1 has form (1) or (2), then its contraction, as specified in the lemma, has a Socratic proof in the relevant $\mathbb{E}$.

Next, we proceed by cases depending on the rule applied to Q_0. Consider:

$$\frac{Q_0 = ?(\ S(\gamma)(\gamma)(\beta) \vdash C\ )}{Q_1 = ?(\ S(\gamma)(\gamma)(\beta_0) \vdash C\ ;\ S(\gamma)(\gamma)(\beta_1) \vdash C\ )}\ L_\beta$$

By applying the IH twice, we get that the sequents $S(\gamma)(\_)(\beta_0) \vdash C$ and $S(\gamma)(\_)(\beta_1) \vdash C$ have Socratic proofs in $\mathbb{E}^{\mathbf{Q}}$. Then, by Lemma 2.3 (Separation), there is a successful Socratic transformation of the question $?(S(\gamma)(\_)(\beta_0) \vdash C; S(\gamma)(\_)(\beta_1) \vdash C)$ in $\mathbb{E}^{\mathbf{Q}}$. Thus, by Corollary 2.7, the sequent $S(\gamma)(\_)(\beta) \vdash C$ has a Socratic proof, as required.

Now, let's consider the following case with the contracted formula active:

$$\frac{Q_0 = ?(\ S(-\delta) \vdash \delta\ )}{Q_1 = ?(\ S(-\delta, \gamma(t)) \vdash \delta\ )}\ L_\gamma$$

By Corollary 2.1, $-\delta$ is a γ-formula, and $\gamma(t)$ is its instance. By the IH, the sequent $S(\gamma(t)) \vdash \delta$ has a Socratic proof in $\mathbb{E}^{\mathbf{Q}}$. By applying Lemma 2.7 (Switching), we conclude that $S(-\delta) \vdash \neg\gamma(t)$ has a Socratic proof in $\mathbb{E}^{\mathbf{Q}}$. If $\neg\gamma(t) \neq \delta(t)$, we additionally apply Lemma 2.5 (double negation) to conclude that $S(-\delta) \vdash \delta(t)$ has a Socratic proof. Finally, $S(\_) \vdash \delta$ is provable by Corollary 2.7.

The reasoning is analogous for the remaining cases. □

Lemma 2.9 (Semi-Admissibility of L^{ls}_{contr}, L^{sw}_{contr}, R_{contr} in $\mathbb{E}$)

1. *If there is a successful Socratic transformation of the question* $Q = ?(\Phi(S(A)(A) \vdash T))$ *in* $\mathbb{E} \in \{\mathbb{E}^{\mathbf{LQ}}, \mathbb{E}^{\mathbf{Q}}\}$*, then there is also a successful Socratic transformation of the question* $?(\Phi(S(A)(\_) \vdash T))$ *in* $\mathbb{E}$.
2. *If there is a successful Socratic transformation of the question* $Q = ?(\Phi(S(-A) \vdash A))$ *in* $\mathbb{E}^{\mathbf{Q}}$*, then there is also a successful Socratic transformation of the question* $?(\Phi(S(\_) \vdash A))$ *in* $\mathbb{E}^{\mathbf{Q}}$.
3. *If there is a successful Socratic transformation of the question* $Q = ?(\Phi(\vdash T(A)(A)))$ *in* $\mathbb{E}^{\mathbf{RQ}}$*, then there is also a successful Socratic transformation of the question* $?(\Phi(\vdash T(A)(\_)))$ *in* $\mathbb{E}^{\mathbf{RQ}}$.

Proof The proof is conducted by induction on the rank r of the formula A, with a subinduction on the length l of the successful Socratic transformation of Q.[22] For the sake of conciseness, we will refer to this as 'r-induction' and 'l-induction, with r-IH and l-IH denoting the respective inductive hypotheses.

To facilitate the reasoning, let ϕ stand for the constituent of Q distinguished in the formulation of our lemma. Throughout the following argument, we will utilize Lemma 2.3 (Separation) to transition between the question and the provability of individual sequents, sometimes implicitly.

[22] More technically, a pair of natural numbers $\langle r, l\rangle$ can be assigned to the question Q. Then the proof proceeds by induction on r, and for a fixed r, by induction on l. This approach is reminiscent on the triple induction applied in [6, pp. 13–14] in the proof of the admissibility of cut.

Let **s** stand for the successful Socratic transformation of Q. In the base case of the r-induction, where $r(A) = 0$, the application of a rule to Q does not affect A. If $l = 1$ (the base case of the l-induction), then each constituent of Q is basic, and thus the contraction of ϕ has a Socratic proof by Corollary 2.15. For the inductive part of the l-induction, we apply l-IH and Corollary 2.7.

Now, suppose that the lemma holds for contraction-formulas of rank up to r and consider the case where $r(A) = r + 1$. In this scenario we can reiterate part of the previous reasoning: if **s** is of length 1, then the result follows directly from Corollary 2.15. Furthermore, if $l > 1$ and the rule acts on a sequent other than ϕ, or if it acts on ϕ but on a formula other than A, then the result follows from l-IH (along with Corollary 2.7 and possibly Lemma 2.3).

The most challenging case arises when one occurrence of A is active.

If A is a γ-formula in the antecedent of ϕ, or a δ-formula in its succedent, the result follows from Lemma 2.8.

For the remainder, we analyze three possible cases, leaving the others for the reader to explore. Let $\mathbf{s} = \langle Q_0, Q_1, \ldots, Q_n \rangle$. Suppose A is of the form $\neg\neg B$, and $\phi = S(-A) \vdash A$, and that Q_0/Q_1 is:

$$\frac{Q_0 = ?(\ \Phi\ (\ S\ (\neg B) \vdash \neg\neg B\ ))}{Q_1 = ?(\ \Phi\ (\ S\ (\neg B) \vdash B\ ))}\ R_\kappa$$

If B is a positive formula, then $-B = \neg B$, and hence by applying the r-IH to Q_1, we obtain a Socratic proof of $S(\_) \vdash B$ in $\mathbb{E}^{\mathbf{Q}}$. If B is negative, then $\neg B$ in the antecedent of the sequent is doubly negated. In this case, by Corollary 2.12 (Proof-theoretic invertibility of κ-rules), or directly by Lemma 2.5 (double negation), there is a Socratic proof of $S(-B) \vdash B$, and again, r-IH applies.

If A is an α- or β-formula, we additionally use Corollary 2.1. For instance, suppose A is a β-formula, and we have:

$$\frac{Q_0 = ?(\ \Phi\ (\ S\ (\alpha) \vdash A\ ))}{Q_1 = ?(\ \Phi\ (\ S\ (\alpha_0, \alpha_1) \vdash A\ ))}\ L_\alpha$$

By clause 5 of Corollary 2.13, the sequent $S(\alpha_0, \alpha_1, -\beta_0) \vdash \beta_1$ has a Socratic proof in $\mathbb{E}^{\mathbf{Q}}$. According to clause 4 of Corollary 2.1, either $\alpha_0 = -\beta_0$, or the equality is achieved by applying Lemma 2.5 (double negation). Therefore, by the r-IH, the sequent $S(\alpha_0, \alpha_1) \vdash \beta_1$ is provable in $\mathbb{E}^{\mathbf{Q}}$. Using the same argument, we conclude that $S(\alpha_0) \vdash \beta_1$ is provable, finally leading us back to the question $?(\Phi(S(\_) \vdash A)$.

Lastly, let A be a δ-formula, and suppose we are dealing with:

$$\frac{Q_0 = ?(\ \Phi\ (\ S\ (\delta)(\delta) \vdash\ ))}{Q_1 = ?(\ \Phi\ (\ S\ (\delta)(\delta(a_i)) \vdash\ ))}\ L_\delta$$

Let a_k represent a constant not present in **s**. By clause 4 of Corollary 2.14 (Proof-theoretic invertibility of rule L_δ), the sequent $S(\delta(a_k))(\delta(a_i)) \vdash$ has a Socratic proof

in $\mathbb{E}^{\mathbf{LQ}}$. By clause 1 of Corollary 2.9, the sequent $S(\delta(a_i))(\delta(a_i)) \vdash$ also has a Socratic proof in $\mathbb{E}^{\mathbf{LQ}}$. Thus, the r-IH applies, and we conclude that $S(\delta(a_i))(\_) \vdash$ is provable.
This finishes the proof. □

Returning to the general form of the contraction rules introduced on Sect. 2.5, we can now state the following result:

Corollary 2.16 (Semi-Admissibility of L_{contr}, R_{contr} in $\mathbb{E}$) *If there is a successful Socratic transformation of the question-conclusion of one of the rules L_{contr} or R_{contr} in $\mathbb{E}$, then there is also a successful Socratic transformation of the question-premise of the rule in $\mathbb{E}$.*

Proof This follows directly from the previous lemma. □

Corollary 2.17 (Eliminability of L_{contr}, R_{contr} in $\mathbb{E}$)

1. *Let $\mathbb{E} \in \{\mathbb{E}^{\mathbf{Q}}, \mathbb{E}^{\mathbf{P}}\}$, and let Q stand for a question of $\mathcal{L}^?_{\vdash\mathsf{L}}$ with both-sided constituents. If there is a successful Socratic transformation of Q in $\mathbb{E} \cup \{L_{contr}\}$, then there is also one in $\mathbb{E}$.*
2. *Let $\mathbb{E} \in \{\mathbb{E}^{\mathbf{RQ}}, \mathbb{E}^{\mathbf{RP}}\}$, and let Q represent a question of $\mathcal{L}^?_{\vdash\mathsf{L}}$ with right-sided constituents. If there is a successful Socratic transformation of Q in $\mathbb{E} \cup \{R_{contr}\}$, then there is also one in $\mathbb{E}$.*
3. *Let $\mathbb{E} \in \{\mathbb{E}^{\mathbf{LQ}}, \mathbb{E}^{\mathbf{LP}}\}$, and let Q stand for a question of $\mathcal{L}^?_{\vdash\mathsf{L}}$ with left-sided constituents. If there is a successful Socratic transformation of Q in $\mathbb{E} \cup \{L_{contr}\}$, then there is also one in $\mathbb{E}$.*

Proof The proof proceeds in the same way as for the analogous Corollary 2.11, which concerns eliminability of weakening. Lemma 2.9 is applied here. □

Let us conclude this section with a note on the rule of cut in the erotetic setting. Generally speaking, this rule:

$$\frac{\Gamma \Rightarrow \Delta, A \quad A, \Theta \Rightarrow \Pi}{\Gamma, \Theta \Rightarrow \Delta, \Pi}\ cut \qquad \text{where } \Gamma, \Theta, \Delta, \Pi \text{ are finite multisets of formulas}$$

is often the focal point of structural analysis (in fact, the theorem stating the eliminability of cut in sequent calculus was famously termed *Hauptsatz* by Gentzen [10]). However, in the completeness proofs presented in this book, the cut rule was not required, which is why it has been omitted in this section. Nonetheless, the cut rule in the context of erotetic logic presents an intriguing topic. It was studied within this framework in [33], and further investigation, such as the erotetic interpretation of the rule, is left for future research.

2.6 Completeness of Erotetic Calculi

In [2], the author employs the technique of consistency properties to establish the completeness of erotetic calculi for classical logic.[23] Here, we present a proof using similar techniques, though with some simplifications. Our proof is more straightforward in certain aspects, mainly due to the absence of parameters, though this leads to some complications in the proof of Lemma 2.10, where we demonstrate that a consistent set can be extended to a 'downward-saturated' one.

The appeal of this type of proof lies in its conceptual simplicity. In its most basic form, the completeness proof refers to the so-called *Hintikka sets*,[24] which are sets of formulas defined purely syntactically, in terms of the structure of the formulas and the membership relationship. However, the definition of a Hintikka set encapsulates the semantics of the underlying logic in such a way that it becomes relatively straightforward to prove that a Hintikka set is satisfiable—i.e., that it has a model. The next step involves connecting unprovability in the relevant proof system to Hintikka sets, demonstrating that for every unprovable formula, there exists a countermodel in the form of a Hintikka set.

Herbrand Models, Downward-Saturated Sets, and L-Consistency Properties

As previously mentioned, in this work, we utilize consistency properties, which are, broadly speaking, families of subsets of Hintikka sets. The advantage of using consistency properties for completeness proofs arises from the fact that subsets, rather than entire Hintikka sets, better align with the incremental stages of constructing a derivation. Our initial goal is to prove that every set belonging to a consistency property is satisfiable. In the case of FOL, we introduce Herbrand models for this purpose. We begin by laying out the necessary definitions (cf. [8, Chap. 5]).

Definition 2.11 (*Herbrand Model*) A pair $\langle U, f \rangle$ is called a *Herbrand model for the language* $\mathcal{L}_{\mathsf{FOL}}$, provided that:

- U is the set of all closed terms of the language $\mathcal{L}_{\mathsf{FOL}}$, and
- For each individual constant a_i, each function symbol F_i^n, and each predicate P_i^n, the function f satisfies the following conditions:
  - $f(a_i) = a_i$;
  - $f(F_i^n)$ is an n-argument function that acts on the elements of U such that $f(F_i^n)(t_1 \ldots, t_n) = F_i^n(t_1 \ldots, t_n)$;
  - $f(P_i^n)$ is an n-term relation on U.

Note that there are at least as many Herbrand models (in the above sense) as there are subsets of U^n—in fact, uncountably many. Herbrand models are often referred to as *linguistic* models because they are constructed from linguistic elements. One enjoyable feature of Herbrand models is that they are *standard*, meaning that every

[23] For the dual erotetic calculi described in [2], the author develops the technique of *refutability properties*, which are dual to consistency properties.

[24] See [11].

object in the universe is assigned a name by the interpretation function. This leads to the following property:

Corollary 2.18 *Let $\mathcal{H}$ be a Herbrand model for the language $\mathcal{L}_{\mathsf{FOL}}$.*

1. *For each γ-formula, $\mathcal{H} \models \gamma$ iff for every closed term t of $\mathcal{L}_{\mathsf{FOL}}$, $\mathcal{H} \models \gamma(t)$.*
2. *For each δ-formula, $\mathcal{H} \models \delta$ iff for some closed term t of $\mathcal{L}_{\mathsf{FOL}}$, $\mathcal{H} \models \delta(t)$.*

Proof For clause 1, let γ stand for an arbitrary γ-formula of $\mathcal{L}_{\mathsf{FOL}}$, and assume the quantifier in γ binds x_i. We proceed through the following chain of equivalences:

$$\begin{aligned}\mathcal{H} \models \gamma \text{ iff } & \text{for each object assignment } \sigma, \mathcal{H} \models \gamma[\sigma] \\ \text{iff } & \text{for each object assignment } \sigma, \text{ for each } t \in U, \mathcal{H} \models \gamma(x_i)[\sigma(x_i/t)].\end{aligned}$$

Since $x_i{}^{\mathcal{H}}[\sigma(x_i/t)] = t$ and $t^{\mathcal{H}}[\sigma(x_i/t)] = f(t) = t$, we have:

$$\mathcal{H} \models \gamma(x_i)[\sigma(x_i/t)] \text{ iff } \mathcal{H} \models \gamma(t)[\sigma(x_i/t)].$$

But now x_i is not a free variable in $\gamma(t)$, therefore:

$$\mathcal{H} \models \gamma(x_i)[\sigma(x_i/t)] \text{ iff } \mathcal{H} \models \gamma(t)[\sigma].$$

Thus, we have shown that $\mathcal{H} \models \gamma$ iff for every object assignment σ, for each $t \in U$, $\mathcal{H} \models \gamma(t)[\sigma]$. This means that $\mathcal{H} \models \gamma$ iff for each $t \in U$, $\mathcal{H} \models \gamma(t)$, which was to be proved, as U is the set of all closed terms of $\mathcal{L}_{\mathsf{FOL}}$.

The reasoning is analogous for δ-formulas. □

Definition 2.12 (*First-Order Consistency Property*) A family $\mathcal{C}$ of sets of sentences of the language $\mathcal{L}_{\mathsf{FOL}}$ is called a *first-order consistency property* if the following conditions hold for each $X \in \mathcal{C}$:

1. For any atomic sentence A, if $A \in X$, then $-A \notin X$;
2. If $\kappa \in X$, then $X \cup \{\kappa_0\} \in \mathcal{C}$;
3. If $\alpha \in X$, then $X \cup \{\alpha_0, \alpha_1\} \in \mathcal{C}$;
4. If $\beta \in X$, then $X \cup \{\beta_0\} \in \mathcal{C}$ or $X \cup \{\beta_1\} \in \mathcal{C}$;
5. If $\gamma \in X$, then for every closed term t of the language $\mathcal{L}_{\mathsf{FOL}}$, $X \cup \{\gamma(t)\} \in \mathcal{C}$;
6. If $\delta \in X$, then for some closed term t of $\mathcal{L}_{\mathsf{FOL}}$, $X \cup \{\delta(t)\} \in \mathcal{C}$.

Propositional consistency properties are structures satisfying clauses 1–4 of the above definition, where in clause 1, 'atomic formula' is replaced by 'propositional variable'. We will use the abbreviation 'L-consistency property,' meaning the propositional version when $\mathsf{L} = \mathsf{CPL}$, and the first-order version otherwise.

It is worth noting that an empty set X satisfies all the conditions in the above definition; so does the empty family $\mathcal{C}$.

In the propositional case, the next step is to prove Theorem 2.3 (the propositional model existence theorem), and in the first-order case, we aim to prove Theorem 2.4 (the first-order model existence theorem). To do so, following Hintikka (see [11]), we introduce the concept of a downward-saturated set, also known as Hintikka set, as previously mentioned. A Hintikka set is one that contains, for every κ-, α-, or β-formula, the relevant component(s) of the formula, and for FOL, every (at least one) instance of a γ-formula (δ-formula), respectively.

Definition 2.13 (*Downward-Saturated Set*) A set X of formulas of the language $\mathcal{L}_\mathsf{L}$ is called *downward-saturated* if the following conditions hold:

1. For any atomic sentence A, if $A \in X$, then $-A \notin X$;
2. If $\kappa \in X$, then $\kappa_0 \in X$;
3. If $\alpha \in X$, then $\alpha_0 \in X$ and $\alpha_1 \in X$;
4. If $\beta \in X$, then $\beta_0 \in X$ or $\beta_1 \in X$;
5. If $\gamma \in X$, then for each closed term t of the language $\mathcal{L}_\mathsf{FOL}$, $\gamma(t) \in X$;
6. If $\delta \in X$, then for some closed term t of $\mathcal{L}_\mathsf{FOL}$, $\delta(t) \in X$.

The first condition does not express the saturation property, but the consistency. Sometimes the saturation property is considered without this clause.

The next step is to define, for any arbitrary set X within an L-consistency property, a downward-saturated superset $Y \supseteq X$. As we will demonstrate, this superset Y will have an L-model. This step forms the core of the entire construction, bridging the gap between syntax and semantics.

However, in the case of FOL, complications arise due to the absence of an additional supply of parameters, requiring careful handling at this stage.

Lemma 2.10 *Let $\mathcal{C}$ be an* L*-consistency property. Every set $X \in \mathcal{C}$ can be extended to a downward-saturated set.*

Proof The argument begins with a Lindenbaum-style construction. Thus assume $X \in \mathcal{C}$, where $\mathcal{C}$ is an L-consistency property. Recall that in the case of FOL, X contains only sentences, while for CPL, the term 'sentence' means 'formula.' We focus exclusively on countable languages,[25] hence all sentences of $\mathcal{L}_\mathsf{L}$ can be ordered in a sequence according to some fixed alphabetical order. Let:

$$F_0, F_1, \ldots, F_k, \ldots \tag{2.19}$$

[25] This is not a necessary limitation. For languages with arbitrary cardinality, the construction would require more direct reference to some form of the Axiom of Choice; see [7, p. 60]. This *is not* to say that the presented proof does not require a counterpart of the Axiom of Choice. It enters the stage every time we assume that there is a union of an infinite, monotonically increasing sequence of sets of sentences. The existence of such a union follows from Kuratowski-Zorn Lemma, which asserts that a partially ordered set containing an upper bound for every chain also contains a maximal element. The Kuratowski-Zorn Lemma is equivalent to the Axiom of Choice.

represent such a sequence of sentences. We define a sequence of sets $(X_k)_\infty$ as follows[26]:

$$X_0 = X \text{ and for each } k \in \mathbb{N}_0 :$$

$$X_{k+1} = \begin{cases} X_k \cup \{F_k\} & \text{if for some } Y \in \mathcal{C},\ X_k \cup \{F_k\} \subseteq Y, \\ X_k & \text{otherwise.} \end{cases} \tag{2.20}$$

By this construction we obtain:

$$X = X_0 \subseteq X_1 \subseteq \cdots \subseteq X_k \subseteq X_{k+1} \subseteq \cdots \tag{2.21}$$

Let $\mathbb{X}$ stand for the union of this sequence, that is:

$$\mathbb{X} = \bigcup_{k=0}^{\infty} X_k.$$

Since the sequence $(X_k)_\infty$ is monotonically increasing, we have:

$$\text{if } k < l, \text{ then } X_k \subseteq X_l. \tag{2.22}$$

Additionally, $X \subseteq \mathbb{X}$, hence $\mathbb{X}$ is a superset of X. Note that by (2.20), each X_k is a subset of some $Y \in \mathcal{C}$. We may occasionally refer to this superset of X_k in $\mathcal{C}$ as 'Y^{X_k}'.

Let us observe that for each formula $A \in \mathbb{X}$, there is a set in the sequence (2.21) where the formula first appears. If $A \in X$, then it appears in X_0. If $A \notin X$, we trace its position in the sequence (2.19), say at position $(k-1)$ for some $k \in \mathbb{N}$. Then, $X_k = X_{k-1} \cup \{A\}$, for otherwise A would never 'jump into' $\mathbb{X}$. Clearly, by (2.22), if $A \in X_0$, then $A \in X_k$ for each $k > 0$. In summary, if $A \in \mathbb{X}$ and A occurs in the sequence (2.19) at position $(k-1)$, then we refer to X_k in (2.21) as *A's set in* (2.21).

Another observation is that if A is a complex formula other than β or δ, and a component, A_0, of A precedes A in (2.19), then the component is already in A's set, say it is X_l. The reason is that, by (2.20), $X_l = X_{l-1} \cup \{A\}$ is a subset of some $Y \in \mathcal{C}$, and since $A \in Y$, $\mathcal{C}$ contains also $Y \cup \{A_0\}$. If A_0 is in (2.19) in position k that precedes $(l-1)$, then by (2.22) we have

$$X_k \cup \{A_0\} \subseteq X_l \cup \{A_0\} \subseteq Y \cup \{A_0\} \in \mathcal{C},$$

---

[26] For readers of [8]: the property of being subset-closed is embedded in this definition by requiring that each X_k be a *subset* of a set in $\mathcal{C}$. The property of finite character is not used here, as it is not required that the downward-saturated superset of X belong to $\mathcal{C}$. While the presented approach might sacrifice some elegance, the ability to use the subset-condition rather than the two traditional properties was compelling enough to adopt this construction. In the modal case, we will proceed in a more standard manner (see the next chapter).

therefore $X_{k+1} = X_k \cup \{A_0\}$. Hence, as we said, A_0 is already there. For this reason, when considering κ-, α-, and γ-formulas, we may restrict ourselves to the situation when the complex formula precedes its component(s) in sequence (2.19). More specifically, the following holds:

(a) If $\kappa \in X_k$ but $\kappa_0 \notin X_k$, then κ_0's position, l, in (2.19) is such that $k - 1 < l$;
(b) If $\alpha \in X_k$ but $\alpha_i \notin X_k$ ($i \in \{0, 1\}$), then α_i's position, l, in (2.19) is such that $k - 1 < l$;
(c) If $\gamma \in X_k$ but $\gamma(t) \notin X_k$ for some closed term t, then $\gamma(t)$'s position, l, in (2.19) is such that $k - 1 < l$.

We start with the first clause of Definition 2.13, the one excluding inconsistency on the level of formulas of rank 0. Suppose, to the contrary, that there is an A and '$-A$' both present in $\mathbb{X}$. From among the two formulas' sets in (2.21), we choose the one that follows the other in (2.21), let us call it X_l, and we conclude that $\{A, -A\} \subseteq X_l$ by (2.22). Hence, there is some $Y \in \mathcal{C}$ such that $\{A, -A\} \subseteq X_l \subseteq Y$, which contradicts clause 1 of Definition 2.12.

Now, we will check that conditions 2, 3 and 5 from Definition 2.13 are satisfied; conditions 4 and 6 present their specific problems, as we shall see.

Suppose that $\kappa \in \mathbb{X}$ and let X_k be κ's set in (2.21). If $\kappa_0 \notin X_k$, then, by (a), κ_0 is present in (2.19) on position l, where $k < l$. By (2.22), we have $\kappa \in X_l$. By (2.20), there is a set $Y \in \mathcal{C}$ such that $X_l \subseteq Y$, and since $\kappa \in X_l \subseteq Y$, we must also have $Y \cup \{\kappa_0\} \in \mathcal{C}$ (by clause 2 of Definition 2.12). Since $X_l \cup \{\kappa_0\} \subseteq Y \cup \{\kappa_0\}$, it follows that $X_{l+1} = X_l \cup \{\kappa_0\}$. Hence, $\kappa_0 \in \mathbb{X}$.

Suppose that $\alpha \in \mathbb{X}$ and let X_k be α's set in (2.21). If $\alpha_i \notin X_k$ (for any $i \in \{0, 1\}$), then, by (b), $k - 1 < l$, where l is the position of α_i in (2.19). Since $X_k \subseteq X_l$ (by (2.22)), we have $\alpha \in X_l$, and since $X_l \subseteq Y^{X_l} \in \mathcal{C}$, we also have $Y^{X_l} \cup \{\alpha_0, \alpha_1\} \in \mathcal{C}$ (clause 3 of Definition 2.12). Since $X_l \cup \{\alpha_i\} \subseteq Y^{X_l} \cup \{\alpha_0, \alpha_1\}$, it follows that $X_{l+1} = X_l \cup \{\alpha_i\}$, which yields $\alpha_i \in \mathbb{X}$. This reasoning holds for both $i \in \{0, 1\}$, so if $\alpha \in \mathbb{X}$, then $\{\alpha_0, \alpha_1\} \subseteq \mathbb{X}$.

For the reminder of this proof, when considering instances $\gamma(t)$ or $\delta(t)$, we will always assume that t is a closed term of the language $\mathcal{L}_{\mathsf{FOL}}$.

Assume that we have $\gamma \in \mathbb{X}$ and that X_k is γ's set in (2.21). We consider an arbitrary instance $\gamma(t)$ of γ; let l stand for the formula's position in (2.19). If $\gamma(t) \notin X_k$, then, by (c), $k - 1 < l$, and hence $\gamma \in X_l$ by (2.22). Thus, for $Y^{X_l} \in \mathcal{C}$, we must have $Y^{X_l} \cup \{\gamma(t)\} \in \mathcal{C}$ (clause 5 of Definition 2.12), which yields that $X_{l+1} = X_l \cup \{\gamma(t)\}$. Thus, $\gamma(t) \in \mathbb{X}$. The instance was arbitrary; therefore, every closed instance of γ belongs to $\mathbb{X}$.

That's it for clauses 1–3 and 5 of Definition 2.13. We proceed to the remaining two. Actually, the β-case is a toy example before the case for δ. We drop the assumption that components come after complex formulas in (2.19) and carefully go through all the details once again.

Assume we have $\beta \in \mathbb{X}$. Let k, l and m stand for the positions in (2.19) of β, β_0 and β_1, respectively. First, we assume that both l and m precede k in (2.19). For $X_{k+1} = X_k \cup \{\beta\}$, there must be $Y^{X_{k+1}} \in \mathcal{C}$ such that $X_{k+1} \subseteq Y^{X_{k+1}}$. Hence, there must also be $Y^{X_{k+1}} \cup \{\beta_i\} \in \mathcal{C}$ for at least one β_i (clause 4 of Definition 2.12). If $i = 0$, then:

$$X_l \cup \{\beta_0\} \subseteq X_{k+1} \cup \{\beta_0\} \subseteq Y^{X_{k+1}} \cup \{\beta_0\} \in \mathcal{C},$$

and then $\beta_0 \in X_{l+1}$. Similarly, if $i = 1$, then:

$$X_m \cup \{\beta_1\} \subseteq X_{k+1} \cup \{\beta_1\} \subseteq Y^{X_{k+1}} \cup \{\beta_1\} \in \mathcal{C},$$

and then $\beta_1 \in X_{m+1}$. In both cases, $\beta_i \in \mathbb{X}$. We have shown that if $l < k$ and $m < k$, then $\beta_i \in \mathbb{X}$ for at least one $i \in \{0, 1\}$.

Assume that $k < l$ or $k < m$, and let $n = \max\{l, m\}$. By (2.22), $\beta \in X_n$; hence there is $Y^{X_n} \supseteq X_n$ such that $Y^{X_n} \cup \{\beta_i\} \in \mathcal{C}$ for some $i \in \{0, 1\}$ (clause 4 of Definition 2.12). In this situation, both

$$X_l \cup \{\beta_0\} \subseteq X_n \cup \{\beta_0\} \subseteq Y^{X_n} \cup \{\beta_0\},$$

and

$$X_m \cup \{\beta_1\} \subseteq X_n \cup \{\beta_1\} \subseteq Y^{X_n} \cup \{\beta_1\}.$$

Since for at least one i, $Y^{X_n} \cup \{\beta_i\}$ is in $\mathcal{C}$, one of the above proves that $\beta_i \in \mathbb{X}$.

Before we proceed, let us stress that the union of sets $\mathbb{X}$, defined above by means of (2.19) and (2.20), satisfies conditions 1–5 of Definition 2.13 *for every sequence of formulas of* $\mathcal{L}_{\mathsf{L}}$, not just for (2.19). In turn, as we shall see, in the δ-case expressed by condition 6, things depend on the initial order (2.19). In other words, $\mathbb{X}$ does not satisfy condition 6 in general, and the remaining part of this proof is devoted to specifying a sequence of formulas that will rectify this.

Assume that $\delta \in \mathbb{X}$, and let X_k be δ's set in (2.21). We obviously have $Y^{X_k} \in \mathcal{C}$ such that $X_k \subseteq Y^{X_k}$; hence, by clause 6 of Definition 2.12, $Y^{X_k} \cup \{\delta(t_1)\} \in \mathcal{C}$ for some instance $\delta(t_1)$. If we assume that $\delta(t_1)$ precedes δ in (2.19), then $\delta(t_1) \in \mathbb{X}$, as in the previous cases. Hence, assume that $\delta(t_1)$ occurs in (2.19) at position l, which is further than k, and consider X_l. We have so far:

$$k - 1 < l \text{ and } \delta(t_1) = F_l \text{ in (2.19)},$$

$$X_k \subseteq Y^{X_k} \cup \{\delta(t_1)\} \in \mathcal{C},$$

and by (2.22),

$$\delta \in X_k \subseteq X_l. \tag{2.23}$$

Unfortunately, we cannot conclude that the set $X_l \cup \{\delta(t_1)\}$ has a superset in $\mathcal{C}$, because for $X_k \subset X_l$, the fact that there is in $\mathcal{C}$ a superset of $X_k \cup \{\delta(t_1)\}$ does not entail that there is one for $X_l \cup \{\delta(t_1)\}$.[27] Hence, we must humbly consider the case with $X_{l+1} = X_l$.

On the other hand, since $\delta \in X_l$ and there is a superset $Y^{X_l} \supseteq X_l$ in $\mathcal{C}$, there must also be $Y^{X_l} \cup \{\delta(t_2)\} \in \mathcal{C}$ for some instance $\delta(t_2)$. Things are getting interesting now. Assume that $\delta(t_2)$ is at position m in (2.19). If $\delta(t_2)$ precedes $\delta(t_1)$ in (2.19), then $\delta(t_2) \in \mathbb{X}$, because in this situation

$$X_m \cup \{\delta(t_2)\} \subseteq X_l \cup \{\delta(t_2)\} \subseteq Y^{X_l} \cup \{\delta(t_2)\} \in \mathcal{C}.$$

Hence, assume that $\delta(t_1)$ precedes $\delta(t_2)$ in (2.19). When we apply the same arguments as above, we shall end up with an analogous conclusion—that it is possible that $X_{m+1} = X_m$, $\delta(t_2) \notin \mathbb{X}$, but then there is another instance of δ, $\delta(t_3)$, which is further than $\delta(t_2)$ in (2.19) and is such that its presence somewhere in a member of $\mathcal{C}$ secures condition 6 of Definition 2.12 with respect to $\mathcal{C}$ (consistency property), while still not preserving clause 6 of Definition 2.13 with respect to $\mathbb{X}$ (downward-saturated set). And so on…

This rather unfortunate situation is illustrated in Fig. 2.2. We used 'X sub $\mathcal{C}$' to indicate that X has a superset in $\mathcal{C}$ and 'X nosub $\mathcal{C}$' to express that this is not the case.

The illustration is not intended to lead to a contradiction, but rather to demonstrate that this unfortunate state of affairs can indeed occur. Each time we attempt to ascertain δ by its instance—an instance that already accompanies δ in a set from $\mathcal{C}$—it turns out that we have made some unfortunate choices such that at a particular stage in (2.21), the considered set cannot accommodate the instance. This analysis shows that such a situation is perfectly possible, and consequently, the union $\mathbb{X}$ can contain a δ-formula without any instance of it.

On the other hand, we must keep in mind that this situation is rather peculiar. We are considering consecutive instances of δ, each of which is present in *some* set of $\mathcal{C}$ together with our δ, but every time, we fail to find the instance in the *required* set of $\mathcal{C}$, thereby remaining unlucky infinitely many times. This unfortunate situation is closely linked to the ordering of the consecutive instances of δ in (2.19). For suppose that we simply swapped the positions of $\delta(t_1)$ and $\delta(t_2)$ in (2.19). In that case, we would have $F_{l_1} = \delta(t_2)$, thus the fact that:

$$X_{l_1} \cup \{\delta(t_2)\} \text{ sub } \mathcal{C}$$

would lead to:

$$X_{l_1+1} = X_{l_1} \cup \{\delta(t_2)\} \text{ and } \delta(t_2) \in \mathbb{X}.$$

[27] To give a more specific illustration of this phenomenon, consider $X = \{\exists x \neg P(x)\}$, a first-order consistency property $\mathcal{C}$ with $\{\exists x \neg P(x)\} \in \mathcal{C}$ and $\{\exists x \neg P(x), \neg P(a)\} \in \mathcal{C}$, but also $\{\exists x \neg P(x), P(a)\} \in \mathcal{C}$. Suppose moreover that $P(a)$ and $\neg P(a)$ occur in (2.19) at positions 0 and 1, respectively. Then $\{\exists x \neg P(x), P(a)\} = X_1$, hence $X_2 = X_1$ and $\neg P(a) \notin \mathbb{X}$. But then there must be another instance of $\exists x \neg P(x)$ that satisfies clause 6 of Definition 2.12. An analogous problem pertains to the β-case, where it is easily solvable, however, as the number of components is finite.

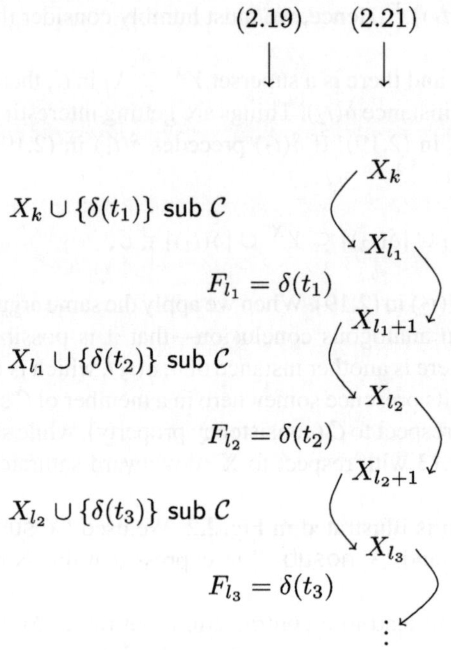

Fig. 2.2 The unfortunate δ-case

Moreover, the opportunity to 'fix' the chosen order of formulas occurs *infinitely many times*, just like the failures described above. Starting with X_k, the formula δ occurs *in every set that follows* in (2.21). For *each such set*, there is an instance of δ—a different instance every time—that, when added to the set, finds a superset in $\mathcal{C}$. In *each case*, the instance occupies a position in (2.19) further than the position of this set in (2.21), and when one attempts to add the instance to an appropriate further set of (2.21), one fails to find a superset in $\mathcal{C}$.

Hence, if we picked *any other pair* of δ-instances traced in this way—by the fact that they occur in a set of $\mathcal{C}$ together with δ—swapping their positions in (2.19) would suffice to ensure that an instance of δ appears in $\mathbb{X}$. Of course, this would alter more than just the instances: the order (2.19), the sequence (2.21), and consequently, the union $\mathbb{X}$ would be different. However, as we shall see shortly, if performed carefully, this instances-swapping trick will preserve at least the initial part of a sequence like (2.21).

So, this is exactly what we are going to do: we apply the swapping trick every time we arrive at the unfortunate δ-case. It is time to get more formal.

We call δ *unlucky in* $\mathbb{X}$ *with respect to* $\mathcal{C}$ iff $\delta \in \mathbb{X}$ and $\delta(t) \notin \mathbb{X}$ for each closed term t of $\mathcal{L}_{\mathsf{FOL}}$. If a δ-formula belongs to $\mathbb{X}$ and is not unlucky in $\mathbb{X}$ with respect to $\mathcal{C}$, then we call it *lucky in* $\mathbb{X}$ *with respect to* $\mathcal{C}$.

As we have seen, the presence of unlucky δ-formulas is possible, but since $\mathcal{C}$ is an L-consistency property, the assumption that no instance of δ belongs to $\mathbb{X}$ yields the unfortunate situation described above. The following holds then:

(d) For each δ that is unlucky in $\mathbb{X}$ with respect to $\mathcal{C}$, there are infinitely many instances, D, of δ such that $\{\delta, D\}$ $\textsf{sub}$ $\mathcal{C}$. Each such instance D of an unlucky δ will be called *favorable*.
(e) If δ is unlucky in $\mathbb{X}$ with respect to $\mathcal{C}$, then each of the favorable instances of δ is further in (2.19) than δ.
(f) If l is a position of a favorable instance, D, of an unlucky δ in (2.19), then $X_{l+1} = X_l$, because, by assumption, $X_l \cup \{D\}$ $\textsf{nosub}$ $\mathcal{C}$.
(g) If δ is unlucky in $\mathbb{X}$ and F_l, F_m are two favorable instances of δ such that $l < m$, then $X_l \cup \{B_m\}$ $\textsf{sub}$ $\mathcal{C}$.

Since the consistency property is settled and it cannot change in the course of this reasoning, we skip the relativization to $\mathcal{C}$ and say simply that a *formula is (un)lucky in* $\mathbb{X}$ (as we did in points (f) and (g) above). Let us observe, however, that the construction of $\mathbb{X}$ depends not only on the initial X in $\mathcal{C}$, but also on (2.19); thus, the concept of a δ-formula (un)lucky in $\mathbb{X}$ hides a relativization to (2.19) as well.

The remaining part of the proof consists of redefining (2.19); each modification will influence (2.21), and thus also the union $\mathbb{X}$.

Let us recall that infinite sequences like (2.19) are bijective functions from $\mathbb{N}_0$ to $FORM_{\textsf{FOL}}$. The swapping operation, formally, consists of setting

$$l_1 \mapsto \delta(t_2) \text{ and } l_2 \mapsto \delta(t_1),$$

where previously we had

$$l_1 \mapsto \delta(t_1) \text{ and } l_2 \mapsto \delta(t_2).$$

This obviously leads to a new sequence. We will write '(2.19)[$l_1 \Leftrightarrow l_2$]' for the result of this operation. We need to perform this operation in some systematic manner, however.

The following scheme, repeating (2.20), will be used. Let

$$A_0, A_1, \ldots, A_k, \ldots \tag{2.24}$$

Assume that set Y_0 of sentences of $\mathcal{L}_{\textsf{L}}$ is provided. For each $k \in \mathbb{N}_0$:

$$Y_{k+1} = \begin{cases} Y_k \cup \{A_k\} & \text{if } Y_k \cup \{A_k\} \textsf{ sub } \mathcal{C}, \\ Y_k & \text{otherwise.} \end{cases} \tag{2.25}$$

In this way, we obtain

$$Y_0 \subseteq Y_1 \subseteq \cdots \subseteq Y_k \subseteq Y_{k+1} \subseteq \cdots \tag{2.26}$$

Finally, let $\mathbb{Y}$ stand for the union of this sequence, that is: $\mathbb{Y} = \bigcup_{k=0}^{\infty} Y_k$.

In the next step, we shall define an infinite sequence of five-tuples. In the base case, the five-tuple is: $\langle s^0, S^0, i_0, \mathbb{Z}_0, Z_0 \rangle$, where

- s^0 is the initially considered sequence (2.19) of sentences of $\mathcal{L}_{\mathsf{L}}$,
- S^0 is the initially considered sequence (2.21) of sets of sentences,
- $i_0 = 0$; this is a counter whose role is to keep track of the largest position in s^0 that *should not* be modified by swapping—the operation can be performed only on formulas that lie further in s^0 than i_0,
- $\mathbb{Z}_0 = \mathbb{X}$, that is, the initially considered union of (2.21),
- $Z_0 = X$.

In the recursive case, we define the $k+1$st five-tuple $\langle s^{k+1}, S^{k+1}, i_{k+1}, \mathbb{Z}_{k+1}, Z_{k+1}\rangle$ for $k \in \mathbb{N}_0$. In each line of (2.27), B_k refers to the sequence s^k of sentences carried by the last defined five-tuple, that is, B_k is the $k+1$st element of s^k. For each $k \in \mathbb{N}_0$, the $k+1$st tuple is given by:

$$\langle s^{k+1}, S^{k+1}, i_{k+1}, \mathbb{Z}_{k+1}, Z_{k+1}\rangle = \tag{2.27}$$

$$= \begin{cases} \langle s^k, S^k, i_{k+1}, \mathbb{Z}_k, Z_k \cup \{B_k\}\rangle & \text{if (i) `safe case'}, \\ \langle s^k, S^k, i_{k+1}, \mathbb{Z}_k, Z_k \cup \{B_k\}\rangle & \text{if (ii) } B_k \text{ is a } \delta\text{-formula} \\ & \text{lucky in } \mathbb{Z}_k \\ \langle s^{k+1}, S^{k+1}, i_{k+1}, \mathbb{Z}_{k+1}, Z_k \cup \{B_k\}\rangle & \text{if (iii) } B_k \text{ is a } \delta\text{-formula} \\ & \text{unlucky in } \mathbb{Z}_k \\ \langle s^k, S^k, i_{k+1}, \mathbb{Z}_k, Z_k\rangle & \text{(iv) otherwise.} \end{cases}$$

'Safe case' in the first line means that B_k is not a δ-formula and $Z_k \cup \{B_k\}$ $\mathsf{sub}\, \mathcal{C}$. In this situation, sequence s^k is not altered, nor are S^k and $\mathbb{Z}_k$. The 'swapping counter' i_{k+1} equals i_k if $i_k \geqslant k+1$, and $i_{k+1} = k+1$ otherwise. Let us explain that the definition of the 'swapping counter' must warrant that if B_k is added to the constructed set, then the counter does not indicate a position smaller than k, for this would open a possibility of losing B_k at some further stage of our construction.

In line (ii), B_k is a δ-formula lucky in $\mathbb{Z}_k$, which means that there is an instance, A_l, of the δ-formula such that (h) $Z_l \cup \{A_l\}$ $\mathsf{sub}\, \mathcal{C}$. As in the previous case, there is no need to modify s^k, S^k, or $\mathbb{Z}_k$. As for the swapping counter, i_{k+1}, let us observe that if $l < k$, then $A_l \in Z_k$ (the reason here is the same as in the argument leading to (a)–(c) above). In this situation, $i_{k+1} = i_k$ if $i_k \geqslant k+1$, and $i_{k+1} = k+1$ otherwise. Conversely, if $l > k$, then $i_{k+1} = i_k$ if $i_k \geqslant l$, and $i_{k+1} = l$ otherwise. This ensures that possible further swappings do not disrupt (h).

For the definition of the third line of (2.27), let B_l and B_m represent the first and second instances of B_k favorable in $\mathbb{Z}_k$ appearing in s_k after position i, i.e., $l, m > i$, $l < m$. Then:

- $s^{k+1} = s^k[l \Leftrightarrow m]$,
- S^{k+1} is a sequence of sets (2.26) defined by scheme (2.25), where we substitute s^{k+1} for (2.24) and set $Y_0 = Z_k$,
- $i_{k+1} = i_k$ if $i_k \geqslant m$, and $i_{k+1} = m$ otherwise,
- $\mathbb{Z}_{k+1}$ is the union of $(S^{k+1})_\infty$.

In the fourth line of (2.27), only the 'swapping counter' can change: $i_{k+1} = i_k$ if $i_k \geqslant k + 1$, and $i_{k+1} = k + 1$ otherwise.

Let us observe that

$$\text{if } k < l, \text{ then } Z_k \subseteq Z_l \tag{2.28}$$

(compare (2.22)). Let $\mathbb{Z}$ be the union of sequence $(Z_k)_\infty$:

$$\mathbb{Z} = \bigcup_{k=0}^{\infty} Z_k.$$

The final step of this proof is to demonstrate that $\mathbb{Z}$ is a downward-saturated superset of X. We begin by establishing the following points:

(I) For each formula $C \in FORM_{\mathsf{FOL}}$, there is a five-tuple $\langle s^k, S^k, i_k, \mathbb{Z}_k, Z_k \rangle$ in the defined sequence such that C is the kth term of sequence s^k.
(II) For each Z_k in the sequence, Z_k sub $\mathcal{C}$.
(III) The swapping counter of the kth five-tuple is never smaller than k.

For the proof of (I), consider the initial sequence (2.19). C occupies some position in (2.19), say l. If no swapping occurred between the base-case five-tuple and the $l + 1$st tuple that affected the position of formula C, then the $l + 1$st tuple satisfies (I) (i.e., $k = l$).

Suppose that swapping was performed, altering the position of C. In this case, the swapping counter is set at a position not smaller than that of C in the new sequence, say the position is m. Further swaps, if performed, cannot change the position of C, hence (I) holds for $k = m$.

(II) follows from the very construction of our sequences, which is based on (2.20) and (2.25). It is also easy to see that (III) holds in each case.

Now, suppose that $\kappa \in \mathbb{Z}$. Assume that for the kth five-tuple $\langle s^k, S^k, i_k, \mathbb{Z}_k, Z_k \rangle$, formula κ is the kth term of s^k—by (I), such a tuple exists. Assume also that for the lth five-tuple $\langle s^l, S^l, i_l, \mathbb{Z}_l, Z_l \rangle$, formula κ_0 is the lth term of s^l—again, by (I), such a tuple exists.

If $l < k$, then (III) ensures that κ_0 precedes κ also in sequence s^k. In this case, formula κ_0 is already in set Z_k, and thus also in $\mathbb{Z}$—the arguments for this conclusion are precisely as in the reasoning leading to (a)–(c) above.

Suppose that $k < l$. Then, since by (II) we have Z_l sub $\mathcal{C}$, and by (2.28) that $\kappa \in Z_l$, we find that $Z_l \cup \{\kappa_0\}$ sub $\mathcal{C}$, which yields $Z_{k+1} = Z_k \cup \{\kappa_0\}$. Thus, we have shown that $\mathbb{Z}$ satisfies clause 2 of Definition 2.13.

Let us stress that the following two circumstances are irrelevant to the conclusiveness of the above argument. First, both κ and κ_0 could have undergone the swapping operation 'before' the construction of the kth and lth five-tuples. (By 'before', we refer to the order of the sequence of five-tuples.) Second, in subsequent five-tuples, the positions of some formulas may change, but these changes can only affect positions greater than k and l. Similarly, these circumstances do not impact the arguments demonstrating that $\mathbb{Z}$ satisfies clauses 1 and 3–5 of Definition 2.13. In the reasoning presented earlier for the initial sum $\mathbb{X}$, we consider the corresponding five-tuple instead of the position of a formula in (2.19); the remainder of the reasoning remains the same.

Finally, let $\delta \in \mathbb{Z}$, and suppose that for the kth five-tuple, formula δ is the kth term of s^k. There are two possible cases: either δ is lucky in $\mathbb{Z}$ or it is not. In the first case, point (h) above guarantees that an instance of δ is in $\mathbb{Z}$. In the second case, the same conclusion follows from (g).

This concludes the proof. □

Model Existence Theorems

Theorem 2.3 (Propositional Model Existence Theorem) *If a set X of formulas of $\mathcal{L}_{\mathsf{CPL}}$ belongs to a propositional consistency property, then X has a* CPL*-model.*

Proof In the proof we make use of the above Lemma 2.10 and of Corollary 2.3.

Hence let $X \in \mathcal{C}$, where $\mathcal{C}$ is a CPL-consistency property. By Lemma 2.10, there is a downward-saturated superset of X, say it is $\mathbb{X}$. As may be expected, we define a CPL-model for the superset. By Corollary 2.3, it will be a CPL-model for X as well.

Just to recall, a CPL-model of a set is a Boolean valuation satisfying all the elements of the set. Let v be such that $v(p_i) = 1$ iff $p_i \in \mathbb{X}$. Let $A \in \mathbb{X}$. We show by induction on rank r of A that $v(A) = 1$.

For the base case, we recall that $\mathbb{X}$ satisfies clause 1 of Definition 2.13, hence if $r(A) = 0$ and A is a propositional variable, then $v(A) = 1$ by the definition of v, and if $A = \neg p_i$, then, by the aforementioned clause, $p_i \notin \mathbb{X}$, which yields that $v(A) = 1$.

The rest of the argument, that is, the inductive step, relies on the remaining clauses of Definition 2.13, and, as a matter of fact, there is nothing interesting to be proved. If we take an arbitrary $A \in \mathbb{X}$ of rank $n + 1$, then its respective component(s): first, belong to $\mathbb{X}$ as well, since $\mathbb{X}$ is downward-saturated; second, are of lower rank, and thus are true under v by IH; hence, third, their truth under v makes A true under v. □

The construction is the same in the first-order case, though some details become more subtle. First of all, we must use Herbrand models.

Theorem 2.4 (First-order Model Existence Theorem) *If a set X of sentences of $\mathcal{L}_{\mathsf{FOL}}$ belongs to a first-order consistency property, then there is a Herbrand model for $\mathcal{L}_{\mathsf{FOL}}$ that is a model of X.*

Proof Assume that $\mathcal{C}$ is a first-order consistency property, and let $X \in \mathcal{C}$. As in the proof of the previous theorem, we show that there is a Herbrand model for the superset $\mathbb{X} \supseteq X$ constructed as in the proof of Lemma 2.10, and then we make use of Corollary 2.3 to conclude that it is a FOL-model for X.

Let $\mathcal{H} = \langle U, f \rangle$ be such that for each predicate P_i^n and each n-tuple of closed terms $\langle t_1, \ldots, t_n \rangle$,

$$\langle t_1, \ldots, t_n \rangle \in f(P_i^n) \text{ iff } P_i^n(t_1, \ldots, t_n) \in \mathbb{X}. \tag{2.29}$$

We show that $\mathcal{H} \models A$ by induction with respect to rank r of formula $A \in \mathbb{X}$.

Base case. Assume that A is an atomic sentence. Let us recall that for atomic formulas, $\mathcal{H} \models A$ iff A is satisfied by every object assignment σ in $\mathcal{H}$. However, formulas in $\mathbb{X}$ are sentences only, therefore the notion of truth in $\mathcal{H}$ does not depend on σ. More technically, for a closed term t, for each σ,

$$t^{\mathcal{H}}[\sigma] = f(t) = t.$$

Therefore, for atomic sentences:

$$\mathcal{H} \models P_i^n(t_1, \ldots, t_n) \text{ iff } f(P_i^n)(t_1, \ldots, t_n)$$

and hence, by (2.29),

$$\mathcal{H} \models P_i^n(t_1, \ldots, t_n) \text{ iff } P_i^n(t_1, \ldots, t_n) \in \mathbb{X}.$$

For sentences of rank 0 of the form '$\neg P_i^n(t_1, \ldots, t_n)$', we recall, as in the proof of Theorem 2.3, that if $\neg P_i^n(t_1, \ldots, t_n) \in \mathbb{X}$, then $P_i^n(t_1, \ldots, t_n) \notin \mathbb{X}$. Thus, in each case, if $A \in \mathbb{X}$ and $r(A) = 0$, then $\mathcal{H} \models A$.

As in the proof of the previous theorem, the inductive part follows from the fact that $\mathbb{X}$ is a downward-saturated set. In the case of γ- and δ-formulas, we apply Corollary 2.18. □

Consistency with Respect to $\mathbb{E}$

The next task in this section is to define the notion of consistency as it pertains to formulas occurring in a sequent. We aim for a notion such that a formula '$-A$' is consistent in this sense iff there is no proof of A. This will work perfectly for sequents with a non-empty accident; that is, we expect a formula '$-A$' to be consistent iff there is no Socratic proof of '$\vdash A$'. If we are to preserve the intuitions we associate with consistency, then moving to the left side of a sequent, we should say that a formula A is consistent iff there is no Socratic proof of '$A \vdash$'. This is in line with Corollary 2.4, stated in Sect. 2.4.

However, one difficulty that arises here is the need to generalize the notion of consistency in order to account for every sequent '$S \vdash T$' and every calculus $\mathbb{E}$. Moreover, the notion must be such that we can smoothly transition between sets in consistency properties and sequences in sequents. For these reasons, we need a custom-made notion of representation between sets of sentences and sequents.

We will need the following auxiliary notion. If X is a set of formulas of $\mathcal{L}_{\mathsf{L}}$, then by '$-X$' we mean:

$$-X = \{-A : A \in X\}.$$

Definition 2.14 (*Set-Representation*) Let ϕ stand for a sequent of the language $\mathcal{L}^?_{\vdash\mathsf{L}}$. We say that X *is a set-representation of sequent* ϕ iff

- $X = \mathsf{set}(S)$ for a left-sided sequent $\phi = S \vdash$;
- $X = -\mathsf{set}(T)$ for a right-sided sequent $\phi = \vdash T$;
- $X = \mathsf{set}(S) \cup \{-A\}$ for a both-sided sequent $\phi = S \vdash A$.

If X is a set-representation of ϕ, then we also say that ϕ *is represented by* X.

The three above cases can be generalized as follows.

Corollary 2.19 *Let '$S \vdash T$' stand for a sequent of the language $\mathcal{L}^?_{\vdash\mathsf{L}}$. X is a set-representation of '$S \vdash T$' iff $X = \mathsf{set}(S) \cup -\mathsf{set}(T)$.*

Let us observe that set-representations of sequents are finite sets, as sequents are finite structures. Moreover,

Corollary 2.20 *Every finite and non-empty set of sentences of the language $\mathcal{L}_\mathsf{L}$ represents a both-sided, a right-sided and a left-sided sequent of the language $\mathcal{L}^?_{\vdash\mathsf{L}}$. Furthermore, the empty set represents no sequent.*

The fact that the empty set represents nothing follows from our prohibition of empty sequents in the language. The easiest way to construct a proof of the corollary is to define an ordering of the sentences of the language $\mathcal{L}_\mathsf{L}$—for example, like (2.19) in the proof of Lemma 2.10—and then arrange the elements of the set according to this order. Afterward, place them into the sequent of the expected form, adding the negation sign when a formula is placed in the succedent. In this way, we would obtain a uniquely determined form of a sequent represented by a given finite set of sentences.

On the other hand, since both the order and the number of occurrences are neglected when moving to sets, a finite set is a set-representation of infinitely many sequents of all three types.

The following lemma expresses the fact that the provability of a sequent depends on its representation, and once this is settled, the specific form of the sequent becomes negligible.

Lemma 2.11 (Representation and Provability) *Let X stand for a finite and non-empty set of sentences of the language $\mathcal{L}_\mathsf{L}$. A both-sided sequent represented by X is provable in $\mathbb{E} \in \{\mathbb{E}^{\mathbf{Q}}, \mathbb{E}^{\mathbf{P}}\}$ iff **every** both-sided sequent represented by X is provable in $\mathbb{E}$.*

*Similarly, a right-sided (left-sided) sequent represented by X is provable in $\mathbb{E} \in \{\mathbb{E}^{\mathbf{RQ}}, \mathbb{E}^{\mathbf{RP}}\}$ (in $\mathbb{E} \in \{\mathbb{E}^{\mathbf{LQ}}, \mathbb{E}^{\mathbf{LP}}\}$) iff **every** right-sided (left-sided) sequent represented by X is provable in the respective $\mathbb{E}$.*

Proof Once the structural analysis from the previous section has been performed, this result follows easily from the lemmas concerning transparency of double negation, weakening, contraction, and switching.

Let $\mathbb{E}$ be fixed, and let X stand for a finite and non-empty set of sentences of the language $\mathcal{L}_\mathsf{L}$. The only-if part is guaranteed by excluding the empty set and by Corollary 2.20. For the if-then direction, let '$S \vdash T$' and '$U \vdash V$' represent two both-sided (right- or left-sided) sequents of the language $\mathcal{L}^{?}_{\vdash\mathsf{L}}$, both represented by X. We will show that if there is a Socratic proof of '$S \vdash T$' in the respective calculus $\mathbb{E}$, then there is also a Socratic proof of '$U \vdash V$' in $\mathbb{E}$.

Hence, suppose that there is a Socratic proof of '$S \vdash T$' in $\mathbb{E}$. The details depend on the differences between '$S \vdash T$' and '$U \vdash V$'. We list the possible cases.

(a) The sequents are left-sided. Since they are both represented by X, it follows that $\mathsf{set}(S) = \mathsf{set}(U)$.
(b) The sequents are right-sided. In this case, $X = -\mathsf{set}(T) = -\mathsf{set}(V)$.
(c) The sequents are both-sided, of the forms '$S \vdash A$' and '$U \vdash B$', and we have: $X = \mathsf{set}(S) \cup \{-A\} = \mathsf{set}(U) \cup \{-B\}$. This gives rise to the following subcases:

 (c.1) $\mathsf{set}(S) = \mathsf{set}(U)$ and $A = B$.
 (c.2) $\mathsf{set}(S) \neq \mathsf{set}(U)$ and $A = B$.
 (c.3) $A \neq B$.

Case (a). The sequent '$S \vdash$' has a Socratic proof. On the other hand, S is a non-empty proper subsequence of the concatenation, '$S\,'\,U$', of S and U. Therefore, by Lemma 2.4 (weakening), there is a Socratic proof of the sequent '$S\,'\,U \vdash$'. Moreover, since the sequences S, U, and '$S\,'\,U$' all share the same set of terms, Corollary 2.16 (admissibility of contraction in the general form) ensures that there is also a Socratic proof of the sequent '$U \vdash$'. The reasoning is the same in case (c.1), where the sequents have the same succedent.

Case (b). The only difficulty here is that the equation $-\mathsf{set}(T) = -\mathsf{set}(V)$ does not necessarily imply $\mathsf{set}(T) = \mathsf{set}(V)$. According to Corollary 2.1, clause 3, the possible difference between these sets is due to doubly negated formulas.

More specifically, if $-\mathsf{set}(T) = -\mathsf{set}(V)$ but $\mathsf{set}(T) \neq \mathsf{set}(V)$, then by Corollary 2.1, clause 3, the two sets can be represented as follows:

$$\mathsf{set}(T) = X_0 \cup \{A_1, \dots, A_k\} \cup \{\neg\neg B_1, \dots, \neg\neg B_n\},$$

$$\mathsf{set}(V) = X_0 \cup \{\neg\neg A_1, \dots, \neg\neg A_k\} \cup \{B_1, \dots, B_n\}.$$

By successive applications of Lemma 2.5 (double negation), we conclude that there is a Socratic proof of a sequent '$\vdash T^*$' such that $\mathsf{set}(T^*) = \mathsf{V}$. The further reasoning follows the same pattern as in case (a).

Subcase (c.2). Here, the sequents take the forms '$S \vdash A$' and '$U \vdash A$'. Since $\mathsf{set}(S) \cup \{-A\} = \mathsf{set}(U) \cup \{-A\}$, the sets $\mathsf{set}(S)$ and $\mathsf{set}(U)$ must differ with respect to $-A$, that is, either $\mathsf{set}(S) \setminus \mathsf{set}(U) = \{-A\}$ or $\mathsf{set}(U) \setminus \mathsf{set}(S) = \{-A\}$.

In the first case, the provable sequent is of the form '$S(-A) \vdash A$'. By Lemma 2.9 (contraction), its sw-left-contraction, '$S(\_) \vdash A$', with no occurrence of '$-A$' on the left, is provable. In the second case, we begin with Lemma 2.4 (weakening) and add '$-A$' to the antecedent of the sequent $S \vdash A$. In both cases, we thus arrive at (c.1).

Subcase (c.3). Further possibilities are: (c.3.1) $-A = -B$, (c.3.2) $-A \neq -B$. In (c.3.1), Corollary 2.1, clause 3, implies that one of A or B is a double negation of the other. Thus, we apply Lemma 2.5 to conclude that '$S \vdash A$' is provable in $\mathbb{E}$, and we arrive at case (c.1).

In (c.3.2), it must be that '$-B$' is a term of S (and '$-A$' is a term of U). Hence, the provable sequent is of the form '$S(-B) \vdash A$'. By Lemma 2.7 (Switching), '$S(-A) \vdash B$' is provable in $\mathbb{E}$, which brings us back to one of the earlier cases. □

Finally, a finite set X of sentences will be called *consistent with respect to* $\mathbb{E}$ iff it is not a set-representation of a sequent provable in $\mathbb{E}$. Since finite sets of sentences do represent some sequents, those that are consistent represent only unprovable sequents. It is important to assume compatibility between the form/sidedness of the sequent and the calculus $\mathbb{E}$. More formally:

Definition 2.15 (*Set Consistent w.r.t* $\mathbb{E}$) Let X stand for a finite set of sentences of the language $\mathcal{L}_{\mathsf{L}}$. We say that X is:

- *consistent with respect to* $\mathbb{E}^{\mathbf{Q}}$ ($\mathbb{E}^{\mathbf{P}}$) iff every both-sided sequent of the language $\mathcal{L}^{?}_{\vdash\mathsf{L}}$ represented by X is unprovable in $\mathbb{E}^{\mathbf{Q}}$ ($\mathbb{E}^{\mathbf{P}}$);
- *consistent with respect to* $\mathbb{E}^{\mathbf{RQ}}$ ($\mathbb{E}^{\mathbf{RP}}$) iff every right-sided sequent of the language $\mathcal{L}^{?}_{\vdash\mathsf{L}}$ represented by X is unprovable in $\mathbb{E}^{\mathbf{RQ}}$ ($\mathbb{E}^{\mathbf{RP}}$);
- *consistent with respect to* $\mathbb{E}^{\mathbf{LQ}}$ ($\mathbb{E}^{\mathbf{LP}}$) iff every left-sided sequent of the language $\mathcal{L}^{?}_{\vdash\mathsf{L}}$ represented by X is unprovable in $\mathbb{E}^{\mathbf{LQ}}$ ($\mathbb{E}^{\mathbf{LP}}$).

It follows that the empty set is consistent with respect to every $\mathbb{E}$.

Finally, the following lemma leads us back to L-consistency properties.

Lemma 2.12 *The family of all finite sets X of sentences of $\mathcal{L}_{\mathsf{L}}$ such that X is consistent with respect to $\mathbb{E}^{\mathbf{Q}}$ ($\mathbb{E}^{\mathbf{P}}$) is an* L*-consistency property. Similarly, the family of all finite sets X of sentences of $\mathcal{L}_{\mathsf{L}}$ such that X is consistent with respect to $\mathbb{E}^{\mathbf{RQ}}$ ($\mathbb{E}^{\mathbf{RP}}$) is an* L*-consistency property. Finally, the family of all finite sets X of sentences of $\mathcal{L}_{\mathsf{L}}$ such that X is consistent with respect to $\mathbb{E}^{\mathbf{LQ}}$ ($\mathbb{E}^{\mathbf{LP}}$) is also an* L*-consistency property.*

Proof We proceed with the proof for $\mathbb{E} \in \{\mathbb{E}^{\mathbf{Q}}, \mathbb{E}^{\mathbf{P}}\}$ in detail, and we will note some minor complications for the right sided case. The proof for the left-sided case is exactly the same as for both-sided sequents.

Let $\mathcal{F}$ denote the family of all sets that are consistent with respect to $\mathbb{E} \in \{\mathbb{E}^{\mathbf{Q}}, \mathbb{E}^{\mathbf{P}}\}$, and assume that $X \in \mathcal{F}$. The empty set is in $\mathcal{F}$, but it trivially satisfies every condition of the definition of an L-consistency property. Assume X is non-empty. We will review the clauses of Definition 2.12 (first-order consistency property). In each case, we use Lemma 2.11, which allows us to define the form of the sequent freely.

Clause 1: Assume that $\{A, -A\} \subseteq X$, where A is a propositional variable or atomic sentence. By Definition 2.15, each sequent represented by X is unprovable in $\mathbb{E}$, and by Corollary 2.20, there exists a sequent ϕ represented by X. Given this assumption, ϕ must be of the form '$S(A)(-A) \vdash B$' or '$S(A) \vdash A$'. It is clear that in both cases, the one-term sequence of questions $\langle ?(\phi) \rangle$ is already a Socratic proof of ϕ in $\mathbb{E}$. Thus, we reach a contradiction. It follows that if $A \in X$, then $-A \notin X$.

For the right-sided case, starting with the assumption that $\{A, -A\} \subseteq X$, and using Definition 2.15, Corollary 2.20, and Lemma 2.11, we conclude that the sequent of the form '$\vdash T(\neg A)(A)$', where $X = -\mathsf{set}(T(\neg A)(A))$, is not provable in $\mathbb{E}$. This also leads to a contradiction.

Clause 2. Assume that (i) $\kappa \in X$, but (ii) $X \cup \{\kappa_0\} \notin \mathcal{F}$. From assumption (ii), it follows that there is a sequent represented by $X \cup \{\kappa_0\}$ that has a Socratic proof in $\mathbb{E}$. Lemma 2.11 then implies that *every sequent* represented by X is provable in $\mathbb{E}$. In particular, a sequent of the form '$S(\kappa_0) \vdash A$', such that κ_0 occurs exactly once in the sequent (and $X \cup \{\kappa_0\} = \mathsf{set}(S(\kappa_0)) \cup \{-A\}$), has a Socratic proof in $\mathbb{E}$.[28] However, the pair $?(S(\kappa) \vdash A)/?(S(\kappa_0) \vdash A)$ is an instance of rule L_κ. Thus, by Corollary 2.7, there is a Socratic proof of '$S(\kappa) \vdash A$' in $\mathbb{E}$. But the sequent '$S(\kappa) \vdash A$' is represented by X. This contradicts the assumption (i). Therefore, (i) implies that $X \cup \{\kappa_0\} \in \mathcal{F}$.

For the right-sided sequents, assumption (ii) implies that there is a sequent of the form '$\vdash T(\neg\kappa_0)$', represented by X, with exactly one occurrence of '$\neg\kappa_0$', and that this sequent is provable in $\mathbb{E}$. The details here depend on '$\neg\kappa_0$'. For example, suppose the κ-formula is of the form '$\forall x_i \neg A$', where x_i is not free in '$\neg A$'. From assumption (ii), the sequent '$\vdash T(\neg\neg A)$' has a Socratic proof in $\mathbb{E}$. By Lemma 2.5 (double negation), the sequent '$\vdash T(A)$' also has a Socratic proof in $\mathbb{E}$. Then, by Corollary 2.7, with rule R_κ applied, the sequent '$\vdash T(\neg\forall x_i \neg A)$' has a Socratic proof in $\mathbb{E}$ as well, but this sequent is represented by X, leading to a contradiction. The other cases proceed similarly.

All the remaining cases are equally easy to get along with. If possible, the considered formulas are assumed to occur in the antecedent, so there is no need to play with the negation, when $-A \in X$. The reasoning for clause 3 of Definition 2.12, concerning α-formulas, is very similar to that for κ-formulas; we skip it.

For clause 4, the both-sided case, assume that (i) $\beta \in X$, but (ii) neither $X \cup \{\beta_0\} \in \mathcal{F}$ (iii) nor $X \cup \{\beta_1\} \in \mathcal{F}$. Assumptions (ii) and (iii) imply that there are sequents '$S(\beta_0) \vdash A$' and '$S(\beta_1) \vdash A$', both with exactly one occurrence of the respective β_i, both provable in $\mathbb{E}$, and such that the sequents are represented by, respectively, $X \cup \{\beta_0\}$ and $X \cup \{\beta_1\}$. By Lemma 2.3 (Separation), there is a successful Socratic transformation of the question '$?(S(\beta_0) \vdash A\ ;\ S(\beta_1) \vdash A)$'. Moreover, the pair

$$?(\ S(\beta) \vdash A\ )\ /\ ?(\ S(\beta_0) \vdash A\ ;\ \ S(\beta_1) \vdash A\ )$$

is an instance of rule L_β. Hence, by Corollary 2.7, there is a Socratic proof of sequent '$S(\beta) \vdash A$' in $\mathbb{E}$. Thus, assumption (i) implies that $X \cup \{\beta_i\} \in \mathcal{F}$ for at least one $i \in \{0, 1\}$. This finishes the propositional case.

Quantifier cases, clause 5: Let (i) $\gamma \in X$ and assume that (ii) there is a closed term t such that $X \cup \{\gamma(t)\} \notin \mathcal{F}$. By assumptions (i), (ii), and Lemma 2.11, there is a sequent '$S(\gamma, \gamma(t)) \vdash A$' provable in $\mathbb{E}^{\mathbf{Q}}$, represented by X. Assume also that $\gamma(t)$ occurs exactly once in the sequent. By Corollary 2.7, using a suitable instance of L_γ, we conclude that the sequent '$S(\gamma) \vdash A$' is provable, which contradicts (i).

---

[28] A minor remark is needed here: since $\kappa \in X$, the set $X \cup \{\kappa_0\}$ has at least two elements, which ensures that it can represent the sequent $S(\kappa_0) \vdash A$.

Finally, clause 6. Let (i) $\delta \in X$, and assume that (ii) for every closed term t, $X \cup \{\delta(t)\} \notin \mathcal{F}$. Assumption (ii) holds, in particular, for every individual constant a_i that is not present in any formula in X. Let us observe that such individual constants exist, as X is a finite set; hence let a_i stand for such a constant (we can be more precise and say it is the first one in some alphabetical order, like we did before). From (ii), we have $X \cup \{\delta(a_i)\} \notin \mathcal{F}$. We take a sequent represented by X with exactly one occurrence of $\delta(a_i)$, and, as before, demonstrate that its provability results in the provability of a sequent represented by X, which contradicts (i).

In the right-sided case, clauses 5 and 6 require, additionally, Corollary 2.1, clause 5, and possibly Lemma 2.5. □

Completeness

We are finally ready to prove completeness theorems.

Theorem 2.5 (Completeness of $\mathbb{E}$ w.r.t the Semantics of Sequents) *If a closed sequent '$S \vdash T$' is correct under every* L*-interpretation of* $\mathcal{L}_{\mathsf{L}}$*, then there exists a Socratic proof of the sequent in the appropriate* $\mathbb{E}$.

Proof Assume '$S \vdash T$' has no Socratic proof in $\mathbb{E}$. Then, by Lemma 2.11, the set-representation

$$X = \mathsf{set}(S) \cup -\mathsf{set}(T)$$

of this sequent belongs to the family $\mathcal{F}$ of all sets consistent with respect to $\mathbb{E}$. By Lemma 2.12, $\mathcal{F}$ is an L-consistency property. Finally, by Theorems 2.3 or 2.4, it has a CPL/a Herbrand model. This contradicts the assumption that '$S \vdash T$' is correct under every L-interpretation of $\mathcal{L}_{\mathsf{L}}$. □

The following theorem is a direct consequence of the previous one.

Theorem 2.6 (Completeness of $\mathbb{E}$ w.r.t the Semantics of L) *If a sentence A of the language* $\mathcal{L}_{\mathsf{L}}$ *is* L*-valid, then:*

1. *there exists a Socratic proof of the sequent '$\vdash A$' in* $\mathbb{E}^{\mathbf{Q}}/\mathbb{E}^{\mathbf{P}}$,
2. *there exists a Socratic proof of the sequent '$\vdash A$' in* $\mathbb{E}^{\mathbf{RQ}}/\mathbb{E}^{\mathbf{RP}}$,
3. *there exists a Socratic proof of the sequent '$-A \vdash$' in* $\mathbb{E}^{\mathbf{LQ}}/\mathbb{E}^{\mathbf{LP}}$.

The results obtained in this and the previous sections sum up to the following corollary expressing the adequacy of $\mathbb{E}$ with respect to the semantics of sequents.

Corollary 2.21 (Adequacy of $\mathbb{E}$ w.r.t the Semantics of Sequents) *A closed sequent of the language* $\mathcal{L}^{?}_{\vdash\mathsf{L}}$ *is correct under every* L*-interpretation of* $\mathcal{L}_{\mathsf{L}}$ *iff there is a Socratic proof of the sequent in the appropriate* $\mathbb{E}$.

Proof By Theorems 2.2 and 2.5. □

Finally, when considering derivability and admissibility, we anticipate the following corollary, Corollary 2.22. However, since all proof-generating mechanisms have been defined for *closed* sequents, the restriction to closed sequents will appear in the corollary as well.

Hence let $\mathbf{r}$ represent an erotetic rule. Recall that, set-theoretically, a rule is a set of pairs (in our case, pairs of questions). Let $\mathbb{Q}^{\mathsf{C}}_{\mathsf{FOL}}$ denote the set of questions of $\mathcal{L}^{?}_{\vdash\mathsf{FOL}}$ whose constituents are exclusively closed sequents of $\mathcal{L}^{?}_{\vdash\mathsf{FOL}}$. By the *closed restriction of* $\mathbf{r}$, denoted $c(\mathbf{r})$, we mean:

$$c(\mathbf{r}) = \mathbf{r} \cap (\mathbb{Q}^{\mathsf{C}}_{\mathsf{FOL}} \times \mathbb{Q}^{\mathsf{C}}_{\mathsf{FOL}}).$$

Naturally, in the case of $\mathsf{L} = \mathsf{CPL}$, $c(\mathbf{r}) = \mathbf{r}$. As we will see, the fact that the second clause of the corollary is proven through semantic arguments (correctness of erotetic rules and completeness of erotetic calculi) requires us to refer to the closed restrictions of rules, whereas clause 1 does not require such restrictions.

Corollary 2.22 *If* $\mathbf{r}$ *is a rule derivable in* $\mathbb{E}$*, then:*

1. $\mathbf{r}$ *is semi-admissible in* $\mathbb{E}$*,*
2. $c(\mathbf{r})$ *is admissible in* $\mathbb{E}$*.*

Proof Suppose an erotetic rule $\mathbf{r}$, of the form Q/Q^*, is derivable in $\mathbb{E}$. Then, there exists a Socratic transformation $\langle Q_1, \ldots, Q_n\rangle$ in $\mathbb{E}$ such that $Q_1 = Q$ and $Q_n = Q^*$. Now assume that $\mathbf{s} = \langle s_1, \ldots, s_k\rangle$ is a successful Socratic transformation of question Q^* in $\mathbb{E}$. The sequence $\langle Q_1, \ldots, Q_n, s_2, \ldots, s_k\rangle$ then forms a successful Socratic transformation of question Q in $\mathbb{E}$. This demonstrates that clause 1 of Definition 2.8 is satisfied, meaning that $\mathbf{r}$ is semi-admissible in $\mathbb{E}$.

The case of clause 2 is more nuanced. Consider the closed restriction $c(\mathbf{r}) = Q/Q^*$ of $\mathbf{r}$. Assume that $\mathbf{s} = \langle Q_1, \ldots, Q_n\rangle$ is a Socratic transformation in $\mathbb{E}$, where $Q_1 = Q$ and $Q_n = Q^*$. Further, assume there exists a successful Socratic transformation of Q in $\mathbb{E}$. Suppose that question Q takes the form '$?(\phi_1, \ldots, \phi_m)$'. By assumption, each ϕ_i, for $i : 1 \leqslant i \leqslant m$, is a closed sequent. By Lemma 2.3 (Separation), for each i, there is a successful Socratic transformation of the question '$?(\phi_i)$'. By the soundness of $\mathbb{E}$ (Theorem 2.2), this implies that for each i, ϕ_i is a sequent that is correct under every L-interpretation. Since only rules from $\mathbb{E}$, which are correct (see Lemmas 2.1 and 2.2), were used in the construction of $\mathbf{s}$, each constituent of question Q_2 is also correct under every L-interpretation. By induction (with respect to n), every constituent ϕ^*_j of question Q^* is a sequent correct under every L-interpretation. Furthermore, since ϕ^*_i is a closed sequent, and by the completeness of $\mathbb{E}$ (Theorem 2.5), each such sequent has a Socratic proof in $\mathbb{E}$. Once again, by Lemma 2.3, there exists a successful Socratic transformation of question Q^* in $\mathbb{E}$. This completes the proof. □

We expect the following to be true:

Corollary 2.23 *If* $\mathbf{r} \in \mathbb{E}$*, then* $\mathbf{r}$ *is both derivable and semi-admissible in* $\mathbb{E}$*; moreover,* $c(\mathbf{r})$ *is admissible in* $\mathbb{E}$*.*

Proof Clearly, if $\mathbf{r} = Q/Q^*$ is a rule of $\mathbb{E}$, then $\langle Q, Q^*\rangle$ is the Socratic transformation that shows that $\mathbf{r}$ is derivable in $\mathbb{E}$. By the previous corollary, $\mathbf{r}$ is also semi-admissible in $\mathbb{E}$, and $c(\mathbf{r})$ is admissible in $\mathbb{E}$. □

2.7 Proof-Search in Erotetic Calculi for FOL

Despite, or rather due to, the undecidability of FOL, any proof system for FOL should provide at least a sketch of a proof-search procedure. We develop such a procedure in this section, using some notions that are well-known from the literature on the subject and are fairly straightforward. First, in order to ensure that the procedure terminates, we use the notion of *quantifier depth* (abbreviated as 'Q-depth'), which is a formal parameter bounding the number of applications of the quantifier rules. It is assumed that the value of Q-depth is supplied by the user of the procedure. The notion was used in this way by Fitting [8, p. 162].

We begin by introducing some necessary terminology. We have already defined basic sequents (see Sect. 2.5). Another useful category is that of a permanently open sequent.[29] A sequent is called *permanently open* if it is not a basic sequent and it does not fall under the schema of a sequent-premise of any rule of the respective erotetic calculus $\mathbb{E}$. In other words, a permanently open sequent is an 'open' (not basic) sequent to which no rule is applicable. When a Socratic transformation starts with a closed sequent of the language $\mathcal{L}^?_{\vdash \mathsf{FOL}}$, the occurrence of a permanently open sequent will be rather rare, but rare does not mean impossible.

The application of quantifier rules must be subject to reasonable and intuitive restrictions. First, if there is a choice between the application of, for example, L_γ and L_δ from $\mathbb{E}^{\mathbf{Q}}$, then L_δ should be applied first. Second, the new constant introduced by the application of L_δ should not be chosen randomly. Third, once the new constants are introduced, rule L_γ should reuse them. The goal of this section is to formalize these restrictions into Algorithm 2.3.

From now on, we assume that the individual constants of the language $\mathcal{L}_{\mathsf{FOL}}$ are given in alphabetical order, e.g., as follows:

$$a_1, a_2, \ldots, a_n, \ldots \tag{2.30}$$

In Table 2.7, we present additional erotetic rules that will be used in Algorithm 2.3. Condition (con) expresses a proviso of the applicability of the rules. As we can see, these rules generalize the applications of the quantifier rules considered previously.

The first rule, L^*_γ, is to be added to $\mathbb{E}^{\mathbf{Q}}$. Rule L^*_γ for calculus $\mathbb{E}^{\mathbf{LQ}}$ is obtained from the presented L^*_γ by dropping 'C' from the succedents of sequents. Rule R^*_δ is designed for calculus $\mathbb{E}^{\mathbf{Q}}$, and rule R^{**}_δ for calculus $\mathbb{E}^{\mathbf{RQ}}$.

We show that each of the additional rules is derivable in the respective calculus $\mathbb{E}$. Corollary 2.22 implies that the closed restrictions of these rules are also admissible in $\mathbb{E}$.

Lemma 2.13 *The generalized quantifier rules are derivable and admissible in the respective erotetic calculi. More specifically:*

---

[29] We have borrowed this term from [32], where the notion of 'permanently unsuccessful sequent' was used in the completeness proof of the original $\mathbf{E}^{\mathbf{PQ}}$.

Table 2.7 Generalized quantifier rules

$$\frac{?(\ \Phi\ (\ S(\gamma)\ \vdash C\ ))}{?(\ \Phi\ (\ S(\gamma, \gamma(t_n), \ldots, \gamma(t_1), \gamma(c^*))\ \vdash C\ ))}\ L^*_\gamma$$

$$\frac{?(\ \Phi\ (\ S(\_)\ \vdash\ \delta\ ))}{?(\ \Phi\ (\ S(-\delta, -\delta(t_n), \ldots, -\delta(t_1))\ \vdash\ \delta(c^*)\ ))}\ R^*_\delta$$

$$\frac{?(\ \Phi\ (\vdash S(\delta)\ ))}{?(\ \Phi\ (\vdash S(\delta, \delta(t_n), \ldots, \delta(t_1), \delta(c^*))\ ))}\ R^{**}_\delta$$

(con) $t_1, \ldots, t_n$ are *all (and only)* the closed terms which occur in the sequent-premise, and c^* is the first constant in the alphabetical order (2.30) which is new to the sequent-premise.

1. *L^*_γ is derivable in $\mathbb{E}^{\mathbf{Q}}$ and hence semi-admissible in $\mathbb{E}^{\mathbf{Q}}$; moreover, $c(L^*_\gamma)$ is admissible in $\mathbb{E}^{\mathbf{Q}}$;*
2. *L^*_γ, with empty-succedent sequents, is derivable in $\mathbb{E}^{\mathbf{LQ}}$, semi-admissible in $\mathbb{E}^{\mathbf{LQ}}$, and $c(L^*_\gamma)$ is admissible in $\mathbb{E}^{\mathbf{LQ}}$;*
3. *R^*_δ is derivable in $\mathbb{E}^{\mathbf{Q}}$, semi-admissible in $\mathbb{E}^{\mathbf{Q}}$, and $c(R^*_\delta)$ is admissible in $\mathbb{E}^{\mathbf{Q}}$; and*
4. *R^{**}_δ is derivable in $\mathbb{E}^{\mathbf{RQ}}$, semi-admissible in $\mathbb{E}^{\mathbf{RQ}}$, and $c(R^{**}_\delta)$ is admissible in $\mathbb{E}^{\mathbf{RQ}}$.*

Proof We will consider only the third case; the remaining three cases are analogous. Suppose that Q is a question of $\mathcal{L}^?_{\vdash\mathsf{FOL}}$ of the following form:

$$?(\ \Phi\ (\ S(\_)\ \vdash\ \delta\ ))$$

We apply the original rule R_δ of $\mathbb{E}^{\mathbf{Q}}$ using the individual constant c^* specified in condition (con) and obtain:

$$?(\ \Phi\ (\ S(-\delta)\vdash\ \delta(c^*)\ ))$$

Let $t_1, \ldots, t_n$ be all (and only) the closed terms that occur in the sequent '$S(\_)\ \vdash\ \delta$'. By clause 5 of Corollary 2.1, the formula '$-\delta$' is a γ-formula such that $\gamma(t_i) \in \{-\delta(t_i), \neg\delta(t_i)\}$. Thus, we apply rule L_γ n times, starting with the closed term t_1 and continuing up to t_n, arriving at:

$$?(\ \Phi\ (\ S(-\delta, \gamma(t_n), \ldots, \gamma(t_1))\ \vdash\ \delta(c^*)\ )).$$

If $\gamma(t_i) = -\delta(t_i)$, then we are fine. Suppose that $\gamma(t_i) = \neg\delta(t_i) \neq -\delta(t_i)$, then $\neg\delta(t_i) = \neg\neg - \delta(t_i)$. We apply L_κ to the last question:

$$\frac{?(\ \Phi\ (\ S(-\delta, \gamma(t_n), \ldots, \neg\neg - \delta(t_i), \ldots, \gamma(t_1))\ \vdash\ \delta(c^*)\ ))}{?(\ \Phi\ (\ S(-\delta, \gamma(t_n), \ldots, -\delta(t_i), \ldots, \gamma(t_1))\ \vdash\ \delta(c^*)\ ))}\ L_\kappa$$

We repeat this step until we arrive at:

$$?(\ \Phi\ (\ S(-\delta, -\delta(t_n), \ldots, -\delta(t_1)) \vdash \delta(c^*)\ )\ )$$

which finishes the derivability part. Semi-admissibility and admissibility of the closed restrictions follow from Corollary 2.22. □

We continue with the terminology needed for Algorithm 2.3. The rules R_β of $\mathbb{E}^{\mathbf{RQ}}$, L_α and R_β of $\mathbb{E}^{\mathbf{Q}}$, L_α of $\mathbb{E}^{\mathbf{LQ}}$, and all the R_κ, L_κ rules will be called *simple non-branching rules*, or *SNB-rules* for short. The SNB-rules either perform no operations on quantifiers or deal with empty quantification. The branching rules R_α of $\mathbb{E}^{\mathbf{RQ}}$, R_α and L_β of $\mathbb{E}^{\mathbf{Q}}$, and L_β of $\mathbb{E}^{\mathbf{LQ}}$ are called *B-rules*, for short. The remaining rules are non-branching quantifier rules. Among them, the rules R_γ of $\mathbb{E}^{\mathbf{RQ}}$, R_γ and L_δ of $\mathbb{E}^{\mathbf{Q}}$, and L_δ of $\mathbb{E}^{\mathbf{LQ}}$ will be called *N-rules* since they always introduce a new constant. The remaining quantifier rules are the original R_δ of $\mathbb{E}^{\mathbf{RQ}}$, R_δ and L_γ of $\mathbb{E}^{\mathbf{Q}}$, and L_γ of $\mathbb{E}^{\mathbf{LQ}}$, as well as the additional admissible rules presented in Table 2.7. These will be called *O-rules*, where 'O' stands for 'old'. The basic task of O-rules is to 'use' the already introduced closed terms; however, as we know, it may happen that the O-rules need to be used with a new constant (see Example 2.4).

We assume that when an N-rule is applied, the new constant introduced into the sequent-conclusion is not arbitrary, but is exactly the first one in the alphabetical order (2.30) that is new to the sequent-premise. We also assume that the rules L^*_γ, R^*_δ, and R^{**}_δ are used instead of the original O-rules. An application of any of these rules results in the introduction of a new constant which, again, is the first one in the alphabetical order (2.30) that is new to the relevant sequent.

The algorithm uses three procedures described below: SNB, B, and N-O. We begin with SNB(sequent, ϕ). The first parameter, 'sequent', is provided, and 'ϕ' is to be bound.

Procedure SNB(sequent, ϕ)

Data: sequent
$\phi \leftarrow$ sequent
while *ϕ is a sequent-premise of an SNB-rule* **r do**
| $\phi \leftarrow$ sequent-conclusion of **r**
end

All active formulas of SNB-rules are found before the while-loop exits.

In the B procedure, n represents the number of constituents of the current question (corresponding to the number of branches in a derivation). The variable r tracks the number of individual constants introduced so far during the derivation of a given sequent. Each constituent of the transformed question is assigned a number, which is recorded as a pair (sequent, number). The variable R stores the right context of the currently analyzed sequent. In 'B((sequent, Q-depth), R, n, (ϕ, r), R, n)', the pair '(sequent, Q-depth)' is provided, as well as the second and third parameter; the remaining parameters are determined when the procedure runs. It is important to note that a B-rule is applied at most once during the execution of the procedure.

Procedure B((sequent, Q-depth), $R, n, (\phi, r), R, n$)

```
Data: sequent
φ ← sequent
if φ is a sequent-premise of a B-rule r then
    φ ← the left sequent-conclusion of r
    ψ ← the right sequent-conclusion of r
    n ← n + 1
    r ← Q-depth
    append(⟨(ψ, r)⟩, R, NewR)
    R ← NewR
end
```

As mentioned previously, to address the undecidability of FOL, we incorporate an elegant technique from [8], used by Fitting in a Prolog theorem-prover for FOL. To ensure a terminating algorithm, we introduce the concept of *Q-depth*, which limits the number of individual constants that may be introduced when transforming a question. The provability of a given sequent is always tested at a specific Q-depth.

In the case of the N-O procedure, the 'sequent' and 'numeral' are provided, while 'ϕ', 'r' are bound during execution. The 'numeral' tracks the number of individual constants introduced thus far and is later used to check if the Q-depth limit has been exceeded.

Procedure N-O((sequent, numeral), (ϕ, r))

```
   Data: sequent, numeral
 1 φ ← sequent
 2 r ← numeral
 3 while φ is a sequent-premise of an N-rule do
 4     φ ← sequent-conclusion of r
 5     r ← r + 1
 6 end
 7 m ← the number of formulas' occurrences in φ (on both sides of the turnstile)
 8 for j = m downto 1 do
 9     if j-th formula occurrence in φ is active in an O-rule r then
10         φ ← sequent-conclusion of r
11         r ← r + 1
12     end
13 end
```

As with Procedure SNB, once the first while-loop (lines 3–6) exits, no formulas-premises of N-rules remain. The value of r increments (line 5) with every application of an N-rule. Afterward, the O-rules are applied. We use a for-loop controlled by the number of occurrences of formulas in the initial sequent, starting the O-rule applications from the last formula. When new formulas are introduced to the right of the active formula (via an application of **r**), we simply move one step to the left. This ensures that all formulas in the original sequent are checked for possible O-rule applications. The for-loop design also prevents repeated applications of O-rules on the same active formulas. Each application of an O-rule introduces a new constant, which is accounted for in line 11.

Algorithm 2.3, incorporates the following principles:

(i) Economy—branching rules are applied only when necessary, after verifying the applicability of other rules.
(ii) Restrictions on quantifier rules—applications of quantifier rules follow intuitive restrictions mentioned earlier.
(iii) Negative answers—the algorithm distinguishes between two scenarios when a negative result is returned:
   - The occurrence of a permanently open sequent, indicating the actual decision that no Socratic proof exists.
   - A situation where no Socratic proof was found within the use of q distinct individual constants, where q is the Q-depth set by the user.

For instance, the sequent '$\neg\exists x(P(x) \to \forall x P(x)) \vdash$' (see Example 2.4) is not provable in $\mathbb{E}^{\mathbf{LPQ}}$ with a Q-depth of 1, but it is provable with a Q-depth greater than 2. A Q-depth of 2 is insufficient because after applying the N-rule L^{*}_{δ}, the rule L^{*}_{γ} is also applied, increasing r to 3 before there is a chance to run Procedure SNB again.

Algorithm 2.3: Proof-search algorithm for $\mathbb{E} \in \{\mathbb{E}^{\mathbf{Q}}, \mathbb{E}^{\mathbf{LQ}}, \mathbb{E}^{\mathbf{RQ}}\}$

Data: question $?(\phi)$, where ϕ is a closed sequent of $\mathcal{L}^{?}_{\vdash\text{FOL}}$, value q of Q-depth
Result: finite Socratic transformation of question $?(\phi)$

1 $n \leftarrow 1$
2 $i \leftarrow 1$
3 $r \leftarrow 0$
4 $R \leftarrow \varnothing$
5 **while** $r \leqslant q$ **do**
6   **while** *ϕ is a sequent-premise of a rule of* $\mathbb{E}$ **and** $r \leqslant q$ **do**
7   SNB(ϕ, ϕ)
8   N-O($\langle\phi, r\rangle$, $\langle\phi, r\rangle$)
9   B($\langle\phi, r\rangle$, R, n, $\langle\phi, r\rangle$, R, n)
10   **end**
11   **if** *ϕ is a basic sequent* **then**
12   **if** $i = n$ **then**
13   **return** 'Sequent ϕ has a Socratic proof in $\mathbb{E}$.'
14   **else**
15             $\langle\phi, r\rangle \leftarrow$ the first element of R
16             $R \leftarrow$ sequence containing the remaining elements of R, without $\langle\phi, r\rangle$
17             $i \leftarrow i + 1$
18   **end**
19   **end**
20 **end**
21 **if** *ϕ is permanently open* **then**
22   **return** 'Sequent ϕ is not provable in $\mathbb{E}$.'
23 **else**
24   **return** 'Sequent ϕ is not provable in $\mathbb{E}$ with Q-depth q.'
25 **end**

References

1. Avron, A. (1996). The method of hypersequents in the proof theory of propositional non-classical logics. In W. Hodges, M. Hyland, C. Steinhorn, & J. Truss (Eds.), *Logic: Foundations to applications* (pp. 1–32). Oxford Science Publications.
2. Chlebowski, S. (2018). *Canonical and dual erotetic calculi for first-order logic (Doctoral dissertation)*. Adam Mickiewicz University.
3. Cormen, T. H., Leiserson, C. E., Rivest, R. L., & Stein, C. (2009). *Introduction to algorithms* (3rd ed.). The MIT Press.
4. Curry, H. B. (1950). *A theory of formal deducibility*. University of Notre Dame.
5. Došen, K. (1985). Sequent-systems for modal logic. *The Journal of Symbolic Logic, 50*(1), 149–168.
6. Dragalin, A. G. (1988). *Mathematical intuitionism*. Introduction to Proof Theory: American Mathematical Society.
7. Fitting, M. (1983). *Proof methods for modal and intuitionistic logics*. Springer.
8. Fitting, M. (1990). *First-order logic and automated theorem proving*. Springer.
9. Gentzen, G. (1969). Investigations into logical deduction. In M. E. Szabo (Ed.), *The collected papers of Gerhard Gentzen* (pp. 68–131). North-Holland Publishing Company.
10. Genzten, G. (1935). Untersuchungen über das logische Schließen [The English translation by Manfred E. Szabo: (9)]. *Mathematische Zeitschrift, 39*, 176–210.
11. Hintikka, J. (1955). Form and content in quantification theory. *Acta Philosophica Fennica, 8*, 3–55.
12. Indrzejczak, A. (n.d.). Natural Deduction. In *The internet encyclopedia of philosophy*. https://www.iep.utm.edu/nat-ded/
13. Indrzejczak, A. (2010). *Natural deduction, hybrid systems and modal logics* (Vol. 30). Springer.
14. Indrzejczak, A. (2014). A survey of nonstandard sequent calculi. *Studia Logica, 102*, 1295–1322. https://doi.org/doi.org/10.1007/s11225-014-9567-y
15. Indrzejczak, A. (2016). Linear time in hypersequent framework. *The Bulletin of Symbolic Logic, 22*(1), 121–144. https://doi.org/10.1017/bsl.2016.2
16. Kleene, S. C. (1952). *Introduction to metamathematics*. North-Holland Publishing Co., P. Noordhoff N.V.
17. Legris, J. (2014). Paul Hertz's systems of propositions as a proof-theoretical conception of logic. In L. C. Pereira, E. H. Haeusler, & V. de Paiva (Eds.), *Advances in natural deduction. A celebration of Dag Prawitz's work* (pp. 93–101). Springer Netherlands.
18. Leszczyńska-Jasion, D. (2021). The method of Socratic proofs. From the logic of questions to proof theory. In M. Cordes (Ed.), *Asking and answering. rivalling approaches to interrogative methods* (pp. 183–198). Narr Francke Attempto Verlag.
19. Negri, S., & von Plato, J. (2001). *Structural proof theory*. Cambridge University Press.
20. Negri, S., & von Plato, J. (2011). *Proof analysis*. A contribution to Hilbert's last problem: Cambridge University Press.
21. Negri, S., & von Plato, J. (1998). Cut elimination in the presence of axioms. *The Bulletin of Symbolic Logic, 4*(4), 418–435.
22. Poggiolesi, F. (2011). *Gentzen calculi for modal propositional logic*. Springer.
23. Pottinger, G. (1983). Uniform cut-free formulations of T, S4 and S5 [abstract of a talk given at Annual Meeting of the Association for Symbolic Logic]. *The Journal of Symbolic Logic, 48*, 900.
24. Shoesmith, D. J., & Smiley, T. J. (1978). *Multiple-conclusion logic*. Cambridge University Press.
25. Smullyan, R. M. (1968). *First-order logic*. Springer.
26. Troelstra, A. S., & Schwichtenberg, H. (2000). *Basic proof theory* (2nd ed.). Cambridge University Press.
27. von Plato, J. (2016). The development of proof theory. In E. N. Zalta (Ed.), *The Stanford Encyclopedia of Philosophy* (2016). https://plato.stanford.edu/archives/win2016/entries/proof-theory-development/

28. Wiśniewski, A. (2004). Socratic proofs. *Journal of Philosophical Logic, 33*(3), 299–326.
29. Wiśniewski, A. (2006). *A right-sided Socratic calculus for classical logic (Research Report)*. Poznań: Adam Mickiewicz University.
30. Wiśniewski, A. (2013). *Questions, inferences, and scenarios*. College Publications.
31. Wiśniewski, A. (2021). Entailment, transmission of truth, and minimality. *Journal of Applied Logics, 8*(10), 2471–2522.
32. Wiśniewski, A., & Shangin, V. (2006). Socratic proofs for quantifiers. *Journal of Philosophical Logic, 35*(2), 147–178. https://doi.org/10.1007/s10992-005-9000-0
33. Wiśniewski, A., & Shangin, V. (2007). Некоторые допустимые правила дла системы сократического вывода (Some admissible rules in the system of Socratic proofs). *Вестник Московского Университета. Серия, 7*(5), 77–88.

Chapter 3
Erotetic Calculi for Propositional Modal Logics

In this chapter, we present the method of Socratic proofs for a number of propositional modal logics called *basic*. In [11], the author presented erotetic calculi for the class of basic modal propositional logics. By *basic modal logics*, we mean K and all of its proper extensions, characterized by any combination of the following properties of the accessibility relation: seriality (also called 'extendability'), reflexivity, transitivity, symmetry, and Euclideanness. These five semantic properties can be combined in $2^5 = 32$ ways, including the empty one, but some combinations of properties entail others. As a result, there are exactly fifteen different basic modal logics. The properties will be defined in Table 3.2, Sect. 3.1, and the relations between logics are displayed in Fig. 3.1, p. 124.

We use the notation for the names of the logics after Rajeev Goré[1]: K, D, T, KB, DB, B, K4, D4, S4, KB4, S5, K5, D5, K45, and D45. In [11], the logics K, D, T, K4, KB, S4 and S5 were discussed in detail, together with soundness and completeness proofs. The author also presented erotetic calculi for the other basic modal logics. In [12], another six modal logics were 'erotetized': S4.2, S4.3, S4F, S4R, S4M, and G. Nevertheless, there are several differences between the version presented here and in [11], which will be indicated below.

In this chapter, we will demonstrate the soundness and completeness of erotetic calculi for the aforementioned fifteen logics with respect to the relevant Kripke semantics. Our account extends the one presented in [11] by capturing global entailment relation. In this respect, our approach is based on ideas found in [4], where Melvin Fitting develops numerous tableau systems with both local and global assumptions. As far as the erotetic aspects are concerned, we will place the well-known basic modalities in the context of questions, assuming both their meaning (semantics) and their potential role in erotetic inferences. In fact, we will not analyze

[1] See [6].

D. Leszczyńska-Jasion, *The Method of Socratic Proofs*, Trends in Logic 64,
https://doi.org/10.1007/978-3-031-82451-7_3

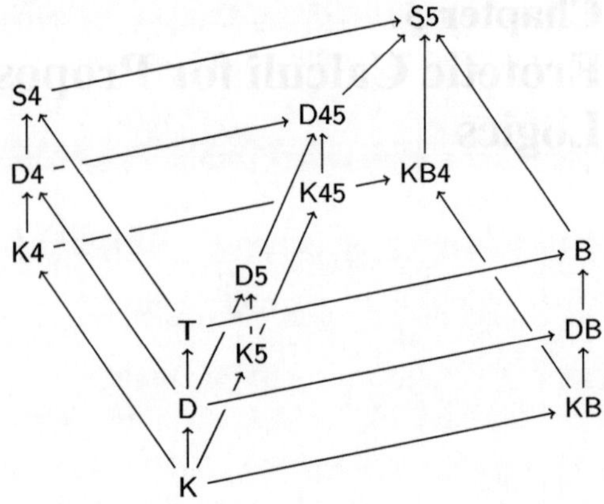

Fig. 3.1 Hasse diagram of the Lattice of basic modal logics

the latter issue but will instead focus on developing the technical machinery that may serve this purpose. This approach will be further developed in Sect. 5.4 of Chap. 5, where we provide some elements of Minimal Erotetic Semantics for languages with questions and modalities.

The reader may wish to compare our account with the inquisitive one in [2, Chaps. 6–7] or [3, Chap. 8], where modalities are defined semantically, and their behavior, as well as their role in questions, is studied on a more intuitive level.

As we will see, erotetic calculi for modal logics make use of languages with indexed formulas, which allow the expression of semantic information. This is a common approach in modal logics; see, e.g. [9, 15, 18]. For this reason, a modal erotetic calculus can be considered a labeled deductive system, cf. [4, 5, 22–24]. On the other hand, since the constituents of questions transformed by erotetic rules are *sequents*, the proof-theoretical relatives of the method of Socratic proofs are proof methods based on sequent calculi.

Among these are 'ordinary' sequent systems, formulated in a 'pure' language, i.e., sequent calculi in which no semantic information is represented directly. Following [6], we may call this systems *implicit*, as opposed to *explicit* systems, where semantic information is explicitly represented.

Implicit sequent systems for modal logics were developed as early as the 1950s (e.g. [16, 17]; see [21] or [25] for an overview). A characteristic feature of implicit systems is their independence of the underlying semantics. In [1], the author postulates this feature as a defining characteristic of a 'good' general proof-theoretical framework.[2]

In the next chapter, we extend the framework developed for modal logics to propositional intuitionistic logic. In the remaining part of this monograph—after Chap. 5, which is devoted to strictly erotetic issues—we will return to the analysis of

[2] See also [21] on this feature of proof systems.

the relationships between explicit and implicit sequent systems in basic modal logics that are neither symmetric nor Euclidean. As we shall see, erotetic calculi for these modal logics can be used as a proof-search tool in implicit sequent systems.

It is worth noting how the diversity of sequent types is employed in building erotetic calculi for non-classical logics. In the previous chapter, we saw three different variants of erotetic calculi for classical logic, each deriving from a syntactically distinct type of sequent, with each sequent type having its own semantic interpretation, as indicated in Corollary 2.4, Sect.2.4. In this and the next chapter, we leverage the diversity of sequent types in constructing erotetic calculi for non-classical logics. The original version of erotetic calculi, as presented in [11], was based on right-sided sequents. Here, we relax the restriction on the antecedent, allowing for both-sided, multisuccedent sequents, thereby enabling us to account for the relation of global entailment. In the case of intuitionistic logic, as we shall see in the next chapter, both-sided sequents can be used to express the basic semantic properties of a sentence being derivable or non-derivable (the intuitionistic counterparts of true/not true), and the sequents are multisuccedent as well.

In Chap. 7, we will return to the topic of erotetic calculi for modal logics, but we will take a different approach to defining modal sequents. The aim of Chap. 7 will be to define a translation between Socratic proofs and sequent-calculi proofs. Since standard sequent calculi for modal logics use both-sided, multisuccedent sequents, we will use analogous sequents on the erotetic side; global entailment will be then neglected.

In this chapter, the proofs of soundness for the erotetic calculi for modal logics follow the proofs contained in [11], whereas the proofs of completeness are new.

Until the end of this chapter, unless stated otherwise, L represents the basic modal logics.

3.1 Language $\mathcal{M}$ and Its Semantics

Language $\mathcal{M}$ is the language of modal propositional logics built upon $\mathcal{L}_{\mathsf{CPL}}$ with '$\Box$' (it is necessary that) and '$\Diamond$' (it is possible that). The notion of *formula of* $\mathcal{M}$ is defined by the following BNF-grammar:

$$A ::= p \mid \neg A \mid \Box A \mid \Diamond A \mid A \wedge A \mid A \vee A \mid A \rightarrow A$$

$FORM_{\mathcal{M}}$ stands for the set of all formulas of language $\mathcal{M}$.

We apply the notation conventions regarding metavariables introduced in the previous chapter. Together with the α-, β-notation for formulas of propositional language we use also the π-, ν-notation, as introduced in [4]. Table 3.1 presents the notation. In the classical case we have defined the uniform notation with the use of complement instead of negation (see Sect.2.3 for the notion of complement and Table 2.1 for uniform notation), we do the same here.

Table 3.1 π-, ν-notation

ν	ν_0	π	π_0
$\Box A$	A	$-\Box A$	$-A$
$-\Diamond A$	$-A$	$\Diamond A$	A

As in the classical case, doubly negated formulas constitute a category of its own—the class of κ-formulas of language $\mathcal{M}$. However, since there are no other κ-formulas, it seems more natural to write explicitly '$\neg\neg A$' in the rules.

The following is an analogue of Corollary 2.1, a modal version for language $\mathcal{M}$. The proof for the classical cases is the same as in Corollary 2.1, hence we shall not repeat it. We repeat, however, the content of corollary for ease of reference.

Corollary 3.1 *For each formula F of language $\mathcal{M}$ the following holds.*

1. *If F is positive, then $\neg F = -F$. If F is negative, then $F = \neg - F$, and hence also $\neg F = \neg\neg - F$.*
2. *If F is of the form '$\neg\neg A$' (doubly negated), then $- - F = A$; otherwise $- - F = F$.*
3. *If $-F = -A$, then*
   (a) *either $F = A$, or*
   (b) *$F = \neg\neg A$, where A is positive, or*
   (c) *$A = \neg\neg F$, where F is positive.*
4. *If F is an α-formula, then '$-F$' is a β-formula and conversely: if F is a β-formula, then '$-F$' is an α-formula. Moreover, it then holds for each $i \in \{0, 1\}$ that $\alpha_i \in \{-\beta_i, \neg\beta_i\}$ and $\beta_i \in \{-\alpha_i, \neg\alpha_i\}$.*
5. *If F is a ν-formula, then '$-\nu$' is a π-formula, and conversely: if F is a π-formula, then '$-\pi$' is a ν-formula. Moreover, it then holds that $\nu_0 \in \{-\pi_0, \neg\pi_0\}$ and $\pi_0 \in \{-\nu_0, \neg\nu_0\}$.*

Proof We consider one case for ν.

Suppose that $\nu = \Box A$, then $-\nu = \neg\Box A$ is a π-formula. The component ν_0 is A, and π_0 is '$-A$'. By clauses 2 and 1 of this corollary, we have either $A = - - A$, when A is not doubly negated, or $A = \neg - A$, if A is negative. Clearly, every formula A is either not doubly negated or is negative. Hence $\nu_0 \in \{-\pi_0, \neg\pi_0\}$. □

We shall use the standard Kripke semantics for modal logics as presented for example in [8]. Thus by a *modal frame* (or simply *frame*) we mean an ordered pair $\langle W, R\rangle$ such that W is a non-empty set whose elements are called *possible worlds*, and $R \subseteq W \times W$. We will use $R(w)$ for the image of w, which is the set $\{w^* : \langle w, w^*\rangle \in R\}$. As is commonly accepted, we write 'wRw^*' instead of '$\langle w, w^*\rangle \in R$'.

Definition 3.1 A *valuation on a frame* $\langle W, R\rangle$ is a function

$$V : FORM_{\mathcal{M}} \times W \longrightarrow \{0, 1\}$$

satisfying, for every $w \in W$, the following conditions.

1. $V(\neg A, w) = 1$ iff $V(A, w) = 0$.
2. $V(A \wedge B, w) = 1$ iff $V(A, w) = 1$ and $V(B, w) = 1$.
3. $V(A \vee B, w) = 1$ iff $V(A, w) = 1$ or $V(B, w) = 1$.
4. $V(A \rightarrow B, w) = 1$ iff $V(A, w) = 0$ or $V(B, w) = 1$.
5. $V(\Box A, w) = 1$ iff for every $w^* \in R(w)$, $V(A, w^*) = 1$.
6. $V(\Diamond A, w) = 1$ iff for some $w^* \in R(w)$, $V(A, w^*) = 1$.

A triple $\langle W, R, V\rangle$, where $\langle W, R\rangle$ is a frame and V is a valuation on it, is called a *model based on frame* $\langle W, R\rangle$.

For every model $\langle W, R, V\rangle$, and every world $w \in W$, the following holds true.

Corollary 3.2 (ν-formulas, π-formulas).

1. $V(\nu, w) = 1$ *iff for every* $w^* \in R(w)$, $V(\nu_0, w^*)$.
2. $V(\pi, w) = 1$ *iff for some* $w^* \in R(w)$, $V(\pi_0, w^*)$.

A formula A of language $\mathcal{M}$ is called *true in a world* w of model $\langle W, R, V\rangle$ iff $V(A, w) = 1$. It is *true in a model* $\langle W, R, V\rangle$ iff it is true in every world of the model. And finally, a formula true in every model is called K-*valid.*

Modal Logics in an Axiomatic Setting

Modal logics are usually presented in an axiomatic setting. Here we follow, roughly, the presentation of [8, Chap. 2]. For modal logic K this is as follows. First, every formula of language $\mathcal{L}_{\mathsf{CPL}}$ which is CPL-valid, is an *axiom of* $\mathbb{A}\mathsf{x}^{\mathsf{K}}$. This semantic expression can be reformulated by adopting any proof system sound and complete with respect to Boolean semantics for classical propositional logic. We would say then that every thesis of this proof system is an axiom of $\mathbb{A}\mathsf{x}^{\mathsf{K}}$. Second, the following single formula called K,

$K \quad \Box(p \rightarrow q) \rightarrow (\Box p \rightarrow \Box q)$

is also an axiom of $\mathbb{A}\mathsf{x}^{\mathsf{K}}$ (there are no other axioms). The notion of *proof of a formula, A, in system* $\mathbb{A}\mathsf{x}^{\mathsf{K}}$ is defined in a standard manner: as a finite sequence of formulas ending with A and such that each of its elements is either an axiom of $\mathbb{A}\mathsf{x}^{\mathsf{K}}$, or can be derived from the axioms by the following rules, *cf.* [8, p. 25]:

US (The *Rule of Uniform Substitution*): The result of uniformly replacing any variable or variables $p_1, \ldots, p_n$, in a theorem of $\mathbb{A}\mathsf{x}^{\mathsf{K}}$ by any formula (of language $\mathcal{M}$) $A_1, \ldots, A_n$ respectively is itself a theorem of $\mathbb{A}\mathsf{x}^{\mathsf{K}}$.

MP (The *Rule of Modus Ponens*, sometimes also called the *Rule of Detachment*): If A and '$A \rightarrow B$' are theorems of $\mathbb{A}\mathsf{x}^{\mathsf{K}}$, so is B.

N (The *Rule of Necessitation*): If A is a theorem of $\mathbb{A}\mathsf{x}^{\mathsf{K}}$, so is '$\Box A$'.

Stronger basic modal logics are built axiomatically by adding one or more of the following formulas to the axiomatic basis of K, see also Table 3.2:

Table 3.2 Axioms of L and L-properties

L	Axioms	L-properties	Definition
K	K	None	
D	K, D	Seriality	$\forall x \exists y (xRy)$
T	K, T	Reflexivity	$\forall x (xRx)$
KB	K, B	Symmetry	$\forall x, y (xRy \to yRx)$
K4	$K, 4$	Transitivity	$\forall x, y, z (xRy \wedge yRz \to xRz)$
K5	$K, 5$	Euclideanness	$\forall x, y, z (xRy \wedge xRz \to yRz)$
DB	K, D, B	Seriality and symmetry	
B	K, T, B	Reflexivity and symmetry	
D4	$K, D, 4$	Seriality and transitivity	
S4	$K, T, 4$	Reflexivity and transitivity	
KB4	$K, B, 4$	Symmetry and transitivity	
S5	$K, T, 5$	Reflexivity and Euclideanness	
D5	$K, D, 5$	Seriality and Euclideanness	
K45	$K, 4, 5$	Transitivity and Euclideanness	
D45	$K, D, 4, 5$	Seriality, transitivity and Euclideanness	

D $\Box p \to \Diamond p$
T $\Box p \to p$
B $p \to \Box \Diamond p$
4 $\Box p \to \Box \Box p$
5 (also called E) $\Diamond p \to \Box \Diamond p$

The relation of derivability is defined in a standard manner, that is:

Definition 3.2 (*Derivability in Axiomatic System for* L) Let A stand for a formula of the language $\mathcal{M}$, and let X be a set of formulas of this language. We say that A *is derivable from* X *in axiomatic system,* $\mathbb{A}\mathsf{x}^{\mathsf{L}}$, *for logic* L iff there exists a finite sequence $\langle F_1, \ldots, F_n \rangle$ of formulas of $\mathcal{M}$ such that $F_n = A$ and each F_i, $1 \leqslant i \leqslant n$, is either an axiom of $\mathbb{A}\mathsf{x}^{\mathsf{L}}$, an element of X or results from some previous formula(s) of this sequence by one of *US, MP, N*.

Semantically, the same logics are defined by requiring the accessibility relation to have some properties; these are presented in Table 3.2. In the table we recall also the definitions of the properties. If $\langle W, R \rangle$ is a frame such that R has the L-properties, then $\langle W, R \rangle$ is called an L-*frame*. Similarly, any $\langle W, R, V \rangle$, where R has the L-properties, is called an L-*model*. A formula A of language $\mathcal{M}$ is called L-*valid* iff it is true in every L-model.

The richness of modal semantics results in two important entailment relations called the *local* and *global* ones. In fact, a third relation is also possible, but it is quite exotic. Local entailment amounts to transmission of truth in a world of a model, whereas global entailment amounts to transmission of truth in a model.[3] Naturally, relativization to L is required. More formally:

Definition 3.3 Let X stand for a set of formulas of the language $\mathcal{M}$ and let A be a single formula of this language. We say that X *locally* L*-entails* A iff

- for every L-model $\langle W, R, V\rangle$ and every possible world $w \in W$, if each element of X is true in w, then A is true in w.

We say that X *globally* L*-entails* A iff

- for every L-model $\langle W, R, V\rangle$, if each element of X is true in $\langle W, R, V\rangle$, then A is true in $\langle W, R, V\rangle$.

The local entailment corresponds to implication in modal logics, and most of popular proof systems for modal logics capture the local entailment (by proving formulas of the form of an implication). However, the relation of derivability, as defined in an axiomatic system, corresponds to the global version of entailment (see [4, Chaps. 3, 4]). For this reason we shall use '$\models$' for global entailment. More specifically, by:

$$X \models_{\mathsf{L}} A$$

we mean that X globally L-entails A.

It is also worth noticing that the split between local and global entailment brings interesting complications into deduction theorems for modal logics. For the local version the semantic deduction theorem is just like the classical one, but for the global version deduction theorems depend on L. We will not elaborate on this issue here, see [19, 20] for further information.

The notions of satisfiability also come in two versions.

Definition 3.4 Let X stand for a set of formulas of the language $\mathcal{M}$, and let $\langle W, R, V\rangle$ be a Kripke model.

- We say that X *is locally satisfied in* $\langle W, R, V\rangle$ *by world* w, if each element of X is true in w. We also say that X *is locally satisfied in* $\langle W, R, V\rangle$, if there exists a world $w \in W$ such that X is locally satisfied in $\langle W, R, V\rangle$ by w.
- We say that X *is globally satisfied in* $\langle W, R, V\rangle$, if it is locally satisfied by each world $w \in W$.

Both notions of satisfiability have the following obvious property.

Corollary 3.3 *If X is locally (globally) satisfied in $\langle W, R, V\rangle$, then each subset of X is also locally (globally) satisfied in $\langle W, R, V\rangle$.*

---

[3] The exotic one is then about transmission of truth in frames; the notion is not recursively enumerable and it cannot be captured by any set of finitary rules of derivation. See [4, p. 69] for more information.

3.2 Language $\mathcal{M}^?_\vdash$

Language $\mathcal{M}^?_\vdash$ of erotetic calculi $\mathbb{E}^\mathsf{L}$ for modal logic L is built upon language $\mathcal{M}$. The construction is analogous to that of $\mathcal{L}^?_{\vdash\mathsf{L}}$, which was built upon $\mathcal{L}_\mathsf{L}$ (for $\mathsf{L} \in \{\mathsf{CPL}, \mathsf{FOL}\}$, see the previous chapter). However, except for the vocabulary of $\mathcal{M}$, the comma, the semicolon and the signs: ?, $\vdash$, *ng*, &, language $\mathcal{M}^?_\vdash$ contains also countably infinitely many numerals: 0, 1, 2, ... Language $\mathcal{M}^?_\vdash$ has well-formed formulas (*wffs*, for short) of three categories: indexed formulas, declarative wffs and questions. *Indexed formulas of* $\mathcal{M}^?_\vdash$ are expressions of the form:

$$(A)^{i_1,\ldots,i_n} \tag{3.1}$$

where A is a formula of $\mathcal{M}$ and $i_1, \ldots, i_n$ is an n-term ($n \geqslant 1$) sequence of numerals. The sequence will be called an *index of wff* (3.1). The numerals are meant to indicate possible worlds, and the order in which the numerals occur in an index presents a partial description of the accessibility relation. The last numeral of an index will play a special role, therefore we will use the inscription: 'ς/i' to indicate that the last term of sequence ς is i. Hence an expression of the form:

$$(A)^{\varsigma/i}$$

represents an arbitrary indexed formula whose index ends with numeral i. Lowercase Latin letters $i, j, \ldots$ will be used for numerals and lowercase Greek letters ς, ρ, possibly with subscripts, will be used for indices.

Sequents of language $\mathcal{M}^?_\vdash$ are expressions of the form:

$$A_1, \ldots, A_k \vdash (B_1)^{\varsigma_1/i_1}, \ldots, (B_n)^{\varsigma_n/i_n} \tag{3.2}$$

where $A_1, \ldots, A_k$ is a k-term ($k \geqslant 0$) sequence of formulas of $\mathcal{M}$ and $(B_1)^{\varsigma_1/i_1}, \ldots, (B_n)^{\varsigma_n/i_n}$ is an n-term ($n \geqslant 1$) sequence of indexed formulas of $\mathcal{M}^?_\vdash$ (as before, we omit the angled brackets around sequences). In the sequel, we will sometimes omit parentheses and write (3.2) rather as:

$$A_1, \ldots, A_k \vdash B_1^{\varsigma_1/i_1}, \ldots, B_n^{\varsigma_n/i_n}$$

Generally, it will depend on the readability of the expression. If $k = 0$, then (3.2) is represented as follows:

$$\vdash B_1^{\varsigma_1/i_1}, \ldots, B_n^{\varsigma_n/i_n} \tag{3.3}$$

In almost all versions of erotetic calculi for modal logics considered so far, [10–13], only sequents of the form (3.3) were concerned. In [14, Chap. 4] multisuccedent sequents with indexed formulas on both sides of sequents were taken into account; in the last chapter of this book we will also consider such sequents.

Until the end of this chapter we use ϕ, ψ for sequents of language $\mathcal{M}^?_\vdash$. *D-wffs of language* $\mathcal{M}^?_\vdash$ are defined in the same way as those of $\mathcal{L}^?_{\vdash\mathsf{L}}$: every sequent of $\mathcal{M}^?_\vdash$ is a d-wff of $\mathcal{M}^?_\vdash$; if $\mathfrak{t}$ is a d-wff of $\mathcal{M}^?_\vdash$, then so is '$ng\mathfrak{t}$'; if $\mathfrak{t}$ and $\mathfrak{u}$ are d-wffs of $\mathcal{M}^?_\vdash$, then so is '$(\mathfrak{t}\&\mathfrak{u})$'. Finally, there are no other d-wffs of $\mathcal{M}^?_\vdash$ than those already mentioned. *Questions of* $\mathcal{M}^?_\vdash$ are expressions of the form:

$$?(\phi_1; \ldots; \phi_n)$$

where '$\phi_1; \ldots; \phi_n$' is an n-term ($n \geqslant 1$) sequence of sequents of $\mathcal{M}^?_\vdash$. As in the previous chapter (see Sect.2.2), we shall sometimes use the term 'seqsequent' for a finite sequence of sequents (the term comes from [26]).

Metavariables $A, B, \ldots$ will be used for formulas of modal language $\mathcal{M}$. Further in this chapter we shall consider finite (possibly empty) sequences of both: formulas of $\mathcal{M}$ and indexed formulas of $\mathcal{M}^?_\vdash$; we need two types of metavariables for that. Hence we will use metavariables $S, T, \ldots$ for the former and $\Gamma, \Delta, \Theta, \ldots$ for the latter. Greek upper-case letters Φ, Ψ will be used for finite (possibly empty) sequences of sequents of $\mathcal{M}^?_\vdash$.

3.3 Erotetic Calculi for Modal Logics, Soundness

If ϕ is a sequent of $\mathcal{M}^?_\vdash$ of the form (3.2), then by '$W(\phi)$' we shall mean the set of all numerals occurring in the indices of formulas of ϕ. More precisely:

$$W(\phi) = \{j : j \text{ is a term of some '}\varsigma_l/i_l\text{'}, 1 \leqslant l \leqslant n\}$$

Obviously, indices occur only right of the turnstile. Hence if $\phi = S \vdash \Gamma$, then we shall often write '$W(\Gamma)$' instead of '$W(\phi)$'. By '$R(\phi)$' (or, consequently, by '$R(\Gamma)$') we mean the set of ordered pairs constituting the formulas' indices in ϕ, formally:

$$R(\phi) = \{\langle j, j^* \rangle : j \text{ immediately precedes } j^* \text{ in some '}\varsigma_l/i_l\text{'}, 1 \leqslant l \leqslant n\}$$

In the presentation of the erotetic rules we apply the notation conventions applied previously, see Sect. 2.3. Table 3.3 presents the rules of erotetic calculi $\mathbb{E}^{\mathsf{L}}$ called *local*. We reuse the names R_α, R_β for simplicity, this should not lead to any confusion. Let us also observe that nothing happens here left of the turnstile, the antecedent is simply rewritten.

Analogously to the notion of an active formula we will use that of an *active indexed formula* of a rule **r**, which is the indexed formula whose schema is distinguished in the sequent-premise of **r**.

Rule R_ν may be applied provided that j is new with respect to sequent $S \vdash \Gamma(\nu^{\varsigma/i})$. By '$\varsigma/i, j$' we will mean the concatenation of sequences 'ς/i' and the

Table 3.3 Local rules of modal $\mathbb{E}^{\mathsf{L}}$

$$\frac{?(\Phi\,(\,S\vdash\Gamma(\,\alpha^{\varsigma}\,)\,)}{?(\Phi\,(\,S\vdash\Gamma(\,\alpha_0{}^{\varsigma}\,)\,;\ S\vdash\Gamma(\,\alpha_1{}^{\varsigma}\,)\,)}\ R_\alpha$$

$$\frac{?(\Phi\,(\,S\vdash\Gamma(\,\beta^{\varsigma}\,)\,)}{?(\Phi\,(\,S\vdash\Gamma(\,\beta_0{}^{\varsigma},\ \beta_1{}^{\varsigma}\,)\,)}\ R_\beta \qquad \frac{?(\Phi\,(\,S\vdash\Gamma(\,\neg\neg A^{\varsigma}\,)\,)}{?(\Phi\,(\,S\vdash\Gamma(\,A^{\varsigma}\,)\,)}\ R_{\neg\neg}$$

$$\frac{?(\Phi\,(\,S\vdash\Gamma(\,\nu^{\varsigma/i}\,)\,)}{?(\Phi\,(\,S\vdash\Gamma(\,\nu_0{}^{\varsigma/i,j}\,)\,)}\ R_\nu \qquad \frac{?(\Phi\,(\,S\vdash\Gamma(\,\pi^{\varsigma/i}\,)\,)}{?(\Phi\,(\,S\vdash\Gamma(\,\pi^{\varsigma/i},\ \pi_0{}^{j}\,)\,)}\ R_\pi$$

one-term sequence '$\langle j\rangle$'. Observe that the last term of sequence $\varsigma/i, j$ is j. For convenience, we usually assume that:

$$j = max(W(\Gamma(\nu^{\varsigma/i}))) + 1 \tag{3.4}$$

which guarantees that the new j is exactly the first (in $\leqslant$) numeral new with respect to the sequent in question. It is convenient to follow this restriction in practice, but as we shall see in Sect. 3.4, it is also necessary to relax it for meta-level (structural) analysis.

The proviso of applicability of rule R_π depends on L and is given in Table 3.4. The following terminology is used.

Definition 3.5 (*R-chain*) Let R be a non-empty binary relation defined in a set W. By *R-chain* we mean a finite, at least two-term sequence $\langle w_1, \ldots, w_n\rangle$ of elements of W such that for each k $(1 \leqslant k < n)$: either $\langle w_k, w_{k+1}\rangle \in R$ or $\langle w_{k+1}, w_k\rangle \in R$.

Table 3.4 Provisos of applicability of rule R_π (ϕ represents the sequent-premise of the rule)

Calculus	Proviso of applicability of rule R_π
$\mathbb{E}^{\mathsf{K}}$, $\mathbb{E}^{\mathsf{D}}$	$\langle i, j\rangle \in R(\phi)$
$\mathbb{E}^{\mathsf{T}}$	$\langle i, j\rangle \in R(\phi)$ **or** $j = i$
$\mathbb{E}^{\mathsf{KB}}$, $\mathbb{E}^{\mathsf{DB}}$	$\langle i, j\rangle \in R(\phi)$ **or** $\langle j, i\rangle \in R(\phi)$
$\mathbb{E}^{\mathsf{B}}$	$\langle i, j\rangle \in R(\phi)$ **or** $\langle j, i\rangle \in R(\phi)$ **or** $j = i$
$\mathbb{E}^{\mathsf{K4}}$, $\mathbb{E}^{\mathsf{D4}}$	There is a directed $R(\phi)$-$[i, j]$-chain
$\mathbb{E}^{\mathsf{S4}}$	There is a directed $R(\phi)$-$[i, j]$-chain **or** $j = i$
$\mathbb{E}^{\mathsf{KB4}}$	There is an $R(\phi)$-$[i, j]$-chain
$\mathbb{E}^{\mathsf{S5}}$	There is an $R(\phi)$-$[i, j]$-chain **or** $j = i$
$\mathbb{E}^{\mathsf{K5}}$, $\mathbb{E}^{\mathsf{D5}}$	$\langle i, j\rangle \in R(\phi)$ **or** there are: a directed $R(\phi)$-$[0, i]$-chain and a directed $R(\phi)$-$[0, j]$-chain
$\mathbb{E}^{\mathsf{K45}}$, $\mathbb{E}^{\mathsf{D45}}$	there is a directed $R(\phi)$-$[i, j]$-chain **or** there are: a directed $R(\phi)$-$[0, i]$-chain and a directed $R(\phi)$-$[0, j]$-chain

An R-chain, $\langle w_1, \ldots, w_n \rangle$, is called *directed*, if $\langle w_k, w_{k+1} \rangle \in R$ for each k ($1 \leqslant k < n$).

Finally, by a *(directed) R-$[w, z]$-chain* we mean a (directed) R-chain whose first term is w and whose last term is z.

Let us observe that, by the above definitions, each R-$[w, z]$-chain is also an R-$[z, w]$-chain.

In Table 3.4, ϕ stands for '$S \vdash \Gamma((\pi)^{\varsigma/i})$', that is, the sequent-premise of the rule. In most cases the proviso is a disjunction, hence for the proviso to hold it is sufficient if at least one of the disjuncts is satisfied.

For those that find the provisos not in sufficient correspondence with the semantic properties here comes what follows. Given a universe (a set) U, the p-closure of a relation $R \subseteq U \times U$, where p is one of the properties: reflexivity, symmetry, transitivity, Euclideanness, is the intersection

$$R^p = \bigcap_i R_i^\star$$

of all relations $R_i^\star$, $R \subseteq R_i^\star \subseteq U \times U$ that have the property p. This definition is correct provided that the intersection exists, and this is warranted by the fact that the full relation, $U \times U$, has all the considered properties. This account is obviously unsuccessful in case of such properties like irreflexivity. More generally, defining closures of relations by the 'intersection condition' is successful, if the property's definition has the form of an implication, where both the antecedent and succedent are conjunctions of positive conditions: stating that relation holds between some objects (the conjunctions can have one conjunct only, as in the case of symmetry, and the antecedent can be empty, as in the case of reflexivity). This warrants that the closure can be always obtained by adding some missing ordered pairs, and also that the intersection is actually *the smallest* superset of R that has the considered property.

We will write R^{ref}, R^{sym}, R^{tr}, R^{Eu} for the respective closures of R. We can generalize the above definition of a closure by saying that, e.g., by *transitive and Euclidean* closure of R we mean the smallest binary relation $R^\star$ such that $R \subseteq R^\star$ and $R^\star$ is both transitive and Euclidean. We shall write $R^{tr,Eu}$ for such a closure, and similarly for any other combination of the four properties.

There are subtleties, however. Since a relation is both transitive and Euclidean iff it is both Euclidean and transitive, relation $R^{tr,Eu}$ is the same relation as $R^{Eu,tr}$. On the other hand, $(R^{tr})^{Eu}$ and $(R^{Eu})^{tr}$ need not be the same thing. Let us consider.

Example 3.1 Let $U = \{0, 1, 2, 3\}$ and $R = \{(0, 1), (2, 1), (2, 3)\}$. We have $R^{tr} = R$, and thus

$$\left(R^{tr}\right)^{Eu} = R^{Eu} = R \cup \{(1, 1), (1, 3), (3, 1), (3, 3)\}.$$

It follows that $(R^{Eu})^{tr} = R^{Eu} \cup \{(0, 3)\}$, therefore

$$\left(R^{tr}\right)^{Eu} \neq \left(R^{Eu}\right)^{tr}.$$

On the other hand, $((R^{tr})^{Eu})^{tr} = R^{tr,Eu}$.

Thus we can see that, in general, for two properties p_1, p_2 of a binary relation, R^{p_1,p_2} is not simply $(R^{p_1})^{p_2}$.

Lets go more formal. Assume a universe U, a binary relation $R \subseteq U \times U$, and properties $p_1, \ldots, p_n$, all defined by implications with both antecedent and succedent being conjunctions of positive conditions (with antecedent possibly empty). By '$R^{p_1,\ldots,p_n}$' we mean the intersection:

$$R^{p_1,\ldots,p_n} = \bigcap_i R_i^\star$$

of all relations $R_i^\star$ such that $R \subseteq R_i^\star \subseteq U \times U$ and $R_i^\star$ has the properties $p_1,\ldots, p_n$.

Relations having a collection of properties definable in terms of the 'inclusion condition' (or, equally, the 'smallest-subset condition') can be also defined by the following approximation. For any finite set $\{p_1, \ldots, p_n\}$ of such single properties, we consider the monotonically increasing sequence:

$$R \subseteq R^{p_1} \subseteq (R^{p_1})^{p_2} \subseteq \ldots \subseteq (\cdots((R^{p_1})^{p_2})\cdots)^{p_n} \tag{3.5}$$

which is finite and its sum is $(\cdots((R^{p_1})^{p_2})\cdots)^{p_n}$. If the sum does not own all of $p_1, \ldots, p_n$, then the approximation is repeated, that is, we assign

$$R \leftarrow (\cdots((R^{p_1})^{p_2})\cdots)^{p_n}$$

and consider (3.5) once again. As far as finite relations are concerned, this process must lead to $R^{p_1,\ldots,p_n}$ in a finite number of steps, because the full relation $U \times U$ is finite. More specifically, if the field of the initial relation R has k elements, then $R^{p_1,\ldots,p_n}$ must be obtained after at most k^2 (3.5)-steps (assuming, for the worst-case analysis, that only one pair is added in each (3.5)-step).

Now let ϕ stand for '$S \vdash \Gamma(\pi\, ^{\varsigma/i})$', again. Table 3.5 presents the provisos of applicability of rule R_π in more familiar terminology.

We can see clearly how the provisos correspond to the semantic definition of L. However, the definitions presented in Table 3.5 have a serious defect, since we are not told how to compute the respective closure. Mainly for this reason the original version of the provisos, that expressed in Table 3.4, is used in this book. We will also show that the two accounts are equivalent, so we can use the two versions in meta-level proofs, depending on our convenience. Before we do this, however, we shall finish the presentation of erotetic calculi.

Table 3.5 Provisos of applicability of rule R_π (ϕ represents the sequent-premise of the rule)

Calculus	Proviso of applicability of rule R_π
$\mathbb{E}^{\mathsf{K}}$, $\mathbb{E}^{\mathsf{D}}$	$\langle i, j\rangle \in R(\phi)$
$\mathbb{E}^{\mathsf{T}}$	$\langle i, j\rangle \in R(\phi)^{ref}$
$\mathbb{E}^{\mathsf{KB}}$, $\mathbb{E}^{\mathsf{DB}}$	$\langle i, j\rangle \in R(\phi)^{sym}$
$\mathbb{E}^{\mathsf{B}}$	$\langle i, j\rangle \in R(\phi)^{ref,sym}$
$\mathbb{E}^{\mathsf{K4}}$, $\mathbb{E}^{\mathsf{D4}}$	$\langle i, j\rangle \in R(\phi)^{tr}$
$\mathbb{E}^{\mathsf{S4}}$	$\langle i, j\rangle \in R(\phi)^{ref,tr}$
$\mathbb{E}^{\mathsf{KB4}}$	$\langle i, j\rangle \in R(\phi)^{sym,tr} = R(\phi)^{sym,Eu}$
$\mathbb{E}^{\mathsf{S5}}$	$\langle i, j\rangle \in R(\phi)^{ref,sym,tr} = R(\phi)^{ref,Eu}$
$\mathbb{E}^{\mathsf{K5}}$, $\mathbb{E}^{\mathsf{D5}}$	$\langle i, j\rangle \in R(\phi)^{Eu}$
$\mathbb{E}^{\mathsf{K45}}$, $\mathbb{E}^{\mathsf{D45}}$	$\langle i, j\rangle \in R(\phi)^{tr,Eu}$

For logics which are serial but not reflexive, i.e., $\mathsf{L} \in \{\mathsf{D}, \mathsf{DB}, \mathsf{D4}, \mathsf{D5}, \mathsf{D45}\}$, the corresponding calculus $\mathbb{E}^{\mathsf{L}}$ contains additionally the following local rule:

$$\frac{?(\Phi\ (S\ \vdash \Gamma\ (\pi^{\ \varsigma/i})))}{?(\ \Phi\ (S\ \vdash \Gamma\ (\pi^{\ \varsigma/i},\ \pi_0^{\ i,j}\ )))}\ R_{\pi\mathsf{D}}$$

with the proviso that j is new with respect to sequent '$S \vdash \Gamma(\pi^{\ \varsigma/i})$'. Actually, we shall put a restriction on j as in (3.4), see Sect. 3.2.

Finally, the global rule allows to pick a formula from the antecedent and place its complement in an arbitrary world indicated in the succedent:

$$\frac{?(\Phi\ (S\ (A) \vdash \Gamma\ (\_)))}{?(\Phi\ (S\ (A) \vdash \Gamma\ (-A^i)))}\ R_G$$

where i is an arbitrary numeral from $W(\Gamma(\_))$. The following definition summarizes the presentation of erotetic calculi from this section.

Definition 3.6 (*Erotetic Calculus for Modal Logic* L) An *erotetic calculus* $\mathbb{E}^{\mathsf{L}}$ *for modal logic* L is the smallest set of erotetic rules comprising:

- rules: R_α, R_β, $R_{\neg\neg}$, R_ν as presented in Table 3.3,
- rule R_π with proviso of its applicability as defined in Table 3.5,
- rule R_G,
- if $\mathsf{L} \in \{\mathsf{D}, \mathsf{DB}, \mathsf{D4}, \mathsf{D5}, \mathsf{D45}\}$, then also rule $R_{\pi\mathsf{D}}$.

Socratic transformations are defined as in the classical case: as sequences of questions regulated by the erotetic rules (see Definition 2.2 in Sect. 2.3). A Socratic proof in $\mathbb{E}^{\mathsf{L}}$ starts with a question of the form $?(S \vdash (A)^0)$, where the numeral is to represent an arbitrary possible world of an arbitrary Kripke model. Formally,

Definition 3.7 (*Socratic Proof in* $\mathbb{E}^{\mathsf{L}}$) Let ϕ be a sequent of the language $\mathcal{M}^{?}_{\vdash}$ of the form '$S \vdash (A)^{0}$'. A *Socratic proof of* ϕ *in* $\mathbb{E}^{\mathsf{L}}$ is a finite Socratic transformation **s** of the question '$?(\phi)$' in $\mathbb{E}^{\mathsf{L}}$ such that each constituent of the last question of **s** has the following form:

(*e*) $S \vdash \Gamma(B^{\varsigma_1/i})(-B^{\varsigma_2/i})$.

Let us stress that 'ς_1/i' and 'ς_2/i' may be *different* sequences. Since possible worlds are indicated by single numerals (not by their sequences) it is not necessary that B and '$-B$' have the same index but that their indices end with the same numeral i. Another observation is that the formulas in sequence S are not changed during the transformation.

Here are some examples. First, with the antecedent empty.

Example 3.2 A Socratic proof of sequent '$\vdash (\Box(p \to q) \to (\Box p \to \Box q))^{0}$' in $\mathbb{E}^{\mathsf{K}}$. Below A stands for '$\neg\Box(p \to q)$' and B for '$\neg\Box p$'.

$$
\begin{array}{c}
?\,(\vdash (\Box(p \to q) \to (\Box p \to \Box q))^{0}) \\
\hline
?\,(\vdash A^{0}, (\Box p \to \Box q)^{0}) \\
\hline
?\,(\vdash A^{0}, B^{0}, (\Box q)^{0}) \\
\hline
?\,(\vdash A^{0}, B^{0}, (q)^{0,1}) \\
\hline
?\,(\vdash A^{0}, (\neg(p \to q))^{1}, B^{0}, (q)^{0,1}) \\
\hline
?\,(\vdash A^{0}, (\neg(p \to q))^{1}, B^{0}, (\neg p)^{1}, (q)^{0,1}) \\
\hline
?\,(\vdash A^{0}, (p)^{1}, B^{0}, (\neg p)^{1}, (q)^{0,1} \,;\, \vdash A^{0}, (\neg q)^{1}, B^{0}, (\neg p)^{1}, (q)^{0,1})
\end{array}
\quad
\begin{array}{l}
R_{\beta} \\ R_{\beta} \\ R_{\nu} \\ R_{\pi} \\ R_{\pi} \\ R_{\alpha}
\end{array}
$$

Example 3.3 A Socratic proof of sequent '$\vdash A^{0}$' in $\mathbb{E}^{\mathsf{D45}}$, where $A = \Diamond(\neg\Diamond(p \wedge q) \to \Box\Diamond\neg q \vee \Box\Box\neg p)$. In the constituents of the last question we skip the inner contexts.

$$
\begin{array}{c}
?(\vdash (\Diamond(\neg\Diamond(p \wedge q) \to \Box\Diamond\neg q \vee \Box\Box\neg p))^{0}) \\
\hline
?(\vdash A^{0}, (\neg\Diamond(p \wedge q) \to \Box\Diamond\neg q \vee \Box\Box\neg p)^{0,1}) \\
\hline
?(\vdash A^{0}, (\Diamond(p \wedge q))^{0,1}, (\Box\Diamond\neg q \vee \Box\Box\neg p)^{0,1}) \\
\hline
?(\vdash A^{0}, (\Diamond(p \wedge q))^{0,1}, (\Box\Diamond\neg q)^{0,1}, (\Box\Box\neg p)^{0,1}) \\
\hline
?(\vdash A^{0}, (\Diamond(p \wedge q))^{0,1}, (\Diamond\neg q)^{0,1,2}, (\Box\Box\neg p)^{0,1}) \\
\hline
?(\vdash A^{0}, (\Diamond(p \wedge q))^{0,1}, (\Diamond\neg q)^{0,1,2}, (\Box\neg p)^{0,1,3}) \\
\hline
?(\vdash A^{0}, (\Diamond(p \wedge q))^{0,1}, (\Diamond\neg q)^{0,1,2}, (\neg p)^{0,1,3,4}) \\
\hline
?(\vdash A^{0}, (\Diamond(p \wedge q))^{0,1}, (p \wedge q)^{4}, (\Diamond\neg q)^{0,1,2}, (\neg p)^{0,1,3,4}) \\
\hline
?(\vdash A^{0}, (\Diamond(p \wedge q))^{0,1}, (p \wedge q)^{4}, (\Diamond\neg q)^{0,1,2}, (\neg q)^{4}, (\neg p)^{0,1,3,4}) \\
\hline
?(\vdash \ldots, (p)^{4}, \ldots, (\neg p)^{0,1,3,4} \,;\, \vdash \ldots, (q)^{4}, \ldots, (\neg q)^{4}, \ldots)
\end{array}
\quad
\begin{array}{l}
R_{\pi}\mathsf{D} \\ R_{\beta} \\ R_{\beta} \\ R_{\nu} \\ R_{\nu} \\ R_{\nu} \\ R_{\pi} \\ R_{\pi} \\ R_{\alpha}
\end{array}
$$

Here comes a justification of applications of rule R_{π} in the above example.

- the first application of R_{π}: $\langle 1, 3, 4 \rangle$ is a directed $R(\phi)$-[1, 4]-chain, where ϕ stands for '$\vdash A^{0}, (\Diamond(p \wedge q))^{0,1}, (\Diamond\neg q)^{0,1,2}, (\neg p)^{0,1,3,4}$',

- the second application of R_π: $\langle 0, 1, 2\rangle$ and $\langle 0, 1, 3, 4\rangle$ are both directed $R(\psi)$-chains, where $\psi = \vdash A^0, (\Diamond(p \wedge q))^{0,1}, (p \wedge q)^4, (\Diamond\neg q)^{0,1,2}, (\neg p)^{0,1,3,4}$.

Here is also a justification in terms of relation properties. The diagram on the left illustrates the relation between numerals in the above example. After closing the relation with respect to transitivity and Euclideanes it becomes universal in the set $\{1, 2, 3, 4\}$ (on the right).

The difference between local and global entailment in modal logics is exemplified by Socratic transformations of questions '$?(\vdash (p \to \Box p)^0)$' and '$?(p \vdash (\Box p)^0)$'. The second is a Socratic proof of sequent '$p \vdash (\Box p)^0$', sequent '$\vdash (p \to \Box p)^0$' is not provable in $\mathbb{E}^{\mathsf{K}}$.

Example 3.4 Socratic transformations of '$?(\vdash (p \to \Box p)^0)$' and '$?(p \vdash (\Box p)^0)$', both in $\mathbb{E}^{\mathsf{K}}$.

$$\dfrac{\dfrac{\vdash (p \to \Box p)^0}{\vdash (\neg p)^0, (\Box p)^0}\ R_\beta}{\vdash (\neg p)^0, (p)^{0,1}}\ R_\nu \qquad\qquad \dfrac{\dfrac{p \vdash (\Box p)^0}{p \vdash (p)^{0,1}}\ R_\nu}{p \vdash (\neg p)^1, (p)^{0,1}}\ R_G$$

The following lemma shows that in the context of Socratic proofs the two versions of provisos—the semantic one given in Table 3.5 and the one expressed in terms of $R(\phi)$-chains in Table 3.4—are equivalent.

Lemma 3.1 *If ϕ occurs in a Socratic proof, then relation $R(\phi)$ satisfies proviso described in Table 3.5 iff it satisfies proviso described in Table 3.4.*

The assumption that ϕ occurs in a Socratic proof is essential in Euclidean logics, where we need chains starting with numeral 0. While in the case of some single properties, like reflexivity or symmetry, the result is straightforward, things are not that obvious in some other cases. We prove Lemma 3.1 for $\mathsf{L} = \{\mathsf{KB4}, \mathsf{K5}, \mathsf{D5}, \mathsf{K45}, \mathsf{D45}\}$, and we skip the remaining, simpler cases.

Assume that ϕ occurs in a Socratic proof constructed in calculus $\mathbb{E}^{\mathsf{L}}$.

Proof of Lemma 3.1 for $\mathsf{L} = \mathsf{KB4}$ We need to show that $\langle i, j\rangle \in R(\phi)^{sym,tr}$ iff there is an $R(\phi)$-$[i, j]$-chain. The universe in this and the other cases is $W(\phi)$.

The if-then direction. By assumption, relation $R(\phi)^{sym,tr}$ is the smallest symmetric and transitive superset of $R(\phi)$, hence it is enough to show that the set of all pairs of numerals $\langle k, l\rangle$ such that there exists an $R(\phi)$-$[k, l]$-chain is itself a binary, symmetric and transitive relation, and a superset of $R(\phi)$ and it will thereafter follow that $R(\phi)^{sym,tr}$ must be a subset of this set.

Thus let X stand for this set of pairs. That $R(\phi) \subseteq X$ follows from the fact that for each $\langle k, l\rangle \in R(\phi)$, the very pair $\langle k, l\rangle$ is an $R(\phi)$-chain. Moreover, by Definition 3.5,

$R(\phi)$-$[k, l]$-chain is also an $R(\phi)$-$[l, k]$-chain, hence X is symmetric. And finally, if there is an $R(\phi)$-$[k, l]$-chain, $\langle k_1, \ldots, k_n \rangle$, and an $R(\phi)$-$[l, m]$-chain, $\langle l_1, \ldots, l_o \rangle$, then sequence $\langle k_1, \ldots, k_n, l_2, \ldots, l_o \rangle$ is an $R(\phi)$-$[k, m]$-chain, hence the relation is transitive. Thus we have shown that $R(\phi)^{sym,tr} \subseteq X$.

For the only-if implication assume that there exists an $R(\phi)$-$[k, l]$-chain. By Definition 3.5, there is a sequence $\langle k_1, \ldots, k_n \rangle$ $(1 < n)$ of numerals such that $k = k_1$, $l = k_n$, and for each m $(1 \leqslant m < n)$: either $\langle k_m, k_{m+1} \rangle \in R(\phi)$ or $\langle k_{m+1}, k_m \rangle \in R(\phi)$. We also make use of the fact that $R(\phi) \subseteq R(\phi)^{sym,tr}$. Our goal is to show that $\langle k_1, k_n \rangle$, that is, $\langle k, l \rangle$, belongs to $R(\phi)^{sym,tr}$. The argument is by induction running through $k_1, \ldots, k_n$.

Hence, for the base case we observe that either $\langle k_1, k_2 \rangle \in R(\phi)$ or $\langle k_2, k_1 \rangle \in R(\phi)$, and since $R(\phi)^{sym,tr}$ is symmetric, in both cases $\langle k_1, k_2 \rangle \in R(\phi)^{sym,tr}$ holds.

Assume (IH) that for m $(1 \leqslant m < n)$: $\langle k_1, k_m \rangle \in R(\phi)^{sym,tr}$. By Definition 3.5, one of $\langle k_m, k_{m+1} \rangle, \langle k_{m+1}, k_m \rangle$ must be in $R(\phi)$. Again, $R(\phi)^{sym,tr}$ is symmetric, hence the two pairs are in $R(\phi)^{sym,tr}$. Since the relation is also transitive, $\langle k_1, k_{m+1} \rangle \in R(\phi)^{sym,tr}$. It follows by induction that $\langle k, l \rangle \in R(\phi)^{sym,tr}$, which finishes the proof. □

***Proof of Lemma 3.1 for* K5 *and* D5** We need to show that $\langle k, l \rangle \in R(\phi)^{Eu}$ iff

(a) $\langle k, l \rangle \in R(\phi)$ or
(b) there are: a directed $R(\phi)$-$[0, k]$-chain and a directed $R(\phi)$-$[0, l]$-chain.

This time we refer to some properties of Socratic proofs. Since each Socratic proof starts with a question of the form $?(S \vdash (A)^0)$ and no rule can introduce a numeral *in front of* 0, for each sequent ψ that occurs in the Socratic proof, the relation $R(\psi)$ has a tree-like structure with 0 being the root. In other words, for each sequent ψ that occurs in a Socratic proof, for each numeral $m \in W(\psi)$ other than 0, there exists a directed $R(\psi)$-$[0, m]$-chain.

Another useful observation is that if sequent ψ occurs in a Socratic proof, then there is no pair of the form $\langle k, k \rangle$ in $R(\psi)$. Hence if $\langle k, l \rangle \in R(\psi)$, then, first, $l \neq 0$, second, by the above argument, there is a directed $R(\psi)$-$[0, l]$-chain, but, third, it is not necessary that there is also a directed $R(\psi)$-$[0, k]$-chain, as k can equal 0, which shows that (a) is not superfluous.

For the first implication we show, similarly to the proof for KB4, that the set, X, of pairs of numerals $\langle k, l \rangle$ that satisfy the proviso, that is (a) or (b), is itself a Euclidean relation and a superset of $R(\phi)$. The fact that $R(\phi) \subseteq X$ is secured by (a). For Euclideanness suppose that $\langle k, l \rangle \in X$ and $\langle k, m \rangle \in X$. There are four possibilities, depending on which part of the proviso is satisfied by the two pairs. First, $\langle k, l \rangle \in R(\phi)$ and $\langle k, m \rangle \in R(\phi)$. We can see that $l \neq 0$ and $m \neq 0$, hence there are: a directed $R(\phi)$-$[0, l]$-chain and a directed $R(\phi)$-$[0, m]$-chain, which shows that also $\langle l, m \rangle \in X$. Second, $\langle k, l \rangle \in R(\phi)$ and for the pair $\langle k, m \rangle$ there are the two directed chains, a directed $R(\phi)$-$[0, m]$-chain in particular. From $\langle k, l \rangle \in R(\phi)$ it follows that there is a directed $R(\phi)$-$[0, l]$-chain, hence, again, $\langle l, m \rangle \in X$. The arguments are similar in the remaining two cases.

For the second implication assume that $\langle k, l\rangle \in X$. Since $R(\phi) \subseteq R(\phi)^{Eu}$, we also assume that $\langle k, l\rangle \notin R(\phi)$. It follows that there are: a directed $R(\phi)$-$[0, k]$-chain: $\langle i_0, \ldots, i_p, \ldots, i_m\rangle$ and a directed $R(\phi)$-$[0, l]$-chain: $\langle j_0, \ldots, j_q, \ldots, j_n\rangle$. Similarly as in the proof for KB4, we conduct the rest of the proof by induction, showing that for $p \geqslant 1, q \geqslant 1$, both $\langle i_p, j_q\rangle \in R(\phi)^{Eu}$ and $\langle j_q, i_p\rangle \in R(\phi)^{Eu}$. Induction is used here with respect to the sum $p + q$.

Base step, $p = q = 1$. Both $\langle 0, i_1\rangle \in R(\phi)$ and $\langle 0, j_1\rangle \in R(\phi)$ by definition of directed chains, hence also both $\langle i_1, j_1\rangle \in R(\phi)^{Eu}$ and $\langle j_1, i_1\rangle \in R(\phi)^{Eu}$.

Assume (IH) that $\langle i_p, j_q\rangle \in R(\phi)^{Eu}$ and $\langle j_q, i_p\rangle \in R(\phi)^{Eu}$. Increasing the value of $p + q$ with 1 implies that either p increases its value or q increases its value, therefore we need to show that (c) $\langle i_{p+1}, j_q\rangle, \langle j_q, i_{p+1}\rangle, \langle i_p, j_{q+1}\rangle, \langle j_{q+1}, i_p\rangle \in R(\phi)^{Eu}$. By definition of directed chains, both $\langle i_p, i_{p+1}\rangle$ and $\langle j_q, j_{q+1}\rangle$ are in $R(\phi)$, and thus also in $R(\phi)^{Eu}$. This, together with (IH) and Euclideanness of $R(\phi)^{Eu}$ entails (c), as required. □

***Proof of Lemma 3.1 for* K45 *and* D45** Let X stand for the set of pairs $\langle k, l\rangle$ satisfying the relevant proviso, that is, such that:

(a) there is a directed $R(\phi)$-$[k, l]$-chain, or
(b) there are: a directed $R(\phi)$-$[0, k]$-chain and a directed $R(\phi)$-$[0, l]$-chain.

We show that $R(\phi)^{tr,Eu} = X$, where X is defined as in the previous cases.

If $\langle k, l\rangle \in R(\phi)$, then the pair is a directed $R(\phi)$-$[k, l]$-chain, hence $\langle k, l\rangle \in X$ by (a), and thus $R(\phi) \subseteq X$.

For transitivity of X assume that $\langle k, l\rangle$ and $\langle l, m\rangle$ are in X. There are four cases. First, for $\langle k, l\rangle$ there is a directed $R(\phi)$-$[k, l]$-chain and for $\langle l, m\rangle$ there is a directed $R(\phi)$-$[l, m]$-chain. Then by the concatenation of the first sequence with a fragment of the second (we do not want to repeat numeral l) we obtain a directed $R(\phi)$-$[k, m]$-chain, which shows, by (a), that $\langle k, m\rangle \in X$. Second case: assume that for $\langle k, l\rangle$ there is a directed $R(\phi)$-$[k, l]$-chain, whereas for $\langle l, m\rangle$ there are a directed $R(\phi)$-$[0, l]$-chain and a directed $R(\phi)$-$[0, m]$-chain. If $k = 0$, then the last chain is already a directed $R(\phi)$-$[k, m]$-chain, hence $\langle k, m\rangle \in X$ by (a). If $k \neq 0$, then there is a directed $R(\phi)$-$[0, k]$-chain (see the beginning of the previous proof, for $\mathsf{L} \in \{$K5, D5$\}$, for justification). The existence of a directed $R(\phi)$-$[0, k]$-chain and a directed $R(\phi)$-$[0, m]$-chain proves, by (b), that $\langle k, m\rangle \in X$. The reasoning is similar in the remaining two cases. Hence we have shown that X is a transitive relation.

For Euclideanness assume that $\langle k, l\rangle$ and $\langle k, m\rangle$ are in X. There are, again, four possible cases. Assume that there is a directed $R(\phi)$-$[k, l]$-chain and a directed $R(\phi)$-$[k, m]$-chain. If $k = 0$, then, by (b), $\langle l, m\rangle \in X$. If $k \neq 0$, then there is a directed $R(\phi)$-$[0, k]$-chain. By arguments analogous to those already used, there is also a directed $R(\phi)$-$[0, l]$-chain and a directed $R(\phi)$-$[0, m]$-chain, hence $\langle l, m\rangle \in X$ by (b). Second case: assume that there is (i) a directed $R(\phi)$-$[k, l]$-chain and for $\langle k, m\rangle$ there are (ii) a directed $R(\phi)$-$[0, k]$-chain and (iii) a directed $R(\phi)$-$[0, m]$-chain. Then we combine chains (ii) and (i) to obtain a directed $R(\phi)$-$[0, l]$-chain which, together with (iii), shows that $\langle l, m\rangle \in X$. Again, the reasoning is similar in the remaining two cases.

We have shown that $R(\phi)^{tr,Eu} \subseteq X$. For the converse inclusion let $\langle k, l\rangle \in X$. Assume first that (a) holds with respect to this pair, that is, there is a directed $R(\phi)$-$[k, l]$-chain $\langle i_1, \ldots, i_m, \ldots, i_n\rangle$. We show by induction with respect to m that $\langle k, i_m\rangle \in R(\phi)^{tr,Eu}$—this is a simplified version of the argument for KB4. If (b) holds, then the reasoning is as in the case of $\mathsf{L} \in \{\mathsf{K5}, \mathsf{D5}\}$. □

Soundness of Erotetic Calculi for Basic Modal Logics

In the modal case, the denotational interpretation of sequents is not a straightforward one. We can see that the explicit representation of semantic information in the sequents makes proving in the system easier, but complicates the meta-analysis.

For this reason in [11], where the proofs of soundness and completeness with respect to Kripke semantics were presented, the author used the so-called *interpretation functions* which map the semantic information encoded in a sequent into Kripke frames. In fact, for a sequent ϕ of language $\mathcal{M}^?_{\vdash}$ and a Kripke frame $\langle W, R\rangle$, interpretation functions are special cases of homomorphisms from $\langle W(\phi), R(\phi)\rangle$ to $\langle W, R\rangle$.[4] By and large, it is not the case that every sequent can be interpreted in every Kripke frame, but a sequent of the form '$\vdash (A)^0$' can, and under such an interpretation function it represents the formula A in an arbitrary world of a Kripke frame.

The above pertains to erotetic calculi for the local account of modal logics, as presented in [11]. Fortunately, the interpretation-function tool need not adjustments to the global account, since there are no numerals on the left side of a sequent.

Definition 3.8 Let ϕ be a sequent of $\mathcal{M}^?_{\vdash}$ and let $\langle W, R\rangle$ be a frame. By an *interpretation of ϕ in $\langle W, R\rangle$* we mean a function $f : W(\phi) \longrightarrow W$ such that

(hm) if $\langle i, j\rangle \in R(\phi)$, then $\langle f(i), f(j)\rangle \in R$.

We say that a *sequent is interpretable in a frame* iff there exists an interpretation function of the sequent in the frame.

It can happen that a sequent is not interpretable in a frame; for example, when $R(\phi) \neq \varnothing$ and $\langle W, R\rangle$ has R empty, then no interpretation exists. However, as we have observed, any sequent of the form '$S \vdash (A)^0$' is interpretable in every frame.

The fact that interpretation functions are homomorphisms from $\langle W(\phi), R(\phi)\rangle$ to $\langle W, R\rangle$ entails also the following lemma. By '$R(\phi)^{\mathsf{L}}$', used in the formulation of the lemma, we mean the closure of $R(\phi)$ under the suitable collection of properties among reflexivity, symmetry, transitivity, Euclideanness, which semantically characterize logic L.

4 This type of functions, acting between relational structures, are called *graph homomorphisms*, which, roughly speaking, are homomorphisms preserving edges. In computer science graph homomorphisms have been studied under the name 'interpretations', see [7, p. 108].

Lemma 3.2 *Let ϕ be a sequent occurring in a Socratic proof of some sequent in $\mathbb{E}^{\mathsf{L}}$, and let f be an interpretation function of sequent ϕ in $\langle W, R\rangle$. If $\langle i, j\rangle \in R(\phi)^{\mathsf{L}}$ and $\langle W, R\rangle$ is an L-frame, then also $\langle f(i), f(j)\rangle \in R$.*

Before the proof, let us observe that the essence of this lemma is that homomorphisms preserve properties like reflexivity, symmetry, transitivity and Euclideanness. As we have already observed, it is important that each definition of a single L-property has the form of an implication, with both antecedent and succedent expressing a positive condition. This also means that the lemma can be easily generalized *at least* to any finite number of this kind of properties.

Proof of Lemma 3.2 We assume that we deal with a collection of the four properties, $\{p_1, \ldots, p_l\}$ ($l \leqslant 4$), let us call it L. Let us recall that $R(\phi)$ is finite. We will consider $R(\phi)^{\mathsf{L}}$ as being approximated by monotonically increasing sequences of relations, as in (3.5):

$$R(\phi) \subseteq R(\phi)^{p_1} \subseteq \ldots \subseteq (^{\cdots}(R(\phi)^{p_1})^{\cdots})^{p_l} \subseteq$$
$$\subseteq ((^{\cdots}(R(\phi)^{p_1})^{\cdots})^{p_l})^{p_1} \subseteq \ldots \text{ repeated (3.5)-steps} \ldots \subseteq R(\phi)^{\mathsf{L}} \quad (3.6)$$

Then (3.6) can be used to build an argument by induction on the consecutive sets approximating $R(\phi)^{\mathsf{L}}$. To simplify the notation for the relations, let us use 'R_0^{ϕ}' for $R(\phi)$ and 'R_n^{ϕ}' for the n-th closure in (3.6).

In the base step we observe that, by Definition 3.8 of an interpretation function, it holds that if $\langle i, j\rangle \in R_0^{\phi}$, then $\langle f(i), f(j)\rangle \in R$.

Assume that for the n-th closure, R_n^{ϕ}, in sequence (3.6) it holds that:

(IH) if $\langle i, j\rangle \in R_n^{\phi}$, then $\langle f(i), f(j)\rangle \in R$.

We consider the next closure, R_{n+1}^{ϕ}, and we assume that $\langle i, j\rangle \in R_{n+1}^{\phi}$. Let us divide the relation as follows:

$$R_{n+1}^{\phi} = R_n^{\phi} \cup \left(R_{n+1}^{\phi} \setminus R_n^{\phi}\right)$$

If $\langle i, j\rangle \in R_n^{\phi}$, then we apply IH. Assume that:

(a) $\langle i, j\rangle \in R_{n+1}^{\phi} \setminus R_n^{\phi}$.

The difference between the two consecutive relations in (3.6) is that for some property p, $R_{n+1}^{\phi} = (R_n^{\phi})^p$. Since in our case $p \in$ {reflexivity, symmetry, transitivity, Euclideanness}, we consider the four cases separately.

The case of reflexivity. (a) holds only if $i = j$, and since f is an interpretation of ϕ in the frame, $f(i) \in W$. But R is reflexive, thus $\langle f(i), f(j)\rangle \in R$.

The case of symmetry. Case (a) holds only if $\langle j, i\rangle \in R_n^{\phi}$. By IH, $\langle f(j), f(i)\rangle \in R$, and since R is symmetric, $\langle f(i), f(j)\rangle \in R$.

Transitivity. Relation R_{n+1}^{ϕ} is a transitive closure of R_n^{ϕ}. From Lemma 3.1 we have learned that (a) holds only if there is a directed R_n^{ϕ}-$[i, j]$-chain. By Definition 3.5, $\langle i_k, i_{k+1}\rangle \in R_n^{\phi}$, where k runs through the indices of the elements of the chain. By

IH, $\langle f(i_k), f(i_{k+1})\rangle \in R$, and since R is transitive, also, finally, $\langle f(i), f(j)\rangle \in R$. A proper argument should go by induction with respect to k, but since we have dealt with these in the proof of Lemma 3.1, we can skip it now.

Euclideanness. Relation R^{ϕ}_{n+1} is a Euclidean closure of R^{ϕ}_{n}, and from Lemma 3.1 we have learned that case (a) holds only if there is a directed R^{ϕ}_{n}-[0, i]-chain and a directed R^{ϕ}_{n}-[0, j]-chain. This time we go through the details of the inductive argument. It will be a subinduction with respect to the main induction. Thus let:

- $\langle i_0, \ldots, i_p, \ldots, i_k\rangle$ be the directed R^{ϕ}_{n}-[0, i]-chain ($i_0 = 0, i_k = i$),
- $\langle j_0, \ldots, j_q, \ldots, j_m\rangle$ be the directed R^{ϕ}_{n}-[0, j]-chain ($j_0 = 0, j_m = j$).

In the base step (of subinduction) we show that:

(b) $\langle f(i_1), f(j_1)\rangle \in R$ and $\langle f(j_1), f(i_1)\rangle \in R$.

Indeed, by IH, $\langle f(0), f(i_1)\rangle \in R$ and $\langle f(0), f(j_1)\rangle \in R$, and hence (b) follows by Euclideanness of R. Suppose that (subinduction hypothesis):

(c) $\langle f(i_p), f(j_q)\rangle \in R$ and $\langle f(j_q), f(i_p)\rangle \in R$.

Since $\langle i_p, i_{p+1}\rangle \in R^{\phi}_{n}$ and $\langle j_q, j_{q+1}\rangle \in R^{\phi}_{n}$, by IH, $\langle f(i_p), f(i_{p+1})\rangle \in R$ and $\langle f(j_q), f(j_{q+1})\rangle \in R$. This, together with (c) and Euclideanness of R, leads to the conclusion that $\langle f(i_{p+1}), f(j_q)\rangle$, $\langle f(j_q), f(i_{p+1})\rangle \in R$, and also that $\langle f(i_p), f(j_{q+1})\rangle$, $\langle f(j_{q+1}), f(i_p)\rangle \in R$, as required. This finishes the proof. □

We proceed with the semantics for sequents of $\mathcal{M}^{?}_{\vdash}$.

Definition 3.9 Assume that $\langle W, R, V\rangle$ is a Kripke model. Let

$$\phi = A_1, \ldots, A_k \vdash (B_1)^{\varsigma_1/i_1}, \ldots, (B_n)^{\varsigma_n/i_n}$$

be a sequent of $\mathcal{M}^{?}_{\vdash}$, and let f be an interpretation of ϕ in frame $\langle W, R\rangle$. We say that ϕ *is true in* $\langle W, R, V\rangle$ *under* f, provided that:

(⋆) if each of $A_1, \ldots, A_k$ is true in $\langle W, R, V\rangle$, then for some j, $1 \leqslant j \leqslant n$, $V(B_j, f(i_j)) = 1$.

It follows that if $k \geqslant 1$ and at least one of $A_1, \ldots, A_k$ is not true in a model, then the whole sequent is true in the model under a given interpretation. For the special case whit $k = 0$ one obtains the following version of truth under f, which was used in [11].

Definition 3.10 Let $\langle W, R, V\rangle$ be a model, let $\phi = \vdash (B_1)^{\varsigma_1/i_1}, \ldots, (B_n)^{\varsigma_n/i_n}$ be a sequent of $\mathcal{M}^{?}_{\vdash}$, and let f be an interpretation of ϕ in frame $\langle W, R\rangle$. We say that ϕ *is true in* $\langle W, R, V\rangle$ *under* f, provided that for some j, $1 \leqslant j \leqslant n$, $V(B_j, f(i_j)) = 1$.

Further, we will say that a sequent, ϕ, is *valid in a frame* $\langle W, R\rangle$ *under an interpretation* f, provided that for every valuation V on the frame, ϕ is true in $\langle W, R, V\rangle$ under f. Thus validity of a sequent is relativized both to a frame and to an interpretation of the sequent in the frame. Finally, we drop the interpretations and say that a sequent, ϕ, is *valid in a frame* $\langle W, R\rangle$ iff ϕ is valid in $\langle W, R\rangle$ under every interpretation of ϕ in $\langle W, R\rangle$. It follows from the last generalization that.

Corollary 3.4 *If a sequent is not interpretable in a frame, then it is valid in the frame.*

We end the enumeration of semantic properties pertaining to sequents of $\mathcal{M}_{\vdash}^{?}$ by saying that a sequent, ϕ, is K-*valid* iff ϕ is valid in every frame. The notion is easily extended to any L defined above, thus a sequent is said to be L-*valid* iff ϕ is valid in every L-frame.

The following lemma shows that the turnstile symbol in an L-valid sequent coincides with global L-entailment.

Lemma 3.3 *Let S stand for a finite, possibly empty, sequence of formulas of $\mathcal{M}$ and let A be a formula of this language. Sequent of the form '$S \vdash (A)^0$' of the language $\mathcal{M}_{\vdash}^{?}$ is L-valid iff* $\mathsf{set}(S) \models_{\mathsf{L}} A$.

Proof The proof is by contraposition.

Assume that sequent '$S \vdash (A)^0$' is not L-valid. By the above definitions it follows that there is an L-frame $\langle W, R\rangle$ and an interpretation f of the sequent in the frame such that '$S \vdash (A)^0$' is not valid in $\langle W, R\rangle$ under f. Further, it follows that for some V, the sequent is not true in model $\langle W, R, V\rangle$ under f. Hence, and from Definition 3.9, it follows that each formula that occurs in sequence S is true in the model, but $V(A, f(0)) = 0$. But this just means that A is not globally L-entailed by $\mathsf{set}(S)$.

Assume that A is not globally L-entailed by $\mathsf{set}(S)$. By a similar chain of definitions we arrive at the conclusion that there is an L-model $\langle W, R, V\rangle$ where each element of $\mathsf{set}(S)$ is true, but A is false at some world $w \in W$. For establishing an interpretation of '$S \vdash (A)^0$' in $\langle W, R\rangle$ it is sufficient to put $f(0) = w$, since there are no other numerals in the sequent. It follows that the sequent is not L-valid. □

The next lemma states that each rule of $\mathbb{E}^{\mathsf{L}}$ preserves L-validity of sequents in both directions: top-down and bottom-up. It means that the rules are both correct and semantically invertible.

Lemma 3.4 (Correctness and Invertibility of Rules of $\mathbb{E}^{\mathsf{L}}$). *Let $?(\Phi)/?(\Psi)$ be an instance of a rule of $\mathbb{E}^{\mathsf{L}}$. Each sequent in Φ is L-valid iff each sequent in Ψ is L-valid.*

Proof The case of R_α. Assume that '$S \vdash \Gamma(\alpha^{\varsigma/i})$' is not L-valid. Then the sequent is not valid in some L-frame $\langle W, R\rangle$, and hence there is an interpretation f of the sequent in the frame, and a valuation V on the frame such that the sequent is not true in $\langle W, R, V\rangle$ under f. By Definition 3.9, each formula in S is true in $\langle W, R, V\rangle$, but no indexed formula from the sequent's succedent satisfies the relevant condition. Since i is the last term of α's index, it follows that $V(\alpha, f(i)) = 0$, and also the other elements of Γ are false in the respective worlds. But then either sequent '$S \vdash \Gamma({\alpha_0}^{\varsigma/i})$' is not true in $\langle W, R, V\rangle$ under f, or sequent '$S \vdash \Gamma({\alpha_1}^{\varsigma/i})$' is not true in $\langle W, R, V\rangle$ under f. The reasoning is analogous for the converse direction.

We skip the cases of R_β and $R_{\neg\neg}$.

The case of R_ν. Suppose that sequent '$S \vdash \Gamma(\nu^{\,\varsigma/i})$' is not L-valid. Then there is an L-frame $\langle W, R\rangle$, an interpretation f of the sequent in $\langle W, R\rangle$ and a valuation V on the frame such that each element of S is true in $\langle W, R, V\rangle$ under f, but no formula from Γ is true in the respective world indicated by the last numeral of its index, in particular, $V(\nu, f(i)) = 0$. By clause 1 of Corollary 3.2 it yields the existence of $w^* \in R(f(i))$ such that $V(\nu_0, w^*) = 0$. We now use function f to define an interpretation g of sequent '$S \vdash \Gamma({\nu_0}^{\varsigma/i,j})$' in $\langle W, R\rangle$, by putting $g(j) = w^*$ and $g(k) = f(k)$ for $k \neq j$. We can do this without violating condition (hm) from Definition 3.8, as j does not occur in '$S \vdash \Gamma(\nu^{\,\varsigma/i})$'. Clearly, sequent '$S \vdash \Gamma({\nu_0}^{\varsigma/i,j})$' is not L-valid, since it is not true in $\langle W, R, V\rangle$ under g.

The converse direction. Assume that sequent '$S \vdash \Gamma({\nu_0}^{\varsigma/i,j})$' is not L-valid. Then we have an L-frame $\langle W, R\rangle$, an interpretation f and a valuation V on the frame such that each formula in S is true in $\langle W, R, V\rangle$, and $V(\nu_0, f(j)) = 0$. Since $\langle i, j\rangle \in R(\Gamma({\nu_0}^{\varsigma/i,j}))$, by condition (hm) from Definition 3.8, $\langle f(i), f(j)\rangle \in R$, therefore $V(\nu, f(i)) = 0$. We obtain interpretation function g of sequent '$S \vdash \Gamma(\nu^{\,\varsigma/i})$' in $\langle W, R\rangle$ by restricting f to $W(\Gamma(\nu^{\,\varsigma/i}))$. The sequent is not true in $\langle W, R, V\rangle$ under g.

The case of R_π. Assume that '$S \vdash \Gamma(\pi^{\,\varsigma/i})$' is not true in an L-model $\langle W, R, V\rangle$ under some f. Then $V(\pi, f(i)) = 0$. The proviso of applicability of the rule warrants that $\langle i, j\rangle \in R(\Gamma(\pi^{\,\varsigma/i}))^{\mathsf{L}}$. By Lemma 3.2, $\langle f(i), f(j)\rangle \in R$, and hence $V(\pi_0, f(j)) = 0$ by clause 2 of Corollary 3.2. This is enough to show that L-validity is preserved bottom-up. For the top-down direction the property is saved by the fact that indexed formula '$\pi^{\,\varsigma/i}$' is rewritten in the conclusion.

In the case of $R_{\pi\mathsf{D}}$ the reasoning is analogous to that for R_ν.

The case of R_G. Suppose that '$A_1, \ldots, A_n \vdash \Gamma(\_)$' is not L-valid, choose $k \in \{1, \ldots, n\}$. Then for some L-model $\langle W, R, V\rangle$ and interpretation f of the sequent in $\langle W, R\rangle$ we have each of $A_1, \ldots, A_n$ true in every world in W, in particular, $V(A_k, f(i)) = 1$. Hence $V(-A_k, f(i)) = 0$ which yields that the sequent-conclusion, '$A_1, \ldots, A_n \vdash \Gamma(-{A_k}^i)$' is not L-valid. The converse direction is equally obvious. □

We note the following corollary without a proof.

Corollary 3.5 *Each sequent of $\mathcal{M}^?_{\vdash}$ of the form*

(*e*) $S \vdash \Gamma(B^{\,\varsigma_1/i})(-B^{\,\varsigma_2/i})$

indicated in Definition 3.7 is L*-valid.*

As in the previous chapter, sequents of language $\mathcal{M}^?_{\vdash}$ of the form (*e*) will be called *basic*.

Theorem 3.1 (Soundness of $\mathbb{E}^{\mathsf{L}}$ wrt Global L-entailment) *If there exists a Socratic proof of sequent '$S \vdash (A)^0$' in $\mathbb{E}^{\mathsf{L}}$, then* $\mathsf{set}(S) \models_{\mathsf{L}} A$.

Proof The proof is organized as that of Theorem 2.2.

Suppose that **p** is a Socratic proof of '$S \vdash (A)^0$' in $\mathbb{E}^{\mathsf{L}}$ and assume that **p** has n terms. By Definition 3.7 and Corollary 3.5, each constituent of the last (nth) term of **p** is L-valid. By Lemma 3.4 also each constituent of the $(n-1)$st term of **p** is L-valid. The argument is analogous in the inductive step, hence by decreasing induction '$S \vdash (A)^0$' is L-valid. By Lemma 3.3, $\mathsf{set}(S) \models_{\mathsf{L}} A$. □

It follows that.

Corollary 3.6 *If there exists a Socratic proof of a sequent of the form '$\vdash (A)^0$' in $\mathbb{E}^{\mathsf{L}}$, then the formula A is L-valid.*

In [11] the author proves the following theorem (there are slight differences in the formulation).

Theorem 3.2 (Completeness of $\mathbb{E}^{\mathsf{L}}$ wrt Kripke Semantics). *If a formula A of the language $\mathcal{M}$ is L-valid, then the sequent $\vdash (A)^0$ is provable in $\mathbb{E}^{\mathsf{L}}$.*

See [11, p. 67, Theorem 3.1]. Completeness is proved there by quite standard counter-model construction on the basis of a *complete* Socratic transformation of question '$?(\vdash (A)^0)$' which is not a Socratic proof of the relevant sequent. A complete Socratic transformation is one constructed according to a certain procedure described in [11]. However, the procedure was developed for the purpose of completeness proof only, and it needs not terminate. The problem of developing a terminating procedure for the case of transitive modal logics was solved by the author in [13].

Here we need to prove completeness of the global version of erotetic calculi for modal logics. We undertake this task in Sect. 3.5.

3.4 Admissible, Derivable and Structural Rules: The Modal Case

The goal of this section is the same as in the case of classical logic: it is mainly the analysis of structural rules of weakening and contraction, that will be of use in the next section concerning completeness. As before, we use:

Definition 3.11 Let Q be a question of the language $\mathcal{M}^?_{\vdash}$. A Socratic transformation of question Q in $\mathbb{E}^{\mathsf{L}}$ is *successful* if it is finite and each constituent of its last question is of the form (e) indicated in Definition 3.7.

Before we proceed to technical results, let us recall the findings from the previous chapter concerning the notions of admissibility and derivability.

Definition 3.12 (*Derivability, Semi-admissibility, Admissibility in* $\mathbb{E}^{\mathsf{L}}$) Let $\mathbf{r} = Q/Q^*$ be an erotetic rule. We will say that **r** is *derivable in* erotetic calculus $\mathbb{E}^{\mathsf{L}}$ iff there exists a finite Socratic transformation **s** of question Q in $\mathbb{E}^{\mathsf{L}}$ such that Q^* is its last term.

We will say that **r** is *semi-admissible in* erotetic calculus $\mathbb{E}^{\mathsf{L}}$ iff the following condition holds:

1. if there exists a successful Socratic transformation of question Q^* in $\mathbb{E}^{\mathsf{L}}$, then there exists a successful Socratic transformation of question Q in $\mathbb{E}^{\mathsf{L}}$.

If, except for clause 1, **r** satisfies also clause 2:

2. if there exists a successful Socratic transformation of question Q in $\mathbb{E}^{\mathsf{L}}$, then there exists a successful Socratic transformation of question Q^* in $\mathbb{E}^{\mathsf{L}}$, then we say that **r** is *admissible in* erotetic calculus $\mathbb{E}^{\mathsf{L}}$.

We move to auxiliary definitions and results. By *length* of a finite Socratic transformation we mean, as in the previous chapter, the number of terms of the Socratic transformation.

Corollary 3.7 *If Q/Q^* is an instance of a rule* **r** *of $\mathbb{E}^{\mathsf{L}}$ and there is a successful Socratic transformation of question Q^* in $\mathbb{E}^{\mathsf{L}}$, which is of length n, then there is also a successful Socratic transformation of question Q in $\mathbb{E}^{\mathsf{L}}$ which is of length $n + 1$.*

Renaming Numerals

Corollary 3.8 (Renaming Numerals: New for New) *Let* **s** *be a finite Socratic transformation of a question in $\mathbb{E}^{\mathsf{L}}$. Assume that rule* $\mathbf{r} \in \{R_\nu, R_{\pi\mathsf{D}}\}$ *was applied in* **s** *and let i be the new numeral introduced in the sequent-conclusion of* **r**. *Let $k \neq i$ be another numeral which does not occur in* **s**. *Then the result of replacing each occurrence of i in* **s** *with an occurrence of k is a finite Socratic transformation of the same question in $\mathbb{E}^{\mathsf{L}}$ and of the same length. Also, if the initial Socratic transformation,* **s**, *is successful, so is the final one.*

Proof The reasoning is analogous to that concerning renaming individual constants in FOL (see Corollary 2.6, Sect. 2.5). Let $\mathbf{s}^*$ stand for the sequence of questions that results from **s** by performing the described operation.

First, it is easy to see that if Q/Q^* is an instance of one of the rules R_α, R_β, $R_{\neg\neg}$, R_G, then after performing the replacement operation the new pair of questions remains an instance of the same rule. Clearly, the same pertains to instances of R_π that precede the considered application of $\mathbf{r} \in \{R_\nu, R_{\pi\mathsf{D}}\}$ in **s**, and hence also in $\mathbf{s}^*$. Some doubts may concern applications of R_π to further questions of $\mathbf{s}^*$: we may wonder whether a pair satisfying an L-proviso in **s** still satisfies the proviso in $\mathbf{s}^*$. The provisos, as expressed in Table 3.4, depend on the structure of $R(\phi)$, where ϕ is the sequent-premise of R_π. Let ϕ^* stand for the result of replacing i with k in ϕ. Then the structure of $R(\phi)$ and $R(\phi^*)$ is the same, as we can easily establish an isomorphism between them setting $h(i) = k$ and $h(l) = l$ for each $l \in W(\phi)$ other than i. Let us take a look; suppose that $\langle i_1, i_2\rangle \in R(\phi)$. It means that in the index of some indexed formula occurring in ϕ numeral i_1 precedes i_2. We replace i with k and we get ϕ^* where the said index is also modified (provided i occurred in the index at all). Clearly, now $h(i_1)$ precedes $h(i_2)$ in the new index, therefore $\langle h(i_1), h(i_2)\rangle \in R(\phi^*)$. We have shown that h is a homomorphism between $\langle W(\phi), R(\phi)\rangle$ and $\langle W(\phi^*), R(\phi^*)\rangle$. The last thing to observe is that for Euclidean logics the relevant proviso may require the existence of a directed chain starting with numeral 0. This numeral, however, cannot be affected by replacing i with k, as i occurs in **s** after application of R_ν, hence $i \neq 0$.

In the case of applications of R_ν, $R_{\pi\mathsf{D}}$ other than the one considered in the formulation of our corollary applicability is preserved in $\mathbf{s}^*$ due to the assumption that k does not occur in the original Socratic transformation **s**. This warrants that each numeral that was new at a given stage of construction of **s** remains new at the corresponding stage of construction of $\mathbf{s}^*$. Also the considered instance of **r** remains an instance of the same rule in $\mathbf{s}^*$, though this one violates restriction (3.4)—the practical one stating that the new numeral should be always the first in $\leqslant$-order. As we have observed, we need to relax this restriction for meta-level purposes. To sum up, we have shown that $\mathbf{s}^*$ is indeed a Socratic transformation in $\mathbb{E}^{\mathsf{L}}$.

Second, the first question of **s** is not altered, as it cannot contain i, hence **s** and $\mathbf{s}^*$ are Socratic transformations of the same question.

Third, a basic sequent remains basic after changing i to k, therefore if **s** is successful, then $\mathbf{s}^*$ is successful as well.

Finally, the length of $\mathbf{s}^*$ must be the same as that of **s**. □

Corollary 3.8 is used to secure applicability of R_ν in the proof of the Separation Lemma, just as applicability of R_γ, L_δ was secured with the use of Corollary 2.6. With this exception, the remaining parts of the proof of the Separation Lemma (see Sect. 2.5 in the previous chapter) do not depend on the character of erotetic rules, hence we state the following without a proof.

Lemma 3.5 (Separation Lemma: the Modal Case) *Let $\phi_1, \ldots, \phi_n$ stand for sequents of $\mathcal{M}^?_{\vdash}$. The following conditions are equivalent for each* L :

1. *There is a successful Socratic transformation of question '$?(\phi_1, \ldots, \phi_n)$' in $\mathbb{E}^{\mathsf{L}}$.*
2. *For each $i : 1 \leqslant i \leqslant n$, there is a successful Socratic transformation of the question '$?(\phi_i)$' in $\mathbb{E}^{\mathsf{L}}$.*

We recall also the following consequence of the Separation Lemma.

Corollary 3.9 (Length-Preserving Separation: the Modal Case) *If there is a successful Socratic transformation of question '$?(\phi_1, \ldots, \phi_n)$' in $\mathbb{E}^{\mathsf{L}}$ of length l, then for each $i : 1 \leqslant i \leqslant n$, there is a successful Socratic transformation of the question '$?(\phi_i)$' of length not greater than l.*

With the above separation result we can prove the following corollary.

Corollary 3.10 *Let ϕ stand for a sequent of the language $\mathcal{M}^?_{\vdash}$ and let i be a numeral in $W(\phi)$ other than 0. If there is a successful Socratic transformation of '$?(\phi)$' in $\mathbb{E}^{\mathsf{L}}$, then for each $k \in W(\phi)$, there is a successful Socratic transformation in $\mathbb{E}^{\mathsf{L}}$ of the question '$?(\phi^*)$', where the sequent ϕ^* results from ϕ by replacing every occurrence of numeral i with an occurrence of numeral k.*

Proof Let ϕ stand for a sequent such that there is a successful Socratic transformation of $?(\phi)$ in $\mathbb{E}^{\mathsf{L}}$. Let us explain at this point that it is essential that we do not assume that ϕ is *provable* in $\mathbb{E}^{\mathsf{L}}$, for this would yield that it is of the form '$S \vdash A^{0}$', whereas we need something more general. Assume that $k, i \in W(\phi), i \neq 0$. If $i = k$, then the thesis is trivial, hence we assume that $i \neq k$. Finally, let ϕ^* stand for the sequent that results from ϕ by replacing every occurrence of i with an occurrence of k. The proof is by induction on the length of a Socratic transformation of $?(\phi)$.

In the base case, the length equals 1 and hence the sequent is basic. It is easy to see that this property cannot be violated by replacing one numeral with another.

Assume that the length of the Socratic transformation of question $?(\phi)$ is $n+1$, $n \geqslant 1$. We consider the first rule applied to this question. Suppose that it is R_G, sequent ϕ is displayed as '$S \vdash \Gamma(\_)$', where '$(\_)$' indicates some position in sequence Γ; moreover, $j \in W(\phi)$.

$$\frac{?(S \vdash \Gamma(\_))}{?(S \vdash \Gamma(A^{j}))}\ R_G$$

If we drop the first question, then the remaining sequence of questions has n elements and is a successful Socratic transformation of $?(S \vdash \Gamma(A^{j}))$. IH applies to this question, so the effect of replacing in the question every occurrence of i with an occurrence of k also has a successful Socratic transformation in $\mathbb{E}^{\mathsf{L}}$. Let ψ^* stand for the sequent that results from '$S \vdash \Gamma(A^{j})$' by the replacing. Then the pair $?(\phi^*)/?(\psi^*)$ is an instance of R_G, hence, by Corollary 3.7, it follows that question $?(\phi^*)$ has a successful Socratic transformation in $\mathbb{E}^{\mathsf{L}}$.

Let us observe that, in general, it is inessential for the above reasoning whether $j = i$ or not. Things are similar in the case of rules $R_{\neg\neg}$, R_β and R_α; we only go through the branching case. Assume that $\phi = S \vdash \Gamma(\alpha^{\,\varsigma/j})$ and the Socratic transformation starts as follows:

$$\frac{?(S \vdash \Gamma(\alpha^{\,\varsigma/j}))}{?(S \vdash \Gamma(\alpha_0^{\,\varsigma/j})\ ;\ S \vdash \Gamma(\alpha_1^{\,\varsigma/j}))}\ R_\alpha$$

Let ψ_0 and ψ_1 stand for the respective sequents-conclusions of R_α. By Corollary 3.9, there is a successful Socratic transformation of $?(\psi_0)$ of length not exceeding n, the same pertains to $?(\psi_1)$. By IH applied twice, questions $?(\psi_0^*)$ and $?(\psi_1^*)$—where sequent ψ_0^* results from ψ_0 by replacing every occurrence of i with an occurrence of k, and similarly for ψ_1^*—both have a successful Socratic transformation in $\mathbb{E}^{\mathsf{L}}$. By Lemma 3.5, question '$?(\psi_0^*;\ \psi_1^*)$' also has a successful Socratic transformation in $\mathbb{E}^{\mathsf{L}}$. Since $?(\phi^*)/?(\psi_0^*;\ \psi_1^*)$ is an instance of R_α, by Corollary 3.7, question $?(\psi^*)$ has a successful Socratic transformation, as required. The situation is similar in the case of:

$$\frac{?(S \vdash \Gamma(\nu^{\,\varsigma/j}))}{?\left(S \vdash \Gamma\left(\nu_0^{\,\varsigma/j,l}\right)\right)}\ R_\nu$$

By proviso of applicability of R_ν, l is new to the sequent-premise, contrary to i and k, hence the proviso of applicability of the rule is not violated after we replace i with k. The reasoning is similar in the case of $R_{\pi\mathsf{D}}$.

Finally, consider:

$$\frac{?(S \vdash \Gamma(\pi^{\varsigma/j}))}{?(S \vdash \Gamma(\pi^{\varsigma/j}, \pi_0{}^{l}))} R_\pi$$

where pair $\langle j, l \rangle$ satisfies the proviso of applicability of R_π in $\mathbb{E}^{\mathsf{L}}$. Let ψ stand for the sequent-conclusion of the rule and let ψ^* stand for the sequent obtained from ψ by replacing every occurrence of i with k. By IH, question $?(\psi^*)$ has a successful Socratic transformation in $\mathbb{E}^{\mathsf{L}}$. The last thing to show is that $?(\phi^*)/?(\psi^*)$ is an instance of R_π, that is, that a relevant pair of numerals satisfies the proviso stated for L. A general argument can be formulated by setting:

$$h(i) = k, \quad h(i_0) = i_0 \text{ for each } i_0 \in W(\phi) \setminus \{i\}$$

which is a homomorphism between $\langle W(\phi), R(\phi)\rangle$ and $\langle W(\phi^*), R(\phi^*)\rangle$; this can be easily verified as in the proof of Corollary 3.8. This time h is not a bijection, but this does not influence our argument: all the provisos are expressed in terms of positive conditions, stating that if some pair(s) belongs to $R(\phi)$, then some other pair can be used for the application of the rule. Positive conditions are not violated by function h: simply, after the replacing k takes the role i played previously. For example, if $j = l$ (L is reflexive), then the equation must hold true after replacing i with k, whether $j = i$ or not.

At this point we can also see the reason for restricting i to differ from 0: the provisos for Euclidean logics require the existence of some directed chain(s) starting with 0, so we cannot replace 0 with other numeral. □

Let us observe that according to the above corollary, if there is a successful Socratic transformation of $?(S \vdash \Gamma(A^{0,1,2}))$ in $\mathbb{E}^{\mathsf{L}}$, then there is also, e.g., a successful Socratic transformation of $?(S \vdash \Gamma(A^{0,1,0}))$ in the same calculus. Clearly, sequent '$S \vdash \Gamma(A^{0,1,0})$' cannot be obtained by rules of $\mathbb{E}^{\mathsf{L}}$ in the context of a Socratic proof, as no rule allows to introduce 0 after 1. For this reason Corollary 3.10 deals with Socratic transformations, not Socratic proofs. Finally, the gist of the above corollary was to have:

Corollary 3.11 (Renaming Numerals: Old for New) *Let $i \neq 0$. If there is a successful Socratic transformation of a question of the form:*

$$?\left(S \vdash \Gamma\left(\nu_0{}^{\varsigma/k}\right)\left(\nu_0{}^{\varsigma/i}\right)\right)$$

in $\mathbb{E}^{\mathsf{L}}$, where the displayed occurrence of i is the only one in the sequent, then there is also a successful Socratic transformation of question

$$?\left(S \vdash \Gamma\left(\nu_0{}^{\varsigma/k}\right)\left(\nu_0{}^{\varsigma/k}\right)\right)$$

in $\mathbb{E}^{\mathsf{L}}$.

Proof Follows from Corollary 3.10. □

Let us observe that 'ν_0' in the formulation of Corollary 3.11 refers to an arbitrary formula of $\mathcal{M}$, as every formula can be preceded with '□'. The formulation was chosen, however, for a direct application of the corollary later in the proof of Lemma 3.9 of semi-admissibility of modal contraction.

Structural Rules: Weakening and Contraction

We are ready to define the rules of weakening and contraction and to examine their properties. The rule of weakening shall become more important in the last chapter, but it plays its role here as well.

The notion of subsequence is understood as previously: roughly speaking, a subsequence of a sequence S is obtained by deleting *some* elements of S without changing the order of the other elements (see Definition 2.9, Sect. 2.5). Here is the rule of weakening.

$$\frac{?(\Phi(S \vdash \Gamma))}{?(Phi(S \vdash \Gamma^*))}\ R_{weak}$$

where Γ^* is a non-empty proper subsequence of Γ. Contraction is defined as follows:

$$\frac{?(\Phi(S \vdash \Gamma^*))}{?(\Phi(S \vdash \Gamma))}\ R_{contr}$$

where Γ^* is a non-empty proper subsequence of Γ and $\mathsf{set}(\Gamma^*) = \mathsf{set}(\Gamma)$.

We use the same names for the rules of right weakening and contraction as in the previous chapter, this shall not lead to a confusion, though.

The property of R_{weak} and R_{contr} that stands out is that the global assumptions stored in the antecedents of sequents are not subject to structural rules. This is not a necessary limitation, but we will not consider here structural rules operating on global assumptions. Let us also note that in the modal version weakening and contraction concerns occurrences of *indexed* formulas of language $\mathcal{M}^?_\vdash$.

We prove the following lemma.

Lemma 3.6 (Semi-admissibility of R_{weak} in $\mathbb{E}^{\mathsf{L}}$). *If there exists a successful Socratic transformation* **s** *of the question-conclusion of* R_{weak} *in* $\mathbb{E}^{\mathsf{L}}$, *and* **s** *is of length n, then there exists a successful Socratic transformation of the question-premise of the rule in* $\mathbb{E}^{\mathsf{L}}$, *also of length n.*

Proof The proof is organized in the same way as that of Lemma 2.4.

As in the classical first-order setting, we have a difficult case here, which is an application of $\mathbf{r} \in \{R_\nu, R_{\pi\mathsf{D}}\}$. The problem is that indexed formulas added by weakening can bring the numeral that is introduced as new by application of **r**.

In the base case **s** has one term: $?(\Phi(S \vdash \Gamma^*))$. If Γ^* is a subsequence of Γ and sequent $S \vdash \Gamma^*$ is basic, then, trivially, sequent $S \vdash \Gamma$ is basic as well.

In the inductive step we assume that $\mathbf{s} = \langle Q_0, Q_1, \ldots, Q_n \rangle$ is an $(n+1)$-term Socratic transformation of question Q_0 in $\mathbb{E}^{\mathsf{L}}$, and that the second question, Q_1, results from $Q_0 = ?(\Phi(S \vdash \Gamma^*))$, by a rule $\mathbf{r}$ of $\mathbb{E}^{\mathsf{L}}$. We also assume that '$S \vdash \Gamma^*$' is the sequent-premise of $\mathbf{r}$, Γ^* is a finite sequence of indexed formulas of $\mathcal{M}$, and it is a proper and non-empty subsequence of some finite sequence Γ of indexed formulas of $\mathcal{M}$. We deal only with the situation of $\mathbf{r} \in \{R_\upsilon, R_{\pi\mathsf{D}}\}$; let i be the new numeral introduced in Q_1. The additional elements of Γ can have i in some index—here is an example. The instance of R_υ indicated on the left is no longer an instance of this rule after weakening the question with '$(D)^1$'.

$$\frac{?(A, B \vdash (\neg\Diamond C)^0)}{?(A, B \vdash (\neg C)^{0,1})}\, R_\upsilon \qquad \frac{?(A, B \vdash (\neg\Diamond C)^0, (D)^1)}{?(A, B \vdash (\neg C)^{0,1}, (D)^1)}$$

Similarly to the case of FOL, we need to carefully *rename* numerals in order to save applicability of R_υ ($R_{\pi\mathsf{D}}$) in such cases. We take the largest numeral, l, among those occurring in indexed formulas of Γ, and we set $k = max\{l, i\} + 1$. By Corollary 3.8, the result of replacing numeral i, introduced by application of $\mathbf{r} \in \{R_\upsilon, R_{\pi\mathsf{D}}\}$ into the sequent-conclusion of $\mathbf{r}$, with numeral k, where the replacing operation is performed on the whole $\mathbf{s}$, is a finite Socratic transformation of Q_0 in $\mathbb{E}^{\mathsf{L}}$ of the same length as that of $\mathbf{s}$. Further we use IH, just as in the proof of Lemma 2.4. The whole proof relies on Corollaries 3.8 and 3.7. □

Let us repeat that previously we have formulated a restriction on the new numeral: that it is always the first new numeral in $\leqslant$-order (see condition (3.4), Sect.3.3). It is useful for practical reasons, but not for the aims of structural analysis, as we can see, for instance, in the above proof. We have dropped this restriction also in the proof of Corollary 3.10.

The following corollary is a straightforward consequence of Lemma 3.6.

Corollary 3.12 (Eliminability of R_{weak}) *Let Q stand for a question of $\mathcal{M}^?_\vdash$. If there is a successful Socratic transformation of Q in $\mathbb{E}^{\mathsf{L}} \cup \{R_{weak}\}$, then there is also one in $\mathbb{E}^{\mathsf{L}}$.*

Let us note that in proving weakening admissibility result we have used the feature of preserving length; it was a part of the accepted strategy. As in the previous chapter, for contraction we use the more difficult route. We start with proof-theoretic invertibility of the rules of $\mathbb{E}^{\mathsf{L}}$ (Lemma 3.7, Corollary 3.13, 3.14).

Lemma 3.7 (Transparency of Double Negation: the Modal Case) *There is a successful Socratic transformation of question '$?(S \vdash \Gamma(\neg\neg A^{\,\varsigma}))$' in $\mathbb{E}^{\mathsf{L}}$ iff there is one of question '$?(S \vdash \Gamma(A^{\,\varsigma}))$'.*

Proof The proof follows that of Lemma 2.5 and we skip some parts of it. We reason by induction with respect to length of a Socratic transformation.

The only-if direction is trivial and we skip it (see the proof of Lemma 2.5). For the if-then direction, we prove what follows.

(S^N) If a question Q has a successful Socratic transformation in $\mathbb{E}^{L}$, then for each constituent of Q which is of the form '$S \vdash \Gamma(\neg\neg A^{\varsigma})$', question '$?(S \vdash \Gamma(A^{\varsigma}))$' has a successful Socratic transformation in $\mathbb{E}^{L}$.

Base case. Let $\phi = S \vdash \Gamma(\neg\neg A^{\varsigma})$, $\psi = S \vdash \Gamma(A^{\varsigma})$, and assume that ϕ is basic, that is, is of the form *(e)* $S \vdash \Gamma(B^{\varsigma_1/i})(-B^{\varsigma_2/i})$ specified in Definition 3.7. There are two cases to consider: (a) $\neg\neg A \notin \{B, -B\}$, and (b) $\neg\neg A \in \{B, -B\}$. In case (a) sequent ψ is basic as well. Similarly, if (b) holds and $\neg\neg A = B$, then the succedent of ψ contains an occurrence of '$(\neg A)^{\varsigma_2/i}$', hence ψ is basic as well. If $\neg\neg A = -B$, then B must be of the form '$\neg\neg\neg A$'. In this situation the following:

$$\frac{?(S \vdash \Gamma(\neg\neg\neg A^{\varsigma_1/i})(A^{\varsigma/i}))}{?(S \vdash \Gamma(\neg A^{\varsigma_1/i})(A^{\varsigma/i}))}\ R_{\neg\neg}$$

shows that ψ has a Socratic proof in $\mathbb{E}^{L}$.

The inductive part goes along the lines of the proof of Lemma 2.5. We assume that $\langle Q_0, Q_1, \ldots, Q_n\rangle$ is a successful Socratic transformation of question

$$Q_0 = ?(\Phi(S \vdash \Gamma((\neg\neg A)^{\varsigma})))$$

in $\mathbb{E}^{L}$. We also assume that the rule applied to Q_0 acts upon the distinguished sequent, '$S \vdash \Gamma((\neg\neg A)^{\varsigma})$', but the distinguished occurrence of '$(\neg\neg A)^{\varsigma}$' is not active. (Otherwise we reason as in Lemma 2.5.) We distinguish two cases.

(a) Q_1 results from Q_0 by a non-branching rule $\mathbf{r} \in \{R_G, R_\beta, R_{\neg\neg}, R_\nu, R_\pi, R_{\pi\mathrm{D}}\}$.
(b) Q_1 results from Q_0 by R_α.

In case (a), IH applies to question $Q_1 = ?(\Phi(S \vdash \Gamma^*(\neg\neg A^{\varsigma})))$, which yields that question '$?(S \vdash \Gamma^*(A^{\varsigma}))$' has a successful Socratic transformation in $\mathbb{E}^{L}$; further we reason as in Lemma 2.5, Corollary 3.7 is needed. For case (b), question Q_1 is of the form:

$$Q_1 = ?(\Phi(S \vdash \Gamma_1(\neg\neg A^{\varsigma});\ S \vdash \Gamma_2(\neg\neg A^{\varsigma}))),$$

and by IH both questions, '$?(S \vdash \Gamma_1(A^{\varsigma}))$' and '$?(S \vdash \Gamma_2(A^{\varsigma}))$', have successful Socratic transformations in $\mathbb{E}^{L}$. We use Lemma 3.5 (Separation) to conclude that there is a successful Socratic transformation of question '$?(S \vdash \Gamma_1(A^{\varsigma});\ S \vdash \Gamma_2(A^{\varsigma}))$' in $\mathbb{E}^{L}$, and hence, by Corollary 3.7, that there is also one of '$?(S \vdash \Gamma(A^{\varsigma}))$'. □

Corollary 3.13 (Proof-Theoretic Invertibility of Rules R_α, R_β) *The following holds:*

1. *If '$?(S \vdash \Gamma(\alpha^{\varsigma}))$' has a successful Socratic transformation in $\mathbb{E}^{L}$, then '$?(S \vdash \Gamma(\alpha_0{}^{\varsigma}))$' has a successful Socratic transformation in $\mathbb{E}^{L}$, and '$?(S \vdash \Gamma(\alpha_1{}^{\varsigma}))$' has a successful Socratic transformation in $\mathbb{E}^{L}$.*
2. *If '$?(S \vdash \Gamma(\beta^{\varsigma}))$' has a successful Socratic transformation in $\mathbb{E}^{L}$, then '$?(S \vdash \Gamma(\beta_0{}^{\varsigma}, \beta_1{}^{\varsigma}))$' has a successful Socratic transformation in $\mathbb{E}^{L}$.*

Proof There is nothing important to be added to the classical case, hence we can restrict ourselves to the assertion that the proof follows that of Corollary 2.12 and

2.13. We shall provide more details in the proof of the next corollary, where we follow the same pattern. □

Corollary 3.14 (Proof-Theoretic Invertibility of Rules R_π, $R_{\pi\mathsf{D}}$, R_ν) *The following holds:*

1. *For each* L, *if* '$?(S \vdash \Gamma((\pi)^{\varsigma/i}))$' *has a successful Socratic transformation in* $\mathbb{E}^{\mathsf{L}}$, *then for each numeral* j, *such that pair* $\langle i, j\rangle$ *satisfies the proviso of applicability of rule* R_π *in* $\mathbb{E}^{\mathsf{L}}$, *question* '$?(S \vdash \Gamma((\pi)^{\varsigma/i}, (\pi_0)^j))$' *has a successful Socratic transformation in* $\mathbb{E}^{\mathsf{L}}$.
2. *For* $\mathsf{L} \in \{\mathsf{D}, \mathsf{D4}, \mathsf{D5}, \mathsf{DB}, \mathsf{D45}\}$, *if* '$?(S \vdash \Gamma((\pi)^{\varsigma/i}))$' *has a successful Socratic transformation in* $\mathbb{E}^{\mathsf{L}}$, *then for each numeral* j, *that does not occur in* $W(\Gamma((\pi)^{\varsigma/i}))$, *question* '$?(S \vdash \Gamma((\pi)^{\varsigma/i}, (\pi_0)^{i,j}))$' *has a successful Socratic transformation in* $\mathbb{E}^{\mathsf{L}}$.
3. *For each* L, *if* '$?(S \vdash \Gamma((\nu)^{\varsigma/i}))$' *has a successful Socratic transformation in* $\mathbb{E}^{\mathsf{L}}$, *then for each numeral* j, *that does not occur in* $W(\Gamma((\nu)^{\varsigma/i}))$, *question* '$?(S \vdash \Gamma((\nu_0)^{\varsigma/i,j}))$' *has a successful Socratic transformation in* $\mathbb{E}^{\mathsf{L}}$.

Proof Clauses 1 and 2 follow from Lemma 3.6 (semi-admissibility of weakening). Clause 3 is proved by induction with respect to length of a Socratic transformation, in a way presented many times above. We prove:

($\mathbf{S}^\pi$) If a question Q has a successful Socratic transformation in $\mathbb{E}^{\mathsf{L}}$, then for each constituent of Q which is of the form '$S \vdash \Gamma((\nu)^{\varsigma/i})$' and for each numeral j that does not occur in Q, question '$?(S \vdash \Gamma((\nu_0)^{\varsigma/i,j}))$' has a successful Socratic transformation in $\mathbb{E}^{\mathsf{L}}$.

Let $\phi = S \vdash \Gamma((\nu)^{\varsigma/i})$, $\psi = S \vdash \Gamma((\nu_0)^{\varsigma/i,j})$ and $j \notin W(\phi)$.
Base case. Assume that question $Q = ?(\Phi(\phi))$ has a Socratic proof in $\mathbb{E}^{\mathsf{L}}$ of length 1. Then each constituent of Q, ϕ in particular, is basic, hence for some formula A, indexed formulas '$(A)^{\varrho_1/k}$' and '$(-A)^{\varrho_2/k}$' occur in '$\Gamma((\nu)^{\varsigma/i})$'. If '$(\nu)^{\varsigma/i}$' is neither of them, then sequent ψ is basic as well. If '$(\nu)^{\varsigma/i}$' is one of them, then, by Corollary 3.1, clause 5, the other indexed formula is of the form '$(\pi)^{\varrho/i}$' or '$(\neg\neg\pi)^{\varrho/i}$'. In view of Lemma 3.7, the latter case can be omitted. Hence let ϕ be of the form:

$$\phi = S \vdash \Gamma\left((\nu)^{\varsigma/i}\right)\left((\pi)^{\varrho/i}\right)$$

where one of ν, π is the complement of the other and $\nu_0 \in \{\neg\pi_0, -\pi_0\}$. Since $\langle i, j\rangle \in R(\Gamma((\nu_0)^{\varsigma/i,j})((\pi)^{\varrho/i}))$, the following:

$$\frac{?\left(S \vdash \Gamma\left((\nu_0)^{\varsigma/i,j}\right)\left((\pi)^{\varrho/i}\right)\right)}{?\left(S \vdash \Gamma\left((\nu_0)^{\varsigma/i,j}\right)\left((\pi)^{\varrho/i}, (\pi_0)^j\right)\right)}\; R_\pi$$

is a successful Socratic transformation of $?(\psi)$ in $\mathbb{E}^{\mathsf{L}}$.
Inductive step. Assume that there is a successful Socratic transformation $\mathbf{s} = \langle Q, Q_1, \ldots, Q_n\rangle$ of question $Q = ?(\Phi(\phi))$ in $\mathbb{E}^{\mathsf{L}}$ which is of length $n + 1$ ($n \geqslant 1$).

If the rule applied to Q acts on a sequent other than ϕ, then the result follows by IH. Suppose that Q/Q_1 is an instance of **r** and that ϕ is the sequent-premise in Q. We list the possible cases:

1. $\mathbf{r} = R_G$;
2. $\mathbf{r} \in \{R_{\neg\neg}, R_\beta, R_\pi\}$;
3. $\mathbf{r} \in \{R_\nu, R_{\pi\mathsf{D}}\}$;
4. $\mathbf{r} = R_\alpha$ (branching case).

In case 1, Q and Q_1 are of the following forms (by '(_)' we indicate a position in Γ, in S the active formula A is displayed):

$$Q = ?(\Phi\ (S(A) \vdash \Gamma((\nu)^{\varsigma/i})(\_)))$$

$$Q_1 = ?(\Phi\ (S(A) \vdash \Gamma((\nu)^{\varsigma/i})((-A)^k)))$$

where $k \in W(\Gamma((\nu)^{\varsigma/i})(\_))$. Since, by assumption, $j \notin W(\phi)$, we know that $k \neq j$. Question Q_1 has a successful Socratic transformation in $\mathbb{E}^{\mathsf{L}}$ of length n, hence, by IH, question '$?(S(A) \vdash \Gamma((\nu_0)^{\varsigma/i,j})((-A)^k))$' has a successful Socratic transformation in $\mathbb{E}^{\mathsf{L}}$. Finally, as Q/Q_1 is an instance of R_G, also

$$\frac{Q^* = ?(S(A) \vdash \Gamma((\nu_0)^{\varsigma/i,j})(\_))}{Q_1^* = ?(S(A) \vdash \Gamma((\nu_0)^{\varsigma/i,j})((-A)^k))}$$

is such an instance. This, by Corollary 3.7, yields that Q^* has a successful Socratic transformation in $\mathbb{E}^{\mathsf{L}}$, as required. Case 2 is proved in the same way and we skip it.

In case 3 we focus on $\mathbf{r} = R_\nu$, the reasoning is analogous for $R_{\pi\mathsf{D}}$. There are two possible subcases, depending on whether the distinguished '$(\nu)^{\varsigma/i}$' is active (case 3.1) or not active (3.2); each of them divides further depending on whether the numeral introduced as new is identical to j (case a) or not (b). The simplest case is 3.2b, when R_ν acts upon an indexed formula other than the already distinguished '$(\nu)^{\varsigma/i}$' and introduces a numeral different than j; the reasoning then goes as in case 1 and 2. If 3.1a holds, then question Q_1 is already of the form '$?(\Phi(S \vdash \Gamma((\nu_0)^{\varsigma/i,j})))$' and we do not need IH, as the result follows by Lemma 3.5 (Separation) alone. For the other two subcases we need Corollary 3.8 (renaming numerals).

In case 3.1b, rule R_ν acts upon '$(\nu)^{\varsigma/i}$' and introduces a numeral k different than j. We have to consider another two possibilities: j is new to the whole **s** or not. If j is not present in **s**, then we apply Corollary 3.8 replacing every occurrence of k with j. The resulting Socratic transformation $\mathbf{s}^* = \langle Q, Q_1^*, \ldots, Q_n^* \rangle$ obtained from **s** is a successful Socratic transformation of question Q in $\mathbb{E}^{\mathsf{L}}$; then sequence $\langle Q_1^*, \ldots, Q_n^* \rangle$ is a successful Socratic transformation of question Q_1^* in $\mathbb{E}^{\mathsf{L}}$, and ψ is a constituent of Q_1^*. Finally, by Lemma 3.5 (Separation), $?(\psi)$ has a successful Socratic transformation in $\mathbb{E}^{\mathsf{L}}$. However, the situation complicates if j is already present in **s** (we assumed only that it is not present in ϕ). In this situation we must first replace it with some another numeral to obtain the first, just described, possible case. Since j does not occur in Q, and is not introduced to Q_1 with the considered application of

R_ν, it follows that it occurs in question Q_2 or further by application of R_ν or $R_{\pi\mathsf{D}}$. We take the first in $\leqslant$-order numeral, l, that does not occur in $\mathbf{s}$ and we replace j with l. By Corollary 3.8, the result is still a successful Socratic transformation of the same question in $\mathbb{E}^{\mathsf{L}}$ and it does not contain any occurrence of j, hence we can use Corollary 3.8 once again to replace k with j. Let us observe that subcase 3.1b does not use IH, hence we do not even need to rely on the preservation of length which is warranted by Corollary 3.8.

In case 3.2a, R_ν acts upon an indexed formula other than the distinguished '$(\nu)^{\varsigma/i}$', but it introduces our numeral j. Let l stand for the first in $\leqslant$-order numeral that does not occur in $\mathbf{s}$. By Corollary 3.8, the result of replacing j with l in the whole $\mathbf{s}$ is a successful Socratic transformation of the same question and of the same length. IH applies now, j is new to the sequent, and the reasoning goes as previously in case 1, 2 and 3.2b.

Case 4 is the one that requires the apparently redundant formulation of $(\mathbf{S}^\pi)$. Questions Q and Q_1 are now of the forms:

$$Q = ?(\Phi\ (S \vdash \Gamma(\nu^{\,\varsigma/i})(\alpha^{\,\varrho})))$$

$$Q_1 = ?(\Phi\ (S \vdash \Gamma(\nu^{\,\varsigma/i})(\alpha_0{}^{\varrho})\ ;\ S \vdash \Gamma(\nu^{\,\varsigma/i})(\alpha_1{}^{\varrho})))$$

IH applies to Q_1 which has a successful Socratic transformation of length n. It follows that both questions: '$?(S \vdash \Gamma(\nu_0{}^{\varsigma/i,j})(\alpha_0{}^{\varrho}))$' and '$S \vdash \Gamma(\nu_0{}^{\varsigma/i,j})(\alpha_1{}^{\varrho})$' have successful Socratic transformations in $\mathbb{E}^{\mathsf{L}}$. By Lemma 3.5 (Separation), question $Q_1^* = ?(S \vdash \Gamma(\nu_0{}^{\varsigma/i,j})(\alpha_0{}^{\varrho})\ ;\ S \vdash \Gamma(\nu_0{}^{\varsigma/i,j})(\alpha_1{}^{\varrho}))$ has a successful Socratic transformation in $\mathbb{E}^{\mathsf{L}}$ as well. Adding a suitable instance of R_α yields, by Corollary 3.7, that there is a successful Socratic transformation of $?(\psi)$ in $\mathbb{E}^{\mathsf{L}}$. □

Let us recall the general formulation of contraction:

$$\frac{?(\Phi\ (S\ \vdash\ \Gamma^*))}{?(\Phi\ (S\ \vdash\ \Gamma))}\ R_{contr}$$

where Γ^* is a non-empty proper subsequence of Γ and $\mathsf{set}(\Gamma^*) = \mathsf{set}(\Gamma)$ (the set-of-terms condition). As in the previous chapter, we shall use a more specific form of this rule in order to make the proof of admissibility more transparent.

$$\frac{?(\Phi\ (S\ \vdash\ \Gamma\ (A^{\,\varsigma})(\_)))}{?(\Phi\ (S\ \vdash\ \Gamma\ (A^{\,\varsigma})(A^{\,\varsigma})))}\ R^*_{contr}$$

We will also call sequent '$S \vdash \Gamma(A^{\,\varsigma})(\_)$' a *contraction of sequent* '$S \vdash \Gamma(A^{\,\varsigma})(A^{\,\varsigma})$'.

Let us observe that here we deal with right-contraction only. Many things simplify when compared to the FOL-case, which is mainly due to the fact that we consider only one version of the calculi, instead of three, and the left side containing global assumptions is not subject to contraction.

Definition 3.13 (*Rank of a Formula of $\mathcal{M}$*) By a *rank of a formula of the language* $\mathcal{M}$ we mean a function $r : FORM_{\mathcal{M}} \longrightarrow \mathbb{N}_0$, such that:

1. For each propositional variable p_i, $r(p_i) = r(-p_i) = 0$;
2. For each formula of $\mathcal{M}$ of the form '$\neg\neg A$', $r(\neg\neg A) = r(A) + 1$;
3. For each formula α of $\mathcal{M}$, $r(\alpha) = r(\alpha_0) + r(\alpha_1) + 1$;
4. For each formula β of $\mathcal{M}$, $r(\beta) = r(\beta_0) + r(\beta_1) + 1$;
5. For each formula ν of $\mathcal{M}$, $r(\nu) = r(\nu_0) + 1$;
6. For each formula π of $\mathcal{M}$, $r(\pi) = r(\pi_0) + 1$.

By a *rank of an indexed formula A^{ς} of the language $\mathcal{M}^?_{\vdash}$*, we mean the rank of A.

The following is a simpler version of Corollary 2.15.

Corollary 3.15 *A contraction of a basic sequent of $\mathcal{M}^?_{\vdash}$ is also a basic sequent of $\mathcal{M}^?_{\vdash}$.*

Proof By the set-of-terms condition. □

The analogue of Lemma 2.8 (contraction of generally quantified formulas) is also provable.

Lemma 3.8 (Contraction of Indexed π-formulas). *If a question '$?(S \vdash \Gamma(\pi^{\varsigma/i})(\pi^{\varsigma/i}))$' has a successful Socratic transformation in $\mathbb{E}^{\mathsf{L}}$, then the question '$?(S \vdash \Gamma(\pi^{\varsigma/i})(\_))$' also has a successful Socratic transformation in $\mathbb{E}^{\mathsf{L}}$.*

Proof The proof is by induction on the length of a successful Socratic transformation of '$?(S \vdash \Gamma(\pi^{\varsigma/i})(\pi^{\varsigma/i}))$'. If this equals 1, then the sequent is basic and we apply Corollary 3.15.

Assume that the length is $n + 1$, for $n \geqslant 1$, and let $\mathbf{s} = \langle Q_0, Q_1, \ldots, Q_n \rangle$ be the successful Socratic transformation of $Q_0 = ?(S \vdash \Gamma(\pi^{\varsigma/i})(\pi^{\varsigma/i}))$ in $\mathbb{E}^{\mathsf{L}}$. Further reasoning depends on the rule applied to Q_0. We shall consider only R_G and R_π.

Let the Socratic transformation start with:

$$\frac{Q_0 = ?(S\,(A) \vdash \Gamma(\pi^{\varsigma/i})(\pi^{\varsigma/i})(\_))}{Q_1 = ?(S\,(A) \vdash \Gamma(\pi^{\varsigma/i})(\pi^{\varsigma/i})((-A)^j))}\; R_G$$

Since question Q_1 has a successful Socratic transformation in $\mathbb{E}^{\mathsf{L}}$ of length n, by IH there is a successful Socratic transformation of $Q_1^* = ?(S(A) \vdash \Gamma(\pi^{\varsigma/i})(\_)((-A)^j))$ in $\mathbb{E}^{\mathsf{L}}$. Since the following:

$$\frac{?(S\,(A) \vdash \Gamma(\pi^{\varsigma/i})(\_)(\_))}{Q_1^* = ?(S\,(A) \vdash \Gamma(\pi^{\varsigma/i})(\_)((-A)^j))}$$

is an instance of R_G, by Corollary 3.7, question '$?(S(A) \vdash \Gamma(\pi^{\varsigma/i})(\_))$' has a successful Socratic transformation in $\mathbb{E}^{\mathsf{L}}$, as required (the absence of additional indication $(\_)$ is inessential).

Assume that $\pi^{\varsigma/i}$ is active in Q_0, that is, the Socratic transformation starts with:

$$\frac{Q_0 = ?(S \vdash \Gamma\ (\pi^{\varsigma/i})(\pi^{\varsigma/i}))}{Q_1 = ?(S \vdash \Gamma\ (\pi^{\varsigma/i}, (\pi_0)^j)(\pi^{\varsigma/i}))}\ R_\pi$$

where numeral j is such that the pair $\langle i, j\rangle$ satisfies the proviso of applicability of R_π in $\mathbb{E}^{\mathsf{L}}$. By IH, there is a successful Socratic transformation of $Q_1^* = ?(S \vdash \Gamma(\pi^{\varsigma/i}, (\pi_0)^j)(\_))$. By a suitable instance of R_π combined with Corollary 3.7, the result follows. □

Lemma 3.9 (Semi-admissibility of R^*_{contr} in $\mathbb{E}^{\mathsf{L}}$). *If a question '$?(S \vdash \Gamma(A^\varsigma)(A^\varsigma))$' has a successful Socratic transformation in $\mathbb{E}^{\mathsf{L}}$, then the question '$?(S \vdash \Gamma(A^\varsigma)(\_))$' also has a successful Socratic transformation in $\mathbb{E}^{\mathsf{L}}$.*

Proof We apply the same strategy as in the proof of Lemma 2.9: the proof is by induction with respect to rank of A with a subinduction on the length l of a successful Socratic transformation of $?(S \vdash \Gamma(A^\varsigma)(A^\varsigma))$. For convenience, we use the self-explanatory terms 'r-induction', 'r-IH', 'l-induction' and 'l-IH'.

Hence assume that there is a successful Socratic transformation of question $Q = ?(\phi)$ in $\mathbb{E}^{\mathsf{L}}$, where $\phi = S \vdash \Gamma(A^\varsigma)(A^\varsigma)$.
Base case of r-induction. Let $r(A) = 0$. A rule, if applied, does not act on A^ς. If $l = 1$, then ϕ is basic, hence by Corollary 3.15, its contraction is basic as well. In the inductive part of l-induction we use l-IH and Corollary 3.7.

Suppose (r-IH) that the lemma holds for contraction-formulas of rank up to r, and let $r(A) = r + 1$. If $l = 1$, then we reason as above. Let $\mathbf{s} = \langle Q_0, Q_1, \ldots, Q_l\rangle$ stand for a successful Socratic transformation of $Q_0 = Q$. Further reasoning depends on the rule $\mathbf{r}$ applied to Q_0 and in most cases it goes according to the scheme presented in the previous chapter: we use Corollaries 3.13, 3.14 concerning proof-theoretic invertibility of rules of $\mathbb{E}^{\mathsf{L}}$ and r-IH. We skip the cases of $\mathbf{r} \in \{R_{\neg\neg}, R_\beta, R_\alpha\}$. If $\mathbf{r} = R_G$, then the reasoning is as in the proof of the previous lemma (where we used induction on l which is also available in this proof). If A^ς is an indexed π-formula active in Q_0, then we apply Lemma 3.8.

For the last case to consider, suppose that A is a ν-formula, and that $\mathbf{s}$ starts with

$$\frac{Q_0 = ?\left(S \vdash \Gamma\ \left(\nu^{\varsigma/i}\right)\left(\nu^{\varsigma/i}\right)\right)}{Q_1 = ?\left(S \vdash \Gamma\ \left((\nu_0)^{\varsigma/i,k}\right)\left(\nu^{\varsigma/i}\right)\right)}\ R_\nu$$

where k is new to the sequent-conclusion. Let j be a numeral that does not occur in Q_1. By clause 3 of Corollary 3.14, question '$?(S \vdash \Gamma((\nu_0)^{\varsigma/i,k})((\nu_0)^{\varsigma/i,j}))$' has a successful Socratic transformation in $\mathbb{E}^{\mathsf{L}}$. By Corollary 3.11 (renaming numerals), question '$?(S \vdash \Gamma((\nu_0)^{\varsigma/i,k})((\nu_0)^{\varsigma/i,k}))$' has a successful Socratic transformation in $\mathbb{E}^{\mathsf{L}}$. By r-IH, question '$?(S \vdash \Gamma((\nu_0)^{\varsigma/i,k})(\_))$' has a successful Socratic transformation in $\mathbb{E}^{\mathsf{L}}$. Finally, since $?(S \vdash \Gamma(\nu^{\varsigma/i})(\_))/?(S \vdash \Gamma((\nu_0)^{\varsigma/i,k})(\_))$ is an instance of R_ν, the result follows by Corollary 3.7. □

We can now go back to the general form of contraction rule, R_{contr}. Lemma 3.9 yields the following two corollaries, which we state without proofs.

Corollary 3.16 (Semi-admissibility of R_{contr} in $\mathbb{E}^{\mathsf{L}}$) *If there is a successful Socratic transformation of the question-conclusion of R_{contr} in $\mathbb{E}^{\mathsf{L}}$, then there is a successful Socratic transformation of the question-premise of R_{contr} in $\mathbb{E}^{\mathsf{L}}$.*

Corollary 3.17 (Eliminability of R_{contr} in $\mathbb{E}^{\mathsf{L}}$) *Let Q stand for a question of $\mathcal{M}^{?}_{\vdash}$. If there is a successful Socratic transformation of Q in $\mathbb{E}^{\mathsf{L}} \cup \{R_{contr}\}$, then there is also one in $\mathbb{E}^{\mathsf{L}}$.*

In the classical case we have used permutation to deal with some problematic cases in the proof of Lemma 2.7 concerning switching, and this, in turn, played its role in corollaries needed for semi-admissibility of contraction. Permutation and switching appear at the stage together with both-sided, single-succedent sequents. In the modal case we did not need permutation to have semi-admissible structural rules. However, once we have both weakening and contraction at our disposal, the two rules can be used to simulate the rule of permutation.

Corollary 3.18 (Permutation) *Assume that A^{ς} is a k-th term of a sequence $\Gamma(A^{\varsigma})(B^{\varrho})$ and B^{ϱ} is its n-th term. If there is a successful Socratic transformation of the question '$?(\Phi(S \vdash \Gamma(A^{\varsigma})(B^{\varrho})))$' in $\mathbb{E}^{\mathsf{L}}$, then there is also a successful Socratic transformation of the question '$?(\Phi(S \vdash \Gamma(B^{\varrho})(A^{\varsigma})))$' in $\mathbb{E}^{\mathsf{L}}$, where B^{ϱ} is the k-th term of the succedent of the displayed sequent, and A^{ς} is its n-th term.*

Proof Suppose that there is a successful Socratic transformation of question $?(\Phi(S \vdash \Gamma(A^{\varsigma})(B^{\varrho})))$ in $\mathbb{E}^{\mathsf{L}}$. Then by Lemma 3.6 (weakening), there is a successful Socratic transformation of question $?(\Phi(S \vdash \Gamma(A^{\varsigma}, B^{\varrho})(A^{\varsigma}, B^{\varrho})))$ in $\mathbb{E}^{\mathsf{L}}$. Finally, by Corollary 3.16 (contraction), there is a successful Socratic transformation of $?(\Phi(S \vdash \Gamma(B^{\varrho})(A^{\varsigma})))$ in $\mathbb{E}^{\mathsf{L}}$. □

Rule R^{*}_{contr} and its generalization R_{contr} pertain to contraction of indexed formulas of an arbitrary form. Is there a specifically modal version? As we have observed before, the rules of contraction secure information; the following Lemmas 3.10–3.13 pertain to specifically modal cases of securing information.

Lemma 3.10 *Let $\mathsf{L} \in \{\mathsf{K4}, \mathsf{S4}, \mathsf{K45}, \mathsf{KB4}\}$. If a sequent $\phi = S \vdash \Gamma(\pi^{\varsigma/i})(\pi^{j})$ occurs in a Socratic proof conducted in $\mathbb{E}^{\mathsf{L}}$ and $\langle i, j\rangle \in R(\phi)$, then there is a successful Socratic transformation of the question '$?(S \vdash \Gamma(\pi^{\varsigma/i})(\_))$' in $\mathbb{E}^{\mathsf{L}}$.*

Let us recall that the use of '(_)' means that the occurrence of an indexed formula is removed from the sequent.

Let us comment on Lemma 3.10 before we start the proof. The assumption that ϕ occurs in a Socratic proof is to guarantee that the transformation starts with a sequent of the form '$S \vdash (A)^{0}$', which entails that the structure of relation $R(\Gamma(\pi^{\varsigma/i})(\pi^{j}))$ is a tree with 0 in the root. This is relevant when considering proviso of applicability of rule R_{π} in $\mathsf{K45}$.

Lemma 3.10 states that the following rule, applicable provided that $\langle i, j\rangle \in R(\Gamma(\pi^{\varsigma/i}))$, is semi-admissible in $\mathbb{E}^{\mathsf{L}}$ for L transitive but not serial.

$$\frac{?(\Phi\,(S \vdash \Gamma(\pi^{\varsigma/i})(\_)))}{?(\Phi\,(S \vdash \Gamma(\pi^{\varsigma/i})(\pi^{j})))}$$

The proviso of applicability of the rule warrants that a world described by j is accessible from that indicated by i. If we add to this picture the fact that on the right side of a sequent π-formulas 'semantically behave' like formulas preceded with '$\Box$', then we can see that the above rule allows to secure strictly modal information in accessible worlds.

We shall see further that the lemma holds for $\mathsf{L} = \{\mathsf{D4}, \mathsf{D45}\}$ as well. Unfortunately, the presence of rule $R_{\pi\mathsf{D}}$ significantly complicates the proof of this fact. We start with the more elegant case.

Proof of Lemma 3.10 Suppose that Q is a question occurring in a Socratic proof of some sequent in $\mathbb{E}^{\mathsf{L}}$, $\mathsf{L} \in \{\mathsf{K4}, \mathsf{S4}, \mathsf{K45}, \mathsf{KB4}\}$. We prove that for each constituent of Q which is of the form '$S \vdash \Gamma(\pi^{\varsigma/i})(\pi^{j})$', and is such that $\langle i, j\rangle \in R(S \vdash \Gamma(\pi^{\varsigma/i})(\pi^{j}))$, question '$?(S \vdash \Gamma(\pi^{\varsigma/i})(\_))$ has a successful Socratic transformation in $\mathbb{E}^{\mathsf{L}}$.

Thus let $\phi = S \vdash \Gamma(\pi^{\varsigma/i})(\pi^{j})$ and $\phi^{*} = S \vdash \Gamma(\pi^{\varsigma/i})(\_)$, where $\langle i, j\rangle \in R(\phi)$, and suppose that ϕ occurs in a question of a Socratic proof conducted in $\mathbb{E}^{\mathsf{L}}$. Then by Lemma 3.5 (Separation), there is a successful Socratic transformation $\mathbf{s} = \langle Q_1, \ldots, Q_n\rangle$ of $?(\phi)$ in $\mathbb{E}^{\mathsf{L}}$. Let us also observe that $R(\phi) = R(\phi^{*})$. Further reasoning is by induction on n.
Base case: $n = 1$, hence sequent ϕ is basic. If, at the same time, sequent ϕ^{*} is not basic, then it must be the case that ϕ is basic due to the presence of π^{j} in Γ, which means that either '$-\pi^{\varrho/j}$' or '$\neg\pi^{\varrho/j}$' is present in this sequent too (where it can happen to be the same formula). We consider the case with the complement. By assumption, $-\pi^{\varrho/j}$ occurs in ϕ^{*} as well. In this situation the following is a successful Socratic transformation of $?(\phi^{*})$ in $\mathbb{E}^{\mathsf{L}}$.

$$\cfrac{\cfrac{?(S \vdash \Gamma(\pi^{\varsigma/i})(-\pi^{\varrho/j})(\_))}{?(S \vdash \Gamma(\pi^{\varsigma/i})(\nu_0^{\varrho/j,k})(\_))}\, R_{\nu}}{?(S \vdash \Gamma(\pi^{\varsigma/i}, \pi_0^{k})(\nu_0^{\varrho/j,k})(\_))}\, R_{\pi}$$

where k occurs as new, and ν_0 is a component of $-\pi$, which is a ν-formula by Corollary 3.1. By the same corollary, one of π_0, ν_0 is the complement of the other, hence the displayed sequent of the last question is basic. Rule R_{π} is applicable to the second question, as $\langle i, j\rangle \in R(\phi^{*}) = R(\phi)$.
Inductive case: Suppose a rule $\mathbf{r}$ was applied to Q_1. Let us start with the following situation:

$$\frac{Q_1 = ?((S \vdash \Gamma(\pi^{\varsigma/i})(\pi^{j}))}{Q_2 = ?((S \vdash \Gamma(\pi^{\varsigma/i})(\pi^{j}, \pi_0^{k}))}\, R_{\pi}$$

where pair $\langle j, k\rangle$ satisfies the L-proviso. IH applies to the question-conclusion, hence there is a successful Socratic transformation of $?(S \vdash \Gamma(\pi^{\varsigma/i})(\pi_0{}^k))$ in $\mathbb{E}^{\mathsf{L}}$. By Corollary 3.18 (permutation), also $?(S \vdash \Gamma(\pi^{\varsigma/i}, \pi_0{}^k)(\_))$ has a successful Socratic transformation in $\mathbb{E}^{\mathsf{L}}$. Moreover, no pair of numerals disappears when we drop the indexed formula π^j, therefore $\langle j, k\rangle$ still satisfies the L-proviso in sequent '$S \vdash \Gamma(\pi^{\varsigma/i})(\pi_0{}^k)$'. Since, by assumption, $\langle i, j\rangle \in R(\phi)$, also pair $\langle i, k\rangle$ satisfies proviso of applicability of R_π, as the considered logics are transitive. This is easy to see, if we use the version of L-proviso presented in Table 3.5. However, in the case of Euclidean logics, the fact that ϕ occurs in a Socratic proof is necessary to use this version; see also Lemma 3.1.

To sum up, the following is a correct application of R_π:

$$\frac{?(S \vdash \Gamma(\pi^{\varsigma/i})(\_))}{?(S \vdash \Gamma(\pi^{\varsigma/i}, \pi_0{}^k)(\_))}\ R_\pi$$

which yields that the question-premise has a successful Socratic transformation in $\mathbb{E}^{\mathsf{L}}$.

If **r** acts upon an indexed formula other than π^j, then we use IH to Q_2, together with an instance of the same rule followed by Corollary 3.7. □

As we mentioned, the result extends to serial transitive logics. We shall need the following auxiliary notion.

Definition 3.14 Let ϕ be a sequent of the language $\mathcal{M}^?_\vdash$ and let $i \in W(\phi)$. We call X *a set-representation of* i *in* ϕ iff X is the set of all formulas A of $\mathcal{M}$, such that indexed formula of the form '$(A)^{\varsigma/i}$' occurs in ϕ.

Lemma 3.11 *Let* $\mathsf{L} \in \{\mathsf{D4}, \mathsf{D45}\}$. *If a sequent* $\phi = S \vdash \Gamma(\pi^{\varsigma/i})(\pi^j)$ *occurs in a Socratic proof conducted in* $\mathbb{E}^{\mathsf{L}}$ *and* $\langle i, j\rangle \in R(\phi)$, *then there is a successful Socratic transformation of the question* '$?(S \vdash \Gamma(\pi^{\varsigma/i})(\_))$' *in* $\mathbb{E}^{\mathsf{L}}$.

Proof The general idea of the proof is like in the previous case. However, this time we can have an application of $R_{\pi\mathsf{D}}$ acting upon indexed formula π^j. This problem forces us to adapt a much more complicated strategy in filling in the general scheme with details.

Hence let $\phi = S \vdash \Gamma(\pi^{\varsigma/i})(\pi^j)$, $\phi^* = S \vdash \Gamma(\pi^{\varsigma/i})(\_)$, where $\langle i, j\rangle \in R(\phi)$, and suppose that ϕ occurs in a question of a Socratic proof conducted in $\mathbb{E}^{\mathsf{L}}$. By Lemma 3.5 (Separation), there is a successful Socratic transformation **s** of question $?(\phi)$ in $\mathbb{E}^{\mathsf{L}}$. Let $\mathbf{s} = \langle Q_1, \ldots, Q_n\rangle$. We shall construct a Socratic transformation $\mathbf{s}^*$ of $Q_1^* = ?(\phi^*)$, with induction running through $1 \ldots n$. In the previous proof it was sufficient to analyze Q_1/Q_2 and apply IH. This time, at each step, we check that $\mathbf{s}^*$ has properties (a)–(d) listed below.

(a) For $h : 1 \leqslant h \leqslant n$, question Q_h of **s** and question Q_h^* of $\mathbf{s}^*$ have the same number of constituents.

Questions Q_1 and Q_1^* have the same number of constituents, and further the property will be assured by applying branching rule in **s*** every time it was applied in **s**. If ψ is an m-th constituent of question Q_h of **s**, then the m-th constituent of the h-th question of **s*** will be called *sequent corresponding to* ψ *in* **s***, or simply a *corresponding sequent*, where this does not lead to a confusion.

Another property is that for $h : 1 \leqslant h \leqslant n$, for each constituent ψ of Q_h:

(b) $W(\psi) = W(\psi^*)$, where ψ^* is the sequent corresponding to ψ in **s***.

We verify that Q_1^* has this property: one occurrence of numeral j present in ϕ is missing in ϕ^*, but the assumption that $\langle i, j\rangle \in R(\phi)$ warrants that there is at least one more occurrence of j in ϕ^*, therefore $W(\phi) = W(\phi^*)$. It will be easy to see that this holds at further stages. The situation is more complicated with respect to relations $R(\psi)$ and $R(\psi^*)$. If rule $R_{\pi\mathrm{D}}$ is applied in **s** with respect to $\pi^{\,j}$, then in **s*** it will be applied to $\pi^{\,\varsigma/i}$. In **s** it results in pair $\langle j, k\rangle$ added to $R(\psi)$, whereas in **s*** we will have $\langle i, k\rangle$ added to $R(\psi^*)$. However, the following property will be secured.

(c) For $h : 1 \leqslant h \leqslant n$, for each constituent ψ of Q_h and its corresponding sequent ψ^*, there is a finite, possibly empty, set K of numerals such that

$$R(\psi)\cup\{\langle i, k\rangle : k\in K\}=\{\langle j, k\rangle : k\in K\}\cup R(\psi^*).$$

Property (c) states that the only possible difference between $R(\psi)$ and $R(\psi^*)$ is in pairs starting with j. One can think of K as the set of worlds to which modal information must be 'redirected'. In the case of question Q_1^*, $R(\phi) = R(\phi^*)$ and $K = \varnothing$. Here is the last property.

(d) For $h : 1 \leqslant h \leqslant n$ and ψ, ψ^* defined as previously, for each $k \in W(\psi)\backslash\{j\}$, if X is the set-representation of k in ψ and X^* is the set-representation of k in ψ^*, then $X = X^*$. Moreover, if X is the set-representation of j in ψ and X^* is the set-representation of j in ψ^*, then $X = X^* \cup \{\pi\}$.

Property (d) warrants that as far as sets-representations are concerned, the only possible difference is in formula π, which is missing in j. This is obviously true with respect to set-representations of numerals in sequents ϕ and ϕ^*.

We have shown that (a)–(d) hold with respect to the first element of **s***. If question Q_1 is the last one in **s**, then we arrive at the base case and we reason as in the proof of the previous lemma.

In the inductive part we consider only the modal cases.

Suppose that we deal with an instance of R_ν and let ψ stand for its sequent-premise.

$$\frac{Q_h = ?\left(\Psi\ \left(S \vdash \Delta\ \left(\nu^{\varrho/l}\right)\right)\right)}{Q_{h+1} = ?\left(\Psi\left(S \vdash \Delta\ \left(\nu_0^{\varrho/l,l^*}\right)\right)\right)}\ R_\nu$$

By IH, Q_h^* has property (b), therefore numeral l^* is new to sequent ψ^* corresponding to ψ. By the same reason, Q_h^* has also property (d), therefore formula ν belongs to the set-representation of l in ψ^*, that is, there is an indexed formula $\nu^{\varrho^*/l}$ (for some ϱ^*) in sequent ψ^*. This warrants that R_ν can be applied to Q_h^* as well:

$$\frac{Q_h^* = ?\left(\Psi^* \left(S \vdash \Theta \left(\nu^{\varrho^*/l}\right)\right)\right)}{Q_{h+1}^* = ?\left(\Psi^* \left(S \vdash \Theta : \left(\nu_0^{\varrho^*/l,l^*}\right)\right)\right)} R_\nu$$

Again, it is easy to see that each constituent of Q_{h+1}^* has properties (a)–(d). Let us verify (c). By IH, for some K

$$R(\psi) \cup \{\langle i, k\rangle : k \in K\} = \{\langle j, k\rangle : k \in K\} \cup R(\psi^*).$$

Trivially, if we add pair $\langle l, l^*\rangle$ to both sides of the equation, we arrive at the following true equation

$$R(\psi) \cup \{\langle l, l^*\rangle\} \cup \{\langle i, k\rangle : k \in K\} = \{\langle j, k\rangle : k \in K\} \cup \{\langle l, l^*\rangle\} \cup R(\psi^*)$$

which yields immediately that the sequents-conclusions of questions Q_{h+1} and Q_{h+1}^* satisfy (c). (Set K is not altered in this step.)

Suppose we consider an application of rule $R_{\pi\mathsf{D}}$ with respect to an indexed π-formula of the form '$(A)^{\varrho/l}$'. Let ψ and ψ^* stand for, respectively, the sequent premise of Q_h and Q_h^*. If $l \neq j$, then we reason as in the previous case of an application of R_ν. Moreover, if $l = j$, but A is not the particular π-formula considered in this lemma, then in $\mathbf{s}^*$ we apply $R_{\pi\mathsf{D}}$ similarly as in $\mathbf{s}$—that is, with respect to an indexed formula '$(A)^{\varrho^*/j}$' whose occurrence in ψ^* is guaranteed by (d).

Finally, assume that we deal with the troublesome case with $R_{\pi\mathsf{D}}$ applied with respect to π^j. Indexed formula '$\pi^{\varsigma/i}$' is not active here, but it is present in the sequent, as indexed π-formulas are not eliminated in Socratic transformations. We display it below for further reference.

$$\frac{Q_h = ?\left(\Psi \left(S \vdash \Delta \left(\pi^{\varsigma/i}\right) \left(\pi^j\right)\right)\right)}{Q_{h+1} = ?\left(\Psi \left(S \vdash \Delta \left(\pi^{\varsigma/i}\right) \left(\pi^j, \pi_0^{j,k^*}\right)\right)\right)} R_{\pi\mathsf{D}}$$

In this case, k^* is new with respect to the sequent-premise, and hence it is new with respect to the corresponding sequent in Q_h^* (IH, property (b)). However, 'π^j' is possibly missing, therefore in $\mathbf{s}^*$ we apply $R_{\pi\mathsf{D}}$ to '$\pi^{\varsigma/i}$' as follows.

$$\frac{Q_h^* = ?\left(\Psi^* \left(S \vdash \Theta \left(\pi^{\varsigma/i}\right) (\_)\right)\right)}{Q_{h+1}^* = ?\left(\Psi^* \left(S \vdash \Theta \left(\pi^{\varsigma/i}, \pi_0^{i,k^*}\right) (\_)\right)\right)} R_{\pi\mathsf{D}}$$

Thus in **s** we introduce new numeral k^* after j, whereas in $\mathbf{s}^*$ it goes after i. By IH, for some K, (c) holds, that is

$$R(\psi) \cup \{\langle i, k\rangle : k \in K\} = \{\langle j, k\rangle : k \in K\} \cup R(\psi^*).$$

Let χ, χ^* represent the sequent-conclusion of, respectively, Q_{h+1} and Q^*_{h+1}. Let also $K^* = K \cup \{k^*\}$. After the application of $R_{\pi\mathsf{D}}$ in **s**, we have $R(\chi) = R(\psi) \cup \{\langle j, k^*\rangle\}$, and analogously in $\mathbf{s}^*$: $R(\chi^*) = R(\psi^*) \cup \{\langle i, k^*\rangle\}$. But this entails

$$R(\chi) \cup \{\langle i, k\rangle : k \in K^*\} = \{\langle j, k\rangle : k \in K^*\} \cup R(\chi^*),$$

hence property (c) is secured for K^*. It is easy to see that (d) holds as well.

As a summary of the applications of $R_{\pi\mathsf{D}}$ let us emphasize that the set K described in (c) is altered only when the rule is applied with respect to '$\pi^{\,j}$'.

Before we, finally, examine the applications of R_π, we first prove condition ($\star$) formulated below. Let ψ stand for a constituent of Q_h, and let ψ^* represent its corresponding sequent in Q^*_h. In addition, let us assume that an indexed π-formula of the form '$(A)^{\varrho/l_1}$' occurs in ψ.

($\star$) If there is a directed $R(\psi)$-$[l_1, l_2]$-chain, then either (i) there is a directed $R(\psi^*)$-$[l_1, l_2]$-chain, or (ii) A is the formula π considered in this lemma, $l_1 = j$ and there is a directed $R(\psi^*)$-$[i, l_2]$-chain.

The condition warrants that if R_π is applicable to ψ under a given proviso, then the proviso is satisfied by ψ^* as well, though in one special case it is satisfied for numeral i rather than j. Let us observe that, by IH, (c) holds for ψ, ψ^* and some K. Let K_i stand for $\{\langle i, k\rangle : k \in K\}$ and K_j for $\{\langle j, k\rangle : k \in K\}$. From the said (c) it follows that

($\star_{1\subseteq}$) $R(\psi) \subseteq K_j \cup R(\psi^*)$, and
($\star_{2\subseteq}$) $K_i \subseteq R(\psi^*)$.

The proof of ($\star$) is by induction on the length of $R(\psi)$-$[l_1, l_2]$-chain. In the base case, $\langle l_1, l_2\rangle \in R(\psi)$. By ($\star_{1\subseteq}$), $\langle l_1, l_2\rangle \in K_j \cup R(\psi^*)$. If $\langle l_1, l_2\rangle \in R(\psi^*)$, then we arrive at case (i) of ($\star$). Assume that $\langle l_1, l_2\rangle \notin R(\psi^*)$. Then $\langle l_1, l_2\rangle \in K_j$, $l_1 = j$ and $l_2 \in K$. This situation happens only in the special case with $\pi^{\,j}$ involved. Hence it follows that $\langle i, l_2\rangle \in K_i$ and thus, by ($\star_{2\subseteq}$), $\langle i, l_2\rangle \in R(\psi^*)$, which is case (ii) of ($\star$), as required.

Inductive step: suppose that $\langle i_1, \ldots, i_o\rangle$, $2 < o$ is a directed $R(\psi)$-$[l_1, l_2]$-chain, and assume the thesis ($\star$) holds for i_p, $2 \leqslant p < o$, that is, (i) there is a directed $R(\psi^*)$-$[i_1, i_p]$-chain, or (ii) $i_1 = j$ and there is a directed $R(\psi^*)$-$[i, i_p]$-chain (this is our internal IH). In addition, we know that $\langle i_p, i_{p+1}\rangle \in R(\psi)$ by the definition of a chain. We use ($\star_{1\subseteq}$), and obtain the following possible subcases:

(iii) $\langle i_p, i_{p+1}\rangle \in K_j$, but $\langle i_p, i_{p+1}\rangle \notin R(\psi^*)$.
The fact that $R(\psi)$ and $R(\psi^*)$ differ with respect to pair $\langle i_p, i_{p+1}\rangle$ yields that this is the special case with $A = \pi$. By ($\star_{1\subseteq}$), $i_p = j$ and $i_{p+1} \in K$, hence

$\langle i, i_{p+1}\rangle \in K_i$. By $(\star_{2\subseteq})$, $\langle i, i_{p+1}\rangle \in R(\psi^*)$. The pair is a directed chain required in (ii) of $(\star)$.

(iv) $\langle i_p, i_{p+1}\rangle \in R(\psi^*)$ and part (i) of the internal IH holds with respect to i_p. We obtain directed $R(\psi^*)$-$[i_1, i_{p+1}]$-chain, hence (i) of $(\star)$ holds with respect to i_{p+1}.

(v) $\langle i_p, i_{p+1}\rangle \in R(\psi^*)$ and part (ii) of the internal IH holds with respect to i_p. We obtain directed $R(\psi^*)$-$[i, i_{p+1}]$-chain, hence (ii) of $(\star)$ holds with respect to i_{p+1}.

This finishes the subinduction proof of $(\star)$.

Now we can consider an application of R_π. Let ψ and ψ^* be understood as previously. The burden of this case is in showing that the rule is still applicable in $\mathbf{s}^*$, however, we have moved most of the burden to $(\star)$, and only some cases will need additional arguments.

Thus let A be a π-formula (possibly, but not necessarily, the one discussed in this lemma), and let A_0 stand for its component. Assume that R_π is applied as follows.

$$\frac{Q_h = ?\left(\Psi\left(S \vdash \Delta\left(A^{\varrho/l}\right)\right)\right)}{Q_{h+1} = ?\left(\Psi\left(S \vdash \Delta\left(A^{\varrho/l}, A_0^{k^*}\right)\right)\right)}\ R_\pi$$

As a special case, the above scheme captures the following troublesome case with 'π^j' active.

$$\frac{Q_h = ?\left(\Psi\left(S \vdash \Delta\left(\pi^{\varsigma/i}\right)\left(\pi^j\right)\right)\right)}{Q_{h+1} = ?\left(\Psi\left(S \vdash \Delta\left(\pi^{\varsigma/i}\right)\left(\pi^j, \pi_0^{k^*}\right)\right)\right)}\ R_\pi$$

The fact of applying R_π entails that the proviso of applicability of the rule is satisfied by pair $\langle l, k^*\rangle$ in sequent ψ (in the special case $l = j$). We examine the possible cases for pair $\langle l, k^*\rangle$ satisfying the L-proviso in sequent ψ, and we show that in each case one of the pairs $\langle l, k^*\rangle$, $\langle i, k^*\rangle$ satisfies the proviso in sequent ψ^*, so application of R_π to Q_h^* in a way that adds A_0 to the set-representation of k^* is legitimate. In our special case, this application is:

$$\frac{Q_h^* = ?\left(\Psi^*\left(S \vdash \Theta\left(\pi^{\varsigma/i}\right)(\_)\right)\right)}{Q_{h+1}^* = ?\left(\Psi^*\left(S \vdash \Theta\left(\pi^{\varsigma/i}, \pi_0^{k^*}\right)(\_)\right)\right)}\ R_\pi$$

Let us also explain that if $l = j$ but $A \neq \pi$, then R_π will be applied in $\mathbf{s}^*$ just as in the cases where $l \neq j$. Applications of R_π do not influence condition (c), hence only (d) will be checked.

Here are the possible cases for $\langle l, k^*\rangle$:

(vi) there is a directed $R(\psi)$-$[l, k^*]$-chain,

(vii) there are: a directed $R(\psi)$-$[0, l]$-chain and a directed $R(\psi)$-$[0, k^*]$-chain, (for $\mathsf{L} = \mathsf{D45}$).

In the case (vi), condition $(\star)$ warrants that either there is a directed $R(\psi^*)$-$[l, k^*]$-chain, or we deal with the special case with $l = j$ and there is a directed $R(\psi)$-

$[i, k^*]$-chain. In both cases R_π is applicable, as required, and thus property (d) holds for the resulting sequents-conclusions.

In the case (vii): let us recall that we started with a Socratic proof, hence since $\langle i, j\rangle \in R(\psi)$, it follows that $j \neq 0$. For this reason ($\star$) entails that there is a directed $R(\psi^*)$-$[0, l]$-chain and a directed $R(\psi^*)$-$[0, k^*]$-chain. If $l \neq j$, or $l = j$ but $A \neq \pi$, then by IH, (d) holds for ψ and ψ^*, therefore in ψ^* there occurs $A^{\varrho^*/l}$ (for some ϱ^*). In these cases we simply apply R_π with respect to this indexed formula.

Suppose that $l = j$ and $A = \pi$. In this last case we need to prove that R_π is applicable with respect to $\pi^{\varsigma/i}$. If it happens that $i = 0$, then the existence of a directed $R(\psi^*)$-$[0, k^*]$-chain satisfies the D45-proviso. If $i \neq 0$, then since at the beginning we deal with a Socratic proof, there must exist a directed $R(\psi^*)$-$[0, i]$-chain, hence the Euclidean part of D45-proviso is satisfied. Thus in each case we can apply R_π to '$\pi^{\varsigma/i}$'.

We have indicated the way to construct a Socratic transformation **s*** of question $?(\phi^*)$ satisfying conditions (a)–(d). Property (d) yields that the last question of **s*** is based on basic sequents—as the last question of **s** is—with one possible exception: that when a sequent in **s** is basic due to the presence of π^j. In this situation we reason as in the proof of the previous lemma. This finishes the proof. □

The next lemma concerns the semi-admissibility of the following rule in $\mathbb{E}^{\mathsf{L}}$:

$$\frac{?\left(\Phi\ \left(S \vdash \Gamma\left(\nu_0^{\varsigma/i}\right)(\_)\right)\right)}{?\left(\Phi\ \left(S \vdash \Gamma\left(\nu_0^{\varsigma/i}\right)(\nu^j)\right)\right)}$$

where $\langle i, j\rangle \in R(\Gamma(\nu_0^{\varsigma/i})(\nu^j))$, or, for $\mathsf{L} \in \{\mathsf{B}, \mathsf{S5}\}$, $i = j$. L is one of KB, DB, B, S5, KB4.

Lemma 3.12 *Let* $\mathsf{L} \in \{\mathsf{KB}, \mathsf{DB}, \mathsf{B}, \mathsf{S5}, \mathsf{KB4}\}$. *If a sequent* $\phi = S \vdash \Gamma(\nu_0^{\varsigma/i})(\nu^j)$ *occurs in a Socratic proof conducted in* $\mathbb{E}^{\mathsf{L}}$ *and* $\langle i, j\rangle \in R(\phi)$, *or, for* $\mathsf{L} \in \{\mathsf{B}, \mathsf{S5}\}$, $i = j$, *then there is a successful Socratic transformation of the question '*$?(S \vdash \Gamma(\nu_0^{\varsigma/i})(\_))$*' in* $\mathbb{E}^{\mathsf{L}}$.

Proof The proof is organized as that of Lemma 3.11, but properties (a)–(d) receive a different formulation.

Let $\phi = S \vdash \Gamma(\nu_0^{\varsigma/i})(\nu^j)$, $\phi^* = S \vdash \Gamma(\nu_0^{\varsigma/i})(\_)$, where $\langle i, j\rangle \in R(\phi)$ or, for $\mathsf{L} \in \{\mathsf{B}, \mathsf{S5}\}$, $i = j$, and suppose that ϕ occurs in a question of a Socratic proof conducted in $\mathbb{E}^{\mathsf{L}}$. By Lemma 3.5 (Separation), there is a successful Socratic transformation $\mathbf{s} = \langle Q_1, \dots, Q_n\rangle$ of $Q_1 = ?(\phi)$ in $\mathbb{E}^{\mathsf{L}}$. Based on **s** we describe how to construct a successful Socratic transformation **s*** of $?(\phi^*)$.

Base case, $n = 1$. If ϕ^* is not a basic sequent, then ϕ must be basic due to the presence of indexed formula '$(-\nu)^{\varrho/j}$', for some ϱ. In the following instance of R_π, π_0 is the component of ν_0 (see Corollary 3.1, we skip possible application of Lemma 3.7).

$$\frac{Q_1^* = ?\left(S \vdash \Gamma\left(\nu_0^{\varsigma/i}\right)\left((-\nu)^{\varrho/j}\right)(\_)\right)}{Q_2^* = ?\left(S \vdash \Gamma\left(\nu_0^{\varsigma/i}\right)\left((-\nu)^{\varrho/j}, \pi_0{}^{i}\right)(\_)\right)} R_\pi$$

We need to justify the application of R_π. If $\mathsf{L} \in \{\mathsf{B}, \mathsf{S5}\}$ and $i = j$, then the application satisfies reflexivity-part of the L-proviso. If $\mathsf{L} \in \{\mathsf{KB}, \mathsf{DB}, \mathsf{B}, \mathsf{S5}, \mathsf{KB4}\}$ and $\langle i, j\rangle \in R(\phi) = R(\phi^*)$, then the symmetry-part holds. In each possible case, we arrive at a successful Socratic transformation of Q_1^*, as required.

Let us analize a possible course of events before we give a definition of conditions (a)–(d)—this will clearly explain what we aim at. Hence suppose that **s** starts as follows.

$$\frac{Q_1 = ?\left(S \vdash \Gamma\left(\nu_0^{\varsigma/i}\right)\left(\nu^{j}\right)\right)}{Q_2 = ?\left(S \vdash \Gamma\left(\nu_0^{\varsigma/i}\right)\left(\nu_0^{j,k}\right)\right)} R_\nu$$

In **s*** nothing will correspond to this step. Let χ stand for the sequent-conclusion of Q_2. As in the proof of the previous lemma, it is useful to have the notion of a corresponding sequent at our disposal, hence in this particular case we fix that χ in Q_2 corresponds to ϕ^* in Q_1^* (so does ϕ). Since, by assumption, $\langle i, j\rangle \in R(\phi)$, we have $W(\phi) = W(\phi^*)$ and $R(\phi) = R(\phi^*)$. On the other hand, $W(\chi) = W(\phi^*) \cup \{k\}$ and $R(\chi) = R(\phi^*) \cup \{\langle j, k\rangle\}$. Possible applications of rules with respect to '$\nu_0^{j,k}$' or its subformulas will be mimicked in **s*** by applications of the same rules with respect to '$\nu_0^{\varsigma/i}$', hence this difference between $W(\chi)$, $R(\chi)$ on the one hand and $W(\phi^*)$, $R(\phi^*)$ on the other is acceptable.

Until the end of this proof k will stand for the new numeral introduced by application of R_ν with respect to '(ν^{j})' (if there was such an application). Suppose that $\nu_0 = A \wedge B$ and that further in **s** rule R_α is applied with respect to $(A \wedge B)^{j,k}$.

$$\frac{Q_2 = ?\left(S \vdash \Gamma\left((A \wedge B)^{\varsigma/i}\right)\left((A \wedge B)^{j,k}\right)\right)}{Q_3 = ?\left(S \vdash \Gamma\left((A \wedge B)^{\varsigma/i}\right)\left(A^{j,k}\right) \,;\, S \vdash \Gamma\left((A \wedge B)^{\varsigma/i}\right)\left(B^{j,k}\right)\right)} R_\alpha$$

In this situation in **s*** we apply R_α as follows.

$$\frac{Q_1^* = ?\left(S \vdash \Gamma\left((A \wedge B)^{\varsigma/i}\right)(\_)\right)}{Q_2^* = ?\left(S \vdash \Gamma\left(A^{\varsigma/i}\right)(\_) \,;\, S \vdash \Gamma\left(B^{\varsigma/i}\right)(\_)\right)} R_\alpha$$

Let χ_1, χ_2 stand for the constituents of Q_3, and ψ_1^*, ψ_2^*—for those of Q_2^*. We set: χ_1 corresponds to ψ_1^* and χ_2 corresponds to ψ_2^*. In the sequents of **s*** there is no k, but the set-representation of k in a sequent of **s*** is always a subset of the set-representation of i in the corresponding sequent. Again, we shall see that this situation suits us.

In the next possible scenario we want to obtain the same relation between the set-representation of k in a sequent of **s** and the set-representation of i in its corresponding sequent. Suppose that R_α is applied to Q_2 in the following way.

$$\frac{Q_2 = ?\left(S \vdash \Gamma\left((A \wedge B)^{\varsigma/i}\right)\left((A \wedge B)^{j,k}\right)\right)}{Q_3 = ?\left(S \vdash \Gamma\left(A^{\varsigma/i}\right)\left((A \wedge B)^{j,k}\right) \; ; \; S \vdash \Gamma\left(B^{\varsigma/i}\right)\left((A \wedge B)^{j,k}\right)\right)} \, R_\alpha$$

In this situation in $\mathbf{s}^*$ we first apply R_{contr} (see Corollary 3.16) and then R_α.

$$\cfrac{\cfrac{Q_1^* = ?\left(S \vdash \Gamma\left((A \wedge B)^{\varsigma/i}\right)(\_)\right)}{Q_2^* = ?\left(S \vdash \Gamma\left((A \wedge B)^{\varsigma/i}, (A \wedge B)^{\varsigma/i}\right)(\_)\right)} \, R_{contr}}{Q_3^* = ?\left(S \vdash \Gamma\left(A^{\varsigma/i}, (A \wedge B)^{\varsigma/i}\right)(\_) \; ; \; S \vdash \Gamma\left(B^{\varsigma/i}, (A \wedge B)^{\varsigma/i}\right)(\_)\right)} \, R_\alpha$$

Now we set: χ_1 (χ_2) of Q_3 corresponds to the first (the second) constituent of Q_3^*. The application of R_{contr} warrants that the set-representation of k in χ_1 (χ_2) is included in the set-representation of i in the corresponding sequent.

Now we define (a)–(d). For each $h : 1 \leqslant h \leqslant n$,

(a) Corresponding questions have the same number of constituents.

For each constituent ψ of question Q_h, if a constituent, ψ^*, of a question, Q^*, is a sequent corresponding to ψ, then

(b) $W(\psi) = W(\psi^*)$ or $W(\psi) = W(\psi^*) \cup \{k\}$, (where k is the numeral introduced by application of R_ν with respect to ν^j).
(c) If $W(\psi) = W(\psi^*)$, then also $R(\psi) = R(\psi^*)$. If $W(\psi) = W(\psi^*) \cup \{k\}$, then for some finite, possibly empty, set L of numerals

$$R(\psi) \cup \{\langle i, l\rangle : l \in L\} = \{\langle k, l\rangle : l \in L\} \cup \{\langle j, k\rangle\} \cup R(\psi^*).$$

(d) For each $l \in W(\psi)$, if X_l is the set-representation of l in ψ and X_l^* is the set-representation of l in ψ^*, then

(d1) if $l \notin \{i, j\}$, then $X_l = X_l^*$,
(d2) $X_j^* \cup \{\nu\} = X_j$,
(d3) $X_i \subseteq X_i^*$,
(d4) $X_k \subseteq X_i^*$.

The first question of $\mathbf{s}$ corresponds to the first question of $\mathbf{s}^*$, and ϕ corresponds to ϕ^*. The verification that (a)–(d) are satisfied is straightforward.

Suppose that the construction is carried on up to question Q_h of $\mathbf{s}$ ($h < n$). IH states that for each ψ of Q_h, its corresponding sequent ψ^* satisfies (a)–(d).

If Q_h/Q_{h+1} is an instance of R_G that introduces A^k for some A, then in $\mathbf{s}^*$ we introduce A^i; in other cases R_G is applied in the same manner. If Q_h/Q_{h+1} is an instance of rule $\mathbf{r}$ acting upon indexed formula '$(A)^{\varrho/l}$', where:

- $l \notin \{i, j\}$, then by (d1) in ψ^* there is an indexed formula $(A)^{\varrho^*/l}$ (for some ϱ^*), hence we can apply $\mathbf{r}$ to the same effect.

What we mean by 'the same effect' is that in the sequent-conclusion the same formula appears, with a possibly different index, but with the same numeral at its end. The verification of (a)–(d) in this case is easy.

– $l = i$, then by (d3) (IH) we can apply the same rule, but in order to satisfy (d4) (in the inductive step), we precede the application of **r** with one application of R_{contr}, as in the example analyzed above. Then we set: Q_{h+1} corresponds to the question-conclusion of **r**.
– $l = k$, then (d4) warrants that in ψ^* there is an indexed formula $(A)^{\varrho^*/i}$.

If $\mathbf{r} = \{R_\nu, R_{\pi\mathsf{D}}\}$ and l^* is the new numeral introduced by this application of **r**, then since in the case of ψ and ψ^*, by IH, (c) holds for some L, now it holds for $L^* = L \cup \{l^*\}$.

– $l = j$, then if $(A)^{\varrho/l}$ is ν^j, then nothing corresponds to this step in $\mathbf{s}^*$, just as we explained above. Questions Q_h and Q_{h+1} both correspond to the same question in $\mathbf{s}^*$—the last one constructed. If $(A)^{\varrho/l}$ is not ν^j, then we reason as in the first case.

In the above description we have assumed that R_π is applicable in $\mathbf{s}^*$, whenever it was applicable in **s**. We need to justify this assumption. Strictly speaking, we do not need R_π to be applicable in $\mathbf{s}^*$ in exactly the same way it was in **s**—it is sufficient if it is applicable to the effect of adding the same formulas to the set-representations of the respective numerals, and in the case of numeral k—to the set-representation of i.

Suppose that R_π is applicable to ψ in Q_h by reflexivity, and assume, in addition, that it is not applicable to the corresponding question in $\mathbf{s}^*$. The only possible difference between $R^{ref}(\psi)$ and $R^{ref}(\psi^*)$ is in pair $\langle k, k\rangle$. By IH, the active π-formula belongs to the set-representation of i in ψ^*, hence the rule is applicable to ψ^* with respect to the π-formula with an indexed ended by i.

Suppose that the applicability of R_π to sequent ψ depends on the pair $\langle j, k\rangle$, which does not belong to $R(\psi^*)$. Then if we deal with a symmetric logic, then we use pair $\langle i, j\rangle$, as $\langle i, j\rangle, \langle j, i\rangle \in R^{sym}(\psi^*)$. Intuitively speaking, whatever can be introduced to, or from, k by rule R_π in sequents of **s**, will be introduced in the corresponding sequents to, or from, i. □

Lemma 3.13 *Let* $\mathsf{L} \in \{\mathsf{K5}, \mathsf{D5}, \mathsf{S5}, \mathsf{K45}, \mathsf{D45}\}$. *If* $\phi = S \vdash \Gamma(\nu^{\varsigma/i})(\nu^j)$ *occurs in a Socratic proof conducted in* $\mathbb{E}^{\mathsf{L}}$ *and* $\langle i, j\rangle \in R(\phi)$, *then there is a successful Socratic transformation of the question* '$?(S \vdash \Gamma(\nu^{\varsigma/i})(\_))$' *in* $\mathbb{E}^{\mathsf{L}}$.

Proof The proof is organized as that of Lemma 3.12 with the same formulation of (a)–(d). We discuss only the points that differ.

Base case with $n = 1$. If ϕ^* is not basic although ϕ is, then we use the following transformation:

$$\cfrac{\cfrac{?(S \vdash \Gamma(\nu^{\varsigma/i})((-\nu)^{\varrho/j})(\_))}{?(S \vdash \Gamma({\nu_0}^{\varsigma/i,l})((-\nu)^{\varrho/j})(\_))}\ R_\nu}{?(S \vdash \Gamma({\nu_0}^{\varsigma/i,l})((-\nu)^{\varrho/j}, \pi_0^{\,l})(\_))}\ R_\pi$$

Numeral l in the second question is new. R_π is applicable to the second question, as there are: a directed $R(\Gamma({\nu_0}^{\varsigma/i,l})((-\nu)^{\varrho/j}))$-$[0, l]$-chain and a directed $R(\Gamma({\nu_0}^{\varsigma/i,l})((-\nu)^{\varrho/j}))$-$[0, j]$-chain.

In the previous proof nothing corresponded in $\mathbf{s}^*$ to the application of R_ν to a constituent of question Q_h with respect to ν^j, as ν_0 was already present in the set-representation of i. In the present case, if in $\mathbf{s}$ we have

$$\frac{Q_h = ?\left(\Phi\left(S \vdash \Delta\left(\nu^{\,\varsigma/i}\right)\left(\nu^{\,j}\right)\right)\right)}{Q_{h+1} = ?\left(\Phi\left(S \vdash \Delta\left(\nu^{\,\varsigma/i}\right)\left(\nu_0^{\,j,k}\right)\right)\right)}\ R_\nu$$

then in $\mathbf{s}^*$ we will use

$$\frac{\dfrac{Q^* = ?\left(\Psi\left(S \vdash \Theta\left(\nu^{\,\varsigma/i}\right)(\_)\right)\right)}{Q^{**} = ?\left(\Psi\left(S \vdash \Theta\left(\nu^{\,\varsigma/i}, \nu^{\,\varsigma/i}\right)(\_)\right)\right)}\ R_{contr}}{Q^{***} = ?\left(\Psi\left(S \vdash \Theta\left(\nu^{\,\varsigma/i}, \nu_0^{\,\varsigma/i,k}\right)(\_)\right)\right)}\ R_\nu$$

As in the proof of Lemma 3.12, the application of R_{contr} has no independent proof-theoretic meaning; we use it only for the sake of the adopted strategy of proving our lemma. (Condition (d) would break without it.) The possibility to use the same numeral k in $\mathbf{s}^*$ will be guaranteed by condition (b). In setting the correspondence between elements of $\mathbf{s}$ and $\mathbf{s}^*$ question Q^{**} is skipped, as previously—Q_{h+1} corresponds to Q^{***} and the sequent-conclusion of Q_{h+1} corresponds to that of Q^{***}.

As far as other rules are concerned, if R_G is applied to ψ to the effect of adding A^k, then in $\mathbf{s}^*$ we introduce the same indexed formula (in the proof of Lemma 3.12 we changed the numeral by introducing A^i). Similarly, if a rule applied to ψ in Q_h acts upon an indexed formula $A^{\varrho/k}$, then we apply the same rule in the same way with respect to $A^{\varrho^*/k}$, whose occurrence in ψ^* (for some ϱ^*) is guaranteed by condition (d). In all the other cases, we act as described in the proof of Lemma 3.12.

Condition (a) is exactly as in the proof of Lemma 3.12. Conditions (b)–(d) are simplified. For each $h : 1 \leqslant h \leqslant n$, for each constituent ψ of question Q_h, if a constituent, ψ^*, of a question, Q^*, is a sequent corresponding to ψ, then

(b) $W(\psi) = W(\psi^*)$.
(c) $R(\psi) = R(\psi^*)$ or $R(\psi) \cup \{\langle i, k\rangle\} = \{\langle j, k\rangle\} \cup R(\psi^*)$.

The first possibility holds, if R_π was not applied to ψ with respect to 'ν^j'.

(d) For each $l \in W(\psi)$, if X_l is the set-representation of l in ψ and X_l^* is the set-representation of l in ψ^*, then

(d1) if $l \neq j$, then $X_l = X_l^*$,
(d2) $X_j^* \cup \{\nu\} = X_j$.

The verification that our construction of $\mathbf{s}^*$ satisfies (a)–(d) is like in the previous proof—we skip it.

Possible difference in the structure of $R(\psi)$ and $R(\psi^*)$ is in pair $\langle j, k\rangle$, which belongs to $R(\psi)$, but not to $R(\psi^*)$. On the other hand, pair $\langle i, k\rangle$ belongs to $R(\psi^*)$ (but not to $R(\psi)$), which warrants that there is a directed $R(\psi^*)$-[0, j]-chain and a directed $R(\psi^*)$-[0, k]-chain. This entails that $\langle j, k\rangle \in R^{Eu}(\psi^*)$. Since we deal with Euclidean logics, applicability of R_π is saved in $\mathbf{s}^*$.

Finally, if the last question of $\mathbf{s}^*$ contains a sequents which is not basic, then we apply rules R_ν and R_π in a way presented at the beggining of this proof. □

In the last part of this section we have shown the semi-admissibility, in the respective $\mathbb{E}^{\mathsf{L}}$, of the following rules. The right column indicates proviso of applicability of the respective rule.

$$\frac{?(\Phi(S \vdash \Gamma(\pi^{\varsigma/i})(\_)))}{?(\Phi(S \vdash \Gamma(\pi^{\varsigma/i})(\pi^j)))}\ R^{\pi(tr)}_{contr} \qquad \begin{array}{l}\langle i, j\rangle \in R(\Gamma(\pi^{\varsigma/i}))\\ \text{transitive } \mathsf{L}\end{array}$$

$$\frac{?(\Phi(S \vdash \Gamma(\nu_0{}^{\varsigma/i})(\_)))}{?(:\Phi(S \vdash \Gamma(\nu_0{}^{\varsigma/i})(\nu^j)))}\ R^{\nu(sym)}_{contr} \qquad \begin{array}{l}\langle i, j\rangle \in R(\Gamma(\nu_0{}^{\varsigma/i})(\nu^j))\\ \text{or, for } \mathsf{L} \in \{\mathsf{B}, \mathsf{S5}\}, i = j\\ \text{symmetric } \mathsf{L}\end{array}$$

$$\frac{?(Phi(S \vdash \Gamma(\nu^{\varsigma/i})(\_)))}{?(\Phi(S \vdash \Gamma(\nu^{\varsigma/i})(\nu^j)))}\ R^{\nu(Eu)}_{contr} \qquad \begin{array}{l}\langle i, j\rangle \in R(\Gamma(\nu^{\varsigma/i}))\\ \text{Euclidean } \mathsf{L}\end{array}$$

3.5 Completeness of $\mathbb{E}^{\mathsf{L}}$

In this section we extend the techniques used to prove completeness of classical logic to the modal cases. The ideas are borrowed from [4], but all minor adjustments are ours.

L-consistency Properties

We start with the notion of consistency property adjusted to K. Let:

$$(X)^\sharp = \{\nu_0 : \nu \in X\}. \tag{3.7}$$

In the sequel we will use the notation $(X)^\sharp$ quite frequently, later we shall also change its meaning to capture some of the L-properties. However, until stated otherwise, $(X)^\sharp$ is understood as given in (3.7).

The following properties of function $(\bullet)^\sharp : 2^{FORM_{\mathcal{M}}} \longrightarrow 2^{FORM_{\mathcal{M}}}$ will be useful in this section. The result is obvious and we state it without a proof.

Corollary 3.19 *Let X and Y be sets of formulas of $\mathcal{M}$. The following holds:*

1. *If $X \subseteq Y$, then $(X)^\sharp \subseteq (Y)^\sharp$.*
2. $(X \cup Y)^\sharp = (X)^\sharp \cup (Y)^\sharp$.
3. *If $(X)^\sharp$ is finite, then so is $\{\nu : \nu \in X\}$.*

One comment is in order here. Set X is not uniquely determined by set $(X)^\sharp$, and it can obviously happen that X is not finite though $(X)^\sharp$ is. For example, $X = VAR \cup \{\Box p\}$, $(X)^\sharp = \{p\}$. Another example shows that also for finite sets, $(X)^\sharp$ does not determine X. For $(X)^\sharp = \{p, \neg q\}$, X can be one of $\{\Box p, \Box\neg q, r\}$, $\{\neg\Diamond\neg p, \neg\Diamond q, \neg s\}$ (and many others). However, if $(X)^\sharp$ is finite, then the number of ν-formulas 'producing' the elements of $(X)^\sharp$ must be finite.

Definition 3.15 (K-*consistency Property*) Let $\mathcal{C}$ stand for a family of sets of formulas of $\mathcal{M}$. We say that $\mathcal{C}$ *is a* K-*consistency property* if each set $X \in \mathcal{C}$ satisfies the following conditions:

1. If $p_i \in X$, then '$-p_i$' $\notin X$.
2. If $\neg\neg A \in X$, then $X \cup \{A\} \in \mathcal{C}$.
3. If $\alpha \in X$, then $X \cup \{\alpha_0, \alpha_1\} \in \mathcal{C}$.
4. If $\beta \in X$, then $X \cup \{\beta_0\} \in \mathcal{C}$ or $X \cup \{\beta_1\} \in \mathcal{C}$.
5. If $\pi \in X$, then $(X)^\sharp \cup \{\pi_0\} \in \mathcal{C}$.

Conditions 1–4 are classical. The fifth condition is strictly modal and it says that if there is a π-formula in a world represented by set X, then there must be another world, containing π_0, but then it takes also all the ν_0-components of ν-formulas from X.

The above definition is based on Fitting's account. The next notion is also taken from [4, p. 51].

Definition 3.16 (Z-*compatibility*) Let $\mathcal{C}$ be a K-consistency property and let Z stand for a set of formulas of $\mathcal{M}$. We say that $\mathcal{C}$ *is* Z-*compatible* if, for each $X \in \mathcal{C}$ and for each $A \in Z$, $X \cup \{A\} \in \mathcal{C}$.

Definition 3.17 (*Subset Closed Properties*) A family $\mathcal{F}$ of sets is called *subset closed* iff for each $X \in \mathcal{F}$, every subset of X is also in $\mathcal{F}$.

The following lemma proves that each K-consistency property can be extended to a subset closed K-consistency property (cf. [4, Lemma 5.4, p. 52]).

Lemma 3.14 *Let $\mathcal{C}$ stand for a* K-*consistency property, and let $\mathcal{C}^*$ be a family containing all subsets of sets from $\mathcal{C}$. Then $\mathcal{C}^*$ is a subset closed* K-*consistency property. Moreover, if $\mathcal{C}$ is Z-compatible, then $\mathcal{C}^*$ is Z-compatible.*

Proof Let $\mathcal{C}^*$ be defined as in the statement of the lemma.

If $X^* \in \mathcal{C}^*$, then $X^* \subseteq X$ for some $X \in \mathcal{C}$. If $Y \subseteq X^*$, then also $Y \subseteq X$, which yields that Y is in $\mathcal{C}^*$. Hence $\mathcal{C}^*$ is subset closed.

The rest of the proof consists in checking that the conditions stated in Definition 3.15 and Definition 3.16 hold. Most of this is straightforward, hence we only go through conditions 4, 5 of Definition 3.15 and Z-compatibility.

Let $X^* \in \mathcal{C}^*$. We know that $X^* \subseteq X$ for some $X \in \mathcal{C}$. Assume that $\beta \in X^*$, then also $\beta \in X$, and hence $X \cup \{\beta_i\} \in \mathcal{C}$ for some $i \in \{0, 1\}$. Each subset of $X \cup \{\beta_i\}$ is in $\mathcal{C}^*$, including $X^* \cup \{\beta_i\}$, which was to be proved.

Assume that $\pi \in X^* \in \mathcal{C}^*$. There is $X \in \mathcal{C}$ such that $X^* \subseteq X$, hence $\pi \in X$ and there is $(X)^\sharp \cup \{\pi_0\} \in \mathcal{C}$. By clause 1 of Corollary 3.19, the following holds:

$$(X^*)^\sharp \subseteq X^\sharp, \text{ hence } (X^*)^\sharp \cup \{\pi_0\} \subseteq X^\sharp \cup \{\pi_0\} \in \mathcal{C},$$

and for this reason $(X^*)^\sharp \cup \{\pi_0\} \in \mathcal{C}^*$.

Let $A \in Z$ and let $X^* \in \mathcal{C}^*$ be arbitrary. There is $X \supseteq X^*$ in $\mathcal{C}$, and there must be also $X \cup \{A\} \in \mathcal{C}$, since $\mathcal{C}$ is Z-compatible. As $X^* \cup \{A\} \subseteq X \cup \{A\}$, we have $X^* \cup \{A\} \in \mathcal{C}^*$, as required. □

Definition 3.18 (*Properties of Finite Character*) We say that a family $\mathcal{F}$ of sets is *of finite character* if the following holds: $X \in \mathcal{F}$ iff every finite subset of X is in $\mathcal{F}$.

Corollary 3.20 *Every property of finite character is subset closed.*

Proof Let $\mathcal{F}$ stand for a property of finite character, let $X \in \mathcal{F}$ and $Y \subseteq X$. We consider an arbitrary finite subset Z of Y. The aim is to show that $Z \in \mathcal{F}$, since then $Y \in \mathcal{F}$ by Definition 3.18. But this is quite obvious, as $Z \subseteq Y$ and $Y \subseteq X$, hence Z is a finite subset of X, which yields that $Z \in \mathcal{F}$ by the same Definition 3.18. It follows that $\mathcal{F}$ is subset closed. □

Corollary 3.21 *Every property of finite character is closed under chain unions.*

Proof Let $\mathcal{F}$ stand for a family of sets that is of finite character. Let

$$X_1 \subseteq \ldots \subseteq X_k \subseteq \ldots$$

be a chain of elements of $\mathcal{F}$ and let $\mathbb{X}$ stand for its union. Suppose that X is an arbitrary finite subset of $\mathbb{X}$; as in the proof of the previous lemma, our aim is to show that $X \in \mathcal{F}$.

Indeed, let $X = \{x_1, \ldots, x_n\}$. Each x_i occurs in some X_{k_i} building the chain. We consider all of $X_{k_1}, \ldots, X_{k_n}$ and we choose the largest one (in the $\subseteq$-order), say it is X_{max}. Then $X = \{x_1, \ldots, x_n\} \subseteq X_{max}$. Since X is finite and X_{max} is in the property $\mathcal{F}$ which is of finite character, also $X \in \mathcal{F}$, which was to be proved. □

In the next lemma we extend subset closed K-consistency properties to properties of finite character.

Lemma 3.15 *Let $\mathcal{C}$ stand for a subset closed* K*-consistency property, and let $\mathcal{C}^*$ be a family, such that $X \in \mathcal{C}^*$ iff every finite subset of X belongs to $\mathcal{C}$. $\mathcal{C}^*$ is a* K*-consistency property of finite character. Moreover, if $\mathcal{C}$ is Z-compatible, then $\mathcal{C}^*$ is Z-compatible.*

Proof First, the finite character. Suppose that $X \in \mathcal{C}^*$. Then every finite subset of X is in $\mathcal{C}$. But every finite subset of any of these finite subsets is also in $\mathcal{C}$, since $\mathcal{C}$ is subset closed. Henceforth, every finite subset of X is in $\mathcal{C}^*$. The other way round: if for every finite subset Y of X, $Y \in \mathcal{C}^*$, then every finite subset of Y is in $\mathcal{C}$, in particular, $Y \in \mathcal{C}$, as Y is finite. It means that every finite subset of X is in $\mathcal{C}$, hence $X \in \mathcal{C}^*$. We have shown that $\mathcal{C}^*$ is of finite character.

Now, as in the proof of Lemma 3.14, we check conditions of Definitions 3.15 and 3.16. This time the reasoning is slightly more complicated. Let $X \in \mathcal{C}^*$.

If $p_i \in X$, then assuming that $\neg p_i \in X$ we obtain a contradiction immediately, since $\{p_i, \neg p_i\}$ is a finite subset of X, hence it would be in $\mathcal{C}$. It follows that if $\mathcal{C}$ is a K-consistency property, then $\mathcal{C}^*$ satisfies clause 1 of Definition 3.15.

Assume that $\neg\neg A \in X$. The way to show that $X \cup \{A\} \in \mathcal{C}^*$ is by proving that every finite subset of $X \cup \{A\}$ is in $\mathcal{C}$. Hence let Y be an arbitrary finite subset of $X \cup \{A\}$. $Y \backslash \{A\}$ is a finite subset of X alone, therefore $Y \backslash \{A\} \in \mathcal{C}$ by the construction of $\mathcal{C}^*$. Since $\neg\neg A \in X$, the following

$$Y^* = (Y \backslash \{A\}) \cup \{\neg\neg A\}$$

is definitely also a finite subset of X, thus it belongs to $\mathcal{C}$, which is a K-consistency property, and hence also

$$Y^* \cup \{A\} = Y \cup \{A, \neg\neg A\} \in \mathcal{C}.$$

Independently of whether A or $\neg\neg A$ is in Y or not, we have $Y \in \mathcal{C}$, as $\mathcal{C}$ is subset closed. Thus we have shown that $X \cup \{A\} \in \mathcal{C}^*$.

We skip the case of α-formulas. For clause 4 of Definition 3.15 we reason by contradiction. Assume that $\beta \in X$, and that neither $X \cup \{\beta_0\}$ nor $X \cup \{\beta_1\}$ is in $\mathcal{C}^*$. Then there must be (i) a finite subset $Y_0 \subseteq X \cup \{\beta_0\}$ that *is not* in $\mathcal{C}$, and (ii) a finite subset $Y_1 \subseteq X \cup \{\beta_1\}$ that *is not* in $\mathcal{C}$. Since, by the construction of $\mathcal{C}^*$, each finite subset of X is in $\mathcal{C}$, it must be the case that $\beta_0 \in Y_0$ and $\beta_1 \in Y_1$. On the other hand, both $Y_i \backslash \{\beta_i\}$ are finite subsets of X, and so is their sum, hence

$$(Y_0 \backslash \{\beta_0\}) \cup (Y_1 \backslash \{\beta_1\}) \in \mathcal{C}.$$

But as $\beta \in X$, the following also holds:

$$Y^* = ((Y_0 \backslash \{\beta_0\}) \cup (Y_1 \backslash \{\beta_1\})) \cup \{\beta\} \in \mathcal{C}.$$

Since $\mathcal{C}$ is a K-consistency property, one of the two must be the case: either (iii) $Y^* \cup \{\beta_0\} \in \mathcal{C}$ or (iv) $Y^* \cup \{\beta_1\} \in \mathcal{C}$. As $\mathcal{C}$ is subset closed, (iii) yields $Y_0 \in \mathcal{C}$ and (iv) yields $Y_1 \in \mathcal{C}$, a contradiction with (i) and (ii). Hence we have shown that $\mathcal{C}^*$ satisfies condition 4 of Definition 3.15.

Assume that $X \in \mathcal{C}^*$ and $\pi \in X$. Again, we need to show that every finite subset of $(X)^\sharp \cup \{\pi_0\}$ belongs to $\mathcal{C}$, hence let $Y \subseteq (X)^\sharp \cup \{\pi_0\}$ be finite. It follows that for each $A \in Y$, either $A = \pi_0$, or there is $\nu \in X$ such that $A = \nu_0$. Let Y^* be defined as follows,

(Y^* def) $B \in Y^*$ iff B is a ν-formula, $B \in X$ and the component of B, ν_0, is in Y.

The point is that Y^* is such that $(Y^*)^\sharp = Y$ or $(Y^*)^\sharp \cup \{\pi_0\} = Y$ (depending on whether $\pi_0 \in Y$ or not). By clause 3 of Corollary 3.19, Y^* is finite. Moreover, it is a subset of X and so is $Y^* \cup \{\pi\}$, hence $Y^* \cup \{\pi\} \in \mathcal{C}$. Since $\mathcal{C}$ is a K-consistency property, it follows that $(Y^*)^\sharp \cup \{\pi_0\} \in \mathcal{C}$. If it happens that $(Y^*)^\sharp \cup \{\pi_0\} \neq Y$ because $\pi_0 \notin Y$, then we make one more round through subset closed property of $\mathcal{C}$. Thus we have shown that $\mathcal{C}^*$ satisfies condition 5 of Definition 3.15.

Assume that $\mathcal{C}$ is Z-compatible and that $A \in Z$. Let Y stand for an arbitrary finite subset of $X \cup \{A\}$. Since $Y \backslash \{A\}$ is a finite subset of X, $Y \backslash \{A\} \in \mathcal{C}$, which is sufficient for us, if $A \notin Y$. Otherwise, we observe that $\mathcal{C}$ is Z-compatible, therefore $Y \cup \{A\} \in \mathcal{C}$. This finishes the proof. □

Finally, we obtain what follows.

Corollary 3.22 *For each Z-compatible* K*-consistency property $\mathcal{C}$, there is a Z-compatible* K*-consistency property $\mathcal{C}^{**}$ of finite character, which is such that $\mathcal{C} \subseteq \mathcal{C}^{**}$.*

Proof Let $\mathcal{C}$ be a (Z-compatible) K-consistency property. By Lemma 3.14, there is a (Z-compatible) K-consistency subset closed property $\mathcal{C}^*$ such that every subset of a set from $\mathcal{C}$ is in $\mathcal{C}^*$. Since $X \subseteq X$, this yields that if $X \in \mathcal{C}$, then $X \in \mathcal{C}^*$, that is, $\mathcal{C} \subseteq \mathcal{C}^*$.

Further, by Lemma 3.15, there is a (Z-compatible) K-consistency property $\mathcal{C}^{**}$ of finite character such that $X \in \mathcal{C}^{**}$ iff each finite subset of X is in $\mathcal{C}^*$. If $X \in \mathcal{C}^*$, then each finite subset of X is in $\mathcal{C}^*$, as this one is subset closed. Hence also $X \in \mathcal{C}^{**}$. We have shown that $\mathcal{C} \subseteq \mathcal{C}^{**}$. □

Now we need to show that every set in a K-consistency property is satisfiable. The two notions of satisfiability—that of local and global satisfiability (see Definition 3.4)—occur in one theorem, called *Strong Model Existence Theorem* (the name used by Fitting). As we shall see, the fact that $X \in \mathcal{C}$ will correspond to local satisfiability of X, and Z-compatibility of $\mathcal{C}$ will correspond to global satisfiability of Z. Before we proceed to the theorem, we first show that each set in a consistency property can be extended to a downward-saturated one, where only classical conditions of saturation are considered at the moment. The idea of the proof is very similar to that presented in the classical case, but this time we use consistency properties of finite character, which simplifies the proof. Besides, we won't get unlucky this time, fortunately.

Definition 3.19 (*Downward-Saturated Set: the Classical Case*) A set X of formulas of the language $\mathcal{M}$ is *downward-saturated* iff

1. If $p_i \in X$, then '$-p_i$' $\notin X$;
2. If '$\neg\neg A$' $\in X$, then $A \in X$;
3. If $\alpha \in X$, then $\alpha_0 \in X$ and $\alpha_1 \in X$;
4. If $\beta \in X$, then $\beta_0 \in X$ or $\beta_1 \in X$.

Lemma 3.16 *Let $\mathcal{C}$ be a* K*-consistency property of finite character. For each set $X \in \mathcal{C}$, there is a downward-saturated set $\mathbb{X} \in \mathcal{C}$, such that $X \subseteq \mathbb{X}$. Also, if $\mathcal{C}$ is Z-compatible, then $X \cup Z \subseteq \mathbb{X}$.*

Proof Let $X \in \mathcal{C}$, where $\mathcal{C}$ is a K-consistency property of finite character. The initial part of the construction reminds that used in the classical case. Let:

$$F_0, F_1, \ldots, F_k, \ldots \tag{3.8}$$

stand for an ordering of all formulas of $\mathcal{M}$. We define sequence $(X_k)_\infty$ of sets as follows:

$X_0 = X$ and for each $k \in \mathbb{N}_0$:

$$X_{k+1} = \begin{cases} X_k \cup \{F_k\} & \text{if } X_k \cup \{F_k\} \in \mathcal{C}, \\ X_k & \text{otherwise.} \end{cases}$$

Let:

$$\mathbb{X} = \bigcup_{k=0}^{\infty} X_k.$$

Obviously, $X \subseteq \mathbb{X}$, and, by Corollary 3.21, $\mathbb{X} \in \mathcal{C}$. The first condition of Definition 3.19 follows from this fact and the fact that $\mathcal{C}$ is a K-consistency property.

For the remaining part of the proof it is sufficient to show that $\mathbb{X}$ is maximal in $\mathcal{C}$ with respect to $\subseteq$. Indeed, if $\mathbb{X}$ is $\subseteq$-maximal in $\mathcal{C}$, then the antecedent of clause 2–4 of Definition 3.19 yields its succedent by Definition 3.15 (K-consistency property). For example, if we assume that $\beta \in \mathbb{X}$, then, by Definition 3.15, $\mathbb{X} \cup \{\beta_i\} \in \mathcal{C}$ for $i = 0$ or $i = 1$, but this, together with $\mathbb{X}$-maximality, yields $\beta_i \in \mathbb{X}$. Similarly, if $\mathcal{C}$ is Z-compatible, then for a $\subseteq$-maximal set $\mathbb{X}$, for each $A \in Z$, A is already in $\mathbb{X}$, thus $Z \subseteq \mathbb{X}$.

Henceforth suppose that A is a formula such that $\mathbb{X} \cup \{A\} \in \mathcal{C}$. Let k stand for A's position in (3.8). Checking the construction of $\mathbb{X}$, we can see that $X_k \cup \{A\} \subseteq \mathbb{X} \cup \{A\}$. Since $\mathcal{C}$ is subset closed (see Corollary 3.20), $\mathbb{X} \cup \{A\} \in \mathcal{C}$ yields $X_k \cup \{A\} \in \mathcal{C}$. For this reason $X_{k+1} = X_k \cup \{A\}$, which entails $A \in \mathbb{X}$.

This finishes the proof. □

Superframes and Model Existence Theorems

The proof of Lemma 3.16 describes a construction assigning a downward-saturated supserset $\mathbb{X} \in \mathcal{C}$ to each $X \in \mathcal{C}$. It means that for a fixed K-consistency property $\mathcal{C}$ and ordering $S =$ (3.8) of formulas of $\mathcal{M}$ we can define a function $\mathbb{S}_S^{\mathcal{C}}$ assigning $X \mapsto \mathbb{X}$. Let us note for further reference:

$$\mathbb{S}_S^{\mathcal{C}} : X \mapsto \mathbb{X}. \tag{3.9}$$

We use this fact in the following definition, which will be further utilised in the proof of Strong Model Existence Theorem.

Definition 3.20 (*Superframe*) Let $\mathcal{C}$ stand for a K-consistency property of finite character, and let S stand for an arbitrary sequence ordering all formulas of $\mathcal{M}$. By a *superframe* $\langle \mathbb{W}, \mathbb{R}\rangle$ *built with reference to S upon* $\mathcal{C}$, we mean the following structure:

$$\mathbb{W} = \{\mathbb{S}_S^{\mathcal{C}}(X) : X \in \mathcal{C}\}, \ \ \mathbb{R} \subseteq \mathbb{W} \times \mathbb{W}, \ \ \mathbb{R} = \{\langle \mathbb{X}, \mathbb{Y}\rangle : \mathbb{X}^{\sharp} \subseteq \mathbb{Y}\}.$$

Let us observe that we could have defined $\mathbb{W}$ by picking up from $\mathcal{C}$ its $\subseteq$-maximal elements. However, it is useful to express the above definition with function $\mathbb{S}_S^{\mathcal{C}}$.

Superframes built upon K-consistency properties of finite character have the following two nice properties, which we shall refer to in the proof of Strong Model Existence Theorem.

Corollary 3.23 *If* $\langle \mathbb{W}, \mathbb{R}\rangle$ *is a superframe built with reference to S upon* $\mathcal{C}$, *then* $\mathbb{W} \subseteq \mathcal{C}$. *Also, if* $\mathcal{C}$ *is Z-compatible, then for each* $\mathbb{X} \in \mathbb{W}$, $Z \subseteq \mathbb{X}$.

Proof Follows from the proof of Lemma 3.16. □

The second property is more sophisticated.

Lemma 3.17 *Let* $\mathcal{C}$ *be a* K*-consistency property of finite character, let S stand for a sequence ordering all formulas of* $\mathcal{M}$, *and let* $\langle \mathbb{W}, \mathbb{R}\rangle$ *stand for a superframe built with reference to S upon* $\mathcal{C}$. *For each* $\mathbb{X} \in \mathbb{W}$, *and for every* $\pi \in \mathbb{X}$, *there is* $\mathbb{Y} \in \mathbb{W}$, *such that* $(\mathbb{X})^{\sharp} \cup \{\pi_0\} \subseteq \mathbb{Y}$.

Proof Let $\pi \in \mathbb{X} \in \mathbb{W}$. Since $\mathbb{X} \in \mathcal{C}$ (by Corollary 3.23) and $\mathcal{C}$ is a K-consistency property, by Definition 3.15, clause 5, the following set:

$$(\mathbb{X})^{\sharp} \cup \{\pi_0\} \tag{3.10}$$

is in $\mathcal{C}$. Then also the following:

$$\mathbb{Y} = \mathbb{S}_S^{\mathcal{C}}((\mathbb{X})^{\sharp} \cup \{\pi_0\}), \tag{3.11}$$

where $\mathbb{S}_S^{\mathcal{C}}$ is a function set by (3.9), belongs to $\mathcal{C}$. This finishes the proof. □

Theorem 3.3 [Strong Model Existence Theorem) *If a set X of formulas of* $\mathcal{M}$ *belongs to a Z-compatible* K*-consistency property, then X is locally satisfied in a Kripke model globally satisfying Z.*

Proof Let $X \in \mathcal{C}$ where $\mathcal{C}$ is a Z-compatible K-consistency property. By Corollary 3.22, $\mathcal{C}$ can be extended to a Z-compatible K-consistency property $\mathcal{C}^{**}$ which is of finite character. Hence we have $X \in \mathcal{C}^{**}$.

Let S stand for an ordering of all formulas of language $\mathcal{M}$, and let $\langle \mathbb{W}, \mathbb{R}\rangle$ be a superframe built with reference to S upon $\mathcal{C}^{**}$. We define valuation function V on the frame, which is such that for each propositional variable, p_i, and each $\mathbb{X} \in \mathbb{W}$,

$$V(p_i, \mathbb{X}) = 1 \text{ iff } p_i \in \mathbb{X}. \tag{3.12}$$

We prove what follows by induction on the rank of formula A.

(a) For each set $\mathbb{X} \in \mathbb{W}$, for each formula $A \in \mathbb{X}$, $V(A, \mathbb{X}) = 1$.

We will also prove:

(b) For each set $\mathbb{X} \in \mathbb{W}$, Z is locally satisfied in $\mathbb{X}$, and thus Z is globally satisfied in $\langle \mathbb{W}, \mathbb{R}, V \rangle$.

Since for the initial set X there is $\mathbb{S}^{\mathcal{C}}_{S}(X) \in \mathbb{W}$, (a) and (b) together with Corollary 3.3, entail our theorem.

We start with (a). Let $\mathbb{X} \in \mathbb{W}$ be arbitrary and assume that $A \in \mathbb{X}$. In the base case, $r(A) = 0$. If A is a propositional variable, p_i, then $V(A, \mathbb{X}) = 1$ by (3.12). If $A = \neg p_i$, then we recall the fact that $\mathbb{W} \subseteq \mathcal{C}^{**}$ and $\mathcal{C}^{**}$ is a K-consistency property, which yields that $\mathbb{X}$ cannot contain both p_i and '$\neg p_i$'. Hence $p_i \notin \mathbb{X}$, and therefore $V(A, \mathbb{X}) = 1$ by (3.12).

IH: for each set $\mathbb{X} \in \mathbb{W}$, for each formula $A \in \mathbb{X}$ of rank up to n, $V(A, \mathbb{X}) = 1$.

Again, let $\mathbb{X} \in \mathbb{W}$ be arbitrary. Among the classical cases we consider only α-formulas. Let $A \in \mathbb{X}$ stand for such an α-formula of rank $n + 1$. Since $\mathbb{X}$ is downward-saturated, also $\{\alpha_0, \alpha_1\} \subseteq \mathbb{X}$. But then IH applies to both components, which yields immediately that $V(A, \mathbb{X}) = 1$.

Let $A \in \mathbb{X}$ be a ν-formula of rank $n + 1$ and assume that $\mathbb{Y} \in \mathbb{R}(\mathbb{X})$. It means that $\nu_0 \in \mathbb{Y}$ by Definition 3.20. IH applies to formulas in $\mathbb{Y}$ of rank up to n, and $r(\nu_0) = n$, hence $V(\nu_0, \mathbb{Y}) = 1$ by IH. Since $\mathbb{Y}$ was an arbitrary set in $\mathbb{R}(\mathbb{X})$, we arrive at $V(\nu, \mathbb{X}) = 1$.

Let $A \in \mathbb{X}$ be a π-formula of rank $n + 1$. By Lemma 3.17, there is $\mathbb{Y} \in \mathbb{W}$ such that $\pi_0 \in \mathbb{Y}$. By IH applied to this set, $V(\pi_0, \mathbb{Y}) = 1$, hence also $V(\pi, \mathbb{X}) = 1$. This finishes the proof of (a).

By Definition 3.20, each $\mathbb{X} \in \mathbb{W}$ is an $\mathbb{S}^{\mathcal{C}}_{S}$-value of some $X \in \mathcal{C}^{**}$. By Lemma 3.16, $Z \subseteq \mathbb{X}$. By (a), Z is locally satisfied in $\mathbb{X}$, but $\mathbb{X}$ was arbitrary, hence (b) follows. □

The Strong Model Existence Theorem entails the weaker version, the Model Existence Theorem.

Theorem 3.4 (Model Existence Theorem) *If a set, X, of formulas of $\mathcal{M}$ belongs to a* K*-consistency property, then X is locally satisfied in a Kripke model.*

Proof The proof is contained in the proof of the previous theorem. □

K-consistency with Respect to $\mathbb{E}^{\mathsf{K}}$

Here comes the part linking the notion of consistency with erotetic calculi. We start with some auxiliary notions.

Definition 3.21 Let Z stand for a set of formulas of $\mathcal{M}$. Let $\phi = S \vdash \Gamma$ be a sequent of the language $\mathcal{M}^{?}_{\vdash}$. If $Z = \mathsf{set}(S)$, then we say that *Z is the set of global assumptions of ϕ*, and we call ϕ a *Z-sequent*, for short.

There are always two difficult moments in the construction of completeness proof using consistency properties: the first one is when we show that syntactically defined consistency property is semantically satisfiable (Theorem 3.3); the second consists in assigning an unprovable sequent to a consistency property. Here comes the second stage.

Let us recall that if X is a set of formulas, then $-X = \{-F : F \in X\}$. Another useful convention is this: let Z be a finite set of formulas of $\mathcal{M}$ and let ϕ be a Z-sequent. By a *(Z, ϕ)-family of sets* we shall mean a family of all sets of the form

$$Y = Z \cup -X$$

where for some $i \in W(\phi)$, X is the set-representation of i in ϕ. That is, Z picks up all the global assumptions and X picks up all the formulas in a world i. This notion is further generalized as follows.

Definition 3.22 (*Z-family*) Let Z be a finite set of formulas of $\mathcal{M}$.

1. For every Z-sequent ϕ, (Z, ϕ)-family of sets is a Z-family of sets.
2. If $\mathcal{F}$ and $\mathcal{G}$ are Z-families of sets, then their sum, $\mathcal{F} \cup \mathcal{G}$, is also a Z-family of sets.

Nothing else is a Z-family.

In what follows we shall focus on sequents ϕ for which there is no successful Socratic transformation of $?(\phi)$ in $\mathbb{E}^{\mathsf{K}}$. Let us recall that the notion of Socratic proof is tailored for sequents of the form '$S \vdash A^{\mathbf{0}}$', whereas we need something more general. In order to simplify the terminology, from now on, by *sequent unprovable in* $\mathbb{E}^{\mathsf{K}}$ we shall mean sequent ϕ of an arbitrary form for which there is no successful Socratic transformation of $?(\phi)$ in $\mathbb{E}^{\mathsf{L}}$.

In order to obtain a K-consistency property we need a Z-family constructed out of *unprovable* sequents only which is *maximal* at the same time, that is, uses all unprovable Z-sequents. The following concept satisfies these requirements.

Definition 3.23 (*K-consistent Z-family*) Let Z stand for a finite set of formulas of $\mathcal{M}$ and let $\mathcal{F}$ be a Z-family. We say that $\mathcal{F}$ is *K-consistent*, provided that the following condition (con^{K}_{Z}) holds:

(con^{K}_{Z}) $Y \in \mathcal{F}$ iff for some Z-sequent ϕ unprovable in $\mathbb{E}^{\mathsf{K}}$, Y belongs to a (Z, ϕ)-family.

We note the following without a proof.

Corollary 3.24 *Every family of sets can be extended to a subset closed family of sets.*

Lemma 3.18 *Every K-consistent Z-family can be extended to a Z-compatible K-consistency property.*

Proof Let $\mathcal{F}$ stand for a K-consistent Z-family. We first extend $\mathcal{F}$ to a subset closed family $\mathcal{F}^*$ of sets (by Corollary 3.24), and we prove that $\mathcal{F}^*$ is a Z-compatible K-consistency property.

We check the conditions from Definition 3.15 of K-consistency property, one by one. Hence let (1) $Y \in \mathcal{F}^*$. By assumption, Y is a subset of set $Z \cup -X \in \mathcal{F}$ where (2) X is the set-representation of some $i \in W(\phi)$, (3) is not provable in $\mathbb{E}^{\mathsf{K}}$, and (4) ϕ is a Z-sequent. Let us observe that Z and X must be finite, by definition. Clearly, $\mathcal{F} \subseteq \mathcal{F}^*$, hence the fact that $Y \subseteq Z \cup -X$ and $Z \cup -X \in \mathcal{F}$ will immediately yield that $\mathcal{F}^*$ is Z-compatible, once we prove that it is a K-consistency property.

Assume that for some propositional variable p_l, $\{p_l, \neg p_l\} \subseteq Y$. There are the following possibilities.

(a) $\{p_l, \neg p_l\} \subseteq Z$;
(b) one of the literals is in Z and the other is in $-X$, more specifically:

- $p_l \in Z, \neg p_l \in -X$, and '$\neg p_l$' is the complement of p_l,
- $p_l \in Z, \neg p_l \in -X$, and '$\neg p_l$' is the complement of '$\neg\neg p_l$',
- $\neg p_l \in Z$, $p_l \in -X$, then p_l must be the complement of '$\neg p_l$';

(c) $\{p_l, \neg p_l\} \subseteq -X$.

In the case (a), ϕ is of the form '$S(p_l)(\neg p_l) \vdash \Gamma$', and by assumption (2) we have $i \in W(\phi)$. We choose an arbitrary position in Γ which we indicate by writing '$\Gamma(\_)$'. The following:

$$\cfrac{\cfrac{?(S(p_l)(\neg p_l) \ \vdash \ \Gamma(\_))}{?(S(p_l)(\neg p_l) \ \vdash \ \Gamma((\neg p_l)^i))}\ R_G}{?(S(p_l)(\neg p_l) \ \vdash \ \Gamma((p_l)^i, (\neg p_l)^i))}\ R_G$$

is a Socratic proof of ϕ in $\mathbb{E}^{\mathsf{K}}$. But this contradicts the assumption (3) concerning ϕ.

Assume that (b) is the case. Depending on the subcase ϕ is of one of the following three forms:

$$\phi \ = \ S(p_l) \ \vdash \ \Gamma((p_l)^{\varsigma/i}),$$

$$\phi \ = \ S(p_l) \ \vdash \ \Gamma((\neg\neg p_l)^{\varsigma/i}),$$

$$\phi \ = \ S(\neg p_l) \ \vdash \ \Gamma((\neg p_l)^{\varsigma/i}).$$

In each case one application of R_G produces a Socratic proof of ϕ in $\mathbb{E}^{\mathsf{K}}$, contrary to assumption (3). Finally, if (c) is the case, then the sequent is basic already.

We proceed to the second clause of Definition 3.15. Let $\neg\neg A \in Y$. Then (a) $\neg\neg A \in Z$ or (b) $\neg\neg\neg A \in X$. If (a) is the case, then ϕ is of the form '$S(\neg\neg A) \vdash \Gamma$'. We indicate an arbitrary position in Γ by writing it as '$\Gamma(\_)$'. We have $i \in W(\phi)$. The following Socratic transformation:

$$\cfrac{?(S(\neg\neg A) \ \vdash \ \Gamma(\_))}{?(S(\neg\neg A) \ \vdash \ \Gamma((\neg A)^i))}\ R_G$$

shows that if there is a Socratic proof of sequent $\psi = S(\neg\neg A) \vdash \Gamma((\neg A)^i)$ in $\mathbb{E}^{\mathsf{K}}$, then ϕ is provable as well, thus sequent ψ is not provable in $\mathbb{E}^{\mathsf{K}}$ by assumption (3). On the other hand, if X is the set representation of $i \in W(\phi)$, then the set representation of $i \in W(\psi)$ is $X \cup \{\neg A\}$. Finally, $-\neg A = A$, hence $Z \cup (-X \cup \{A\})$ belongs to a (Z, ψ)-family, where ψ is not provable in $\mathbb{E}^{\mathsf{K}}$. This yields that $Z \cup (-X \cup \{A\})$ belongs to $\mathcal{F}$, and hence each subset of this set, $Y \cup \{A\}$ in particular, belongs to $\mathcal{F}^*$, which was to be proved.

The second case requires the use of contraction, for otherwise '$\neg\neg A$' can disappear from the considered sequent. This is the first time we shall use Lemma 3.9 (semi-admissibility of contraction). Sequent ϕ is now of the form '$S \vdash \Gamma((\neg\neg\neg A)^{\varsigma/i})$'. The following shows that if there is a Socratic proof of sequent '$S \vdash \Gamma((\neg\neg\neg A)^{\varsigma/i}, (\neg A)^{\varsigma/i})$' in $\mathbb{E}^{\mathsf{K}} \cup \{R_{contr}\}$, then ϕ is provable in $\mathbb{E}^{\mathsf{K}} \cup \{R_{contr}\}$, and hence, by Corollary 3.17 (eliminability of contraction) which follows directly from Lemma 3.9, ϕ is provable in $\mathbb{E}^{\mathsf{K}}$, contrary to assumption (3).

$$\dfrac{\dfrac{?(S \vdash \Gamma((\neg\neg\neg A)^{\varsigma/i}))}{?(S \vdash \Gamma((\neg\neg\neg A)^{\varsigma/i}, (\neg\neg\neg A)^{\varsigma/i}))}\, R_{contr}}{?(S \vdash \Gamma((\neg\neg\neg A)^{\varsigma/i}, (\neg A)^{\varsigma/i}))}\, R_{\neg\neg}$$

To state the obvious, a sequent unprovable in $\mathbb{E}^{\mathsf{K}} \cup \{R_{contr}\}$ is not provable in $\mathbb{E}^{\mathsf{K}}$. Therefore the above Socratic transformation shows that $Z \cup (-X \cup \{A\}) \in \mathcal{F}$, which yields that $Y \cup \{A\} \in \mathcal{F}^*$.

Clause 3. Assume that $\alpha \in Y$. Again, we have two cases: (a) $\alpha \in Z$ and (b) $\alpha \in -X$. In (a) we apply R_G first. By Corollary 2.1, '$-\alpha$' is a β-formula, hence in the second step we can apply R_β.

$$\dfrac{\dfrac{?(S\,(\alpha) \vdash \Gamma(\_))}{?(S\,(\alpha) \vdash \Gamma((-\alpha)^i))}\, R_G}{?(S\,(\alpha) \vdash \Gamma((\beta_0)^i, (\beta_1)^i))}\, R_\beta$$

Since ϕ is not provable in $\mathbb{E}^{\mathsf{K}}$, neither is '$S\,(\alpha) \vdash \Gamma((\beta_0)^i, (\beta_1)^i)$'. If $\beta_k = \neg\alpha_k$ for both $k \in \{0, 1\}$, then the above shows that

$$Y \cup -\{\neg\alpha_0, \neg\alpha_1\} = Y \cup \{\alpha_0, \alpha_1\} \in \mathcal{F}^*.$$

Assume that $\beta_k = -\alpha_k \neq \neg\alpha_k$. Then α_k must be a negative formula. Suppose further that α_k is not doubly negated, that is, it is of the form '$\neg A$' for some positive formula A. Then

$$-\beta_k = - - \alpha_k = - - \neg A = -A = \neg A = \alpha_k.$$

Finally, if α_k is of the form '$\neg\neg A$', then $-\beta_k = - - \alpha_k = A$. Without loss of generality, suppose it happens with respect to α_0 only. In this case we need Lemma 3.7 (transparency of double negation) which entails that sequent

$$S\,(\alpha) \vdash \Gamma((\neg\neg\beta_0)^i, (\beta_1)^i)$$

is not provable in $\mathbb{E}^{\mathsf{K}}$, if '$S(\alpha) \vdash \Gamma((\beta_0)^i, (\beta_1)^i)$' is not. This, in turn, yields that $Y \cup \{\alpha_0, \alpha_1\} \in \mathcal{F}^*$.

If (b) holds, we shall need R_{contr}, as above. Furthermore, $\alpha \in -X$, hence either (i) $\neg\alpha \in X$ or (ii) α is a negative formula '$\neg B$' with $B \in X$ positive. In the first case we have to consider

$$\phi = S \vdash \Gamma((\neg\alpha)^{\varsigma/i})$$

with two further subcases: (i.i) $\neg\alpha$ is a β-formula, or (i.ii) α is of the form '$\neg\neg\beta$'. If (ii) $\alpha = \neg B$, then we consider

$$\phi = S \vdash \Gamma((B)^{\varsigma/i}).$$

B is a β-formula in this case. The solution is the following Socratic transformation:

$$\cfrac{\cfrac{\cfrac{?(S \vdash \Gamma((\neg\neg\beta)^{\varsigma/i}))}{?(S \vdash \Gamma((\beta)^{\varsigma/i}))}\,R_{\neg\neg}}{?(S \vdash \Gamma((\beta)^{\varsigma/i}, (\beta)^{\varsigma/i}))}\,R_{contr}}{?(S \vdash \Gamma((\beta)^{\varsigma/i}, (\beta_0)^{\varsigma/i}, (\beta_1)^{\varsigma/i}))}\,R_{\beta}$$

where the first application of $R_{\neg\neg}$ is relevant only in case (i.ii). Further we show that $Y \cup \{\alpha_0, \alpha_1\} \in \mathcal{F}^*$ using Corollary 3.17 (eliminability of contraction) and the same arguments as in case (a).

Clause 4. Let $\beta \in Y$, possible cases are: (a) $\beta \in Z$, (b) $\beta \in -X$. By Corollary 3.1, '$-\beta$' is an α-formula, hence in (a) we have what follows,

$$\cfrac{\cfrac{?(S(\beta) \vdash \Gamma(\_))}{?(S(\beta) \vdash \Gamma((-\beta)^i))}\,R_G}{?(S(\beta) \vdash \Gamma((\alpha_0)^i);\ S(\beta) \vdash \Gamma((\alpha_1)^i))}\,R_{\alpha}$$

In the case (b), we first apply $R_{\neg\neg}$ to '$?(\Phi(S \vdash \Gamma((\neg\neg\alpha)^{\varsigma/i})))$', if this is necessary; then further:

$$\cfrac{\cfrac{?(S \vdash \Gamma((\alpha)^{\varsigma/i}))}{?(S \vdash \Gamma((\alpha)^{\varsigma/i}, (\alpha)^{\varsigma/i}))}\,R_{contr}}{?(S \vdash \Gamma((\alpha)^{\varsigma/i}, (\alpha_0)^{\varsigma/i});\ S \vdash \Gamma((\alpha)^{\varsigma/i}, (\alpha_1)^{\varsigma/i}))}\,R_{\alpha}$$

Since, by assumption (3), ϕ is not provable in $\mathbb{E}^{\mathsf{K}}$, it follows that, in the case (a), there is no successful Socratic transformation of question:

$$?(S(\beta) \vdash \Gamma((\alpha_0)^{\varsigma/i});\ S(\beta) \vdash \Gamma((\alpha_1)^{\varsigma/i}))$$

in calculus $\mathbb{E}^{\mathsf{K}}$, and in the case (b), there is no successful Socratic transformation of question:

$$?(S \vdash \Gamma\,((\alpha)^{\varsigma/i}, (\alpha_0)^{\varsigma/i});\ \ S \vdash \Gamma\,((\alpha)^{\varsigma/i}, (\alpha_1)^{\varsigma/i}))$$

in $\mathbb{E}^{\mathsf{K}} \cup \{R_{contr}\}$. In the later case we go through Corollary 3.17. By Lemma 3.5 (Separation), in case (a) it follows that at least one of the sequents:

$$S(\beta) \vdash \Gamma((\alpha_0)^{\varsigma/i}),\ \ S(\beta) \vdash \Gamma((\alpha_1)^{\varsigma/i})$$

is not provable in $\mathbb{E}^{\mathsf{K}}$; in case (b) this concerns at least one of:

$$S \vdash \Gamma((\alpha)^{\varsigma/i}, (\alpha_0)^{\varsigma/i}),\ \ S \vdash \Gamma((\alpha)^{\varsigma/i}, (\alpha_1)^{\varsigma/i}).$$

Similarly as for clause 3, we arrive at the conclusion that $Y \cup \{\beta_k\} \in \mathcal{F}^*$ for some $k \in \{0, 1\}$. (To deal with double negation we apply $R_{\neg\neg}$ if necessary, and we use Lemma 3.7 if needed.)

Clause 5, let $\pi \in Y$. Recall that Y is finite. Let $\nu^1, \dots, \nu^n$ stand for all ν-formulas in Y. Let us also recall that the aim is to show that $Y^\sharp \cup \{\pi_0\} = \{\nu_0^1, \dots, \nu_0^n, \pi_0\} \in \mathcal{F}^*$.

We first establish that there is $Z \cup \{\pi_0\} \in \mathcal{F}^*$, then we proceed to ν-formulas in order to show that their components extend $Z \cup \{\pi_0\}$ in $\mathcal{F}^*$. Hence assume that $\pi \in Z$ (this is our (a)-case, (b)-case to be considered later). By Corollary 3.1, $-\pi$ is a ν-formula, hence the following is a Socratic transformation in $\mathbb{E}^{\mathsf{K}}$ showing that (for some j) sequent '$S(\pi) \vdash \Gamma((\nu_0)^{i,j})$' is not provable in $\mathbb{E}^{\mathsf{K}}$.

$$\cfrac{\cfrac{?(S(\pi) \vdash \Gamma(\_))}{?(S(\pi) \vdash \Gamma((-\pi)^i))}\ R_G}{?(S(\pi) \vdash \Gamma((\nu_0)^{i,j}))}\ R_\nu$$

By Corollary 3.1 again, $\nu_0 \in \{-\pi_0, \neg\pi_0\}$. The set $\{\nu_0\}$ is a set-representation of j in sequent '$S(\pi) \vdash \Gamma((\nu_0)^{i,j})$', hence $Z \cup -\{\nu_0\} \in \mathcal{F}^*$. Since $\nu_0 \in \{-\pi_0, \neg\pi_0\}$, it follows that $-\nu_0 \in \{- - \pi_0, -\neg\pi_0\} = \{- - \pi_0, \pi_0\}$. If $-\nu_0 = \pi_0$, then we have $Z \cup \{\pi_0\} \in \mathcal{F}^*$. Otherwise, $-\nu_0 = - - \pi_0$ and $-\nu_0 \neq \pi_0$, which is the case only if π_0 is doubly negated. In this case we make use of Lemma 3.7 (transparency of double negation) which yields that sequent '$S(\pi) \vdash \Gamma((\neg\neg\nu_0)^{i,j})$' is not provable in $\mathbb{E}^{\mathsf{K}}$.

Now let (b) $\pi \in -X$. Then, by Corollary 3.1, $\neg\pi \in X$ or π is of the form '$\neg\nu$' and $\nu \in X$. Either way, the corresponding formula in X is ν or '$\neg\neg\nu$'. Then the following (with the first step possibly omitted):

$$\cfrac{\cfrac{?(S \vdash \Gamma((\neg\neg\nu)^{\varsigma/i}))}{?(S \vdash \Gamma((\nu)^{\varsigma/i}))}\ R_{\neg\neg}}{?(S \vdash \Gamma((\nu_0)^{\varsigma/i,j}))}\ R_\nu$$

shows that $Z \cup \{\pi_0\} \in \mathcal{F}^*$, the detailed arguments are as above.

Among the ν-formulas $\nu^1, \dots, \nu^n \in Y$, some (possibly none, possibly all) are in Z and some (again, possibly none, possibly all) in $-X$. Let $\nu^{l_1}, \dots, \nu^{l_k}$ stand for ν-formulas in Z, and let $\nu^{m_1}, \dots, \nu^{m_o}$ stand for ν-formulas in $-X$. We first consider ν^{l_1}. By Corollary 3.1, $-\nu^{l_1}$ is a π-formula, we will mark it as 'π^{l_1}'. For simplicity, the last sequent considered above, that is, $\phi = S \vdash \Gamma((\nu_0)^{\varsigma/i,j})$, will be now represented by '$S(\nu^{l_1}) \vdash \Gamma(\_)$' (let us also recall that $\langle i, j\rangle \in R(\phi)$). We apply:

$$\cfrac{\cfrac{?(S(\nu^{l_1}) \vdash \Gamma(\_))}{?(S(\nu^{l_1}) \vdash \Gamma((\pi^{l_1})^i))}\, R_G}{?(S(\nu^{l_1}) \vdash \Gamma((\pi^{l_1})^i, (\pi_0^{l_1})^j))}\, R_\pi$$

and we repeat the applications of R_G, R_π with respect to each of $\nu^{l_2}, \ldots, \nu^{l_k}$, until we arrive at:

$$?(S \vdash \Gamma((\pi^{l_1})^i, (\pi_0^{l_1})^j, \ldots, (\pi^{l_k})^i, (\pi_0^{l_k})^j))$$

For simplicity, the ν-formulas occurring in S are not displayed.

Next we continue with $\nu^{m_1} \in -X$. There are the following cases: (i) ν^{m_1} is of the form '$\neg\pi^{m_1}$' and $\pi^{m_1} \in X$ (the π-formula is positive and thus $-\pi^{m_1} = \neg\pi^{m_1}$), or (ii) $\neg\nu^{m_1} \in X$, in this case either '$\neg\nu^{m_1}$' is a π-formula, for which we shall use π^{m_1}, or '$\neg\nu^{m_1}$' is of the form '$\neg\neg\pi^{m_1}$'. We build a Socratic transformation as follows (we omit the first step if not needed):

$$\cfrac{\cfrac{?(S \vdash \Gamma((\neg\neg\pi^{m_1})^{\varsigma_1/i}))}{?(S \vdash \Gamma((\pi^{m_1})^{\varsigma_1/i}))}\, R_{\neg\neg}}{?(S \vdash \Gamma((\pi^{m_1})^{\varsigma_1/i}, (\pi_0^{m_1})^j))}\, R_\pi$$

and analogously for $\nu^{m_2}, \ldots, \nu^{m_o}$, till we arrive at a question whose only constituent contains, *int. al.*, the following formulas:

$$(\pi^{l_1})^i, (\pi_0^{l_1})^j, \ldots, (\pi^{l_k})^i, (\pi_0^{l_k})^j, (\pi^{m_1})^{\varsigma_i/i}, (\pi_0^{m_1})^j, \ldots, (\pi^{m_o})^{\varsigma_o/i}, (\pi_0^{m_o})^j$$

right of the turnstile. If needed, we apply Lemma 3.7 (double negation). From the above it follows that

$$Z \cup \{\pi_0, \nu_0^{l_1}, \ldots, \nu_0^{l_k}, \nu_0^{m_1}, \ldots, \nu_0^{m_1}\} \in \mathcal{F}.$$

Finally, $\mathcal{F} \subseteq \mathcal{F}^*$ and $\mathcal{F}^*$ is subset closed, therefore $Y^\sharp \cup \{\pi_0\} \in \mathcal{F}^*$.

We have shown that $\mathcal{F}^*$ is a (Z-compatible) K-consistency property. □

Theorem 3.5 (Completeness of $\mathbb{E}^{\mathsf{K}}$ wrt Semantics of Sequents). *For every* K*-valid sequent* ϕ *of the language* $\mathcal{M}^?_\vdash$*, there is a successful Socratic transformation of* $?(\phi)$ *in* $\mathbb{E}^{\mathsf{K}}$.

Proof Let $\phi = S \vdash \Gamma$ stand for a sequent of $\mathcal{M}^?_\vdash$ and assume that there is no required Socratic transformation of $?(\phi)$ in $\mathbb{E}^{\mathsf{K}}$. We will show that ϕ is not K-valid.

Let $\mathsf{set}(S) = Z$, let $0, \ldots, n$ represent all numerals occurring in indices of indexed formulas of Γ. Then:

$$\mathcal{F} = \{Z \cup -X_0, \ldots, Z \cup -X_n\}$$

is a $(Z, S \vdash \Gamma)$-family. Since '$S \vdash \Gamma$' is not provable in $\mathbb{E}^{\mathsf{K}}$, by condition (con_Z^{K}) of Definition 3.23, $\mathcal{F}$ is a subset of a K-consistent Z-family $\mathcal{F}^*$ which, by Lemma 3.18, can be extended to a Z-compatible K-consistency property $\mathcal{F}^{**}$. Further, by Theorem 3.3 (strong model existence theorem), for each set $Y_i = Z \cup -X_i$ (where $0 \leqslant i \leqslant n$), Y_i is locally satisfied in a Kripke model $\langle W_i, R_i, V_i\rangle$ globally satisfying Z. Below we indicate how to join the models in order to obtain one that falsifies the whole sequent '$S \vdash \Gamma$'.

We make an assumption that for $i \neq j$, $W_i \cap W_j = \varnothing$. If the assumption is not met, then we redefine one of the models, say it is $\langle W_j, R_j, V_j\rangle$, in the following way. For each $w \in W_i \cap W_j$ we take a 'copy' of w—let us call it w^*—such that $w^* \notin W_i \cup W_j$, and we swap w in W_j for w^*, leaving 'the original' w in W_i. By calling w^* a 'copy' we mean that the truth-conditions of all formulas in w^* are the same as those in w. Assuming that this operation can always be performed, we suppose, without loss of generality, that $W_i \cap W_j = \varnothing$ for each $i \neq j$. Now we can build one model from those mentioned. Let $\langle W, R\rangle$ stand for a frame such that:

$$W = \bigcup_{i=0}^{n} W_i, \quad R = \bigcup_{i=0}^{n} R_i.$$

The valuation function, viewed set-theoretically, could be defined by sum as well. We can also set it as follows: let $w \in W_i \subseteq W$ and $A \in \mathit{FORM}_{\mathcal{M}}$, then $V(A, w) = V_i(A, w)$.

Let us now recall Theorem 3.3, which guarantees that for each Y_i, there is a world in the model $\langle W_i, R_i, V_i\rangle$ in which all the elements of Y_i are true. Let us call the world 'w^i'. It follows that set Y_i is locally satisfied in model $\langle W, R, V\rangle$ as well. What is more, set Z is globally satisfied in $\langle W, R, V\rangle$.

Finally, we indicate an interpretation function of sequent '$S \vdash \Gamma$' in frame $\langle W, R\rangle$ (recall Definition 3.8, Sect. 3.3). Simply, let $f : W(S \vdash \Gamma) \longrightarrow W$ be such that for each $i \in W(S \vdash \Gamma)$, $f(i) = w^i$. It follows that sequent '$S \vdash \Gamma$' is not true in the frame $\langle W, R\rangle$ under the considered interpretation f, which contradicts the assumption that the sequent is K-valid. □

The following theorem is a straightforward corollary from the preceding one.

Theorem 3.6 (Completeness of $\mathbb{E}^{\mathsf{K}}$ wrt Semantics of $\mathcal{M}$). *If a sequent of the form '$S \vdash (A)^{0}$' is* K*-valid, then there is a Socratic proof of the sequent in* $\mathbb{E}^{\mathsf{K}}$.

Finally, we end this subsection with the following theorem.

Theorem 3.7 (Soundness and Completeness of $\mathbb{E}^{\mathsf{K}}$ wrt Global K-entailment). *A sequent of* $\mathcal{M}^?_{\vdash}$ *of the form '$S \vdash (A)^{0}$' has a Socratic proof in* $\mathbb{E}^{\mathsf{K}}$ *iff* $\mathsf{set}(S) \models_{\mathsf{K}} A$.

Proof The if-then direction follows from Theorem 3.1 (soundness of $\mathbb{E}^{\mathsf{L}}$ wrt global entailment). For the only-if direction we use Lemma 3.3, expressing the connection between L-validity of sequents and global L-entailment, and then we use the previous theorem: if $\mathsf{set}(S) \models_{\mathsf{K}} A$, then sequent '$S \vdash (A)^{0}$' is K-valid, hence it is provable in $\mathbb{E}^{\mathsf{K}}$. □

Extensions to $\mathbb{E}^{\mathsf{L}}$

As could be expected, we start with a redefinition of $(X)^{\sharp}$ for logics above K.[5] An explicit relativization to L is now added to this notation, with the understanding that $(X)^{\sharp\mathsf{K}}$ is just $(X)^{\sharp}$. Intuitively speaking, $(X)^{\sharp\mathsf{L}}$ is the content of set X that must be carried to a world L-accessible from X. The definition of $(\bullet)^{\sharp\mathsf{L}}$ is provided in Table 3.6. For a given logic L and set X, the content of $(\bullet)^{\sharp\mathsf{L}}$ is defined in terms of four sets displayed in the first row of the table. The plus-minus notation is self-explanatory.

The definition of $(\bullet)^{\sharp\mathsf{L}}$ captures transitivity, symmetry and Euclideanness; the other two properties: seriality and reflexivity, will be captured later, in the definition of an L-consistency property.

We shall also see later, in the proof of Lemma 3.22, how the component-sets of $(X)^{\sharp\mathsf{L}}$ work in preserving the L-properties of R-accessibility relation. By and large, set $\{\nu : \nu \in X\}$ corresponds to transitivity of R, set $\{\pi : \pi_0 \in X\}$ corresponds to its symmetry, and set $\{\pi : \pi \in X\}$—to its Euclideanness.

For some logic we can choose the $(\bullet)^{\sharp\mathsf{L}}$-description. For example, KB4 and KB5 (and hence also KB45) is the same logic for which different definitions of $(\bullet)^{\sharp\mathsf{L}}$ can be given. However, the above account is compatible with the axiomatic and semantic descriptions presented at the beginning of this chapter.

Clauses 1 and 2 of Corollary 3.19 apply to all definitions of function $(\bullet)^{\sharp\mathsf{L}}$. Clause 3 needs a reformulation.

Corollary 3.25 (L-version of Corollary *3.19*) *Let X and Y be sets of formulas of $\mathcal{M}$. The following holds for each* L.

1. *If $X \subseteq Y$, then $(X)^{\sharp\mathsf{L}} \subseteq (Y)^{\sharp\mathsf{L}}$.*
2. $(X \cup Y)^{\sharp\mathsf{L}} = (X)^{\sharp\mathsf{L}} \cup (Y)^{\sharp\mathsf{L}}$.
3. *If $(X)^{\sharp\mathsf{L}}$ is finite, then:*
   (a) $\{\nu : \nu \in X\}$ *is finite for every* L,
   (b) *for* L ∈ {K5, D5, S5, K45, D45}, *also* $\{\pi : \pi \in X\}$ *is finite,*
   (c) *for* L ∈ {KB, DB, B, S5, KB4}, *set X is finite.*

Table 3.6 Classes of logics and their $(\bullet)^{\sharp\mathsf{L}}$ definition

L	For $X \subseteq FORM_{\mathcal{M}}$, $(X)^{\sharp\mathsf{L}}$ is the sum of			
	$\{\nu_0 : \nu \in X\}$	$\{\nu : \nu \in X\}$	$\{\pi : \pi_0 \in X\}$	$\{\pi : \pi \in X\}$
K, D, T	+	−	−	−
K4, D4, S4	+	+	−	−
KB, DB, B	+	−	+	−
K5, D5	+	−	−	+
KB4	+	+	+	−
K45, D45	+	+	−	+
S5	+	−	+	+

[5] Compare [4, Sect. 5 of Chap. 2, Sect. 4 of Chap. 5].

Proof Clause 3 (c) follows from the fact that every element of X can be viewed as a π_0-component of some π-formula. □

Definition 3.24 (L-*consistency Property*) Let $\mathcal{C}$ stand for a family of sets of formulas of $\mathcal{M}$. We say that $\mathcal{C}$ *is an* L-*consistency property* if each set $X \in \mathcal{C}$ satisfies the following conditions.

1. If $p_i \in X$, then '$-p_i$' $\notin X$.
2. If $\neg\neg A \in X$, then $X \cup \{A\} \in \mathcal{C}$.
3. If $\alpha \in X$, then $X \cup \{\alpha_0, \alpha_1\} \in \mathcal{C}$.
4. If $\beta \in X$, then $X \cup \{\beta_0\} \in \mathcal{C}$ or $X \cup \{\beta_1\} \in \mathcal{C}$.
5. If $\pi \in X$, then $(X)^{\sharp_{\mathsf{L}}} \cup \{\pi_0\} \in \mathcal{C}$.
6. For $\mathsf{L} \in \{\mathsf{D}, \mathsf{DB}, \mathsf{D4}, \mathsf{D5}, \mathsf{D45}\}$, $(X)^{\sharp_{\mathsf{L}}} \in \mathcal{C}$.
7. For $\mathsf{L} \in \{\mathsf{T}, \mathsf{B}, \mathsf{S4}, \mathsf{S5}\}$, $X \cup (X)^{\sharp_{\mathsf{L}}} \in \mathcal{C}$.

The notion of Z-compatibility applies as previously (see Definition 3.16).

Seriality is captured by clause 6, and reflexivity by clause 7 of Definition 3.24. As we have observed, the remaining properties are captured by the definition of $(\bullet)^{\sharp_{\mathsf{L}}}$. However, for symmetric and Euclidean logics one more trick is needed—we shall use cut classes.[6]

Definition 3.25 (*Cut Class*) Let $X \subseteq FORM_{\mathcal{M}}$. We call X a *cut class* if

1. X is closed under subformulas, that is, if $A \in X$ and B is a subformula of A, then $B \in X$.
2. X is closed under prefixing with modal operators, that is, if $A \in X$, then also $\{\Box A, \neg\Box A, \Diamond A, \neg\Diamond A\} \subseteq X$.

Let us observe that the set $FORM_{\mathcal{M}}$ of all formulas of the language $\mathcal{M}$ is a cut class. For the following definition compare [4, p. 205–206].

Definition 3.26 (L-*consistency Property with Cut Class* X) Let $\mathcal{C}$ be an L-consistency property. We will say that $\mathcal{C}$ is an L-*consistency property with cut class* X if for each $Y \in \mathcal{C}$, and for every $A \in X$,

$$Y \cup \{A\} \in \mathcal{C} \text{ or } Y \cup \{-A\} \in \mathcal{C}.$$

If $X = FORM_{\mathcal{M}}$, then we call $\mathcal{C}$, simply, a *cut-class* L-*consistency property*.

To sum up, cut class L-consistency properties are L-consistency properties with cut class $FORM_{\mathcal{M}}$. We shall use them further to deal with symmetric and Euclidean logics. The cut class could be sensibly restricted, as in the original account [4], but we shall stay with the general version, as further considerations simplify when we take the whole set $FORM_{\mathcal{M}}$ as X.

Further in this section we repeat the results proved for K and check the parts of the proofs that are L-sensitive.

[6] For the original definition of a cut class see [4, p. 202].

Lemma 3.19 (L-version of Lemma *3.14*). *Let $\mathcal{C}$ stand for an* L*-consistency property, and let $\mathcal{C}^*$ be a family containing all subsets of sets from $\mathcal{C}$. Then $\mathcal{C}^*$ is a subset closed* L*-consistency property. Moreover, if $\mathcal{C}$ is Z-compatible (cut-class) then $\mathcal{C}^*$ is Z-compatible (cut-class).*

Proof Let $\mathcal{C}^*$ be defined as in the statement of the lemma. Clauses 1–5 and Z-compatibility are proved as in Lemma 3.14; we use clause 1 of Corollary 3.25 instead of clause 1 of Corollary 3.19.
Clause 6. Let $\mathsf{L} \in \{\mathsf{D}, \mathsf{DB}, \mathsf{D4}, \mathsf{D5}, \mathsf{D45}\}$ and $X^* \in \mathcal{C}^*$. $X^* \subseteq X \in \mathcal{C}$ for some X, hence $(X)^{\sharp_{\mathsf{L}}} \in \mathcal{C}$ by clause 6 of Definition 3.24. On the other hand, $(X^*)^{\sharp_{\mathsf{L}}} \subseteq (X)^{\sharp_{\mathsf{L}}}$ (by clause 1 of Corollary 3.25), therefore $(X^*)^{\sharp_{\mathsf{L}}} \in \mathcal{C}^*$, which was to be proved. Clause 7 is equally obvious.

Suppose $\mathcal{C}$ is a cut-class L-consistency property. Let $X^* \in \mathcal{C}^*$ and $A \in FORM_{\mathcal{M}}$. Since $X^* \subseteq X$ for some $X \in \mathcal{C}$, and $\mathcal{C}$ fulfils Definition 3.26, $X \cup \{A\} \in \mathcal{C}$ or $X \cup \{-A\} \in \mathcal{C}$. Since $X^* \cup \{A\} \subseteq X \cup \{A\}$, and similarly for '$-A$', it follows that also $X^* \cup \{A\} \in \mathcal{C}^*$ or $X^* \cup \{-A\} \in \mathcal{C}^*$, that is, $\mathcal{C}^*$ is a cut-class L-consistency property.
□

Lemma 3.20 (L-version of Lemma *3.15*). *Let $\mathcal{C}$ stand for a subset closed* L*-consistency property, and let $\mathcal{C}^*$ be a family such that $Y \in \mathcal{C}^*$ iff every finite subset of Y belongs to $\mathcal{C}$. $\mathcal{C}^*$ is an* L*-consistency property of finite character. Moreover, if $\mathcal{C}$ is Z-compatible (cut-class), then $\mathcal{C}^*$ is Z-compatible (cut-class).*

Proof Let $\mathcal{C}^*$ be as in the statement of the lemma and assume that $X \in \mathcal{C}^*$. Clauses 1–4, finite character and Z-compatibility are proved as in Lemma 3.15 (with Corollary 3.25 used instead of Corollary 3.19).

For clause 5, we assume that $\pi \in X$ and that Y is an arbitrary finite subset of $(X)^{\sharp_{\mathsf{L}}} \cup \{\pi_0\}$; adjustments are needed in definition (Y^* def) of set Y^*. We recall the version for K (see the proof of Lemma 3.15):

(Y^* def) $B \in Y^*$ iff (i) B is a ν-formula, $B \in X$ and the component of B, ν_0, is in Y.

The result is such that $Y^* \subseteq X \in \mathcal{C}$ (this is guaranteed by $B \in X$), so that clause 5 of Definition 3.24 applies to Y^*, and at the same time $(Y^*)^{\sharp_{\mathsf{L}}} = Y$. However, the construction depends on $(\bullet)^{\sharp_{\mathsf{L}}}$, hence if the definition of $(X)^{\sharp_{\mathsf{L}}}$ includes:

(a) set $\{\nu : \nu \in X\}$, then $B \in Y^*$ iff (i) holds as above, or (ii) B is a ν-formula such that $B \in X$ and $B \in Y$;
(b) set $\{\pi : \pi_0 \in X\}$, then $B \in Y^*$ iff (i) holds as above, or (iii) B is a component, π_0, of a π-formula such that $B \in X$ and the π-formula belongs to Y;
(c) set $\{\pi : \pi \in X\}$, then $B \in Y^*$ iff (i) holds as above, or (iv) B is a π-formula such that $B \in X$ and $B \in Y$.

If definition of $(X)^{\sharp_{\mathsf{L}}}$ includes more than one of (a)–(c), then the respective conditions (ii)–(iv) are combined as well. We use clause 3 of Corollary 3.25 (instead of Corollary 3.19) to show that Y^* is finite. Except for these adjustments, the reasoning for clause 5 is the same as for K.

Clause 6. Let $\mathsf{L} \in \{\mathsf{D}, \mathsf{DB}, \mathsf{D4}, \mathsf{D5}, \mathsf{D45}\}$ and $X \in \mathcal{C}^*$. Let Y be a finite subset of $(X)^{\sharp_\mathsf{L}}$; the goal is to show that $Y \in \mathcal{C}$. For this reason let

$$Y^* = \{\nu : \nu \in X \text{ and } \nu_0 \in Y\} \pm \{\nu : \nu \in X \text{ and } \nu \in Y\}$$
$$\pm \{\pi_0 : \pi_0 \in X \text{ and } \pi \in Y\} \pm \{\pi : \pi \in X \text{ and } \pi \in Y\}.$$

Whether the consecutive sets after '$\pm$' are added to $\{\nu : \nu \in X \text{ and } \nu_0 \in Y\}$ depends on L and its definition of $(\bullet)^{\sharp_\mathsf{L}}$. By clause 3 of Corollary 3.25, Y^* is finite. On the other hand, $Y^* \subseteq X$, hence by definition of $\mathcal{C}^*$, $Y^* \in \mathcal{C}$. By clause 6 of Definition 3.24, $(Y^*)^{\sharp_\mathsf{L}} \in \mathcal{C}$, and $(Y^*)^{\sharp_\mathsf{L}} = Y$.
Clause 7. Let $\mathsf{L} \in \{\mathsf{T}, \mathsf{B}, \mathsf{S4}, \mathsf{S5}\}$ and $X \in \mathcal{C}^*$. The goal is to show that $X \cup (X)^{\sharp_\mathsf{L}} \in \mathcal{C}^*$. Hence let Y be a finite subset of $X \cup (X)^{\sharp_\mathsf{L}}$. Y can be presented as a sum $X_1 \cup X_2$, where:

$$X_1 = X \cap Y, \ \ X_2 = (X)^{\sharp_\mathsf{L}} \cap Y.$$

For X_2 we define set X_2^* just as Y^* was defined for Y in the reasoning for clause 6. Then it holds that $X_2^* \subseteq X$, and also that (d) $(X_2^*)^{\sharp_\mathsf{L}} = X_2$. Further, since $X_1 \cup X_2^*$ is a finite subset of X, we have $X_1 \cup X_2^* \in \mathcal{C}$. By clause 7, the following set is also an element of $\mathcal{C}$:

$$(X_1 \cup X_2^*) \cup (X_1 \cup X_2^*)^{\sharp_\mathsf{L}}$$

By clause 2 of Corollary 3.25, and by (d), this equals:

$$(X_1 \cup X_2^*) \cup (X_1)^{\sharp_\mathsf{L}} \cup X_2.$$

Since $\mathcal{C}$ is subset closed, also $X_1 \cup X_2 = Y \in \mathcal{C}$, as required.

Assume that $\mathcal{C}$ is a cut-class L-consistency property, let $X \in \mathcal{C}^*$ and $A \in \mathit{FORM}_\mathcal{M}$. We need to show that $X \cup \{A\} \in \mathcal{C}^*$ or $X \cup \{-A\} \in \mathcal{C}^*$. For this reason suppose that $X \cup \{A\} \notin \mathcal{C}^*$. By definition of $\mathcal{C}^*$, there must be a finite subset of $X \cup \{A\}$ that is not in $\mathcal{C}$, and since every finite subset of X belongs to $\mathcal{C}$, there must be then a finite subset X_0 of X such that $X_0 \cup \{A\} \notin \mathcal{C}$. But $\mathcal{C}$ is a cut-class L-consistency property, thus $X_0 \cup \{-A\} \in \mathcal{C}$.

Now let X_f be an arbitrary finite subset of $X \cup \{-A\}$. Set $(X_0 \cup X_f)\backslash\{-A\}$ is also a finite subset of X, thus it is extended with A or $-A$ in $\mathcal{C}$. But it cannot be A that extends the set in $\mathcal{C}$, for $\mathcal{C}$ is subset closed, and hence if $(X_0 \cup X_f) \cup \{A\} \in \mathcal{C}$, then $X_0 \cup \{A\}$ would be in $\mathcal{C}$, contrary to what we have established. Thus $(X_0 \cup X_f) \cup \{-A\} \in \mathcal{C}$. Again, $\mathcal{C}$ is subset closed, hence also $X_f \cup \{-A\} \in \mathcal{C}$, and $X_f \in \mathcal{C}$ (relevant, if $-A \notin X_f$). It follows that every finite subset of $X \cup \{-A\}$ is in $\mathcal{C}$. We have shown that if $X \cup \{A\} \notin \mathcal{C}^*$, then $X \cup \{-A\} \in \mathcal{C}^*$, as required: $\mathcal{C}^*$ is a cut-class L-consistency property. □

As in the case of $\mathsf{L} = \mathsf{K}$, we have:

Corollary 3.26 (L-version of Corollary *3.22*) *For each (Z-compatible, cut-class)* L-*consistency property* $\mathcal{C}$, *there exists a (Z-compatible, cut-class)* L-*consistency property* $\mathcal{C}^{**}$ *of finite character, such that* $\mathcal{C} \subseteq \mathcal{C}^{**}$.

The proof is the same as that of Corollary 3.22.

The next corollary generalizes clause 1 of definitions of consistency properties to formulas of arbitrary rank. As far, we did not need to state this consistency condition explicitly, but we need it now with cut-class properties.

Corollary 3.27 *If* $\mathcal{C}$ *is a cut-class* L-*consistency property, then for each* $A \in FORM_{\mathcal{M}}$ *and every* $X \in \mathcal{C}$, *if* $A \in X$, *then* $-A \notin X$.

Proof The proof goes by induction on A's rank. The base case follows directly from clause 1 of Definition 3.24.

IH has the following formulation: for every set $X \in \mathcal{C}$, for each formula A of rank up to n, if $A \in X$, then $-A \notin X$.

The argument is by contradiction. Assume that for some $X \in \mathcal{C}$ and a formula A of rank equal to $n+1$, $\{A, -A\} \subseteq X$.

If $A = \neg\neg B$, then by clause 2 of Definition 3.24 also $X \cup \{B\} \in \mathcal{C}$. Obviously, $r(B) = n$ and $-A = \neg B$. Formula '$\neg B$' belongs to $X \cup \{B\}$. If B is positive, then $-B = \neg B$, hence the set contradicts IH. If $B = \neg C$, then $-A = \neg\neg C$. By clause 2 again, $(X \cup \{B\}) \cup \{C\} \in \mathcal{C}$, and then this set contradicts IH.

If A is an α-formula, then '$-A$' is a β-formula. By clause 3 of Definition 3.24, $X \cup \{\alpha_0, \alpha_1\} \in \mathcal{C}$, where α_0, α_1 are the components of A. On the other hand, $-A \in X \cup \{\alpha_0, \alpha_1\}$, hence, by clause 4 of Definition 3.24, $(X \cup \{\alpha_0, \alpha_1\}) \cup \{\beta_i\} \in \mathcal{C}$ for one of $i \in \{0, 1\}$. By clause 4 of Corollary 3.1, one of α_i, β_i is the complement of the other, both are of rank up to n, hence set $(X \cup \{\alpha_0, \alpha_1\}) \cup \{\beta_i\}$ contradicts IH. The reasoning is analogous, if A is a β-formula.

If A is a ν- or π-formula, then we use clause 5 of Definition 3.24 and clause 5 of Corollary 3.1 to show that set $(X)^{\sharp_{\mathsf{L}}} \cup \{\pi_0\}$ contradicts IH. □

Also the following holds.

Lemma 3.21 (L-version of Lemma *3.16*). *Let* $\mathcal{C}$ *be an* L-*consistency property of finite character. For each set* $X \in \mathcal{C}$, *there is a downward-saturated set* $\mathbb{X} \in \mathcal{C}$, *such that:*

1. $X \subseteq \mathbb{X}$,
2. *if* $\mathcal{C}$ *is Z-compatible, then* $X \cup Z \subseteq \mathbb{X}$,
3. *if* $\mathcal{C}$ *has the cut-class property, then for each* $A \in FORM_{\mathcal{M}}$, $A \in \mathbb{X}$ *or* $-A \in \mathbb{X}$,
4. *if* $\mathcal{C}$ *has the cut-class property, then if* $A \in \mathbb{X}$, *then also* $\neg\neg A \in \mathbb{X}$.

Proof For clauses 1 and 2 the proof is basically the same as that of Lemma 3.16, where we have sketched a quasi-procedure of constructing the downward-saturated supersets with reference to an order of formulas of $\mathcal{M}$. The proof relies, *int.al.*, on the fact that finite character properties are closed under chain unions (see Corollary 3.21), which yields that the downward-saturated sets are elements of an L-consistency property maximal with respect to $\subseteq$.

Clause 3 follows immediately from this last observation: if for each $A \in FORM_{\mathcal{M}}$, $\mathbb{X} \cup \{A\} \in \mathcal{C}$ or $\mathbb{X} \cup \{-A\} \in \mathcal{C}$, and $\mathbb{X}$ is $\subseteq$-maximal in $\mathcal{C}$, then for each $A \in FORM_{\mathcal{M}}$, $A \in \mathbb{X}$ or $-A \in \mathbb{X}$.

The last clause 4 is a technical addendum needed later to simplify arguments concerning formulas and their complements. It obviously follows from the previous clause and Corollary 3.27: if $\neg\neg A \notin \mathbb{X}$, then $\neg A \in \mathbb{X}$, but then also $A \notin \mathbb{X}$. □

As previously, the following will be used in the proof of Strong Model Existence Theorem for L.

Definition 3.27 (L-superframe) Let $\mathcal{C}$ stand for an L-consistency property of finite character. For symmetric and Euclidean logics, i.e. L ∈ {KB, DB, B, KB4, K5, D5, K45, D45, S5}, we require $\mathcal{C}$ to be cut-class as well. Further, let S stand for an arbitrary sequence ordering all formulas of $\mathcal{M}$, and let $\mathbb{S}^{\mathcal{C}}_{S}$ be a function defined on $\mathcal{C}$ by (3.9) $\mathbb{S}^{\mathcal{C}}_{S} : W \mapsto \mathbb{W}$.

By an L-*superframe* $\langle \mathbb{W}, \mathbb{R} \rangle$ *built with reference to* S *upon* $\mathcal{C}$ we mean the following structure:

$$\mathbb{W} = \{\mathbb{S}^{\mathcal{C}}_{S}(X) : X \in \mathcal{C}\}, \quad \mathbb{R} \subseteq \mathbb{W} \times \mathbb{W}, \quad \mathbb{R} = \{\langle \mathbb{X}, \mathbb{Y} \rangle : \mathbb{X}^{\sharp_{\mathsf{L}}} \subseteq \mathbb{Y}\}.$$

For short, by an L-*superframe* we mean an L-superframe built with reference to some ordering S upon some L-consistency property $\mathcal{C}$; S and $\mathcal{C}$ will be clear from the context. Moreover, a K-*superframe* is simply a superframe as defined in Definition 3.20.

The following holds as before.

Corollary 3.28 (L-version of Corollary *3.23*) *If* $\langle \mathbb{W}, \mathbb{R} \rangle$ *is an* L-*superframe built with reference to* S *upon* $\mathcal{C}$, *then* $\mathbb{W} \subseteq \mathcal{C}$. *Also, if* $\mathcal{C}$ *is* Z-*compatible, then for each* $\mathbb{X} \in \mathbb{W}$, $Z \subseteq \mathbb{X}$.

Proof Follows from the proof of Lemma 3.21 (Lemma 3.16). □

What is more, the following lemma is true.

Lemma 3.22 *Every* L-*superframe* $\langle \mathbb{W}, \mathbb{R} \rangle$ *is an* L-*frame, that is,* $\mathbb{R}$ *has the* L-*properties.*

Proof Let $\langle \mathbb{W}, \mathbb{R} \rangle$ be an L-superframe, that is an L-superframe built with reference to some S and some L-consistency property $\mathcal{C}$.

Let L ∈ {D, DB, D4, D5, D45}. If $\mathbb{X} \in \mathbb{W}$, then $\mathbb{X} \in \mathcal{C}$ by Corollary 3.28. Hence $(\mathbb{X})^{\sharp_{\mathsf{L}}} \in \mathcal{C}$ by clause 6 of Definition 3.24, and thus also $\mathbb{S}^{\mathcal{C}}_{S}((\mathbb{X})^{\sharp_{\mathsf{L}}}) \in \mathbb{W}$. Trivially, $(\mathbb{X})^{\sharp_{\mathsf{L}}} \subseteq \mathbb{S}^{\mathcal{C}}_{S}((\mathbb{X})^{\sharp_{\mathsf{L}}})$, hence $\langle \mathbb{X}, \mathbb{S}^{\mathcal{C}}_{S}((\mathbb{X})^{\sharp_{\mathsf{L}}}) \rangle \in \mathbb{R}$, so $\mathbb{R}$ is serial.

Let L ∈ {T, B, S4, S5} and $\mathbb{X} \in \mathbb{W}$. We have $\mathbb{X} \in \mathcal{C}$, and by clause 7 of Definition 3.24, $\mathbb{X} \cup (\mathbb{X})^{\sharp_{\mathsf{L}}} \in \mathcal{C}$. Since $\mathbb{X}$ is maximal in $\mathcal{C}$, it follows that $(\mathbb{X})^{\sharp_{\mathsf{L}}} \subseteq \mathbb{X}$, and therefore $\langle \mathbb{X}, \mathbb{X} \rangle \in \mathbb{R}$. Hence $\mathbb{R}$ is reflexive.

Let $\mathsf{L} \in \{\mathsf{KB}, \mathsf{DB}, \mathsf{B}, \mathsf{KB4}\}$, assume that $\langle \mathbb{X}, \mathbb{Y} \rangle \in \mathbb{R}$, and let $A \in (\mathbb{Y})^{\sharp\mathsf{L}}$. The aim is to show that $A \in \mathbb{X}$, since $(\mathbb{Y})^{\sharp\mathsf{L}} \subseteq \mathbb{X}$ yields that $\mathbb{R}$ is symmetric. From the assumption that $\langle \mathbb{X}, \mathbb{Y} \rangle \in \mathbb{R}$, for each $\mathsf{L} \in \{\mathsf{KB}, \mathsf{DB}, \mathsf{B}, \mathsf{KB4}\}$, we have:

(a) $(\mathbb{X})^{\sharp\mathsf{L}} \subseteq \mathbb{Y}$, hence $\{\nu_0 : \nu \in \mathbb{X}\} \cup \{\pi : \pi_0 \in \mathbb{X}\} \subseteq \mathbb{Y}$,

which we shall use in a moment. If $\mathsf{L} \in \{\mathsf{KB}, \mathsf{DB}, \mathsf{B}\}$, then A could be (i) the component ν_0, of a ν-formula that belongs to $\mathbb{Y}$, or (ii) a π-formula such that its component, π_0, belongs to $\mathbb{Y}$. If $\mathsf{L} = \mathsf{KB4}$, then there is also another possibility: (iii) A is a ν-formula that belongs to $\mathbb{Y}$. We assume, for a proof by contradiction, that $A \notin \mathbb{X}$. Since for the considered logics $\mathcal{C}$ is a cut-class L-consistency property, this assumption yields:

(b) $-A \in \mathbb{X}$.

If (i) holds, then A is the component, ν_0, of a ν-formula in $\mathbb{Y}$. By clause 5 of Corollary 3.1, $-\nu$ is a π-formula such that $\pi_0 \in \{-A, \neg A\}$. If $\pi_0 = \neg A$ and $\neg A \neq -A$, then also $\neg\neg - A \in \mathbb{X}$ by clause 4 of Lemma 3.21. Either way, (b) $-A \in \mathbb{X}$ entails $\pi_0 \in \mathbb{X}$. By (a), formula π, which equals $-\nu$, belongs to $\mathbb{Y}$; but this contradicts Corollary 3.27.

In the case (ii), A is a π-formula such that $\pi_0 \in \mathbb{Y}$. By (b), formula ν, which is the complement of A, belongs to $\mathbb{X}$. Then by (a), $\nu_0 \in \mathbb{Y}$. By clause 5 of Corollary 3.1, one of ν_0, π_0 is the complement of the other, hence follows a contradiction with Corollary 3.27.

If (iii) holds, then '$-A$' is a π-formula in $\mathbb{X}$. In $\mathsf{KB4}$, $\{\pi : \pi_0 \in \mathbb{X}\} \subseteq (\mathbb{X})^{\sharp\mathsf{L}}$. Since $(\mathbb{X})^{\sharp\mathsf{L}} \subseteq \mathbb{Y}$, also $\Diamond - A \in \mathbb{Y}$. By clause 5 of Definition 3.24, $(\mathbb{Y})^{\sharp\mathsf{L}} \cup \{-A\} \in \mathcal{C}$. Let $\mathbb{Z} = \mathbb{S}^{\mathcal{C}}_{S}((\mathbb{Y})^{\sharp\mathsf{L}} \cup \{-A\})$. Since A is a ν-formula in $\mathbb{Y}$, also $A \in \mathbb{Z}$—a contradiction with Corollary 3.27. To sum up, we have shown by cases that $(\mathbb{Y})^{\sharp\mathsf{L}} \subseteq \mathbb{X}$, so for $\mathsf{L} \in \{\mathsf{KB}, \mathsf{DB}, \mathsf{B}, \mathsf{KB4}\}$, $\mathbb{R}$ is symmetric.

Let $\mathsf{L} \in \{\mathsf{K4}, \mathsf{D4}, \mathsf{S4}, \mathsf{KB4}, \mathsf{K45}, \mathsf{D45}\}$, $\langle \mathbb{X}, \mathbb{Y} \rangle \in \mathbb{R}$ and $\langle \mathbb{Y}, \mathbb{Z} \rangle \in \mathbb{R}$. Let $A \in (\mathbb{X})^{\sharp\mathsf{L}}$; we need to show that $A \in \mathbb{Z}$. If (i) A is a component, ν_0, of a ν-formula in $\mathbb{X}$, then the ν-formula is in $\mathbb{Y}$ as well, and thus $A \in \mathbb{Z}$. Similarly, if (ii) A is a ν-formula in $\mathbb{X}$, then also $A \in \mathbb{Y}$ and $A \in \mathbb{Z}$. This finishes the argument for $\mathsf{L} \in \{\mathsf{K4}, \mathsf{D4}, \mathsf{S4}\}$. For $\mathsf{KB4}$, (iii) A could be a π-formula such that its component, π_0, belongs to $\mathbb{X}$. We use the fact that for this logic $\mathcal{C}$ is a cut-class $\mathsf{KB4}$-consistency property and we reason by contradiction. Assume that $A \notin \mathbb{Z}$, then $-A \in \mathbb{Z}$ by clause 3 of Lemma 3.21. From $\langle \mathbb{X}, \mathbb{Y} \rangle \in \mathbb{R}$, we have $A \in \mathbb{Y}$. From $\langle \mathbb{Y}, \mathbb{Z} \rangle \in \mathbb{R}$, $\Diamond A \in \mathbb{Z}$. By clause 5 of Definition 3.24, $(\mathbb{Z})^{\sharp\mathsf{L}} \cup \{A\} \in \mathcal{C}$. Since '$-A$' is a ν-formula (clause 5 of Corollary 3.1), it is carried to $(\mathbb{Z})^{\sharp\mathsf{L}} \cup \{A\}$, and we have a contradiction with Corollary 3.27. Finally, for logics $\mathsf{K45}, \mathsf{D45}$, assume that (iv) A is a π-formula such that $A \in \mathbb{X}$. This case is as straightforward as (ii), since π-formulas are carried by $\mathbb{R}$-accessibility to $\mathbb{Z}$. We have shown that for the considered logics $\mathbb{R}$ is transitive.

Let $\mathsf{L} \in \{\mathsf{K5}, \mathsf{D5}, \mathsf{K45}, \mathsf{D45}, \mathsf{S5}\}$, (a) $\langle \mathbb{X}, \mathbb{Y} \rangle \in \mathbb{R}$ and (b) $\langle \mathbb{X}, \mathbb{Z} \rangle \in \mathbb{R}$. We assume that $A \in (\mathbb{Y})^{\sharp\mathsf{L}}$ and aim to show that $A \in (\mathbb{Z})^{\sharp\mathsf{L}}$. For the considered logics, by the definition of $(\bullet)^{\sharp\mathsf{L}}$, assumptions (a) and (b) entail

(c) $\{\pi : \pi \in \mathbb{X}\} \subseteq \mathbb{Y}$ and $\{\pi : \pi \in \mathbb{X}\} \subseteq \mathbb{Z}$.

By (c), and by the fact that $\mathcal{C}$ is a cut-class L-consistency property, if there is a ν-formula in $\mathbb{Y}$, then it belongs to $\mathbb{X}$ as well, for otherwise we would have $\nu, -\nu \in \mathbb{Y}$. This establishes that if A is the component, ν_0, of a ν-formula in $\mathbb{Y}$, then $\nu \in \mathbb{X}$ and hence $A \in \mathbb{Z}$. Further reasoning is by contradiction; the details depend on L and in each case there are two more possibilities. Assume that

(d) $A \notin \mathbb{Z}$.

If $\mathsf{L} \in \{\mathsf{K5}, \mathsf{D5}, \mathsf{S5}\}$, then A is a π-formula such that (i) $\pi_0 \in \mathbb{Y}$ (only for S5), or (ii) $\pi = A \in \mathbb{Y}$. From (d), $-A \in \mathbb{Z}$ (cut-class), and since $-A$ is a ν-formula, also $-A \in \mathbb{X}$ (argument as previously). In the case (i) it suffices to observe that the component of $-A$ belongs to $\mathbb{Y}$ (by (a)) which yields a contradiction with $\pi_0 \in \mathbb{Y}$. In the case (ii), we use the fact that $-A \in \mathbb{X}$ entails $\Diamond - A \in \mathbb{Y}$. By clause 5 of Definition 3.24, there is $(\mathbb{Y})^{\sharp_\mathsf{L}} \cup \{-A\}$ in the underlying $\mathcal{C}$, but since A is a π-formula in $\mathbb{Y}$, a contradiction follows.

Let $\mathsf{L} \in \{\mathsf{K45}, \mathsf{D45}\}$. There are the following possible cases: (iii) A is a ν-formula and $A \in \mathbb{Y}$, (iv) A is a π-formula and $A \in \mathbb{Y}$. In these logics ν-formulas are carried by the $\mathbb{R}$-accessibility relation forward and backwards. Hence in (iii) from $A \in \mathbb{Y}$ follows $A \in \mathbb{X}$, and thus also $A \in \mathbb{Z}$. In (iv) we use (d) again—from (d) $-A \in \mathbb{Z}$, hence $-A \in \mathbb{X}$ and $-A \in \mathbb{Y}$—a contradiction.

We have shown that for all Euclidean logics relation $\mathbb{R}$ is Euclidean.

This finishes the proof. □

Let us observe that the cut-class property is crucial for proving symmetry and Euclideanness. What is more, having cut-class L-consistency properties is needed *only* for this purpose. The proof of the above lemma reveals also that carrying the set $\{\nu : \nu \in X\}$ to an R-accessible world warrants transitivity of R, carrying the set $\{\pi : \pi_0 \in X\}$ warrants symmetry of R, whereas in the case of Euclideanness the property is warranted by carrying $\{\pi : \pi \in X\}$ together with at least one of the other two mentioned sets. Obviously, in the detailed arguments $\{\nu_0 : \nu \in X\}$ is used too, but this set warrants standard behavior of π and ν.

Similarly to the case of K, the following lemma is true.

Lemma 3.23 *Let $\mathcal{C}$ be an L-consistency property of finite character, S be a sequence ordering all formulas of $\mathcal{M}$, and assume $\langle \mathbb{W}, \mathbb{R} \rangle$ is an L-superframe built with reference to S upon $\mathcal{C}$. For each $\mathbb{X} \in \mathbb{W}$, and for every $\pi \in \mathbb{X}$, there is $\mathbb{Y} \in \mathbb{W}$, such that $(\mathbb{X})^{\sharp_\mathsf{L}} \cup \{\pi_0\} \subseteq \mathbb{Y}$.*

Proof The proof relies on: Definition 3.27, Corollary 3.28, and clause 5 of Definition 3.24, and it is basically the same as that of Lemma 3.17. □

We proceed to:

Theorem 3.8 (Strong Model Existence Theorem for L). *If a set X of formulas of $\mathcal{M}$ belongs to a Z-compatible L-consistency property, then X is locally satisfied in a Kripke L-model globally satisfying Z.*

Proof The proof is analogous to that of Theorem 3.3.

Let $X \in \mathcal{C}$, where $\mathcal{C}$ is a Z-compatible L-consistency property. We extend $\mathcal{C}$ to a Z-compatible L-consistency property $\mathcal{C}^{**}$ which is of finite character; also if $\mathcal{C}$ is a cut-class L-consistency property, then so is $\mathcal{C}^{**}$ (see Corollary 3.26). Next, we choose some ordering S of formulas of $\mathcal{M}$ and we build an L-superframe with reference to S upon $\mathcal{C}^{**}$ (Definition 3.27). Let $\langle \mathbb{W}, \mathbb{R} \rangle$ stand for the frame. We define valuation function on $\mathbb{W}$ by requiring that for each p_i and each $\mathbb{X} \in \mathbb{W}$,

$$V(p_i, \mathbb{X}) = 1 \text{ iff } p_i \in \mathbb{X}. \tag{3.13}$$

Finally, we conduct the proof by induction: we show that for each set $\mathbb{X} \in \mathbb{W}$ and for each formula $A \in \mathbb{X}$, $V(A, \mathbb{X}) = 1$; induction is with respect to rank of formula A. The reasoning is exactly as in the proof of Theorem 3.3; even in the case of modal formulas, the burden of the proof has already been moved to some previous lemmas and corollaries. Therefore, for $\pi \in \mathbb{X}$, in order to show that $V(\pi, \mathbb{X}) = 1$ we use Lemma 3.23 which guarantees that there is a set $\mathbb{Y}$ that is $\mathbb{R}$-accessible from $\mathbb{X}$, and such that π_0 is true in $\mathbb{Y}$. In the case of $\nu \in \mathbb{X}$, for each $\mathbb{Y}$ that is $\mathbb{R}$-accessible from $\mathbb{X}$, we have $(\mathbb{X})^{\sharp_{\mathsf{L}}} \subseteq \mathbb{Y}$ by definition of $\mathbb{R}$, and this yields that $\nu_0 \in \mathbb{Y}$ for each definition of $(\bullet)^{\sharp_{\mathsf{L}}}$. This finishes the part showing that $V(A, \mathbb{X}) = 1$ for every formula A and each set $\mathbb{X} \in \mathbb{W}$. For set $X \in \mathcal{C}$, there is $\mathbb{S}^{\mathcal{C}}_{S}(X) \in \mathbb{W}$, hence X is locally satisfied in $\langle \mathbb{W}, \mathbb{R}, V \rangle$. Moreover, by Corollary 3.28, Z is satisfied in each world of the model, hence the model satisfies Z globally. Finally, by Lemma 3.22, $\mathbb{R}$ has the appropriate L-properties, therefore $\langle \mathbb{W}, \mathbb{R}, V \rangle$ is an L-model. This finishes the proof. □

The strong model existence theorem yields the following weaker version.

Theorem 3.9 *If a set X of formulas of $\mathcal{M}$ belongs to an* L*-consistency property, then X is locally satisfied in a Kripke* L*-model.*

Here comes the second stage of the completeness enterprise. We build the notion of consistency for erotetic calculi, just as we did with K. Definition 3.14 (set-representation of numeral i in a sequent ϕ) and Definition 3.21 (Z-sequent) apply in the same form. Further, for a finite set Z of formulas of $\mathcal{M}$ and a Z-sequent we introduce the notion of *(Z, ϕ)-family of sets*, which is a family of all sets of the form:

$$Y = Z \cup -X$$

where for some $i \in W(\phi)$, X is the set-representation of i in ϕ. This notion is generalized over all Z-sequents in the definition of Z-family which applies without changes (see Definition 3.22). Next, we construe L-consistent Z-families, in order to deal with unprovable sequents only.

Definition 3.28 (L-consistent Z-family) Let Z stand for a finite set of formulas of $\mathcal{M}$, and let $\mathcal{F}$ be a Z-family. We say that $\mathcal{F}$ is L*-consistent*, provided that the following condition holds:

(con^{L}_{Z}) $Y \in \mathcal{F}$ iff for some Z-sequent ϕ unprovable in $\mathbb{E}^{\mathsf{L}}$, Y belongs to a (Z, ϕ)-family.

As previously, now we prove:

Lemma 3.24 (L-version of Lemma *3.18*). *Every* L*-consistent Z-family can be extended to a Z-compatible* L*-consistency property.*

Proof The proof is constructed as that of Lemma 3.18. For an L-consistent Z-family $\mathcal{F}$ we first extend $\mathcal{F}$ to a subset closed family $\mathcal{F}^*$ of sets, and then we show that $\mathcal{F}^*$ has all the required properties.

For clause 1 of Definition 3.24 the reasoning is exactly as in the proof of Lemma 3.18; we skip this part, as well as most of the non-modal clauses. For the record, in verifying that $\mathcal{F}^*$ satisfies clause 2 of the definition we use Lemma 3.9 (semi-admissibility of contraction) in the respective subcase. Similarly, in clause 3 (for α-formulas) we need Lemma 3.9, and for some subcases also Lemma 3.7 (double negation). In the reasoning for β-formulas (clause 4) we use also Lemma 3.5 (Separation).

Clause 5 of Definition 3.24 considers π-formulas. Let $Y \in \mathcal{F}^*$ and $\pi \in Y$; the aim is to show that $Y^{\sharp_{\mathsf{L}}} \cup \{\pi_0\} \in \mathcal{F}^*$. By assumption, $Y \subseteq Z \cup -X \in \mathcal{F}$, X is the set-representation of some $i \in W(\phi)$ and ϕ is a Z-sequent unprovable in $\mathbb{E}^{\mathsf{L}}$. The part of the reasoning showing that if $\pi \in Y \in \mathcal{F}^*$, then $Z \cup \{\pi_0\} \in \mathcal{F}^*$ is exactly the same as in the proof of Lemma 3.18, only further reasoning is L-sensitive. Hence let us assume that we have already shown that, if (a) $\pi \in Z$, then the following sequent is unprovable in $\mathbb{E}^{\mathsf{L}}$ (we assume that S is such that $\mathsf{set}(S) = Z$):

$$S\,(\pi) \vdash \Gamma\,({\nu_0}^{i,j})$$

where one of ν_0, π_0 is the complement of the other; and if (b) $\pi \in -X$, then it is a sequent of the form

$$S \vdash \Gamma\,({\nu_0}^{\varsigma/i,j}).$$

We need to show, by successive applications of modal erotetic rules, that set $Z \cup \{\pi_0\}$ can be extended to $(Y)^{\sharp_{\mathsf{L}}} \cup \{\pi_0\} \in \mathcal{F}^*$. In the case with $\nu \in Y$ and $\nu_0 \in (Y)^{\sharp_{\mathsf{L}}}$ the reasoning is as in the proof of Lemma 3.18. There are three more cases to consider, according to the definition of $(\bullet)^{\sharp_{\mathsf{L}}}$ given in Table 3.6, and in these cases additional semi-admissible rules will be used (Lemmas 3.10–3.13). Let ψ stand for the last obtained unprovable sequent, with '$(\nu_0)^{\varsigma/i,j}$' in the succedent.

Case 1: $\nu \in Y$ and $\nu \in (Y)^{\sharp_{\mathsf{L}}}$. Then $\mathsf{L} \in \{\mathsf{K4}, \mathsf{D4}, \mathsf{S4}, \mathsf{K45}, \mathsf{D45}, \mathsf{KB4}\}$. We have two subcases depending on whether (a) $\nu \in Z$, or (b) $\nu \in -X$. We need to put $-\nu \in X^j$, where X^j is the set-representation of j in ψ. In (a) we simply consider an instance of R_G which puts '$(-\nu)^j$' in the succedent of an unprovable sequent. In subcase (b), assume that the last considered unprovable sequent is of the form:

$$\psi = S \;\vdash\; \Gamma((\nu_0)^{\varsigma/i,j})((-\nu)^{\varrho/i}).$$

Lemmas 3.10 and 3.11 yield that if ψ is not provable in $\mathbb{E}^{\mathsf{L}}$, then the following is neither:

$$\chi = S \vdash \Gamma((\nu_0)^{\varsigma/i,j})((-\nu)^{\varrho/i})((-\nu)^j).$$

Case 2: $\pi_0 \in Y$ and $\pi \in (Y)^{\sharp_{\mathsf{L}}}$. Then we have $\mathsf{L} \in \{\mathsf{KB}, \mathsf{DB}, \mathsf{B}, \mathsf{S5}, \mathsf{KB4}\}$. In subcase (a) $\pi_0 \in Z$; then R_G puts '$(-\pi_0)^i$' in the succedent and then we reason as in subcase (b) discussed below.

In subcase (b): $\pi_0 \in -X$. Sequent ψ is now of the form (modulo double negation):

$$\psi = S \vdash \Gamma((\nu_0)^{\varsigma/i,j})((-\pi_0)^{\varrho/i})$$

By Lemma 3.12, the following is not provable in $\mathbb{E}^{\mathsf{L}}$, as ψ is not.

$$\chi = S \vdash \Gamma((\nu_0)^{\varsigma/i,j})((-\pi_0)^{\varrho/i})((-\pi)^j).$$

Case 3: let $\pi \in Y$ and $\pi \in (Y)^{\sharp_{\mathsf{L}}}$. Then $\mathsf{L} \in \{\mathsf{K5}, \mathsf{D5}, \mathsf{S5}, \mathsf{K45}, \mathsf{D45}\}$. If $\pi \in Z$ (subcase (a)), then we reason as above. Suppose that (b) $\pi \in -X$. Sequent ψ is of the form:

$$\psi = S \vdash \Gamma((\nu_0)^{\varsigma/i,j})((-\pi)^{\varrho/i})$$

By Lemma 3.13, the following sequent is not provable in $\mathbb{E}^{\mathsf{L}}$, since ψ is not:

$$\chi = S \vdash \Gamma((\nu_0)^{\varsigma/i,j})((-\pi)^{\varrho/i})((-\pi)^j).$$

We are left with clauses 6 and 7 of Definition 3.24. Again, let $Y \in \mathcal{F}^*$. In the reasoning we run across the main case, where we consider set $\{\nu_0 : \nu \in Y\}$. Except for the main case there are three further cases, each divided into two subcases (a) and (b) as above.

Case 6. $\mathsf{L} \in \{\mathsf{D}, \mathsf{DB}, \mathsf{D4}, \mathsf{D5}, \mathsf{D45}\}$, we need to show that $(Y)^{\sharp_{\mathsf{L}}} \in \mathcal{F}^*$. We first need to introduce a new j, the set-representation of whose will be the required $(Y)^{\sharp_{\mathsf{L}}} \in \mathcal{F}^*$; then another scheme of action is needed, showing how to add further elements of the set, if its creation was already initiated somehow.

Hence assume that $\nu \in Y$ and $\nu_0 \in (Y)^{\sharp_{\mathsf{L}}}$ (this is the main case mentioned in the previous paragraph). To start the construction of the required set: if (a) $\nu \in Z$, then we apply R_G introducing $(-\nu)^i$, so next we are free to apply $R_{\pi\mathsf{D}}$ which brings $(\pi_0)^{i,j}$ (with $\pi_0 = -\nu_0$ modulo double negation). If (b) $\nu \in -X$, then we just apply $R_{\pi\mathsf{D}}$ to $(-\nu)^{\varsigma/i}$ and also end up with $(\pi_0)^{i,j}$. To continue the construction of the set: the pair $\langle i, j\rangle$ is already there in the last obtained (unprovable) sequent. If (a) $\nu \in Z$, then we apply R_G introducing $(-\nu)^i$, and then R_π. If (b) $\nu \in -X$, then this is just a single application of R_π.

We have started the reasoning in the previous paragraph with the assumption that $\nu \in Y$. What if there is no ν-formula in Y? If we deal with $\mathsf{L} \in \{\mathsf{D}, \mathsf{D4}\}$, then $(Y)^{\sharp_{\mathsf{L}}}$ is just empty in such cases. Trivially, the empty set belongs to $\mathcal{F}^*$, as it is subset closed.

Assume that $\nu \in Y$ and $\nu \in (Y)^{\sharp \mathsf{L}}$. (In this case $\mathsf{L} \in \{\mathsf{D4}, \mathsf{D45}\}$). If we do not have $\langle i, j \rangle$ in the last sequent already, then we act as in the previous case to start the construction of the set. In the proceeding process of construction, if (a) $\nu \in Z$, then we just apply R_G once again and put $(-\nu)^j$ in the succedent. If (b) $\nu \in -X$, then we need Lemma 3.11. Below π_0 is $-\nu_0$ modulo double negation. Lemma 3.11 warrants that if

$$\psi = S \vdash \Gamma\,(-\nu^{\varsigma/i}, \pi_0{}^{i,j})(\_)$$

is not provable in $\mathbb{E}^{\mathsf{L}}$, then:

$$\chi = S \vdash \Gamma\,(-\nu^{\varsigma/i}, \pi_0{}^{i,j})(-\nu^j)$$

is neither.

Assume that $\pi_0 \in Y$ and $\pi \in (Y)^{\sharp \mathsf{L}}$. We have $\mathsf{L} = \mathsf{DB}$. If there is any other formula in the sequent that can be used to trigger the building process, then we use it. Once we have $\langle i, j \rangle$ in a sequent (with j introduced as new), we use Lemma 3.12, which yields that we can bring $(-\pi)^j$ to the succedent.

Assume that $\pi \in Y$ and $\pi \in (Y)^{\sharp \mathsf{L}}$. $\mathsf{L} \in \{\mathsf{D5}, \mathsf{D45}\}$. If we do not have $\langle i, j \rangle$ yet, then for the start: if (a) $\pi \in Z$, then we use R_G and R_ν. If (b) $\pi \in -X$, then we use R_ν. Suppose there already is $\langle i, j \rangle$ in the last sequent. If (a) $\pi \in Z$, then use R_G. If (b) $\pi \in -X$, then apply Lemma 3.13. If there is neither a π-formula nor a ν-formula in Y, then $(Y)^{\sharp \mathsf{D5}}$ and $(Y)^{\sharp \mathsf{D45}}$ is empty, and the empty set belongs to $\mathcal{F}^*$.

This finishes the part of the proof showing that $\mathcal{F}^*$ satisfies clause 6 of Definition 3.24.

Clause 7. Let $Y \in \mathcal{F}^*$ and suppose that $\mathsf{L} \in \{\mathsf{T}, \mathsf{B}, \mathsf{S4}, \mathsf{S5}\}$. Assume that $\nu \in Y$ and $\nu_0 \in (Y)^{\sharp \mathsf{L}}$. If $\nu \in Z$, then we apply R_G introducing $(-\nu)^i$, and then R_π with $i = j$ to get $(-\nu_0)^i$. If $\nu \in -X$, then it is sufficient to apply R_π with $i = j$ to get $(-\nu_0)^i$.

Assume that $\nu \in Y$ and $\nu \in (Y)^{\sharp \mathsf{L}}$ ($\mathsf{L} = \mathsf{S4}$). In this case ν already 'extends' Y, hence then there is nothing to be proved.

Assume that $\pi_0 \in Y$ and $\pi \in (Y)^{\sharp \mathsf{L}}$. We have $\mathsf{L} \in \{\mathsf{B}, \mathsf{S5}\}$. If $\pi_0 \in Z$, then apply R_G introducing $(-\pi_0)^i$. If $\pi_0 \in -X$, then $(-\pi_0)^{\varsigma/i}$ is already there. Then we apply Lemma 3.12.

Assume that $\pi \in Y$ and $\pi \in (Y)^{\sharp \mathsf{L}}$, we have $\mathsf{L} = \mathsf{S5}$. As in the case for $\mathsf{S4}$, there is nothing to be proved.

Z-compatibility holds trivially. □

Theorem 3.10 (Completeness of $\mathbb{E}^{\mathsf{L}}$ wrt Semantics of $\mathcal{M}$). *If a sequent of the form '$S \vdash (A)^0$' is* L*-valid, then there is a Socratic proof of the sequent in* $\mathbb{E}^{\mathsf{L}}$.

Proof The proof is organized as that of Theorem 3.5; we assume that $\phi = S \vdash (A)^0$ has no Socratic proof in $\mathbb{E}^{\mathsf{L}}$, and we show that ϕ is not L-valid.

Hence let $\mathsf{set}(S) = Z$. Clearly, $\mathcal{F} = \{Z \cup \{-A\}\}$ is a (Z, ϕ)-family. Since ϕ is not provable in $\mathbb{E}^{\mathsf{L}}$, by condition (con^{L}_Z) of Definition 3.28, $\mathcal{F}$ is a subset of an L-consistent Z-family $\mathcal{F}^*$ which, by Lemma 3.24, can be extended to a Z-compatible

L-consistency property $\mathcal{F}^{**}$. Further, by Theorem 3.8 (Strong Model Existence Theorem for L), set $Z \cup \{-A\}$ is locally satisfied by world w in a Kripke L-model $\langle W, R, V \rangle$ globally satisfying Z. This shows that sequent ϕ is not true in the frame $\langle W, R \rangle$ under interpretation f such that $f(0) = w$, hence ϕ is not L-valid. □

Finally, we end this section with the following theorem.

Theorem 3.11 (Adequacy of $\mathbb{E}^{\mathsf{L}}$ wrt Global L-entailment). *A sequent of $\mathcal{M}^{?}_{\vdash\mathsf{L}}$ of the form '$S \vdash (A)^0$' has a Socratic proof in $\mathbb{E}^{\mathsf{L}}$ iff* $\mathsf{set}(S) \models_{\mathsf{L}} A$.

Proof The if-then direction follows from Theorem 3.1 (soundness of $\mathbb{E}^{\mathsf{L}}$ wrt global entailment). For the only-if direction we use Lemma 3.3, expressing the connection between L-validity of sequents and global L-entailment, and then we use the previous theorem: if $\mathsf{set}(S) \models_{\mathsf{L}} A$, then sequent '$S \vdash (A)^0$' is L-valid, hence it is provable in $\mathbb{E}^{\mathsf{L}}$. □

References

1. Avron, A. (1996). The method of hypersequents in the proof theory of propositional non-classical logics. In W. Hodges, M. Hyland, C. Steinhorn, & J. Truss (Eds.), *Logic: Foundations to applications* (pp. 1–32). Oxford Science Publications.
2. Ciardelli, I. A. (2016). *Questions in logic* (Doctoral dissertation). University of Amsterdam.
3. Ciardelli, I. A. (2022). *Inquisitive Logic. Consequence and Inference in the Realm of Questions*. Springer.
4. Fitting, M. (1983). *Proof methods for modal and intuitionistic logics*. Springer.
5. Gabbay, D. M. (1996). *Labelled deductive systems*. Oxford University Press.
6. Goré, R. (1999). Tableau methods for modal and temporal logics. In M. D'Agostino, D. M. Gabbay, R. Hähnle, & J. Posegga (Eds.), *Handbook of tableau methods* (pp. 297–396). Springer.
7. Hahn, G., & Tardif, C. (1997). Graph homomorphisms: Structure and symmetry. In G. Hahn & G. Sabidussi (Eds.), *Graph symmetry*. NATO ASI Series (Series C: Mathematical and physical sciences) (Vol. 497, pp. 107–166). Springer. https://doi.org/10.1007/978-94-015-8937-6_4
8. Hughes, G. E., & Cresswell, M. J. (1996). *A new introduction to modal logic*. Routledge.
9. Konikowska, B. (1999). Rasiowa-Sikorski deduction systems: A handy tool for computer science logics. In J. L. Fiadeiro (Ed.), *Recent trends in algebraic development techniques*. WADT 1998. Lecture Notes in Computer Science (Vol. 1589, pp. 184–197). Springer.
10. Leszczyńska, D. (2004). Socratic proofs for some normal modal propositional logics. *Logique et Analyse, 47*(185–188), 259–285.
11. Leszczyńska, D. (2007). *The method of Socratic proofs for normal modal propositional logics*. Adam Mickiewicz University Press.
12. Leszczyńska-Jasion, D. (2008). The method of Socratic proofs for modal propositional logics: K5, S4.2, S4.3, S4M, S4F, S4R and G. *Studia Logica, 89*(3), 371–405.
13. Leszczyńska-Jasion, D. (2009). A loop-free decision procedure for modal propositional logics K4, S4 and S5. *Journal of Philosophical Logic, 38*(2), 151–177.
14. Leszczyńska-Jasion, D. (2018). *From questions to proofs*. Between the logic of questions and proof theory: Faculty of Social Science Publishers, Adam Mickiewicz University.
15. Negri, S. (2005). Proof analysis in modal logic. *Journal of Philosophical Logic, 34*, 507–544.
16. Ohnishi, M., & Matsumoto, K. (1957). Gentzen method in modal calculi. *Osaka Mathematical Journal, 9*(2), 113–130.

17. Ohnishi, M., & Matsumoto, K. (1959). Gentzen method in modal calculi. *II. Osaka Mathematical Journal, 11*(2), 115–120.
18. Orłowska, E. (1996). Relational proof systems for modal logics. In H. Wansing (Ed.), *Proof theory of modal logic* (pp. 55–78). Springer.
19. Perzanowski, J. (1972). The first list of the deduction theorems characteristic for several modal calculi formalized after the manner of Lemmon. *Bulletin of the Section of Logic, 1*(4), 21–31.
20. Perzanowski, J. (1973). The deduction theorems for the modal propositional calculi formalized after the manner of Lemmon. *Part I. Reports on Mathematical Logic, 1*, 1–12.
21. Poggiolesi, F. (2011). *Gentzen calculi for modal propositional logic*. Springer.
22. Priest, G. (2008). *An introduction to non-classical logic: From if to is*. Cambridge University Press.
23. Russo, A. (1995). *Modal labelled deductive systems*. Imperial College Research Report DoC 95/7.
24. Viganò, L. (2000). *Labelled non-classical logics*. Springer.
25. Wansing, H. (2002). Sequent systems for modal logics. In D. M. Gabbay & F. Guenthner (Eds.), *Handbook of philosophical logic*, vol. 8, 2nd ed., pp. 61–145). Kluwer Academic Publishers.
26. Wiśniewski, A. (2021). Entailment, transmission of truth, and minimality. *Journal of Applied Logics, 8*(10), 2471–2522.

Chapter 4
Erotetic Calculus for Propositional Intuitionistic Logic

In this chapter, we analyze the method of Socratic proofs for propositional intuitionistic logic (INT, for short). The erotetic calculus for this logic is presented here for the first time. We shall consider Kripke semantics; hence, the construction follows that for modal logics.

It is worth noting how the diversity of sequent types is utilized in building erotetic calculi for non-classical logics. In previous chapters, we have explored three different variants of erotetic calculi for classical logic; in the modal case, we have seen how both-sided, multisuccedent sequents can account for the relation of global entailment in modal logics. For intuitionistic logic, both-sided sequents can be employed, as we shall see, to express the basic semantic properties of a sentence's derivability or non-derivability (the intuitionistic counterparts of true or not true); thus, these sequents are also multisuccedent. However, this is not a departure from the commonly assumed single-succedent format.

The fact that multiple formulas are allowed in the succedent follows from the use of indices, enabling the introduction of a more detailed semantic information. In Chap. 7, we shall examine the relation between the calculus presented here and the more traditional single-succedent version.

4.1 Erotetic Calculus for Intuitionistic Logic

The modal logics considered in the previous chapter are extensions of CPL, as they are expressed in the language $\mathcal{M}$ built upon $\mathcal{L}_{\mathsf{CPL}}$ and adhere to CPL-validity. In contrast, intuitionistic logic is a restriction of CPL that rejects certain laws of classical logic. Although intuitionistic logic is not, strictly speaking, a modal logic, it is closely connected to S4. As in the previous case, we will rely on Kripke semantics.

D. Leszczyńska-Jasion, *The Method of Socratic Proofs*, Trends in Logic 64,
https://doi.org/10.1007/978-3-031-82451-7_4

Table 4.1 Axiomatic basis of INT

$A \supset (B \supset A)$
$(A \supset B) \supset ((A \supset (B \supset C)) \supset (A \supset C))$
$A \supset (B \supset (A \wedge B))$
$(A \wedge B) \supset A$
$(A \wedge B) \supset B$
$A \supset (A \vee B)$
$B \supset (A \vee B)$
$(A \supset C) \supset ((B \supset C) \supset ((A \vee B) \supset C))$
$(A \supset B) \supset ((A \supset \sim B) \supset \sim A)$
$\sim A \supset (A \supset B)$
MP: From $A \supset B$ and A derive B

Since modal operators are not needed in the language, we can return to the simpler language $\mathcal{L}_{\mathsf{CPL}}$ at the lower level. As previously, we use *VAR* for the set of propositional variables. However, we find it appropriate to use distinct symbols for negation and implication; thus, we adopt '$\sim$' instead of '$\neg$' and '$\supset$' in place of '$\rightarrow$'. The language $\mathcal{L}_{\mathsf{INT}}$ is given by the following BNF-grammar:

$$A ::= p \mid \sim A \mid A \wedge A \mid A \vee A \mid A \supset A$$

By $FORM_{\mathsf{INT}}$ we mean the set of all formulas of $\mathcal{L}_{\mathsf{INT}}$.

Intuitionistic Logic in an Axiomatic Setting

'INT' denotes propositional intuitionistic logic defined axiomatically. Table 4.1 provides the axiomatization of the propositional part of first-order intuitionistic logic, as formulated by Kleene [4]. The only rule of inference is *Modus Ponens* (MP).

To construct the erotetic calculus for INT, we employ indices, as our approach is based on Kripke-style semantics. Here, however, we are not concerned with distinguishing between local and global entailment, although both notions can indeed be defined in Kripke semantics as in the modal case. The rationale for this choice lies in the fact that, in intuitionistic logic, the distinction between global and local satisfiability becomes, to some extent, insignificant. Roughly speaking, when a propositional variable holds in a world within an intuitionistic Kripke model, the model's structure ensures that the variable also holds in every world accessible from it. (This property of a model's structure is known as *heredity*.) Given the worlds in which a propositional variable holds, it is easy to construct a Kripke model where the variable holds globally—that is, in every world of the model.[1]

Another reason for not addressing global entailment is that we need the two sides of a sequent for a different purpose. Intuitionistic logic is not a logic of Boolean

[1] This argument is a 'hand-waving' version of the one given in [3, 462–464].

functions in the way that classical logic is. We cannot divide intuitionistic formulas into α- and β-types, nor can we use intuitionistic negation to define a complement operation. To put it differently, while we could attempt this, the defined objects would not be as semantically well-behaved as their classical counterparts.

However, the two sides of a sequent can be used to indicate whether a formula holds or does not hold in a world within a Kripke model. It may be more fitting to interpret the left and right sides of a sequent as designated for formulas that do or do not have a proof. A detailed exploration of the philosophical and mathematical underpinnings of intuitionism is beyond the scope of this book, but readers can consult [2, 6, 8], and the references contained therein for further information.

The language of the intuitionistic erotetic calculus will be denoted as $\mathcal{L}^{?}_{\vdash \mathsf{INT}}$. As in the modal case, the language includes *indexed formulas*, which are expressions of the form:

$$A^{\varsigma/i} \tag{4.1}$$

where A in (4.1) is a formula of $\mathcal{L}_{\mathsf{INT}}$, and '$\varsigma/i$' in (4.1) is a finite, non-empty sequence of numerals, with i as the last element of the sequence. As previously, we will use ς, ϱ, possibly with subscripts, to refer to indices.

Sequents of the language $\mathcal{L}^{?}_{\vdash \mathsf{INT}}$ are expressions of the form:

$$\Gamma \vdash \Delta \tag{4.2}$$

where both Γ and Δ are finite sequences of indexed formulas; each can be empty, but not both simultaneously.

As we see, sequents (4.2) of $\mathcal{L}^{?}_{\vdash \mathsf{INT}}$ are both-sided and multisuccedent. *Atomic d-wffs of* $\mathcal{L}^{?}_{\vdash \mathsf{INT}}$ are simply the sequents of this language, while *complex d-wffs* are constructed as before, using the connectives 'ng' and '&'. Last but not least, *questions of the language* $\mathcal{L}^{?}_{\vdash \mathsf{INT}}$ are expressions of the form:

$$?(\phi_1, \ldots, \phi_n)$$

where $\langle \phi_1, \ldots, \phi_n \rangle$ is a nonempty sequence of sequents (a seqsequent) of $\mathcal{L}^{?}_{\vdash \mathsf{INT}}$.

For the remainder of this section, we use ϕ and ψ to denote sequents of $\mathcal{L}^{?}_{\vdash \mathsf{INT}}$. The other conventions introduced above are applied accordingly (see Sects. 2.2 and 3.2).

The semantics of INT resemble these for the modal logic $\mathsf{S4}$; while an intuitionistic frame aligns with an $\mathsf{S4}$-frame, models are defined through the so-called forcing relation rather than valuations.

Definition 4.1 (*Intuitionistic Frame and Model*) An *intuitionistic frame* is a pair $\langle W, R \rangle$, where $W \neq \varnothing$ and $R \subseteq W \times W$ is a reflexive and transitive relation. An *intuitionistic model* is a triple $\langle W, R, \Vdash \rangle$, where $\langle W, R \rangle$ is an intuitionistic frame and

$$\Vdash \; \subseteq W \times \mathit{FORM}_{\mathsf{INT}}$$

is a relation, called *forcing*, such that each $w \in W$ satisfies the following conditions:

1. $w \Vdash A \wedge B$ iff $w \Vdash A$ and $w \Vdash B$,
2. $w \Vdash A \vee B$ iff $w \Vdash A$ or $w \Vdash B$,
3. $w \Vdash \sim A$ iff for each $w^* \in R(w)$, $w^* \nVdash A$,
4. $w \Vdash A \supset B$ iff for each $w^* \in R(w)$, $w^* \nVdash A$ or $w^* \Vdash B$,

(H) For each propositional variable p_i, if $w \Vdash p_i$ and $w^* \in R(w)$, then $w^* \Vdash p_i$.

Condition (H) is referred to as the *heredity condition*.

Succinctly, if $w \Vdash A$, we say that A is *forced by* w. A formula is considered *true in an intuitionistic model* if it is forced by every world of that model. A formula is called *intuitionistically valid* if it is true in all intuitionistic models.

If X is a set of formulas of the language $\mathcal{L}_{\mathsf{INT}}$, then by $w \Vdash X$, we mean that for each $A \in X$, $w \Vdash A$.

As in the modal case, we need to interpret sequents of $\mathcal{L}^?_{\vdash\mathsf{INT}}$ within intuitionistic frames. As before, we use the set '$W(\phi)$' and the relation '$R(\phi)$' for this purpose. Let ϕ be a sequent of the language $\mathcal{L}^?_{\vdash\mathsf{INT}}$ of the form:

$$A_1^{\varsigma_1/i_1}, \ldots, A_k^{\varsigma_k/i_k} \vdash B_1^{\varrho_1/j_1}, \ldots, B_n^{\varrho_n/j_n} \tag{4.3}$$

Then, we define $W(\phi)$ and $R(\phi)$ as follows:

$$W(\phi) = \{j : j \text{ is a term of the sequence '}\varsigma_l/i_l\text{' for some } 1 \leqslant l \leqslant k,$$

$$\text{or of the sequence '}\varrho_m/j_m\text{' for some } 1 \leqslant m \leqslant n\}$$

$$R(\phi) = \{\langle j, j^*\rangle : j \text{ immediately precedes } j^* \text{ in '}\varsigma_l/i_l\text{' for some } 1 \leqslant l \leqslant k,$$

$$\text{or in '}\varrho_m/j_m\text{' for some } 1 \leqslant m \leqslant n\}.$$

Definition 4.2 (*Interpretation Function*) Let ϕ be a sequent of $\mathcal{L}^?_{\vdash\mathsf{INT}}$, and let $\langle W, R\rangle$ be an intuitionistic frame. An *interpretation of* ϕ *in* $\langle W, R\rangle$ is a function $f : W(\phi) \longrightarrow W$, such that:

(hi) if $\langle i, j\rangle \in R(\phi)$, then $\langle f(i), f(j)\rangle \in R$.

Definition 4.3 Let $\langle W, R, \Vdash\rangle$ be an intuitionistic model. Let ϕ stand for a sequent of $\mathcal{L}^?_{\vdash\mathsf{INT}}$ of the form (4.3) specified above, and let f be an interpretation of ϕ in the frame $\langle W, R\rangle$. ϕ is *true in the model* $\langle W, R, \Vdash\rangle$ *under the interpretation* f *of* ϕ *in* $\langle W, R\rangle$ iff the following condition holds:

($i\star$) if for each l ($1 \leqslant l \leqslant k$), $f(i_l) \Vdash A_l$, then for some m ($1 \leqslant m \leqslant n$), $f(j_m) \Vdash B_m$.

The condition ($i\star$) can be interpreted as corresponding to material interpretation of sequents (compare our comments in Sect. 2.2 of Chap. 2); we need to remember, however, that the interpretation refers to the sequents of the language $\mathcal{L}^{?}_{\vdash\mathsf{INT}}$, containing indexed formulas.

Definition 4.4 Let $\langle W, R\rangle$ stand for an intuitionistic frame. A sequent ϕ of $\mathcal{L}^{?}_{\vdash\mathsf{INT}}$ is *valid in* $\langle W, R\rangle$ *under an interpretation* f *of* ϕ *in* $\langle W, R\rangle$ iff for each forcing relation $\Vdash$ defined on $\langle W, R\rangle$, ϕ is true in $\langle W, R, \Vdash\rangle$ under f. Moreover, the sequent ϕ is *valid in* $\langle W, R\rangle$ iff it is valid in $\langle W, R\rangle$ under every interpretation of ϕ in $\langle W, R\rangle$.

As in the modal case, we have:

Corollary 4.1 *A sequent of* $\mathcal{L}^{?}_{\vdash\mathsf{INT}}$ *is valid in every intuitionistic frame in which it is not interpretable.*

Definition 4.5 A sequent of $\mathcal{L}^{?}_{\vdash\mathsf{INT}}$ is *intuitionistically valid* iff it is valid in every intuitionistic frame.

Corollary 4.2 *A sequent of* $\mathcal{L}^{?}_{\vdash\mathsf{INT}}$ *of the form* '$\vdash A^0$' *is intuitionistically valid iff formula* A *is intuitionitically valid.*

Proof If the sequent '$\vdash A^0$' is intuitionistically valid, then for each intuitionistic frame $\langle W, R\rangle$, '$\vdash A^0$' is valid in the frame under every interpretation f. The sequent contains no pairs of numerals, so it is interpretable in every frame. Indeed, any assignment $f(0) = w \in W$ is an interpretation of '$\vdash A^0$' in $\langle W, R\rangle$. It follows that for each $w \in W$, we can set $f(0) = w$ and have $w \Vdash A$ by Definition 4.3, hence A is intuitionistically valid.

For the only-if direction suppose that A is intuitionistically valid. Let $\langle W, R, \Vdash\rangle$ stand for an intuitionistic model, and let f be an interpretation function of the sequent in $\langle W, R\rangle$. The value $f(0)$ is a world in W, and $f(0) \Vdash A$, as A is intuitionistically valid. This holds for arbitrary forcing relation defined on the frame, hence '$\vdash A^0$' is valid in $\langle W, R\rangle$ under f (by Definition 4.4). As f is arbitrary, the sequent is valid in the frame (by Definition 4.4 again), and since the frame is arbitrary, it is intuitionistically valid by Definition 4.5. □

Rules of Erotetic Calculus $\mathbb{E}^{\mathsf{INT}}$

This section presents the rules of erotetic calculus $\mathbb{E}^{\mathsf{INT}}$ for propositional intuitionistic logic INT.

The first rule captures the heredity condition. All rules here are straightforward when considered under an indirect interpretation: formulas that are forced in a given world are maintained in the antecedent of a sequent, while those that are not forced are placed in the succedent. Under this indirect interpretation, rule R_H states that a propositional variable p_n forced in a world, indicated by i, may 'travel' to another accessible world. This transfer occurs only in the antecedent, with numeral j already present in the sequent-premise.

$$\frac{?(\Phi(\Gamma\ (p_n{}^{\varsigma/i}\ ) \vdash \Delta))}{?(\Phi(\Gamma(p_n{}^{\varsigma/i},\ p_n{}^{j}) \vdash \Delta))}\ R_H$$

As in modal logics, a proviso of applicability for R_H ensures that the worlds denoted by j are indeed accessible from those indicated by i. Reflexivity can be disregarded, as reintroducing a variable in the same world is unnecessary. Rule R_H may be applied as long as condition (4) (referring to the concept of an R-chain introduced in Definition 3.5 in Sect. 3.3) is satisfied.

(4) there is a directed $R(\phi)$-$[i, j]$-chain, where ϕ is the sequent-premise of the rule.

Here are the rules for negation. Recall that '(_)' in the sequent-premise of $L_\sim$ indicates an arbitrary position in the respective sequence of formulas. In turn, the same inscription in the sequent-conclusion of $R_\sim$ means that the indexed formula displayed in the sequent-premise is removed from its position.

$$\frac{?(\Phi(\Gamma((\sim A)^{\varsigma/i}) \vdash \Delta\ (\_)))}{?(\Phi(\Gamma\ ((\sim A)^{\varsigma/i}) \vdash \Delta\ (A^j)))}\ L_\sim \qquad \frac{?(\Phi(\Gamma\ (\_) \vdash \Delta((\sim A)^{\varsigma/i})))}{?(\Phi(\Gamma(A^{\varsigma/i,j}) \vdash \Delta\ (\_)))}\ R_\sim$$

In the case of $L_\sim$, numeral j must satisfy $(t+4)$, where:

$(t+4)$ $i = j$, or there is a directed $R(\phi)$-$[i, j]$-chain, where ϕ is the sequent-premise of the rule.

In the case of $R_\sim$,

(nw) j is new to the sequent-premise of the rule.

Analogously to the modal case, for practical reasons we expect that

$$j = \max\{W(\phi)\} + 1,$$

where ϕ is the sequent-premise, though in the structural analysis we will drop this assumption.

Here are the rules for implication. The left rule, $L_\supset$, is subject to condition $(t+4)$, while in the right rule, $R_\supset$, the numeral j is subject to (nw).

$$\frac{?(\Phi(\Gamma((A \supset B)^{\varsigma/i})(\_) \vdash \Delta\,(\_)))}{?(\Phi(\Gamma((A \supset B)^{\varsigma/i})(\_) \vdash \Delta(A^{j});\ \Gamma((A \supset B)^{\varsigma/i})(B^{j}) \vdash \Delta\,(\_)))}\ L_{\supset}$$

$$\frac{?(\Phi(\Gamma\,(\_) \vdash \Delta((A \supset B)^{\varsigma/i})))}{?(\Phi(\Gamma(A^{\varsigma/i,j}) \vdash \Delta(B^{\varsigma/i,j})))}\ R_{\supset}$$

Finally—the rules for conjunction and disjunction; there are no surprises:

$$\frac{?(\Phi(\Gamma((A \wedge B)^{\varsigma/i}) \vdash \Delta))}{?(\Phi(\Gamma(A^{\varsigma/i}, B^{\varsigma/i}) \vdash \Delta))}\ L_{\wedge} \qquad \frac{?(\Phi(\Gamma \vdash \Delta((A \vee B)^{\varsigma/i})))}{?(\Phi(\Gamma \vdash \Delta(A^{\varsigma/i}, B^{\varsigma/i})))}\ R_{\vee}$$

$$\frac{?(\Phi(\Gamma \vdash \Delta((A \wedge B)^{\varsigma/i})))}{?(\Phi(\Gamma \vdash \Delta(A^{\varsigma/i});\ \Gamma \vdash \Delta(B^{\varsigma/i})))}\ R_{\wedge}$$

$$\frac{?(\Phi(\Gamma((A \vee B)^{\varsigma/i}) \vdash \Delta))}{?(\Phi(\Gamma(A^{\varsigma/i}) \vdash \Delta;\ \Gamma(B^{\varsigma/i}) \vdash \Delta))}\ L_{\vee}$$

The following definition introduces the erotetic calculus for INT.

Definition 4.6 (*Erotetic Calculus for Intuitionistic Propositional Logic*) Erotetic calculus $\mathbb{E}^{\mathsf{INT}}$ is the smallest set of rules containing R_H, $L_{\sim}$, $R_{\sim}$, $L_{\supset}$, $R_{\supset}$, $L_{\wedge}$, $R_{\wedge}$, $L_{\vee}$, $R_{\vee}$, where

- R_H is subject to (4),
- $L_{\sim}$ and $L_{\supset}$ are subject to $(t+4)$, and
- $R_{\sim}$, $R_{\supset}$ are subject to (nw).

As previously, a *Socratic transformation of a question Q of* $\mathcal{L}^{?}_{\vdash\mathsf{INT}}$ is a sequence of questions starting with Q and regulated by the rules of $\mathbb{E}^{\mathsf{INT}}$.

Definition 4.7 (*Socratic proof in* $\mathbb{E}^{\mathsf{INT}}$) Let ϕ be a sequent of the language $\mathcal{L}^{?}_{\vdash\mathsf{INT}}$ of the form '$\Gamma \vdash A^{0}$'. A *Socratic proof of ϕ in* $\mathbb{E}^{\mathsf{INT}}$ is a finite Socratic transformation **s** of the question '$?(\phi)$' in $\mathbb{E}^{\mathsf{INT}}$, such that each constituent of the last question of **s** has the following form:

(f) $\Delta(B^{\varsigma/i}) \vdash \Theta\,(B^{\varrho/i})$.

As in the modal case, indices 'ς/i' and 'ϱ/i' end with the same numeral, but apart from this, they may be distinct.

If the last question of a finite Socratic transformation consists only of constituents of the form (f), we call it *successful*; otherwise, it is termed *unsuccessful*. Let us recall that the distinction between a Socratic proof and a successful Socratic transformation lies in the first question. In a Socratic proof, this question must be of the form '$?(\Gamma \vdash A^{0})$', whereas in a successful Socratic transformation, it can be arbitrary.

Below, we present an example of a Socratic proof and a Socratic transformation of a question that contains an unprovable sequent.

Example 4.1 A Socratic proof of $\vdash (p \supset \sim\sim p)^0$ in $\mathbb{E}^{\mathsf{INT}}$.

$$
\begin{array}{cl}
?(\vdash (p \supset \sim\sim p)^0) & \\ \hline
?(p^{0,1} \vdash (\sim\sim p)^{0,1}) & R_{\supset} \\ \hline
?(p^{0,1}, (\sim p)^{0,1,2} \vdash) & R_{\sim} \\ \hline
?(p^{0,1}, (\sim p)^{0,1,2} \vdash p^2) & L_{\sim} \\ \hline
?(p^{0,1}, p^2, (\sim p)^{0,1,2} \vdash p^2) & R_H
\end{array}
$$

Example 4.2 An unsuccessful Socratic transformation of question $?(\vdash (((p \supset q) \supset p) \supset p)^0)$; starting with 4th question, ψ represents the last constituent '$((p \supset q) \supset p)^{0,1}, p^1 \vdash p^{0,1}$', and starting with 5th question, $A = (p \supset q) \supset p$.

$$
\begin{array}{cl}
?(\vdash (((p \supset q) \supset p) \supset p)^0) & \\ \hline
?(((p \supset q) \supset p)^{0,1} \vdash p^{0,1}) & R_{\supset} \\ \hline
?(((p \supset q) \supset p)^{0,1} \vdash (p \supset q)^1, p^{0,1};\ ((p \supset q) \supset p)^{0,1}, p^1 \vdash p^{0,1}) & L_{\supset} \\ \hline
?(((p \supset q) \supset p)^{0,1}, p^{1,2} \vdash q^{1,2}, p^{0,1};\ \psi) & R_{\supset} \\ \hline
?((A)^{0,1}, p^{1,2} \vdash (p \supset q)^2, q^{1,2}, p^{0,1};\ (A)^{0,1}, p^2, p^{1,2} \vdash q^{1,2}, p^{0,1};\ \psi) & L_{\supset} \\ \hline
?((A)^{0,1}, p^{1,2}, p^{2,3} \vdash q^{2,3}, q^{1,2}, p^{0,1};\ (A)^{0,1}, p^2, p^{1,2} \vdash q^{1,2}, p^{0,1};\ \psi) & R_{\supset}
\end{array}
$$

The Socratic transformation presented in Example 4.2 could be continued with successive applications of rules $L_{\supset}$ and $R_{\supset}$; it is easy to see, however, that this would not lead to a success—it is not possible to obtain p with numeral 1 ending its index in the antecedent.

Soundness of $\mathbb{E}^{\mathsf{INT}}$

In Chap. 3, we proved several results concerning provisos of applicability of modal rules, relation $R(\phi)^{\mathsf{L}}$, and Kripke models that carry over to intuitionistic logic due to the same semantic framework. For convenience of reference, we state these results below.

Lemma 4.1 *Let ϕ stand for a sequent of the language $\mathcal{L}^{?}_{\vdash \mathsf{INT}}$ and let $\langle W, R\rangle$ be an intuitionistic frame.*

1. *There is a directed $R(\phi)$-$[i, j]$-chain iff $\langle i, j\rangle \in R^{tr,ref}$. (Cf Lemma 3.1.)*
2. *If f is an interpretation function of ϕ in $\langle W, R\rangle$ and $\langle i, j\rangle \in R(\phi)^{tr,ref}$, then also $\langle f(i), f(j)\rangle \in R$. (Cf. Lemma 3.2.)*

Proof The proof is contained in those of Lemmas 3.1 and 3.2. □

The reader may observe that the formulation of Lemmas 3.1 and 3.2 is more restrictive than that of Lemma 4.1, as it require that ϕ occurs in a Socratic proof. This restriction, however, pertains specifically to Euclidean logics, where we needed directed $R(\phi)$-chains starting with numeral 0.

We state the following corollary (without a proof), and then we prove that all rules of $\mathbb{E}^{\mathsf{INT}}$ are correct and invertible.

Corollary 4.3 *Each sequent of* $\mathcal{L}^{?}_{\vdash \mathsf{INT}}$ *of the form:*

$(f) \quad \Delta(B^{\varsigma/i}) \vdash \Theta(B^{\varrho/i}),$

indicated in Definition 4.7, is intuitionistically valid.

Sequents of $\mathcal{L}^{?}_{\vdash \mathsf{INT}}$ of the indicated form (f) are called *basic*, as previously.

Lemma 4.2 (Correctness and Invertibility of Rules of $\mathbb{E}^{\mathsf{INT}}$) *Let* $?(\Phi)/?(\Psi)$ *be an instance of a rule of* $\mathbb{E}^{\mathsf{INT}}$. *Each sequent in* Φ *is intuitionistically valid iff each sequent in* Ψ *is intuitionistically valid.*

Proof The cases of rules for conjunction and disjunction are very similar to α- and β-rules in modal logics and we skip them.

Rule R_H. The top-down direction is warranted by 'rewriting' the active indexed formula '$p_n^{\varsigma/i}$' in the sequent-conclusion. For the bottom-up direction assume that there is an intuitionistic frame, $\langle W, R\rangle$, an interpretation f of the sequent-premise '$\Gamma(p_n^{\varsigma/i}) \vdash \Delta$' in $\langle W, R\rangle$, and the forcing relation $\Vdash$ such that (a) the sequent is not true in model $\langle W, R, \Vdash\rangle$ under f. Let us observe that f is also an interpretation of the sequent-conclusion of R_H in the same frame $\langle W, R\rangle$, as the sequents do not differ with respect to pairs of numerals. By assumption (a), each formula occurring in '$\Gamma(p_n^{\varsigma/i})$' is forced by the respective world indicated by the last numeral of its index—in particular, $f(i) \Vdash p_n$. By (4), there is a directed $R(\phi)$-$[i, j]$-chain, where ϕ is for the sequent-premise of R_H. By Lemma 4.1, $\langle f(i), f(j)\rangle \in R$. Finally, by heredity condition, also $f(j) \Vdash p_n$, which yields that the sequent-conclusion is not true in $\langle W, R, \Vdash\rangle$ under f.

Rule $R_{\sim}$. Assume that the sequent-premise, '$\Gamma(\_) \vdash \Delta({\sim} A^{\varsigma/i})$', is not valid in an intuitionistic frame $\langle W, R\rangle$ under f. Then the formulas from $\Gamma(\_)$ are forced in the worlds designated by the respective numerals, and formulas from $\Delta({\sim} A^{\varsigma/i})$ are not, in particular, $f(i) \nVdash {\sim} A$. By Definition 4.1, there is $w^* \in R(f(i))$ such that $w^* \Vdash A$. We define interpretation g of the sequent-conclusion, '$\Gamma(A^{\varsigma/i,j}) \vdash \Delta(\_)$', in $\langle W, R\rangle$ as follows: we put $g(j) = w^*$ and $g(n) = f(n)$ for every $n \neq j$. That g satisfies Definition 4.2 of interpretation function follows from the fact that j is new to the sequent-premise and from the fact that $w^* \in R(f(i))$. It follows that the sequent-conclusion is not true in $\langle W, R, \Vdash\rangle$ under g, which proves the lemma's statement in the bottom-up direction. For the top-down direction we assume that the sequent-conclusion, '$\Gamma(A^{\varsigma/i,j}) \vdash \Delta(\_)$', is not true in $\langle W, R, \Vdash\rangle$ under some f. In order to obtain an interpretation of the sequent-premise in the same frame we need to consider restriction of f to domain $W(\phi)\backslash\{j\}$, where ϕ is the sequent-conclusion. The other details are straightforward. (See also the proof of Lemma 3.4 for analogous modal cases.)

$L_{\supset}$. For simplicity, let $\phi = \Gamma(A \supset B^{\varsigma/i})(\_) \vdash \Delta(\_)$ and:

$$\psi_1 = \Gamma(A \supset B^{\varsigma/i})(\_) \vdash \Delta(A^j)$$

$$\psi_2 = \Gamma(A \supset B^{\varsigma/i})(B^j) \vdash \Delta(\_)$$

The top-down direction is easy, as the key implication is rewritten in both sequent-conclusions, hence we consider only the bottom-up direction. Assume that ϕ is not true in some intuitionistic model $\langle W, R, \Vdash\rangle$ under an interpretation f. From this assumption we have $f(i) \Vdash A \supset B$. The rule satisfies proviso $(t+4)$, hence $f(j) \in R(f(i))$ by Lemma 4.1. By Definition 4.1, $f(j) \nVdash A$ or $f(j) \Vdash B$, and this yields immediately that at least one of ψ_1, ψ_2 is not true in $\langle W, R, \Vdash\rangle$ under f, as required.

The other cases, $L_{\sim}$ and $R_{\supset}$, are similar. □

Theorem 4.1 (Soundness of $\mathbb{E}^{\mathsf{INT}}$). *If there exists a Socratic proof of a sequent in* $\mathbb{E}^{\mathsf{INT}}$*, then the sequent is intuitionistically valid.*

Proof The proof is organized as the proof of Theorem 2.2 in Chap. 2 and that of Theorem 3.1 in the previous chapter. We skip it. □

It follows from Theorem 4.1 that:

Corollary 4.4 *If there exists a Socratic proof of sequent '*$\vdash (A)^0$*' in* $\mathbb{E}^{\mathsf{INT}}$*, then formula A is intuitionistically valid.*

4.2 Admissible, Derivable and Structural Rules: The Intuitionistic Case

The goal of this section mirrors that of classical and modal logic: we analyze additional structural rules and prove that their application cannot lead from a question with a successful Socratic transformation to one without such a transformation.

Before we proceed to the technical results, let us recall the definitions of admissibility and derivability as they pertain to the erotetic format.

Definition 4.8 (Derivability, Semi-admissibility, Admissibility in $\mathbb{E}^{\mathsf{INT}}$) Let $\mathbf{r} = Q/Q^*$ be an erotetic rule. The rule $\mathbf{r}$ is *derivable in* erotetic calculus $\mathbb{E}^{\mathsf{INT}}$ iff there exists a finite Socratic transformation $\mathbf{s}$ of question Q in $\mathbb{E}^{\mathsf{INT}}$, with Q^* as its last term. Moreover, $\mathbf{r}$ is *semi-admissible in* $\mathbb{E}^{\mathsf{INT}}$ iff the following condition holds:

1. If there is a successful Socratic transformation of Q^* in $\mathbb{E}^{\mathsf{INT}}$, then there is a successful Socratic transformation of Q in $\mathbb{E}^{\mathsf{INT}}$.

If, except for the clause 1, **r** satisfies also clause 2:

2. If there is a successful Socratic transformation of Q in $\mathbb{E}^{\mathsf{INT}}$, then there is a successful Socratic transformation of Q^* in $\mathbb{E}^{\mathsf{INT}}$,

then we say that **r** is *admissible in* erotetic calculus $\mathbb{E}^{\mathsf{INT}}$.

By *length* of a finite Socratic transformation we mean, as in the previous chapters, the number of terms of the Socratic transformation.

Corollary 4.5 *If Q/Q^* is an instance of a rule* **r** *of $\mathbb{E}^{\mathsf{INT}}$ and there is a successful Socratic transformation of question Q^* in $\mathbb{E}^{\mathsf{INT}}$, of length n, then there is also a successful Socratic transformation of question Q in $\mathbb{E}^{\mathsf{INT}}$, that is of length $n + 1$.*

Corollary 4.6 (Renaming Numerals: New for New) *Let* **s** *be a finite Socratic transformation of a question in $\mathbb{E}^{\mathsf{INT}}$. Assume rule* $\mathbf{r} \in \{R_{\sim}, R_{\supset}\}$ *was applied in* **s** *and let j be the new numeral introduced in the sequent-conclusion of* **r**. *Further, let $k \neq j$ be another numeral which does not occur in* **s**. *Then the result of replacing each occurrence of j in* **s** *with an occurrence of k is a finite Socratic transformation of the same question in $\mathbb{E}^{\mathsf{INT}}$ and of the same length. Also, if the initial Socratic transformation,* **s**, *is successful, so is the final one.*

Proof The proof follows that of Corollary 3.8. □

Lemma 4.3 (Separation Lemma: the Intuitionistic Case) *Let $\phi_1, \ldots, \phi_n$ stand for sequents of $\mathcal{L}^{?}_{\vdash\mathsf{INT}}$. The following conditions are equivalent:*

1. *There exists a successful Socratic transformation of the question '$?(\phi_1, \ldots, \phi_n)$' in $\mathbb{E}^{\mathsf{INT}}$.*
2. *For each $i : 1 \leqslant i \leqslant n$, there exists a successful Socratic transformation of the question '$?(\phi_i)$' in $\mathbb{E}^{\mathsf{INT}}$.*

Proof Corollary 4.6 is used to secure the applicability of rules $R_{\sim}$ and $R_{\supset}$. In all the other aspects the proof follows that of Lemma 2.3, see also Lemma 3.5. □

From the proof of Lemma 4.3 also the following consequence follows.

Corollary 4.7 (Length-Preserving Separation: the Intuitionistic Case) *If there is a successful Socratic transformation of a question '$?(\phi_1, \ldots, \phi_n)$' in $\mathbb{E}^{\mathsf{INT}}$ of length l, then for each $i : 1 \leqslant i \leqslant n$, there is a successful Socratic transformation of the question '$?(\phi_i)$' of length not greater than l.*

With the above separation result we can prove the following corollary.

Corollary 4.8 *Let ϕ stand for a sequent of the language $\mathcal{L}^{?}_{\vdash\mathsf{INT}}$ and let $i \in W(\phi)$. If there is a successful Socratic transformation of $?(\phi)$ in $\mathbb{E}^{\mathsf{INT}}$, then for each $k \in W(\phi)$, there is also a successful Socratic transformation in $\mathbb{E}^{\mathsf{INT}}$ of the question $?(\phi^*)$, where ϕ^* results from ϕ by replacing every occurrence of numeral i with an occurrence of numeral k.*

Proof The proof follows that of Corollary 3.10. It relies on Corollaries 4.5, 4.7, and Lemma 4.3. □

Corollary 4.8 is further used in the specific formulation of the next corollary.

Corollary 4.9 (Renaming Numerals: Old for New) *If there is a successful Socratic transformation of a question of the form:*

$$?(\Gamma(A^{\varsigma/k})(A^{\varsigma/i}) \vdash \Delta)$$

in $\mathbb{E}^{\mathsf{INT}}$*, where the displayed occurrence of i is the only one in the sequent, then there is also a successful Socratic transformation of the question*

$$?(\Gamma(A^{\varsigma/k})(A^{\varsigma/k}) \vdash \Delta)$$

in $\mathbb{E}^{\mathsf{INT}}$.

Proof Follows from Corollary 4.8. □

We are ready to define the rules of weakening and contraction and to examine their properties.

The notion of subsequence is understood as previously: roughly speaking, a subsequence of a sequence S is obtained by deleting *some* elements of S, without changing the order of the other elements (see Defnition 2.9, Sect. 2.5). Here are the rules of weakening and contraction in the general form.

$$\frac{?(\Phi(\Gamma \vdash \Delta))}{?(\Phi(\Gamma^* \vdash \Delta))}\ L_{weak} \qquad \frac{?(\Phi(\Gamma \vdash \Delta))}{?(\Phi(\Gamma \vdash \Delta^*))}\ R_{weak}$$

where Γ^* is a non-empty proper subsequence of Γ, similarly for Δ^* and Δ.

$$\frac{?(\Phi(\Gamma^* \vdash \Delta))}{?(\Phi(\Gamma \vdash \Delta))}\ L_{contr} \qquad \frac{?(\Phi(\Gamma \vdash \Delta^*))}{?(\Phi(\Gamma \vdash \Delta))}\ R_{contr}$$

where Γ^* is a non-empty proper subsequence of Γ such that $\mathsf{set}(\Gamma^*) = \mathsf{set}(\Gamma)$, and similarly for Δ^*.

Lemma 4.4 (Semi-admissibility of L_{weak}, R_{weak} in $\mathbb{E}^{\mathsf{INT}}$) *If there is a successful Socratic transformation* **s** *of the question-conclusion of* L_{weak}/R_{weak} *in* $\mathbb{E}^{\mathsf{INT}}$*, and* **s** *is of length n, then there is a successful Socratic transformation of the question-premise of the rule in* $\mathbb{E}^{\mathsf{INT}}$*, and it is also of length n.*

Proof The proof is by induction and follows the pattern of the proofs of Lemmas 2.4 and 3.6. In order to secure the applicability of $\mathbf{r} \in \{R_{\sim}, R_{\supset}\}$, we use Corollary 4.6. □

Corollary 4.10 (Eliminability of L_{weak}, R_{weak}) *If there is a successful Socratic transformation of a question in* $\mathbb{E}^{\mathsf{INT}} \cup \{L_{weak}, R_{weak}\}$*, then there is also one in* $\mathbb{E}^{\mathsf{INT}}$.

Proof By Lemma 4.4. □

Lemma 4.5 (Proof-Theoretic Invertibility of the Rules of $\mathbb{E}^{\mathsf{INT}}$)

1. *If there is a successful Socratic transformation of a question* $?(\Gamma(p_n{}^{\varsigma/i}) \vdash \Delta)$ *in* $\mathbb{E}^{\mathsf{INT}}$*, then there is a successful Socratic transformation of the question* $?(\Gamma(p_n{}^{\varsigma/i}, p_n{}^{j}) \vdash \Delta)$ *in* $\mathbb{E}^{\mathsf{INT}}$.
2. *If there is a successful Socratic transformation of a question* $?(\Gamma(A \wedge B^{\varsigma/i}) \vdash \Delta)$ *in* $\mathbb{E}^{\mathsf{INT}}$*, then there is a successful Socratic transformation of the question* $?(\Gamma(A^{\varsigma/i}, B^{\varsigma/i}) \vdash \Delta)$ *in* $\mathbb{E}^{\mathsf{INT}}$.
3. *If there is a successful Socratic transformation of a question* $?(\Gamma \vdash \Delta(A \vee B^{\varsigma/i}))$ *in* $\mathbb{E}^{\mathsf{INT}}$*, then there is a successful Socratic transformation of the question* $?(\Gamma \vdash \Delta(A^{\varsigma/i}, B^{\varsigma/i}))$ *in* $\mathbb{E}^{\mathsf{INT}}$.
4. *If there is a successful Socratic transformation of a question* $?(\Gamma \vdash \Delta(A \wedge B^{\varsigma/i}))$ *in* $\mathbb{E}^{\mathsf{INT}}$*, then there is a successful Socratic transformation of the question* $?(\Gamma \vdash \Delta(A^{\varsigma/i});\ \Gamma \vdash \Delta(B^{\varsigma/i}))$ *in* $\mathbb{E}^{\mathsf{INT}}$.
5. *If there is a successful Socratic transformation of a question* $?(\Gamma(A \vee B^{\varsigma/i}) \vdash \Delta)$ *in* $\mathbb{E}^{\mathsf{INT}}$*, then there is a successful Socratic transformation of the question* $?(\Gamma(A^{\varsigma/i}) \vdash \Delta;\ \Gamma(B^{\varsigma/i}) \vdash \Delta)$ *in* $\mathbb{E}^{\mathsf{INT}}$.
6. *If there is a successful Socratic transformation of a question* $?(\Gamma({\sim} A^{\varsigma/i}) \vdash \Delta(\_))$ *in* $\mathbb{E}^{\mathsf{INT}}$*, then there is a successful Socratic transformation of the question* $?(\Gamma({\sim} A^{\varsigma/i}) \vdash \Delta(A^{j}))$ *in* $\mathbb{E}^{\mathsf{INT}}$.
7. *Let* $\phi = \Gamma(\_) \vdash \Delta({\sim} A^{\varsigma/i})$*. If there is a successful Socratic transformation of a question* $?(\phi)$ *in* $\mathbb{E}^{\mathsf{INT}}$*, then for each numeral* j*, that does not belong to* $W(\phi)$*, there is a successful Socratic transformation of the question* $?(\Gamma(A^{\varsigma/i,j}) \vdash \Delta(\_))$ *in* $\mathbb{E}^{\mathsf{INT}}$.
8. *If there is a successful Socratic transformation of a question* $?(\Gamma(A \supset B^{\varsigma/i})(\_) \vdash \Delta(\_))$ *in* $\mathbb{E}^{\mathsf{INT}}$*, then there is a successful Socratic transformation of the question* $?(\Gamma(A \supset B^{\varsigma/i})(\_) \vdash \Delta(A^{j});\ \Gamma(A \supset B^{\varsigma/i})(B^{j}) \vdash \Delta(\_))$ *in* $\mathbb{E}^{\mathsf{INT}}$.
9. *Let* $\phi = \Gamma(\_) \vdash \Delta(A \supset B^{\varsigma/i})$*. If there is a successful Socratic transformation of a question* $?(\phi)$ *in* $\mathbb{E}^{\mathsf{INT}}$*, then for each numeral* j*, that does not belong to* $W(\phi)$*, there is a successful Socratic transformation of the question* $?(\Gamma(A^{\varsigma/i,j}) \vdash \Delta(B^{\varsigma/i,j}))$ *in* $\mathbb{E}^{\mathsf{INT}}$.

Proof Clauses 1, 6 follow from Lemma 4.4 and clause 8 from Lemmas 4.4 and 4.3. Let us observe that for this reason the clauses can be formulated without referring to provisos of applicability of the rules in question.

Clauses 2–5 are proved as in the classical case (but simpler—there is no double negation and switching), see Corollary 2.13 in Chap. 2.

Clause 7 and 9: the reasoning is analogous to that for the rule R_{ν} in Corollary 3.14. Let $\phi = \Gamma(\_) \vdash \Delta({\sim} A^{\varsigma/i})$, $\phi^{*} = \Gamma(A^{\varsigma/i,j}) \vdash \Delta(\_)$, $j \notin W(\phi)$, and suppose there is a successful Socratic transformation $\mathbf{s} = \langle Q_0, Q_1, \ldots, Q_n \rangle$ of question $Q_0 = ?(\phi)$. We prove the following statement by induction with respect to the length of $\mathbf{s}$.

(S) If a question Q has a successful Socratic transformation in $\mathbb{E}^{\mathsf{INT}}$, then for each constituent of Q, which is of the form $\Gamma(\_) \vdash \Delta(\sim A^{\varsigma/i})$, and for each numeral j, that does not occur in the sequent, there is a successful Socratic transformation of the question $?(\Gamma(A^{\varsigma/i,j}) \vdash \Delta(\_))$ in $\mathbb{E}^{\mathsf{INT}}$.

In the base case, ϕ is a basic sequent. If at the same time ϕ^* is not basic, then it must be the case that ϕ is basic due to the presence of $(\sim A)^{\varsigma^*/i}$ in its antecedent (for some sequence ς^*). Clearly $\langle i, j\rangle$ satisfies proviso of applicability of $L_\sim$; we apply the rule introducing A^j to the succedent of the resulting sequent. In this way we have shown that there is a successful Socratic transformation of question $?(\phi^*)$.

Inductive case. We consider Q_0/Q_1. Suppose it is an instance of $R_\wedge$.

$$\frac{Q_0 = ?\left(\Gamma(\_)\,((B \wedge C)^{\varrho}\,) \;\vdash\; \Delta\left((\sim A)^{\varsigma/i}\right)\right)}{Q_1 = ?\left(\Gamma(\_)\,(B^{\varrho}, C^{\varrho}\,) \;\vdash\; \Delta\left((\sim A)^{\varsigma/i}\,\right)\right)}\; L_\wedge$$

By IH, there is a successful Socratic transformation of question $Q_1^* = ?(\Gamma(A^{\varsigma/i,j})(B^{\varrho}, C^{\varrho}) \vdash \Delta(\_))$. As the following Q_0^*/Q_1^* is an instance of $L_\wedge$,

$$\frac{Q_0^* = ?\left(\Gamma(A^{\varsigma/i,j})\,((B \wedge C)^{\varrho}\,) \;\vdash\; \Delta\,(\_)\right)}{Q_1^* = ?\left(\Gamma(A^{\varsigma/i,j})\,(B^{\varrho}, C^{\varrho}\,) \;\vdash\; \Delta\,(\_)\,\right)}\; L_\wedge$$

there is a successful Socratic transformation of Q_0^*.

Suppose that Q_0/Q_1 is an instance of $L_\supset$.

$$\frac{?(\Gamma((B \supset C)^{\varrho/k})(\_) \vdash \Delta((\sim A)^{\varsigma/i})(\_))}{?(\Gamma((B \supset C)^{\varrho/k})(\_) \vdash \Delta((\sim A)^{\varsigma/i})(B^l)\,;\; \Gamma((B \supset C)^{\varrho/k})(C^l) \vdash \Delta((\sim A)^{\varsigma/i})(\_))}$$

By Corollary 4.7 (separation), there is a successful Socratic transformation of question $?(\Gamma((B \supset C)^{\varrho/k})(\_) \vdash \Delta((\sim A)^{\varsigma/i})(B^l))$ which is of length up to n, hence, by IH, there is a successful Socratic transformation of question $?(\Gamma((B \supset C)^{\varrho/k})(A^{\varsigma/i,j}) \vdash \Delta(\_)(B^l))$ in $\mathbb{E}^{\mathsf{INT}}$. The same arguments pertain to the second constituent of Q_1. By a suitable instance of $L_\supset$, there is a successful Socratic transformation of question $?(\Gamma((B \supset C)^{\varrho/k})(A^{\varsigma/i,j}) \vdash \Delta(\_)(\_))$.

If Q_0/Q_1 is an instance of $R_\supset$, then we need to be careful with the new numeral—if it happens that $R_\supset$ introduces numeral j, then we first apply Corollary 4.6 and change j to some other numeral that does not occur in **s**. We do the same, if j was introduced in some other question of **s**. If the rule applied to Q_0 is $R_\sim$ and it acts upon $(\sim A)^{\varsigma/i}$, then the result follows without the use of IH, we only apply Corollary 4.6 to change the new numeral into j, if the one introduced was different from it. If $R_\sim$ is applied with respect to some other indexed formula, then we reason as in the case with $R_\supset$. We skip all the remaining cases.

Clause 9 is dealt with in analogous manner. □

Definition 4.9 (Rank of Formula of $\mathcal{L}_{\mathsf{INT}}$) A *rank of a formula of the language* $\mathcal{L}_{\mathsf{INT}}$ is a function $r : FORM_{\mathsf{INT}} \longrightarrow \mathbb{N}_0$, such that:

1. for each propositional variable p_i, $r(p_i) = 0$;
2. for each formula of $\mathcal{L}_{\mathsf{INT}}$ of the form '$\sim A$', $r(\sim A) = r(A) + 1$;
3. for each formula of $\mathcal{L}_{\mathsf{INT}}$ of the form '$(A \otimes B)$', where $\otimes \in \{\wedge, \vee, \supset\}$, $r(A \otimes B) = r(A) + r(B) + 1$.

By a *rank of an indexed formula A^{ς} of the language $\mathcal{L}^{?}_{\vdash\mathsf{INT}}$* we mean the rank of A.

As in the previous chapter, we will use more detailed forms of the rules of contraction, which makes the proof of admissibility more transparent.

$$\frac{?(\Phi(\Gamma(A^{\varsigma})(\_) \vdash \Delta))}{?(\Phi(\Gamma(A^{\varsigma})(A^{\varsigma}) \vdash \Delta))}\ L^*_{contr} \qquad \frac{?(\Phi(\Gamma \vdash \Delta(A^{\varsigma})(\_)))}{?(\Phi(\Gamma \vdash \Delta(A^{\varsigma})((A^{\varsigma})))}\ R^*_{contr}$$

We call the sequent '$\Gamma(A^{\varsigma})(\_) \vdash \Delta$' a *left-contraction of the sequent* '$\Gamma(A^{\varsigma})(A^{\varsigma}) \vdash \Delta$', and we call the sequent '$\Gamma \vdash \Delta(A^{\varsigma})(\_)$' a *right-contraction of the sequent* '$\Gamma \vdash \Delta(A^{\varsigma})(A^{\varsigma})$'. Moreover, a left-contraction and a right-contraction of a sequent are simply called *contractions* of this sequent.

It is easy to see that the following corollary is true.

Corollary 4.11 *A contraction of a basic sequent of $\mathcal{L}^{?}_{\vdash\mathsf{INT}}$ is a basic sequent.*

The next lemma is an analogue of Lemma 3.8 and it pertains to indexed formulas with negation or implication as the main operator, occurring in the antecedent of a sequent. Semantically speaking, the lemma concerns a situation analogous to that of generally quantified formulas in FOL or indexed π-formulas in modal L. For this reason, we coined the said formulas 'intuitionistic necessities'.

Lemma 4.6 (Contraction of Intuitionistic Necessities)

1. *If a question $?(\Gamma((\sim A)^{\varsigma/i})((\sim A)^{\varsigma/i}) \vdash \Delta)$ has a successful Socratic transformation in $\mathbb{E}^{\mathsf{INT}}$, then the question $?(\Gamma((\sim A)^{\varsigma/i})(\_) \vdash \Delta)$ has one as well.*
2. *If a question $?(\Gamma((A \supset B)^{\varsigma/i})((A \supset B)^{\varsigma/i}) \vdash \Delta)$ has a successful Socratic transformation in $\mathbb{E}^{\mathsf{INT}}$, then the question $?(\Gamma((A \supset B)^{\varsigma/i})(\_) \vdash \Delta)$ has one as well.*

Proof The proof is organised as that of Lemma 3.8 and there is nothing to be gained in repeating the details. Let us limit ourselves to listing, for the record, corollaries needed for the proof: Corollary 4.11 for the base case, Corollary 4.5 in the inductive case, and for the branching cases—Corollary 4.7 (length-preserving separation). □

Lemma 4.7 (Semi-admissibility of L^*_{contr}, R^*_{contr} in $\mathbb{E}^{\mathsf{INT}}$)

1. *If a question $?(\Gamma(A^{\varsigma})(A^{\varsigma}) \vdash \Delta)$ has a successful Socratic transformation in $\mathbb{E}^{\mathsf{INT}}$, then the question $?(\Gamma(A^{\varsigma})(\_) \vdash \Delta)$ also has a successful Socratic transformation in $\mathbb{E}^{\mathsf{INT}}$.*
2. *If a question $?(\Gamma \vdash \Delta(A^{\varsigma})(A^{\varsigma}))$ has a successful Socratic transformation in $\mathbb{E}^{\mathsf{INT}}$, then the question $?(\Gamma \vdash \Delta(A^{\varsigma})(\_))$ also has a successful Socratic transformation in $\mathbb{E}^{\mathsf{INT}}$.*

Proof We apply the same strategy as in the proof of Lemmas 2.9 and 3.9: the proof is by induction with respect to rank of A with a subinduction on the length l of a successful Socratic transformation of $?(\Gamma(A^{\varsigma})(A^{\varsigma}) \vdash \Delta)$ for the first clause and $?(\Gamma \vdash \Delta(A^{\varsigma})(A^{\varsigma}))$ for the second. In the base case of the main induction we rely on Corollaries 4.11 and 4.5.

In the (main) inductive part we use proof-theoretic invertibility of rules proved in Lemmas 4.5, and 4.6 if A^{ς} is an 'intuitionistic necessity'. The most troublesome case is that when the main operator of A is $\sim$ or $\supset$ and the contraction formula is in the succedent (so it pertains only to clause 2 of the lemma). In this situation we use additionally Corollary 4.9. □

Lemma 4.7 entails the following two corollaries, which we formulate without proofs.

Corollary 4.12 (Semi-admissibility of L_{contr} and R_{contr} in $\mathbb{E}^{\mathsf{INT}}$) *If there is a successful Socratic transformation of the question-conclusion of rule L_{contr}/R_{contr} in $\mathbb{E}^{\mathsf{INT}}$, then there is a successful Socratic transformation of the question-premise of the rule in $\mathbb{E}^{\mathsf{INT}}$.*

Corollary 4.13 (Eliminability of L_{contr} and R_{contr} in $\mathbb{E}^{\mathsf{INT}}$) *Let Q stand for a question of $\mathcal{L}^{?}_{\vdash \mathsf{INT}}$. If there is a successful Socratic transformation of Q in $\mathbb{E}^{\mathsf{INT}} \cup \{L_{contr}, R_{contr}\}$, then there is also one in $\mathbb{E}^{\mathsf{INT}}$.*

The following two lemmas concern contraction in the intuitionistic context. They express the semi-admissibility of the following rules in $\mathbb{E}^{\mathsf{INT}}$.

$$\frac{?(\Phi(\Gamma((A \supset B)^{\varsigma/i})(\_) \vdash \Delta))}{?(\Phi(\Gamma((A \supset B)^{\varsigma/i})((A \supset B)^{j}) \vdash \Delta))}\ L^{\supset}_{contr} \quad \langle i, j\rangle \in R(\Gamma((A \supset B)^{\varsigma/i}) \vdash \Delta)$$

$$\frac{?\Phi(\Gamma((\sim A)^{\varsigma/i})(\_) \vdash \Delta))}{?(\Phi(\Gamma((\sim A)^{\varsigma/i})((\sim A)^{j}) \vdash \Delta))}\ L^{\sim}_{contr} \quad \langle i, j\rangle \in R(\Gamma((\sim A)^{\varsigma/i}) \vdash \Delta)$$

Lemma 4.8 *Let $\phi = \Gamma((A \supset B)^{\varsigma/i})((A \supset B)^{j}) \vdash \Delta$. If there is a successful Socratic transformation of $?(\phi)$ in $\mathbb{E}^{\mathsf{INT}}$ and $\langle i, j\rangle \in R(\phi)$, then there is a successful Socratic transformation of the question $?(\Gamma((A \supset B)^{\varsigma/i})(\_) \vdash \Delta)$ in $\mathbb{E}^{\mathsf{INT}}$.*

Proof The proof of this lemma (and the next one) will follow the proof of Lemma 3.10 from the previous chapter. Hence let $\phi = \Gamma((A \supset B)^{\varsigma/i})((A \supset B)^{j}) \vdash \Delta$, $\phi^* = \Gamma((A \supset B)^{\varsigma/i})(\_) \vdash \Delta$, assume that $\langle i, j\rangle \in R(\phi)$, and that there is a successful Socratic transformation **s** of question $?(\phi)$ in $\mathbb{E}^{\mathsf{INT}}$. We reason by induction with respect to the length of **s**.

Base case. If the length is 1, then ϕ is basic. If ϕ^* is not basic, then ϕ must be basic due to the presence of $(A \supset B)^{\varrho/j}$, for some ϱ, in the succedent. Then the following is a successful Socratic transformation of $?(\phi^*)$. To minimise the size of the transformation, we put $F = A \supset B$.

$$\cfrac{\cfrac{?(\Gamma(F^{\varsigma/i})(\_) \vdash \Delta(F^{\varrho/j}))}{?(\Gamma(F^{\varsigma/i})(\_)(A^{\varrho/j,k}) \vdash \Delta(B^{\varrho/j,k}))}\ R_{\supset}}{?(\Gamma(F^{\varsigma/i})(A^{\varrho/j,k}) \vdash \Delta(B^{\varrho/j,k}, A^{k});\ \Gamma(F^{\varsigma/i})(A^{\varrho/j,k}, B^{k}) \vdash \Delta(B^{\varrho/j,k}))}\ L_{\supset}$$

Rule $L_{\supset}$ is applicable to the second question, as there is a directed $R(\psi)$-$[i,k]$-chain (where ψ is the constituent of the second question).

Inductive case. Suppose that **s** is of length $n+1$, and that it starts with the following application of $L_{\supset}$ (again, $F = A \supset B$).

$$\frac{?(\Gamma(F^{\varsigma/i})(F^j) \vdash \Delta)}{?(\Gamma(F^{\varsigma/i})(F^j) \vdash \Delta(A^j);\ \Gamma(F^{\varsigma/i})(F^j, B^j) \vdash \Delta)}\ L_{\supset}$$

The question-conclusion has a successful Socratic transformation of length n. By Corollary 4.7, each constituent of the question has a successful Socratic transformation of length not exceeding n. IH applies to these questions, therefore there is a successful Socratic transformation of question $?(\Gamma((A \supset B))^{\varsigma/i}) \vdash \Delta(A^j))$ in $\mathbb{E}^{\mathsf{INT}}$, and another one of question $?(\Gamma((A \supset B)^{\varsigma/i})(B^j) \vdash \Delta)$. By Lemma 4.3, there is a successful Socratic transformation of question $Q = ?(\Gamma((A \supset B))^{\varsigma/i}) \vdash \Delta(A^j)\,;\ \Gamma((A \supset B)^{\varsigma/i})(B^j) \vdash \Delta)$ in $\mathbb{E}^{\mathsf{INT}}$. Since $\langle i, j\rangle \in R(\phi) = R(\phi^*)$, $?(\phi^*)/Q$ is an instance of $L_{\supset}$, thus by Corollary 4.5, there is a successful Socratic transformation of $?(\phi^*)$ in $\mathbb{E}^{\mathsf{INT}}$.

In the case of any other application of a rule to the first question of **s**, we use IH (and separation results, if the rule was branching) and then apply the same rule in the same manner to receive a successful Socratic transformation of $?(\phi^*)$ in $\mathbb{E}^{\mathsf{INT}}$. □

Lemma 4.9 *Let $\phi = \Gamma((\sim A)^{\varsigma/i})((\sim A)^j) \vdash \Delta$. If there is a successful Socratic transformation of $?(\phi)$ in $\mathbb{E}^{\mathsf{INT}}$ and $\langle i, j\rangle \in R(\phi)$, then there is a successful Socratic transformation of the question $?(\Gamma((\sim A)^{\varsigma/i})(\_) \vdash \Delta)$ in $\mathbb{E}^{\mathsf{INT}}$.*

Proof The proof goes along the same scheme as the previous one; we skip it. □

4.3 Completeness of $\mathbb{E}^{\mathsf{INT}}$

The key notion to start with in this section is that of intuitionistic consistency property. As in the modal case, for each set $X \subseteq FORM_{\mathsf{INT}}$, we define the set of formulas from X that will be carried to the accessible worlds.

$$(X)^{\sharp\mathsf{INT}} = \{p_n : p_n \in X\} \cup \{A \supset B : A \supset B \in X\} \cup \{\sim A : \sim A \in X\}$$

The analogues of Corollaries 3.19 and 3.25 hold.

Corollary 4.14 *Let X and Y be sets of formulas of $\mathcal{L}_{\mathsf{INT}}$. The following holds:*

1. *If $X \subseteq Y$, then $(X)^{\sharp\mathsf{INT}} \subseteq (Y)^{\sharp\mathsf{INT}}$.*
2. *$(X \cup Y)^{\sharp\mathsf{INT}} = (X)^{\sharp\mathsf{INT}} \cup (Y)^{\sharp\mathsf{INT}}$.*
3. *$(X)^{\sharp\mathsf{INT}} \subseteq X$.*
4. *If $Y \subseteq (X)^{\sharp\mathsf{INT}}$, then $(Y)^{\sharp\mathsf{INT}} = Y$.*

Proof Clauses 3 and 4 follow from the definition of $(\bullet)^{\sharp\mathsf{INT}}$. □

The INT-consistency properties defined below correspond to partitions of language. The intuition is that in the first set we store formulas shown to be true (proved), and in the second set—those shown to be false (disproved).[2] For this reason the following definition concerns families of pairs of sets of formulas. Our solution was inspired by the ideas developed in [3, Chap. 9], where the author uses formulas signed with 'T' and 'F'. The two sets in our consistency properties correspond to the truth-signs.

Definition 4.10 (INT-*consistency Property)* Let $\mathcal{C}$ stand for a family of pairs of sets of formulas of $\mathcal{L}_{\mathsf{INT}}$. We say that $\mathcal{C}$ *is an intuitionistic consistency property*, INT-*consistency property* for short, if for each pair $\langle X_L, X_R \rangle \in \mathcal{C}$ of sets of formulas,

1. If $p_i \in X_L$, then $p_i \notin X_R$;
2. If $A \wedge B \in X_L$, then $\langle X_L \cup \{A, B\}, X_R \rangle \in \mathcal{C}$;
3. If $A \wedge B \in X_R$, then $\langle X_L, X_R \cup \{A\} \rangle \in \mathcal{C}$ or $\langle X_L, X_R \cup \{B\} \rangle \in \mathcal{C}$;
4. If $A \vee B \in X_L$, then $\langle X_L \cup \{A\}, X_R \rangle \in \mathcal{C}$ or $\langle X_L \cup \{B\}, X_R \rangle \in \mathcal{C}$;
5. If $A \vee B \in X_R$, then $\langle X_L, X_R \cup \{A, B\} \rangle \in \mathcal{C}$;
6. If $\sim A \in X_L$, then $\langle X_L, X_R \cup \{A\} \rangle \in \mathcal{C}$;
7. If $A \supset B \in X_L$, then $\langle X_L, X_R \cup \{A\} \rangle \in \mathcal{C}$ or $\langle X_L \cup \{B\}, X_R \rangle \in \mathcal{C}$;
8. If $\sim A \in X_R$, then $\langle (X_L)^{\sharp\mathsf{INT}} \cup \{A\}, \varnothing \rangle \in \mathcal{C}$;
9. If $A \supset B \in X_R$, then $\langle (X_L)^{\sharp\mathsf{INT}} \cup \{A\}, \{B\} \rangle \in \mathcal{C}$.

The first clause of the above definition expresses consistency at the level of propositional variables. Clauses 2–5 are self-explanatory. Clauses 6 and 7 capture reflexivity of the accessibility relation in intuitionistic models, whereas clauses 8 and 9 are analogues of clause 5 in Definitions 3.15 and 3.24 of modal consistency properties—they settle the accessibility relation. The heredity condition and the 'modal' character of our 'intuitionistic necessities' are captured in the definition of $(\bullet)^{\sharp\mathsf{INT}}$.

As in the previous cases, we need to close intuitionistic consistency properties under subsets and finite character. Then both X_L and X_R in a pair could be extended to sets in families closed under chain unions. We need to adjust our terminology to pairs of sets, but the core meaning of the concepts is retained.

Hence as before, the phrase *set X^* (family $\mathcal{F}^*$) extends set X (family $\mathcal{F}$)* means that $X \subseteq X^*$ (respectively, $\mathcal{F} \subseteq \mathcal{F}^*$). Moreover, we shall say that a pair of sets,

---

[2] It is worth to observe that the 'right-sets' X_R of INT-consistency properties correspond to what may be called 'dual Hintikka sets'. An analogue of this term was used already in [5]. In [1] the author uses 'refutability properties' to deal with dual erotetic calculi (see also Sect. 6.3 in Chap. 6 of this monograph). Interestingly, the term 'refutability properties' was used by Raymond Smullyan in his book [7].

$\langle X_L^*, X_R^*\rangle$, $\sqsubseteq$-*extends* another pair of sets, $\langle X_L, X_R\rangle$ iff $X_L \subseteq X_L^*$ and $X_R \subseteq X_R^*$, symbolically

$$\langle X_L, X_R\rangle \sqsubseteq \langle X_L^*, X_R^*\rangle \text{ iff } X_L \subseteq X_L^* \text{ and } X_R \subseteq X_R^*.$$

If $\mathcal{F}$ is a family of pairs of sets, then we say that it is $\sqsubseteq$-*subset closed*, provided that for each pair $\langle X_L, X_R\rangle \in \mathcal{C}$, family $\mathcal{F}^L_{X_R} = \{Y_L : \langle Y_L, X_R\rangle \in \mathcal{F}\}$ is subset closed and family $\mathcal{F}^R_{X_L} = \{Y_R : \langle X_L, Y_R\rangle \in \mathcal{F}\}$ is subset closed. We will write '$\sqsubseteq$-closed' instead of '$\sqsubseteq$-subset closed', for short. From these definitions it follows that:

Corollary 4.15 *A family $\mathcal{F}$ of pairs of sets is $\sqsubseteq$-closed iff if a pair $P_1 \in \mathcal{F}$ $\sqsubseteq$-extends a pair P_2, then $P_2 \in \mathcal{F}$.*

Proof Suppose that $\mathcal{F}$ is $\sqsubseteq$-closed and that a pair $P_1 = \langle X_L, X_R\rangle \in \mathcal{F}$ $\sqsubseteq$-extends a pair $P_2 = \langle Y_L, Y_R\rangle$. Then, first, $Y_L \in \mathcal{F}^L_{X_R}$, as $\mathcal{F}^L_{X_R}$ is subset closed, and hence $\langle Y_L, X_R\rangle \in \mathcal{F}$. Second, set $\mathcal{F}^R_{Y_L}$ is also subset closed, therefore $Y_R \in \mathcal{F}^R_{Y_L}$, and hence $P_2 \in \mathcal{F}$. This finishes the reasoning for the if-then direction.

For the other direction suppose that (a) if $P_1 \in \mathcal{F}$ $\sqsubseteq$-extends P_2, then $P_2 \in \mathcal{F}$. Let $\langle X_L, X_R\rangle \in \mathcal{F}$ be arbitrary, let $Y \in \mathcal{F}^L_{X_R}$ and suppose that $Z \subseteq Y$. Since pair $\langle Y, X_R\rangle$ $\sqsubseteq$-extends pair $\langle Z, X_R\rangle$, the latter pair belongs to $\mathcal{F}$ by (a), which proves that $\mathcal{F}^L_{X_R}$ is subset closed. We reason analogously for $\mathcal{F}^R_{X_L}$. □

Similarly, we say that a family $\mathcal{F}$ of pairs of sets is of *finite $\sqsubseteq$-character*, provided that for each pair $\langle X_L, X_R\rangle \in \mathcal{F}$, family $\mathcal{F}^L_{X_R} = \{Y_L : \langle Y_L, X_R\rangle \in \mathcal{F}\}$ is of finite character and family $\mathcal{F}^R_{X_L} = \{Y_R : \langle X_L, Y_R\rangle \in \mathcal{F}\}$ is of finite character.

Corollary 4.16 *Each family of pairs of sets of finite $\sqsubseteq$-character is $\sqsubseteq$-closed.*

Proof Let $\mathcal{F}$ be a family of pairs of sets which is of finite $\sqsubseteq$-character. Suppose that $\langle X, Y\rangle \in \mathcal{F}$. By assumption, families $\mathcal{F}^L_Y$ and $\mathcal{F}^R_X$ are of finite character. By Corollary 3.20, both families are subset closed. Thus we have shown that $\mathcal{F}$ is $\sqsubseteq$-closed. □

Corollary 4.17 *Each* INT*-consistency property can be extended to an* INT*-consistency property that is $\sqsubseteq$-closed.*

Proof This corollary states that for an INT-consistency property $\mathcal{C}$ we can always indicate an INT-consistency property $\mathcal{C}^*$ such that (i) $\mathcal{C} \subseteq \mathcal{C}^*$, and (ii) for each pair $\langle X_L, X_R\rangle \in \mathcal{C}^*$, if $X \subseteq X_L$ and $Y \subseteq X_R$, then $\langle X, Y\rangle \in \mathcal{C}^*$.

Construction of $\mathcal{C}^*$ is straightforward. For each $\langle X_L, X_R\rangle \in \mathcal{C}$, for every $X \subseteq X_L$ and $Y \subseteq X_R$ we put $\langle X, Y\rangle \in \mathcal{C}^*$. We need to verify that $\mathcal{C}^*$ satisfies Definition 4.10.

If there is a pair $\langle X, Y\rangle \in \mathcal{C}^*$ with $p_i \in X \cap Y$, then for some $\langle X_L, X_R\rangle \in \mathcal{C}$ such that $X_L \supseteq X$ and $X_R \supseteq Y$, also $p_i \in X_L \cap X_R$—a contradiction.

Assume $A \wedge B \in X$, $\langle X, Y\rangle \in \mathcal{C}^*$. Then for some $\langle X_L, X_R\rangle \in \mathcal{C}$, we have $X \subseteq X_L$ and $Y \subseteq X_R$, hence also $\langle X_L \cup \{A, B\}, X_R\rangle \in \mathcal{C}$ by clause 2 of Definition 4.10. Since $X \cup \{A, B\} \subseteq X_L$, also $\langle X \cup \{A, B\}, Y\rangle \in \mathcal{C}^*$ by assumption. The reasoning is very similar for clauses 3–5 and we skip it.

Assume that $\sim A \in X$ where $\langle X, Y\rangle \in \mathcal{C}^*$. Then we have $\sim A \in X_L$, for some $\langle X_L, X_R\rangle \in \mathcal{C}$ $\sqsubseteq$-extending $\langle X, Y\rangle$. Then $\langle X_L, X_R \cup \{A\}\rangle \in \mathcal{C}$ by clause 6 of Definition 4.10. Since $X \subseteq X_L$ and $Y \cup \{A\} \subseteq X_R \cup \{A\}$, also $\langle X, Y \cup \{A\}\rangle \in \mathcal{C}^*$.

Assume that $A \supset B \in X$, $\langle X, Y\rangle \in \mathcal{C}^*$. There is $\langle X_L, X_R\rangle \in \mathcal{C}$ $\sqsubseteq$-extending $\langle X, Y\rangle$, hence $\langle X_L, X_R \cup \{A\}\rangle \in \mathcal{C}$ or $\langle X_L \cup \{B\}, X_R\rangle \in \mathcal{C}$. As $Y \cup \{A\} \subseteq X_R \cup \{A\}$ and $X \cup \{B\} \subseteq X_L \cup \{B\}$, it follows that $\langle X, Y \cup \{A\}\rangle \in \mathcal{C}^*$ or $\langle X \cup \{B\}, Y\rangle \in \mathcal{C}^*$.

Assume that $\sim A \in X$, where $\langle X, Y\rangle \in \mathcal{C}^*$. There is $\langle X_L, X_R\rangle \in \mathcal{C}$ $\sqsubseteq$-extending $\langle X, Y\rangle$, which is such that $\langle (X_L)^{\sharp\mathsf{INT}} \cup \{A\}, \varnothing\rangle \in \mathcal{C}$. Since $X \subseteq X_L$, also $(X)^{\sharp\mathsf{INT}} \cup \{A\} \subseteq (X_L)^{\sharp\mathsf{INT}} \cup \{A\}$ by clause 1 of Corollary 4.14. Needless to say, $\varnothing \subseteq \varnothing$, therefore $\langle (X)^{\sharp\mathsf{INT}} \cup \{A\}, \varnothing\rangle \in \mathcal{C}^*$, as required. The reasoning is similar in the case of clause 9. □

If an INT-consistency property is $\sqsubseteq$-closed, then it is easy to prove that clause 1 of Definition 4.10 extends to all formulas of $\mathcal{L}_{\mathsf{INT}}$.

Corollary 4.18 *If $\mathcal{C}$ is a $\sqsubseteq$-closed INT-consistency property, then for each $\langle X, Y\rangle \in \mathcal{C}$, $X \cap Y = \varnothing$.*

Proof We prove by induction on rank of a formula A, that if $A \in X$, then $A \notin Y$. The base case follows directly from clause 1 of Definition 4.10.

Let $r(A) = r + 1$. By IH, for each pair $\langle X, Y\rangle \in \mathcal{C}$ and every formula F such that $r(F) \leqslant r$, if $F \in X$, then $F \notin Y$.

Assume that A is of the form '$B \vee C$'. Since $A \in X$, $\langle X \cup \{A\}, Y\rangle \in \mathcal{C}$ or $\langle X \cup \{B\}, Y\rangle \in \mathcal{C}$. Suppose that $\langle X \cup \{A\}, Y\rangle \in \mathcal{C}$. By IH, $\langle X \cup \{A\}, Y \cup \{A\}\rangle \notin \mathcal{C}$. As $\mathcal{C}$ is $\sqsubseteq$-closed, also $\langle X \cup \{A\}, Y \cup \{A, B\}\rangle \notin \mathcal{C}$. It follows that if $\langle X \cup \{A\}, Y\rangle \in \mathcal{C}$, then $A \vee B \notin Y$. We show in a similar way that $\langle X \cup \{B\}, Y\rangle \in \mathcal{C}$ implies $A \vee B \notin Y$. The reasoning is analogous for A of the form '$B \wedge C$'.

Assume that A is of the form '$\sim B$'. This time we reason towards a contradiction. If $\sim B \in Y$, then by clause 8 of Definition 4.10, $\langle (X)^{\sharp\mathsf{INT}} \cup \{B\}, \varnothing\rangle \in \mathcal{C}$. However, $\sim B \in (X)^{\sharp\mathsf{INT}}$, therefore by clause 6 of the same definition, $\langle (X)^{\sharp\mathsf{INT}} \cup \{B\}, \{B\}\rangle \in \mathcal{C}$. We arrive at a contradiction with IH, therefore if $\sim B \in X$, then $\sim B \notin Y$. The reasoning for A of the form '$B \supset C$' is very similar and we skip it. □

Let us also observe, on the margin, that Corollaries 4.17 and 4.18 entail that in every INT-consistency property clause 1 of Definition 4.10 extends to all formulas.

In the case of the presented construction the proof of the following Lemma 4.10—extending $\sqsubseteq$-closed INT-consistency properties to INT-consistency properties of finite $\sqsubseteq$-character—becomes more complicated in details, although the underlying ideas are more or less then same as in the previously examined modal cases.

Lemma 4.10 *Each $\sqsubseteq$-closed INT-consistency property can be extended to an INT-consistency property of finite $\sqsubseteq$-character.*

Proof Let $\mathcal{C}$ be a $\sqsubseteq$-closed INT-consistency property. We extend $\mathcal{C}$ in two separate steps. Let $\mathcal{C}^*$ be such that $\langle X, Y\rangle \in \mathcal{C}^*$ iff for every finite subset $X^f \subseteq X$, $\langle X^f, Y\rangle \in \mathcal{C}$. Once we have $\mathcal{C}^*$, let $\mathcal{C}^{**}$ be such that $\langle X, Y\rangle \in \mathcal{C}^{**}$ iff for every finite subset $Y^f \subseteq Y$, $\langle X, Y^f\rangle \in \mathcal{C}^*$.

First, we need to show that $\mathcal{C} \subseteq \mathcal{C}^{**}$, that is, that $\mathcal{C}^{**}$ extends $\mathcal{C}$. Analogously to the cases analyzed in Chaps 2 and 3, it shows up that this is a consequence of the fact that $\mathcal{C}$ is $\sqsubseteq$-closed; the reasoning, however, is more complicated in this case. Assume that $\langle X, Y\rangle \in \mathcal{C}$. Let $X^f \subseteq X$ be finite. It is a subset of X, thus $\langle X, Y\rangle$ $\sqsubseteq$-extends $\langle X^f, Y\rangle$. As $\mathcal{C}$ is $\sqsubseteq$-closed, $\langle X^f, Y\rangle \in \mathcal{C}$ (see Corollary 4.15). This is the case for every finite subset X^f of X, hence $\langle X, Y\rangle \in \mathcal{C}^*$. We have shown that $\mathcal{C} \subseteq \mathcal{C}^*$.

Let $\langle X, Y\rangle \in \mathcal{C}^*$ be arbitrary. By the construction of $\mathcal{C}^*$, $\langle X^f, Y\rangle \in \mathcal{C}$ for every finite subset X^f of X. Since $\mathcal{C}$ is $\sqsubseteq$-closed, for each X^f also $\langle X^f, Y^*\rangle \in \mathcal{C}$ for every subset Y^* of Y. Let Y^f be arbitrary but fixed finite subset of Y. We have established that for each finite $X^f \subseteq X$, $\langle X^f, Y^f\rangle \in \mathcal{C}$. It follows that $\langle X, Y^f\rangle \in \mathcal{C}^*$. However, this reasoning applies to every Y^f. It means that $\langle X, Y\rangle \in \mathcal{C}^{**}$. We have shown that $\mathcal{C}^* \subseteq \mathcal{C}^{**}$.

Second, we show that $\mathcal{C}^{**}$ is of finite $\sqsubseteq$-character. Let $\langle X, Y\rangle \in \mathcal{C}^{**}$ be arbitrary. We need to show that families of sets $\mathcal{C}^{**L}_{Y} = \{Y_L : \langle Y_L, Y\rangle \in \mathcal{C}^{**}\}$ and $\mathcal{C}^{**R}_{X} = \{Y_R : \langle X, Y_R\rangle \in \mathcal{C}^{**}\}$ are of finite character. Hence let (a) $Y^* \in \mathcal{C}^{**R}_{X}$ (the reasoning is simpler for this set). Assumption (a) is equivalent to $\langle X, Y^*\rangle \in \mathcal{C}^{**}$, and by the construction of $\mathcal{C}^{**}$, this holds iff (b) for every finite $Y^f \subseteq Y^*$, $\langle X, Y^f\rangle \in \mathcal{C}^*$. Since $\mathcal{C}^* \subseteq \mathcal{C}^{**}$, the above reasoning shows that if (a), then for every finite $Y^f \subseteq Y^*$, $\langle X, Y^f\rangle \in \mathcal{C}^{**}$, which is equivalent to $Y^f \in \mathcal{C}^{**R}_{X}$. On the other hand, if we assume that (c) for each finite $Y^f \subseteq Y^*$, $Y^f \in \mathcal{C}^{**R}_{X}$, then by the construction of $\mathcal{C}^{**}$, for each finite $Y^f \subseteq Y^*$, for every finite $Y^{ff} \subseteq Y^f$, $\langle X, Y^{ff}\rangle \in \mathcal{C}^*$, but this just means that for each finite $Y^{ff} \subseteq Y^*$, $\langle X, Y^{ff}\rangle \in \mathcal{C}^*$, that is, (b) holds. We have established above that (b) holds iff (a) holds, hence (c) implies (a), which was to be proved.

For the finite character of set $\mathcal{C}^{**L}_{Y}$, we establish in an analogous way that (a) $X^* \in \mathcal{C}^{**L}_{Y}$ iff (b) for each finite $Y^f \subseteq Y$, $\langle X^*, Y^f\rangle \in \mathcal{C}^*$. By the construction of $\mathcal{C}^*$, (b) holds iff (c) for each finite $Y^f \subseteq Y$ and for every finite $X^f \subseteq X^*$, $\langle X^f, Y^f\rangle \in \mathcal{C}$. Since $\mathcal{C} \subseteq \mathcal{C}^*$, the above reasoning shows that if (a), then for an arbitrary but fixed finite $X^f \subseteq X^*$, we have that for each finite $Y^f \subseteq Y$, $\langle X^f, Y^f\rangle \in \mathcal{C}^*$, which entails $\langle X^f, Y\rangle \in \mathcal{C}^{**}$. Therefore (a) implies that (d) for each finite $X^f \subseteq X^*$, $X^f \in \mathcal{C}^{**L}_{Y}$. On the other hand, if we assume (d), then by the construction of $\mathcal{C}^{**}$, (e) for each finite $X^f \subseteq X^*$, for every finite $Y^f \subseteq Y$, $\langle X^f, Y^f\rangle \in \mathcal{C}^*$. In turn, by the construction of $\mathcal{C}^*$, (e) means that (f) for each finite subset X^{ff} of X^f, for each finite $X^f \subseteq X^*$, for every finite $Y^f \subseteq Y$, $\langle X^{ff}, Y^f\rangle \in \mathcal{C}$. As X^f is a finite subset of itself, (f) entails that (g) for every finite $Y^f \subseteq Y$, $\langle X^f, Y^f\rangle \in \mathcal{C}$ holds for each finite $X^f \subseteq X^*$. By the construction of $\mathcal{C}^*$, (g) implies (b). Hence (d) implies (a), as required.

In the next part of the proof we verify clauses of Definition 4.10. Let us recall that $\mathcal{C}$ is a $\sqsubseteq$-closed INT-consistency property.

Suppose that $\langle X, Y\rangle \in \mathcal{C}^{**}$ and $p_i \in X \cap Y$. Then by the construction of $\mathcal{C}^{**}$, $\langle X, \{p_i\}\rangle \in \mathcal{C}^*$, and by the construction of $\mathcal{C}^*$, $\langle p_i, p_i\rangle \in \mathcal{C}$, which is a contradiction.

From among conditions 2–5 of Definition 4.10 we verify only 4—it happens to be the most complicated of the four. We start the reasoning by showing that the desired effect, expressed by (i) below, holds for $\mathcal{C}^*$, and then we shall proceed with $\mathcal{C}^{**}$.

(i) if (a) $A \vee B \in X$, (b) $\langle X, Y\rangle \in \mathcal{C}^*$, and (c) $\langle X \cup \{A\}, Y\rangle \notin \mathcal{C}^*$, then $\langle X \cup \{B\}, Y\rangle \in \mathcal{C}^*$.

From (c) it follows that there is a finite $X^* \subseteq X \cup \{A\}$ such that $\langle X^*, Y\rangle \notin \mathcal{C}$. Let $X^{**} = (X^* \backslash \{A\}) \cup \{A \vee B\}$. Set X^{**} is a finite subset of X, hence, by (b), $\langle X^{**}, Y\rangle \in \mathcal{C}$. By (a), hence also $\langle X^{**} \cup \{A\}, Y\rangle \in \mathcal{C}$ or $\langle X^{**} \cup \{B\}, Y\rangle \in \mathcal{C}$, as $\mathcal{C}$ is an INT-consistency property. But $\mathcal{C}$ is $\sqsubseteq$-closed, therefore if $\langle X^{**} \cup \{A\}, Y\rangle \in \mathcal{C}$, then $\langle X^*, Y\rangle \in \mathcal{C}$. It follows that (d) $\langle X^{**} \cup \{A\}, Y\rangle \notin \mathcal{C}$.

Now let $X^f \subseteq X$ be an arbitrary but fixed, finite subset of X. Set $X^f \cup X^{**}$ is a finite subset of X, hence $\langle X^f \cup X^{**}, Y\rangle \in \mathcal{C}$. Due to (d), $\langle X^f \cup X^{**} \cup \{A\}, Y\rangle \notin \mathcal{C}$, therefore $\langle X^f \cup X^{**} \cup \{B\}, Y\rangle \in \mathcal{C}$. It follows that $\langle X^f \cup \{B\}, Y\rangle \in \mathcal{C}$. As X^f was arbitrary, we have shown that $\langle X \cup \{B\}, Y\rangle \in \mathcal{C}^*$, which finishes the demonstration for (i).

Another thing to prove is:

(ii) If (a) $\langle X, Y\rangle \in \mathcal{C}^*$ and (b) $Y^* \subseteq Y$, then $\langle X, Y^*\rangle \in \mathcal{C}^*$.

By (a), for each finite $X^f \subseteq X$, $\langle X^f, Y\rangle \in \mathcal{C}$. As $\mathcal{C}$ is $\sqsubseteq$-closed, also $\langle X^f, Y^*\rangle \in \mathcal{C}$. This holds for every finite $X^f \subseteq X$, thus $\langle X, Y^*\rangle \in \mathcal{C}^*$, as required.

Now let us assume that $A \vee B \in X$ and $\langle X, Y\rangle \in \mathcal{C}^{**}$. It follows that for each finite $Y^f \subseteq Y$, $\langle X, Y^f\rangle \in \mathcal{C}^*$. Assume that $\langle X \cup \{A\}, Y\rangle \notin \mathcal{C}^{**}$. Then for some finite $Y^* \subseteq Y$, (a) $\langle X \cup \{A\}, Y^*\rangle \notin \mathcal{C}^*$. Let $Y^f \subseteq Y$ be arbitrary but fixed, finite subset of Y. $Y^f \cup Y^* \subseteq Y$, hence $\langle X, Y^f \cup Y^*\rangle \in \mathcal{C}^*$. By (ii) and (a), $\langle X \cup \{A\}, Y^f \cup Y^*\rangle \notin \mathcal{C}^*$. By (i), $\langle X \cup \{B\}, Y^f \cup Y^*\rangle \in \mathcal{C}^*$. By (ii) again, $\langle X \cup \{B\}, Y^f\rangle \in \mathcal{C}^*$. As Y^f was arbitrary, $\langle X \cup \{B\}, Y\rangle \in \mathcal{C}^{**}$. We have shown that $\mathcal{C}^{**}$ satisfies clause 4 of Definition 4.10.

Clause 6. Let $\langle X, Y\rangle \in \mathcal{C}^{**}$ and $\sim A \in X$. It follows that for every finite $Y_f \subseteq Y$, $\langle X, Y^f\rangle \in \mathcal{C}^*$, and hence further, that for every finite $Y_f \subseteq Y$, for each finite $X^f \subseteq X$, $\langle X^f \cup \{\sim A\}, Y^f\rangle \in \mathcal{C}$, as $X^f \cup \{\sim A\}$ is a finite subset of X. As $\mathcal{C}^*$ is an INT-consistency property, it follows that for every finite $Y_f \subseteq Y$, for every finite $X^f \subseteq X$, $\langle X^f \cup \{\sim A\}, Y^f \cup \{A\}\rangle \in \mathcal{C}$. By assumption, $\mathcal{C}$ is $\sqsubseteq$-closed, therefore also for every finite $Y_f \subseteq Y$, for every finite $X^f \subseteq X$, $\langle X^f, Y^f \cup \{A\}\rangle \in \mathcal{C}$. Hence for every finite $Y_f \subseteq Y$, $\langle X, Y^f \cup \{A\}\rangle \in \mathcal{C}^*$. In other words, for every finite $Y_f \subseteq Y \cup \{A\}$, $\langle X, Y^f\rangle \in \mathcal{C}^*$, hence $\langle X, Y \cup \{A\}\rangle \in \mathcal{C}^{**}$.

For clause 7 the reasoning is analogous to that for clause 4, where (i) takes the form:

(i) If $A \supset B \in X$, $\langle X, Y\rangle \in \mathcal{C}^*$, and $\langle X, Y \cup \{A\}\rangle \notin \mathcal{C}^*$, then $\langle X \cup \{B\}, Y\rangle \in \mathcal{C}^*$.

We skip the rest.

Clause 8. We shall be more sketchy this time. Let $\sim A \in Y$ and $\langle X, Y\rangle \in \mathcal{C}^{**}$. Then $\langle X, \{\sim A\}\rangle \in \mathcal{C}^*$. Let $X^f \subseteq (X)^{\sharp \mathsf{INT}}$ be an arbitrary but fixed, finite subset of $(X)^{\sharp \mathsf{INT}}$. By clause 3, Corollary 4.14, $\langle X^f, \{\sim A\}\rangle \in \mathcal{C}$. By clause 8, Definition 4.10, plus clause 4, Corollary 4.14, $\langle X^f \cup \{A\}, \varnothing\rangle \in \mathcal{C}$. As X^f was arbitrary, $\langle (X)^{\sharp \mathsf{INT}} \cup \{A\}, \varnothing\rangle \in \mathcal{C}^*$. Since $\mathcal{C}^* \subseteq \mathcal{C}^{**}$, $\langle (X)^{\sharp \mathsf{INT}} \cup \{A\}, \varnothing\rangle \in \mathcal{C}^{**}$, which was to be proved.

We prove clause 9 in an analogous manner. We assume that $A \supset B \in Y$, $\langle X, Y\rangle \in \mathcal{C}^{**}$, and establish that then for every finite $Y^f \subseteq Y$, for each finite $X^f \subseteq (X)^{\sharp \mathsf{INT}}$, $\langle X^f, Y^f \cup \{A \supset B\}\rangle \in \mathcal{C}$. Hence $\langle X^f \cup \{A\}, \{B\}\rangle \in \mathcal{C}$. As X^f is arbitrary, $\langle (X)^{\sharp \mathsf{INT}} \cup \{A\}, \{B\}\rangle \in \mathcal{C}^* \subseteq \mathcal{C}^{**}$. □

Corollary 4.17 and Lemma 4.10 yield the following result.

Corollary 4.19 *Each* INT*-consistency property can be extended to an* INT*-consistency property of finite* $\sqsubseteq$*-character.*

Definition 4.11 *(Downward-Saturated Sets in* INT*:the Classical Propositional Connectives)* A set X of formulas of the language $\mathcal{L}_{\mathsf{INT}}$ is *downward-left-saturated* iff

1. If $A \wedge B \in X$, then $A \in X$ and $B \in X$, and
2. If $A \vee B \in X$, then $A \in X$ or $B \in X$.

Moreover, X is called *downward-right-saturated* iff

1. If $A \wedge B \in X$, then $A \in X$ or $B \in X$, and
2. If $A \vee B \in X$, then $A \in X$ and $B \in X$.

Downward-left-saturated (downward-right-saturated) sets will be called *dl-saturated* (*dr-saturated*), for short.

Let us recall that according to Corollary 3.21, each family of finite character is closed under chain unions. This fact will be used in the proof of the next lemma.

Lemma 4.11 *Let $\mathcal{C}$ be an* INT*-consistency property of finite* $\sqsubseteq$*-character. For each pair $\langle X, Y\rangle \in \mathcal{C}$ there is a dl-saturated set $\mathbb{X}$ extending X and a dr-saturated set $\mathbb{Y}$ extending Y such that $\langle \mathbb{X}, \mathbb{Y}\rangle \in \mathcal{C}$.*

Proof Suppose that $\mathcal{C}$ is an INT-consistency property of finite $\sqsubseteq$-character, and let $\langle X, Y\rangle \in \mathcal{C}$. Let

$$F_0, F_1, \ldots, F_k, \ldots \tag{4.4}$$

be an ordering of all formulas of $\mathcal{L}_{\mathsf{INT}}$. We define sequence $(\langle X_k, Y_k\rangle)_\infty$ of pairs of sets as follows.

$X_0 = X$, $Y_0 = Y$ and for each $k \in \mathbb{N}_0$:

$$X_{k+1} = \begin{cases} X_k \cup \{F_k\} & \text{if } \langle X_k \cup \{F_k\}, Y_k\rangle \in \mathcal{C}, \\ X_k & \text{otherwise.} \end{cases}$$

$$Y_{k+1} = \begin{cases} Y_k \cup \{F_k\} & \text{if } \langle X_k, Y_k \cup \{F_k\}\rangle \in \mathcal{C}, \\ Y_k & \text{otherwise.} \end{cases}$$

Finally, let:

$$\mathbb{X} = \bigcup_{k=0}^{\infty} X_k \text{ and } \mathbb{Y} = \bigcup_{k=0}^{\infty} Y_k.$$

Naturally, $X \subseteq \mathbb{X}$ and $Y \subseteq \mathbb{Y}$.

We need to show that (1) $\langle \mathbb{X}, \mathbb{Y}\rangle \in \mathcal{C}$, and (2) $\mathbb{X}$ is dl-saturated and $\mathbb{Y}$ is dr-saturated.

Proof of (1). First, consider

$$\mathcal{F}^L_Y = \{X_L : \langle X_L, Y\rangle \in \mathcal{C}\} \text{ and } \mathcal{F}^R_X = \{Y_R : \langle X, Y_R\rangle \in \mathcal{C}\}$$

By Corollary 4.16, for each X_k, $\langle X_k, Y\rangle \in \mathcal{C}$ and for each Y_k, $\langle X, Y_k\rangle \in \mathcal{C}$. By the same token, for each finite $X^f_k \subseteq X_k$, $\langle X^f_k, Y\rangle \in \mathcal{C}$, and analogously for finite subsets of Y_k. $\mathcal{C}$ is of finite $\sqsubseteq$-character, therefore $\mathbb{X} \in \mathcal{F}^L_Y$ and $\mathbb{Y} \in \mathcal{F}^R_X$. It follows that $\langle \mathbb{X}, Y\rangle$, $\langle X, \mathbb{Y}\rangle \in \mathcal{C}$. Let us now consider family $\mathcal{F}^R_{X_k}$. By the above construction, $Y = Y_0 \in \mathcal{F}^R_{X_k}$ and $Y_k \in \mathcal{F}^R_{X_k}$. Also $\langle X_k, Y_k\rangle \sqsubseteq \langle X_{k+1}, Y_{k+1}\rangle$, hence, and by Corollary 4.16, $Y_{k+1} \in \mathcal{F}^R_{X_k}$. It follows, by induction, that for each $l \geqslant 0$, $Y_l \in \mathcal{F}^R_{X_k}$. By the fact that $\mathcal{F}^R_{X_k}$ is of finite character, by Corollary 3.21 (chain unions), $\mathbb{Y} \in \mathcal{F}^R_{X_k}$. But then also $X_k \in \mathcal{F}^L_{\mathbb{Y}}$. As k was arbitrary, we may repeat the same sequence of arguments: induction, the fact that $\mathcal{F}^L_{\mathbb{Y}}$ is of finite character, Corollary 3.21—to the effect that $\mathbb{X} \in \mathcal{F}^L_{\mathbb{Y}}$. Hence it follows that (1) $\langle \mathbb{X}, \mathbb{Y}\rangle \in \mathcal{C}$.

Proof of (2). Suppose that (a) $A \wedge B \in \mathbb{X}$. Formulas $A \wedge B$, A and B occur in ordering (4.4), let $F_k = A \wedge B$, $F_i = A$, $F_j = B$. By (a) $X_{k+1} = X_k \cup \{A \wedge B\}$, that is, $\langle X_k \cup \{A \wedge B\}, Y_k\rangle \in \mathcal{C}$. Since $\mathcal{C}$ is an INT-consistency property, $\langle X_k \cup \{A \wedge B, A, B\}, Y_k\rangle \in \mathcal{C}$. If $i < k$, then $A \in X_k$. If $i > k$, then we consider set X_i: we have $A \wedge B \in X_i$, therefore $\langle X_i \cup \{A, B\}, Y_i\rangle \in \mathcal{C}$. $\mathcal{C}$ is $\sqsubseteq$-closed, therefore also $\langle X_i \cup \{A\}, Y_i\rangle \in \mathcal{C}$. This yields that $X_{i+1} = X_i \cup \{A\}$, and hence $A \in \mathbb{X}$. We show that $B \in \mathbb{X}$ in a similar manner. The reasoning is analogous in the remaining cases. □

The proof of Lemma 4.11 indicates a function that given an INT-consistency property $\mathcal{C}$ of finite $\sqsubseteq$-character and a sequence S ordering all formulas of language $\mathcal{L}_{\mathsf{INT}}$ assigns

$$\mathbb{S}^{\mathcal{C}}_S : \langle X, Y\rangle \mapsto \langle \mathbb{X}, \mathbb{Y}\rangle. \quad (4.5)$$

Definition 4.12 (INT-*superframe)* Let $\mathcal{C}$ be an INT-consistency property of finite $\sqsubseteq$-character and let S stand for a sequence ordering all formulas of the language $\mathcal{L}_{\mathsf{INT}}$. By INT-*superframe built with reference to S upon $\mathcal{C}$* we mean pair $\langle \mathbb{W}, \mathbb{R}\rangle$, such that:

$$\mathbb{W} = \{\mathbb{S}^{\mathcal{C}}_S(\langle X, Y\rangle) : \langle X, Y\rangle \in \mathcal{C}\}$$

and where $\mathbb{R} \subseteq \mathbb{W} \times \mathbb{W}$ is such that it holds between $\langle \mathbb{X}_L, \mathbb{X}_R\rangle$ and $\langle \mathbb{Y}_L, \mathbb{Y}_R\rangle$ iff $(\mathbb{X}_L)^{\sharp\mathsf{INT}} \subseteq \mathbb{Y}_L$.

Let us observe that worlds of an INT-superframe are pairs of sets of formulas. INT-superframes have properties expressed in the next lemmas.

Lemma 4.12 *Let $\mathcal{C}$ be an* INT*-consistency property of finite $\sqsubseteq$-character, let S stand for a sequence ordering all formulas of $\mathcal{L}_{\mathsf{INT}}$, and let $\langle \mathbb{W}, \mathbb{R}\rangle$ stand for an* INT*-superframe built with reference to S upon $\mathcal{C}$. Then each $\langle \mathbb{X}_L, \mathbb{X}_R\rangle \in \mathbb{W}$ satisfies the following conditions.*

1. $\langle \mathbb{X}_L, \mathbb{X}_R \rangle \in \mathcal{C}$.
2. *For every* $\sim A \in \mathbb{X}_R$, *there is* $\langle \mathbb{Y}_L, \mathbb{Y}_R \rangle \in \mathbb{R}(\langle \mathbb{X}_L, \mathbb{X}_R \rangle)$, *such that* $A \in \mathbb{Y}_L$.
3. *For every* $A \supset B \in \mathbb{X}_R$, *there is* $\langle \mathbb{Y}_L, \mathbb{Y}_R \rangle \in \mathbb{R}(\langle \mathbb{X}_L, \mathbb{X}_R \rangle)$, *such that* $A \in \mathbb{Y}_L$ *and* $B \in \mathbb{Y}_R$.
4. *Sets* $\mathbb{X}_L, \mathbb{X}_R$ *are maximal in the following sense: if* $\langle \mathbb{X}_L \cup \{A\}, \mathbb{X}_R \rangle \in \mathcal{C}$, *then* $A \in \mathbb{X}_L$*; similarly if* $\langle \mathbb{X}_L, \mathbb{X}_R \cup \{A\} \rangle \in \mathcal{C}$, *then* $A \in \mathbb{X}_R$.

Proof Clause 1 follows directly from the previous lemma.

The reasoning for clauses 2 and 3 is similar, we consider only 3.

Suppose that $A \supset B \in \mathbb{X}_R$. $\mathcal{C}$ is an INT-consistency property, therefore by 1, $\langle (\mathbb{X}_L)^{\sharp \mathsf{INT}} \cup \{A\}, \{B\} \rangle \in \mathcal{C}$. Then the required pair $\langle \mathbb{Y}_L, \mathbb{Y}_R \rangle$ is the value

$$\mathbb{S}_S^{\mathcal{C}} \left(\langle (\mathbb{X}_L)^{\sharp \mathsf{INT}} \cup \{A\}, \{B\} \rangle \right).$$

For clause 4 suppose that $\langle \mathbb{X}_L \cup \{A\}, \mathbb{X}_R \rangle \in \mathcal{C}$. Assume that formula A occurs in sequence S on position k. As $\mathcal{C}$ is $\sqsubseteq$-closed, $\langle X_k \cup \{A\}, Y_k \rangle \in \mathcal{C}$. It follows that $X_{k+1} = X_k \cup \{A\}$, therefore $A \in \mathbb{X}_L$. The reasoning is similar for $\mathbb{X}_R$. □

Lemma 4.13 *An* INT*-superframe built with reference to an ordering* S *upon a nonempty* INT*-consistency property* $\mathcal{C}$ *is an intuitionistic frame.*

Proof Let $\langle \mathbb{W}, \mathbb{R} \rangle$ be an INT-superframe built with reference to some ordering S upon a nonempty $\mathcal{C}$. Let us explain the nonemptiness condition. The definition of INT-consistency property does not preclude the case of $\mathcal{C} = \varnothing$. If we take $\mathcal{C} = \varnothing$, then Lemma 4.11 is trivially satisfied, hence in this peculiar case we would obtain an INT-superframe with empty $\mathbb{W}$. Obviously,it would not constitute an intuitionistic frame.

Since in our case $\mathbb{W} \neq \varnothing$, we need to deal only with the properties of $\mathbb{R}$. Hence let $\langle \mathbb{X}, \mathbb{Y} \rangle \in \mathbb{W}$. By clause 3 of Corollary 4.14, $(\mathbb{X})^{\sharp \mathsf{INT}} \subseteq \mathbb{X}$, hence $\mathbb{R}$ holds between $\langle \mathbb{X}, \mathbb{Y} \rangle$ and itself—$\mathbb{R}$ is reflexive.

Transitivity is also easy. Let $w_X = \langle \mathbb{X}_L, \mathbb{X}_R \rangle \in \mathbb{W}$, $w_Y = \langle \mathbb{Y}_L, \mathbb{Y}_R \rangle \in \mathbb{W}$ and $w_Z = \langle \mathbb{Z}_L, \mathbb{Z}_R \rangle \in \mathbb{W}$. If $\mathbb{R}$ holds between w_X and w_Y, and between w_Y and w_Z, then (a) $(\mathbb{X}_L)^{\sharp \mathsf{INT}} \subseteq \mathbb{Y}_L$ and (b) $(\mathbb{Y}_L)^{\sharp \mathsf{INT}} \subseteq \mathbb{Z}_L$. By clause 1 of Corollary 4.14, (a) yields (c) $((\mathbb{X}_L)^{\sharp \mathsf{INT}})^{\sharp \mathsf{INT}} \subseteq (\mathbb{Y}_L)^{\sharp \mathsf{INT}}$. By clause 4 of the same corollary, $((\mathbb{X}_L)^{\sharp \mathsf{INT}})^{\sharp \mathsf{INT}} = (\mathbb{X}_L)^{\sharp \mathsf{INT}}$, hence (c) is equivalent to (d) $(\mathbb{X}_L)^{\sharp \mathsf{INT}} \subseteq (\mathbb{Y}_L)^{\sharp \mathsf{INT}}$. Finally, (d) together with (b) entail that $\mathbb{R}$ is transitive. □

Definition 4.13 Let $\langle X, Y \rangle$ be a pair of sets of formulas of $\mathcal{L}_{\mathsf{INT}}$ and let $\langle W, R, \Vdash \rangle$ be an intuitionistic model. Assume also $w \in W$. We say that $\langle X, Y \rangle$ *is satisfied in world* w *of* $\langle W, R, \Vdash \rangle$ iff $w \Vdash X$ and for each $A \in Y$, $w \nVdash A$. We say that $\langle X, Y \rangle$ *is satisfied in model* $\langle W, R, \Vdash \rangle$ iff there is a world $w \in W$, such that $\langle X, Y \rangle$ is satisfied in world w of $\langle W, R, \Vdash \rangle$.

Theorem 4.2 (Model Existence Theorem for INT). *If a pair of sets belongs to an* INT*-consistency property, then it is satisfied in an intuitionistic model.*

Proof Let (a) $\langle X, Y\rangle \in \mathcal{C}$, where $\mathcal{C}$ is an INT-consistency property. By Corollary 4.19, there is an INT-consistency property $\mathcal{C}^* \supseteq \mathcal{C}$ which is of finite $\sqsubseteq$-character. Let S stand for a sequence ordering all formulas of language $\mathcal{L}_{\mathsf{INT}}$, and let $\langle \mathbb{W}, \mathbb{R}\rangle$ be the INT-superframe built with reference to S upon $\mathcal{C}^*$. By assumption (a), $\mathcal{C}$, and hence also $\mathcal{C}^*$, is nonempty, therefore Lemma 4.13 applies: $\langle \mathbb{W}, \mathbb{R}\rangle$ is an intuitionistic frame.

Let us set a relation $P \subseteq \mathbb{W} \times \mathit{FORM}_{\mathsf{INT}}$ as follows: for each propositional variable p_i, and every $\langle \mathbb{X}_L, \mathbb{X}_R\rangle \in \mathbb{W}$, let

$$\langle \mathbb{X}_L, \mathbb{X}_R\rangle \; P \; p_i \text{ iff } p_i \in \mathbb{X}_L.$$

It follows that for each propositional variable p_i, if $w_X = \langle \mathbb{X}_L, \mathbb{X}_R\rangle \; P \; p_i$, then since $p_i \in (\mathbb{X}_L)^{\sharp\mathsf{INT}}$, we will have $p_i \in \mathbb{Y}_L$ for every pair $w_Y = \langle \mathbb{Y}_L, \mathbb{Y}_R\rangle$ such that $\mathbb{R}$ holds between w_X and w_Y. We have just shown that relation P satisfies heredity condition (H) of Definition 4.1.

We may now extend P to forcing $\Vdash$. The last step is to prove that in the resulting intuitionistic model $\langle \mathbb{W}, \mathbb{R}, \Vdash\rangle$, for each $w_X = \langle \mathbb{X}_L, \mathbb{X}_R\rangle \in \mathbb{W}$, for every formula A

(i) if $A \in \mathbb{X}_L$, then $w_X \Vdash A$, and
(ii) if $A \in \mathbb{X}_R$, then $w_X \nVdash A$.

Let us recall that by Corollary 4.18, $\mathbb{X} \cap \mathbb{Y} = \varnothing$. The proof is by induction on the rank r of A. The base case is included in the definition of $P \subseteq \Vdash$.

Suppose that A is of the form '$B \wedge C$' and that $A \in \mathbb{X}_L$. Set $\mathbb{X}_L$ is dl-saturated (Lemma 4.11), hence $B, C \in \mathbb{X}_L$. By IH, $w_X \Vdash B$ and $w_X \Vdash C$, therefore $w_X \Vdash B \wedge C$. If $A \in \mathbb{X}_R$, then since $\mathbb{X}_R$ is dr-saturated, $w_X \nVdash B$ or $w_X \nVdash C$, which entails $w_X \nVdash B \wedge C$, as required. The reasoning is similar for $A =$ '$B \vee C$'.

Suppose that A is of the form '$B \supset C$' and that $A \in \mathbb{X}_L$. Let $w_Y = \langle \mathbb{Y}_L, \mathbb{Y}_R\rangle$ be such that $\mathbb{R}$ holds between w_X and w_Y. Then $(\mathbb{X}_L)^{\sharp\mathsf{INT}} \subseteq \mathbb{Y}_L$, hence $B \supset C \in \mathbb{Y}_L$. As $\mathcal{C}$ is an INT-consistency property, there is $\langle \mathbb{Y}_L, \mathbb{Y}_R \cup \{B\}\rangle \in \mathcal{C}$ or $\langle \mathbb{Y}_L \cup \{C\}, \mathbb{Y}_R\rangle \in \mathcal{C}$. By clause 4 of Lemma 4.12, in the first case $B \in \mathbb{Y}_R$, and in the second case $C \in \mathbb{Y}_L$. Since w_Y was arbitrary, it follows by IH that $w_X \Vdash B \supset C$.

If A is of the form '$B \supset C$' and $A \in \mathbb{X}_R$, then by clause 3 of Corollary 4.12, there is $\langle \mathbb{Y}_L, \mathbb{Y}_R\rangle \in \mathbb{R}(\langle \mathbb{X}_L, \mathbb{X}_R\rangle)$ such that $B \in \mathbb{Y}_L$ and $C \in \mathbb{Y}_R$. By IH, $w_X \nVdash B \supset C$. The reasoning is analogous for formulas of the form '$\sim B$'.

Let us go back to $\langle X, Y\rangle \in \mathcal{C}$. By the construction of INT-superframe $\langle \mathbb{W}, \mathbb{R}\rangle$, $\langle X, Y\rangle \sqsubseteq \mathbb{S}_S^{\mathcal{C}}(\langle X, Y\rangle)$. By (i) and (ii), the pair $\langle X, Y\rangle$ is satisfied in $\mathbb{S}_S^{\mathcal{C}}(\langle X, Y\rangle) \in \mathbb{W}$, which was to be proved. □

Here comes the time to go back to sequents of $\mathcal{L}^?_{\vdash\mathsf{INT}}$ and Socratic transformations.

Definition 4.14 Let ϕ be a sequent of language $\mathcal{L}^?_{\vdash\mathsf{INT}}$ and let $i \in W(\phi)$. We call $\langle X, Y\rangle$ *a set-representation of i in* ϕ iff

1. X is the set of all formulas A of $\mathcal{L}_{\mathsf{INT}}$, such that the indexed formula '$(A)^{\varsigma/i}$' occurs in the antecedent of ϕ, and
2. Y is the set of all formulas A of $\mathcal{L}_{\mathsf{INT}}$, such that the indexed formula '$(A)^{\varsigma/i}$' occurs in the succedent of ϕ.

By *ϕ-family* we mean the family of all pairs of sets $\langle X, Y\rangle$, such that for some $i \in W(\phi)$, the pair $\langle X, Y\rangle$ is the set-representation of i in ϕ.

For simplicity, we shall use the term *sequent unprovable in* $\mathbb{E}^{\mathsf{INT}}$ for a sequent ϕ such that there is no successful Socratic transformation of $?(\phi)$. We used the same terminology in the previous chapter.

Definition 4.15 A pair $\langle X, Y\rangle$ of sets of formulas of the language $\mathcal{L}_{\mathsf{INT}}$ is INT-*consistent* iff for some sequent ϕ unprovable in $\mathbb{E}^{\mathsf{INT}}$, the pair $\langle X, Y\rangle$ belongs to ϕ-family.

Theorem 4.3 *The family of all* INT*-consistent pairs is an* INT*-consistency property.*

Proof Let $\mathcal{F}$ stand for the family of all INT-consistent pairs of sets. The proof consists in verifying that $\mathcal{F}$ satisfies Definition 4.10. Let $\langle X, Y\rangle \in \mathcal{F}$ and let ϕ be an arbitrary sequent such that $\langle X, Y\rangle$ belongs to ϕ-family.

Clause 1. If $p_i \in X \cap Y$, then ϕ is basic—a contradiction.

Clause 2. Let $A \wedge B \in X$. Assume that $\langle X, Y\rangle$ is a set-representation of numeral $i \in W(\phi)$. Then '$(A \wedge B)^{\varsigma/i}$' occurs in the antecedent of ϕ. Let us consider:

$$\cfrac{\cfrac{?(\Gamma((A \wedge B)^{\varsigma/i}) \vdash \Delta)}{?(\Gamma((A \wedge B)^{\varsigma/i}, (A \wedge B)^{\varsigma/i}) \vdash \Delta)}\; L_{contr}}{?(\Gamma((A \wedge B)^{\varsigma/i}, A^{\varsigma/i}, B^{\varsigma/i}) \vdash \Delta)}\; L_{\wedge}$$

Since ϕ is unprovable in $\mathbb{E}^{\mathsf{INT}}$ and by Lemma 4.7 (semi-admissibility of contraction), the second question of the above Socratic transformation does not have a successful Socratic transformation in $\mathbb{E}^{\mathsf{INT}}$. Then by Lemma 4.5 (invertibility of rules), the third question does not have a successful Socratic transformation in $\mathbb{E}^{\mathsf{INT}}$. It follows that $\langle X \cup \{A, B\}, Y\rangle \in \mathcal{F}$.

Clause 3. Suppose that $A \wedge B \in Y$. We present only the relevant Socratic transformation.

$$\cfrac{\cfrac{?(\Gamma \vdash \Delta(A \wedge B^{\varsigma/i}))}{?(\Gamma \vdash \Delta(A \wedge B^{\varsigma/i}, A \wedge B^{\varsigma/i})}\; R_{contr}}{?(\Gamma \vdash \Delta(A \wedge B^{\varsigma/i}, A^{\varsigma/i});\; \Gamma \vdash \Delta(A \wedge B^{\varsigma/i}, B^{\varsigma/i}))}\; R_{\wedge}$$

Additional application of Lemma 4.3 (separation) is needed to derive the conclusion that at least one of sequents-conclusions of the third question is not provable.

Clause 4 is analogous to clause 3, and clause 5 is analogous to clause 2.

Clause 6. Let $\sim A \in X$. The relevant Socratic transformation showing that $\langle X, Y \cup \{A\}\rangle \in \mathcal{F}$ is the following.

$$\frac{?(\Gamma(\sim A^{\varsigma/i}) \vdash \Delta(\_))}{?(\Gamma(\sim A^{\varsigma/i}) \vdash \Delta(A^j))}\ L_\sim$$

with $i = j$.

Clause 7 is similar to clause 6. The relevant Socratic transformation consists in one application of $L_\supset$ with $i = j$.

Clause 8. Let $\sim A \in Y$. The relevant Socratic transformation is the following.

$$\frac{?(\Gamma(\_) \vdash \Delta(\sim A^{\varsigma/i}))}{?(\Gamma(A^{\varsigma/i,j}) \vdash \Delta(\_))}\ R_\sim$$

where j is new to the sequent-premise. The sequent-conclusion is unprovable. We need to show now that all elements of $(X)^{\sharp\mathsf{INT}}$, where X is the set of formulas represented by i in the antecedent, can be carried to j. Propositional variables in the antecedent can be carried to j by R_H. Suppose that $B \supset C$ is in this set. Since we already have $\langle i, j\rangle \in R(\psi)$ (where ψ is the constituent of the last question displayed above), by Lemma 4.8, sequent $\Gamma(B \supset C^{\varsigma/i})((B \supset C)^j) \vdash \Delta(\_)$ is not provable. For $\sim B \in X$ we refer to Lemma 4.9.

Clause 9 is similar to clause 8. The relevant Socratic transformation consists in one application of $R_\supset$. Then we consider applications of R_H and apply Lemmas 4.8, 4.9. □

Theorem 4.4 (Adequacy of $\mathbb{E}^{\mathsf{INT}}$ wrt Semantics of $\mathcal{L}_{\mathsf{INT}}$). *A formula A of the language $\mathcal{L}_{\mathsf{INT}}$ is intuitionistically valid iff there is a Socratic proof of the sequent* '$\vdash (A)^0$' *in* $\mathbb{E}^{\mathsf{INT}}$.

Proof By Corollary 4.2, formula A is intuitionistically valid iff sequent of the form '$\vdash A^0$' is intuitionistically valid. By Corollary 4.4 (following from the soundness Theorem 4.1), if there is a Socratic proof of '$\vdash A^0$' in $\mathbb{E}^{\mathsf{INT}}$, then the sequent is intuitionistically valid. Suppose that there is no Socratic proof of '$\vdash A^0$' in $\mathbb{E}^{\mathsf{INT}}$. Then by Definition 4.15, pair $\langle \varnothing, \{A\}\rangle$ is INT-consistent. By Theorem 4.3, the pair belongs to an INT-consistency property. By Theorem 4.2, pair $\langle \varnothing, \{A\}\rangle$ is satisfied in a Kripke model $\langle W, R, \Vdash\rangle$, that is, for some $w \in W$, $w \nVdash A$.

This finishes the proof. □

References

1. Chlebowski, S. (2018). *Canonical and dual erotetic calculi for first-order logic* (Doctoral dissertation). Adam Mickiewicz University.
2. Dragalin, A. G. (1988). *Mathematical intuitionism. Introduction to proof theory*. American Mathematical Society.

3. Fitting, M. (1983). *Proof methods for modal and intuitionistic logics*. Springer.
4. Kleene, S. C. (1952). *Introduction to metamathematics*. North-Holland Publishing Co., P. Noordhoff N.V.
5. Leblanc, H., & Snyder, D. P. (1972). Duals of Smullyan trees. *Notre Dame Journal of Formal Logic, 13*(3), 387–393.
6. Moschovakis, J. (n.d.). Intuitionistic logic. In E. N. Zalta (Ed.), *The Stanford Encyclopedia of Philosophy* (Winter 2018). https://plato.stanford.edu/archives/win2018/entries/logic-intuitionistic/
7. Smullyan, R. M. (2014). *A beginner's guide to mathematical logic*. Dover Publications.
8. Van Dalen, D. (1986). Intuitionistic logic. In D. Gabbay & F. Guenthner (Eds.), *Handbook of philosophical logic* (pp. 225–339). Springer. https://doi.org/10.1007/978-94-009-5203-4_4

Chapter 5
Erotetic Calculi Are Calculi of Questions

The pivotal characteristic of erotetic calculi, as introduced by Wiśniewski, is that its construction serves both purely proof-theoretical purposes and more philosophical—erotetic—aims. Although this book focuses on proof-theoretical aspects of the method, the picture would certainly not be complete without the analysis of questions.

The idea standing behind the method of Socratic proofs was to model reasoning in which a question concerning derivability in the underlying logic is answered by 'pure questioning', without the reference to an external source of information (see, first of all, the introductions to [21, 26]). Therefore, the construction of 'Socratic' calculi consists in a formalization of rules transforming questions concerning derivability in the underlying logic. One could ask: what is meant by 'derivability' here, but we shall not make an attempt to answer this question until Sect. 5.6. At the time being, we assume that some common intuitions concerning the notion are sufficient.

The process of transformation of questions is viewed as a step-by-step simplification of the structure of the initial problem. The transformations were termed 'Socratic,' as the agent asking whether '$S \vdash A$' holds is led step-by-step through the consecutive questions to a final question, which may be conceived as 'rhetoric,' as the answer to it is 'obvious.' Here, 'obviousness' indicates that the last question of a Socratic transformation concerns basic derivability cases, such as derivability of a formula from a set containing this formula or from a set containing a formula and its negation. In other words, the last 'rhetoric' question of a Socratic transformation concerns reflexivity and/or *ex falso quodlibet* as *obvious* properties of derivability.

Yet, the more important issue is that of the legitimacy of the transitions between questions. The construction of erotetic calculi warrants that a very important relation holds between a question-premise and a question-conclusion: the relation of erotetic implication, a central notion for Inferential Erotetic Logic[1] (as previously, we will use 'IEL' for short). As we explained in the first chapter, the notion of erotetic

[1] See [18–20, 22]. See also the first chapter of this book.

D. Leszczyńska-Jasion, *The Method of Socratic Proofs*, Trends in Logic 64,
https://doi.org/10.1007/978-3-031-82451-7_5

implication has been developed as a semantic criterion for the correctness of erotetic reasoning, a tool dedicated to describing *valid erotetic inferences*. Therefore, once we establish that the relation of erotetic implication holds between an instance of a question-premise and an instance of a question-conclusion of a rule of our erotetic calculus, we are guaranteed that every Socratic transformation constructed within the calculus represents a correct erotetic reasoning.

The aim of this chapter is to provide all the technical details necessary to establish this connection. Therefore, first, we introduce the definition of erotetic implication for a simple language and explain the most important ideas with this example. Second, we introduce the details needed to define the notion of erotetic implication for the more complex languages considered in this book. This provides us with an additional semantic analysis of the questions of the languages presented here and allows us to state that the rules of erotetic calculi formalize an inferential relation between questions, whose semantic counterpart is the relation of erotetic implication.

How the Erotetic Goal Determines Proof-theoretical Properties

However, before we proceed to this task, let us take one more look at some properties of the erotetic calculi determined precisely by the fact that the method originated as a formalization of erotetic reasoning.

The initial motivation standing behind the method of Socratic proofs justifies the use of both-sided, single succedent sequents—that is, sequents with single formula in the succedent—as expressions concerning derivability. This approach aligns with the so-called *operational interpretation* of sequents,[2] as well as with thinking of a sequent calculus as a formal theory of derivability relation. Traditionally, however, sequent calculi for classical logic are formalized with the use of multisuccedent sequents, and it is commonly accepted that this is the feature that characterizes classical logic, as opposed to intuitionistic logic, formalized by the use of single succedent sequents. As shown in [21, 26], as well as in Chap. 2 of this book, the method of Socratic proofs for CPL and FOL is formalized with single succedent sequents. However, this does not contradict the commonly accepted view of the proof-theoretical difference between intuitionistic and classical logic, since—as we have seen in Chap. 2—the property of 'single-succedentness' is somewhat artificially maintained by shifting a formula from the succedent to the antecedent of a sequent whenever an intuitionistically problematic rule is applied, that is, exactly in the cases of β-formulas and δ-formulas on the right side of the turnstile. The paradigmatic cases here are '$S \vdash A \vee B$' and '$S \vdash \exists x_j A$', which are transformed into '$S(-A) \vdash B$' and '$S(-\exists x_j A) \vdash A(t)$', respectively. In other words, the difference between 'the classical' and 'the intuitionistic' in this setting comes down to the possibility of shifting a formula from right to left, which is effectively another way of allowing more than one formula in the succedent.

On the margin, the fact that shifting is allowed in our intuitionistic erotetic calculus is irrelevant to what has been said above, since the intuitionistic calculus is an explicit one, that is, expressed in a richer language with tools to represent semantic

[2] See [9], especially the beginning of Chap. 3, see also Sect. 2.2 of Chap. 2 of this book.

information. This means that the reasoning represented by a Socratic transformation conducted in an intuitionistic erotetic calculus is not intuitionistic reasoning but classical reasoning about intuitionistic semantics.

Despite its attractiveness in formalizing erotetic reasoning, the idea of 'single-succedentness' has not been extended to non-classical logics. Those discussed in the previous chapters, modal logics in [5–8], paraconsistent logics [2, 27], and a relevant logic [12] were formalized using multisuccedent sequents, and only in the case of relevant logic it is clear that shifting is *really* forbidden. In Sect. 5.6, we will have something more to say about using multisuccedent sequents in the context of calculi formalizing erotetic reasoning.

Finally, the notion of a basic sequent is also related to the erotetic issues through the idea of an *obvious answer* to a question concerning derivability. What about derivability in logics that violate the properties of reflexivity and/or *ex falso*? As we have seen in Chap. 2, in the case of classical logic each basic sequent (the presence of which makes the affirmative answer to a question concerning derivability *obvious*) expresses at least one of these two properties. By and large, the criterion for being *basic*, and consequently the notion of an *obvious answer*, depends on the underlying logic and thus must be adapted to the analyzed case. For example, in paraconsistent and relevant logics, *ex falso* holds in a moderated form or does not hold at all. Therefore, in the erotetic calculus for FDE (for 'First Degree Entailment'—one of the primary relevance logics), basic sequents are only those referring to reflexivity (and thus relevance) of derivability. In the paraconsistent case, the logic is embedded in a classical environment, since the language of the erotetic calculus contains, *int.al*, classical negation alongside paraconsistent negation. Therefore, basic sequents can be defined by the same general schemes as those in classical logic, where, obviously, in the case of a formula and its negation in the antecedent of a sequent, the negation is classical.

5.1 Erotetic Implication: Language $\mathcal{L}^?_{\mathsf{CPL}}$

Here we define in a formal way some of the notions introduced in Chap. 1. At the start, we follow Wiśniewski's account, see [22].

We construct a language with questions, built upon the language $\mathcal{L}_{\mathsf{CPL}}$ of CPL, in the following way. First, we add the question mark '?', braces '{','}' and the comma ',' to the language $\mathcal{L}_{\mathsf{CPL}}$. By a *d-wff of the language* $\mathcal{L}^?_{\mathsf{CPL}}$ we mean a formula of the language $\mathcal{L}_{\mathsf{CPL}}$. A *question of the language* $\mathcal{L}^?_{\mathsf{CPL}}$ is an expression of the form:

$$?\{A_1, \dots, A_n\} \tag{5.1}$$

where $n > 1$ and $A_1, \dots, A_n$ are pairwise syntactically distinct d-wffs of $\mathcal{L}^?_{\mathsf{CPL}}$. D-wffs $A_1, \dots, A_n$ are called *direct answers* to question (5.1). By 'dQ' we denote the set of direct answers to question Q.

As previously, we use Boolean valuations (valuations, for short) to construct semantics for CPL. However, a generalization of entailment in CPL will prove useful in this chapter. This relation, known as *multiple-conclusion entailment*, was examined by Shoesmith and Smiley in [10].

Definition 5.1 (*Multiple-Conclusion Entailment in* CPL) Let X and Y stand for sets of formulas of the language $\mathcal{L}_{\mathsf{CPL}}$. We say that X *multiple-conclusion entails* Y *in* $\mathcal{L}_{\mathsf{CPL}}$, symbolically, $X \Vdash_{\mathsf{CPL}} Y$, iff:

- For every Boolean valuation v, if every element of X is true under v, then there is at least one element of Y true under v.

The concept of erotetic implication was introduced by Wiśniewski in [17]. Below, we quote the definition of the concept from [22, p. 67]:

Definition 5.2 (*Erotetic Implication in* $\mathcal{L}^?_{\mathsf{CPL}}$) A *question* Q *implies a question* Q^* *on the basis of a set of d-wffs* X (in symbols: $\mathsf{Im}(Q, X, Q^*)$) iff:

1. For each $A \in dQ$, $X \cup \{A\} \Vdash_{\mathsf{CPL}} dQ^*$, and
2. For each $B \in dQ^*$, there exists a non-empty proper subset Y of dQ, such that $X \cup \{B\} \Vdash_{\mathsf{CPL}} Y$.

If $\mathsf{Im}(Q, X, Q^*)$, then Q is called the *implying question*, or the *initial question* (as Q *initiates* erotetic reasoning), the elements of X are called *declarative premises* (of the reasoning), and Q^* is called the *implied question*, or the *question-conclusion* of the reasoning.

Clause 1 of Definition 5.2 expresses the property of *transmission of soundness/truth into soundness*, analogous to the transmission of truth characterizing the entailment relation between declaratives. As explained in the first chapter, the erotetic counterpart of truth under a valuation v is the *soundness of a question under valuation* v; a question Q is said to be *sound under* v if at least one of its direct answers is true under v. Hence, clause 1 of Definition 5.2 ensures that if the initial question Q is sound and the declarative premises in X are true, then the question-conclusion Q^* is sound as well.

Clause 2 of the definition is designed to express the idea that each answer to the question-conclusion Q^* must be potentially useful in resolving the issue posed by the initial question Q—in other words, it must determine a non-empty, proper subset Y of the set of direct answers to Q, within which one may search for the true answer to Q, whenever Q^* is sound and all declarative premises in X are true. Wiśniewski calls this second property *open-minded cognitive usefulness* or *goal-directedness* (for more information, see, e.g., [22, Chap. 5], [20]).

5.2 Erotetic Implication in $\mathcal{L}^?_{\mathsf{CPL}}$: A Discussion

The second property of erotetic implication amounts to the existence of a proper subset of dQ, where the solution of the initial problem lies. However, for those who would say that the *mere existence* of the proper subset of dQ is less than *determining* the solution, whether partial or proper, we have the following suggestion. (Let us recall that the symbol '2^X' is used as the power set of X.)

Definition 5.3 (*Functional Erotetic Implication*) Let Q and Q^* be questions of $\mathcal{L}^?_{\mathsf{CPL}}$, and let X denote a set of d-wffs of $\mathcal{L}^?_{\mathsf{CPL}}$. *Question Q functionally implies question Q^* on the basis of X* iff:

1. For each $A \in dQ$, $X \cup \{A\} \Vdash_{\mathsf{CPL}} dQ^*$, and
2. There exists a function $f : dQ^* \longrightarrow 2^{dQ} \setminus \{dQ, \varnothing\}$ such that for each $B \in dQ^*$, $X \cup \{B\} \Vdash_{\mathsf{CPL}} f(B)$.

Clause 2 of the above definition not only assumes the existence of a proper subset for each direct answer to Q^* but also the existence of a function that assigns these proper subsets to the direct answers. On the face of it, clause 2 of Definition 5.3 merely rephrases clause 2 of Definition 5.2. However, considering that in general, the sets of answers to questions do not have to be finite (as we have assumed here), this reformulation allows for a potentially interesting enhancement of the original concept. For instance, by imposing additional constraints on the function—such as requiring an explicit counting procedure for its values, ensuring it is recursively enumerable, or even recursive—one can obtain new computational variations of the concept of erotetic implication.

On the other hand, sometimes it is argued that the notion of erotetic implication is a very strong concept, perhaps too strong due to the universal quantifier used in clause 2 of Definition 5.2 (see, for example, [16] and [3, Chap. 1, Sect. 4.2, the book is in Polish] for this view). It is argued that when one looks for instances of reasoning with questions that can be modeled by the notion of erotetic implication, then it is more realistic to use the following weaker concept:

Definition 5.4 (*Weak Erotetic Implication*) Let Q and Q^* be questions of $\mathcal{L}^?_{\mathsf{CPL}}$, and let X stand for a set of d-wffs of $\mathcal{L}^?_{\mathsf{CPL}}$. *Question Q w-implies question Q^* on the basis of X (*in symbols: $\mathsf{Im}_{\mathsf{w}}(Q, X, Q^*)$*)* iff:

1. For each $A \in dQ : X \cup \{A\} \Vdash_{\mathsf{CPL}} dQ^*$, and
2. For some $B \in dQ^*$: there exists a non-empty proper subset Y of dQ such that $X \cup \{B\} \Vdash_{\mathsf{CPL}} Y$.

If the correctness of an erotetic reasoning was appraised by the w-implication, one could say that it is perfectly reasonable to ask the implied question, even if only one of its answers leads one closer to a solution. Yet still, the transmission of soundness and truth into soundness (clause 1) must be retained.

What is important, in [16] the authors show that there is good empirical evidence to claim that when a human agent solves a problem expressed by a question by means of asking other (implied) questions, they tend to do it in a way which should be modeled formally by means of w-implication rather than erotetic implication.

One may also find various reasons to consider the following weakening of the original erotetic implication ([3, Chap. 1, Sect. 4.2] discusses this issue in the context of the philosophy of science.)

Definition 5.5 (*Falsificationist Erotetic Implication*) Let Q and Q^* be questions of $\mathcal{L}^?_{\mathsf{CPL}}$, and let X stand for a set of d-wffs of $\mathcal{L}^?_{\mathsf{CPL}}$. *Question Q f-implies question Q^* on the basis of X* (in symbols: $\mathsf{Im_f}(Q, X, Q^*)$) iff:

1. For each $A \in dQ : X \cup \{A\} \Vdash_{\mathsf{CPL}} dQ^*$, and
2. For some $B \in dQ^* : X \cup \{B\}$ eliminates a direct answer C to Q, that is, for each valuation v, if all the elements of $X \cup \{B\}$ are true under v, then the direct answer C is false under v.

The reader will find more comments concerning the notions explicated in Definitions 5.2 and 5.5, intuitions behind them, and relations between them, in [22, Chap. 7]. For the purpose of this section, it will be sufficient if we note once again that Definitions 5.2–5.5 do not differ with respect to the first clause.

Erotetic implications, both the original one and the modified versions, differ from the relation of entailment between declaratives in one important aspect: they are not transitive. What we mean by non-transitivity here is that it is possible that for a given set X we have $\mathsf{Im}(Q_0, X, Q_1)$ and $\mathsf{Im}(Q_1, X, Q_2)$, but not $\mathsf{Im}(Q_0, X, Q_2)$. Here is an example cited after [22, Sect. 7.3.2]. It holds that:

$$\mathsf{Im}(?\{p, \neg p\}, \varnothing, ?\{p \wedge q, p \wedge \neg q, \neg p\})$$

$$\mathsf{Im}(?\{p \wedge q, p \wedge \neg q, \neg p\}, \varnothing, ?\{q, \neg q\})$$

but it is not the case that $\mathsf{Im}(?\{p, \neg p\}, \varnothing, ?\{q, \neg q\})$. This observation is true as well if we change 'Im' to '$\mathsf{Im_w}$', or '$\mathsf{Im_f}$'.

The remaining part of this section is devoted to showing that whenever the first clause—common to all the four relations defined above in Definitions 5.2, 5.3, 5.4 and 5.5—holds for two question Q, Q^* and a set X, then one can construct an 'interpolant'[3] question that is erotetically implied by Q on the basis of X, and implies Q^* on the basis of X enlarged with a *presupposition of* Q. The enlargement seems to be a small price for the result.

The notion of a presupposition of a question comes from Belnap [1], and is usually understood as a d-wff which is entailed by each direct answer to the question (see the first chapter of this book, or [22, Sect. 4.3]), that is, roughly, a d-wff that must be true if the question is sound. In the present context, we are interested in a special type of presupposition, called *prospective*.

Definition 5.6 (*Prospective Presupposition of a Question of* $\mathcal{L}^?_{\mathsf{CPL}}$) Let $Q = ?\{A_1, \ldots, A_n\}$ be a question of $\mathcal{L}^?_{\mathsf{CPL}}$. By a *prospective presupposition of* Q, symbolically $PP(Q)$, we mean

$$PP(Q) := A_1 \vee \ldots \vee A_n.$$

[3] I owe Szymon Chlebowski the observation of the analogy between Theorem 5.1 and the Craig Interpolation Theorem.

The notion of prospective presupposition is usually understood in a more general way (again, see [22, Sect. 4.3]), but the above definition contains exactly what is needed in the present context.

It is easy to see that the prospective presupposition of a question has the following properties.

Corollary 5.1 *Let Q be a question of the language $\mathcal{L}^?_{\mathsf{CPL}}$.*

1. *For each $A \in dQ$, $\{A\} \models_{\mathsf{CPL}} PP(Q)$, but also*
2. $\{PP(Q)\} \mid\!\models_{\mathsf{CPL}} dQ$.

The property expressed by clause 1 can be used to define the very notion of presupposition of a question, whereas the second clause states that the truth of the prospective presupposition of a question guarantees soundness of the question.

Suppose that $Q = ?\{A_1, \dots, A_n\}$ and $Q^* = ?\{B_1, \dots, B_k\}$ are two questions such that for a certain set X of d-wffs, the first clause of Definition 5.2 (and thus also of Definitions 5.3, 5.4, 5.5) holds, that is,

1. For each $A_i \in dQ : X \cup \{A_i\} \mid\!\models_{\mathsf{CPL}} dQ^*$.

We do not presume anything about the second clause, so we may deal with (functional) erotetic implication, w-implication, f-implication, or no implication at all. The following operation on questions can be coined their *conjunction*.

$$Q \odot Q^* := ?\{A_1 \wedge B_1, \dots, A_1 \wedge B_k,$$

$$A_2 \wedge B_1, \dots, A_2 \wedge B_k, \dots, A_n \wedge B_1, \dots, A_n \wedge B_k\}.$$

That is, operation '$\odot$' applied to questions Q and Q^* produces another question, whose set $d(Q \odot Q^*)$ of all direct answers equals:

$$d(Q \odot Q^*) = \{A_i \wedge B_j : 1 \leqslant i \leqslant n, 1 \leqslant j \leqslant k\}.$$

By *1*, for each $i : 1 \leqslant i \leqslant n$:

$$X \cup \{A_i\} \mid\!\models_{\mathsf{CPL}} \{B_1, \dots, B_k\}.$$

It is easy to see that then also (for each i):

$$X \cup \{A_i\} \mid\!\models_{\mathsf{CPL}} \{A_i \wedge B_1, \dots, A_i \wedge B_k\},$$

and since $\{A_i \wedge B_1, \dots, A_i \wedge B_k\} \subseteq d(Q \odot Q^*)$, for each i:

$$X \cup \{A_i\} \mid\!\models_{\mathsf{CPL}} d(Q \odot Q^*). \tag{5.2}$$

Moreover, for each i and for each j $(1 \leqslant j \leqslant k)$:

$$X \cup \{A_i \wedge B_j\} \Vdash_{\mathsf{CPL}} \{A_i\}. \tag{5.3}$$

Since $\{A_i\}$ is a non-empty proper subset of dQ, from (5.2) and (5.3) it follows that:

$$\mathsf{Im}(Q, X, Q \odot Q^*).$$

On the other hand, for each i $(1 \leqslant i \leqslant n)$, and for each j $(1 \leqslant j \leqslant k)$:

$$X \cup \{A_i \wedge B_j\} \Vdash_{\mathsf{CPL}} \{B_j\}$$

which warrants that also

$$X \cup \{A_i \wedge B_j\} \Vdash_{\mathsf{CPL}} dQ^* \tag{5.4}$$

since $\{B_j\} \subseteq dQ^*$. Let us observe that (5.4) entails the first clause of the definition of erotetic implication. For the second one, we need the prospective presupposition, $PP(Q) = A_1 \vee \ldots \vee A_n$, of Q. The following holds:

$$X \cup \{PP(Q), B_j\} \Vdash_{\mathsf{CPL}} \{A_1 \wedge B_j, \ldots, A_n \wedge B_j\} \tag{5.5}$$

since for every valuation v that makes each d-wff in $X \cup \{PP(Q), B_j\}$ true, at least one of $A_1, \ldots, A_n$ is true under v, and hence also at least one of the listed conjunctions $A_1 \wedge B_j, \ldots, A_n \wedge B_j$ is true under v. As $\{A_1 \wedge B_j, \ldots, A_n \wedge B_j\}$ is a non-empty, proper subset of $d(Q \odot Q^*)$, (5.4) and (5.5) tell us that:

$$\mathsf{Im}(Q \odot Q^*, X \cup \{PP(Q)\}, Q^*).$$

To sum up, if the soundness of the question Q and the truth of declaratives in X ensures the soundness of Q^* (which corresponds to assumption *1*), then $\mathsf{Im}(Q, X, Q \odot Q^*)$, and $\mathsf{Im}(Q \odot Q^*, X \cup \{PP(Q)\}, Q^*)$. More formally, what we have shown above proves:

Theorem 5.1 (Interpolation Theorem for Questions) *Let Q and Q^* be questions of $\mathcal{L}^?_{\mathsf{CPL}}$, and let X stand for a set of d-wffs of $\mathcal{L}^?_{\mathsf{CPL}}$. If questions Q, Q^* and set X are such that for each $A \in dQ$:*

$$X \cup \{A\} \Vdash_{\mathsf{CPL}} dQ^*,$$

then $\mathsf{Im}(Q, X, Q \odot Q^*)$ *and* $\mathsf{Im}(Q \odot Q^*, X \cup \{PP(Q)\}, Q^*)$.

The fact that the set of declaratives must be enriched with $PP(Q)$ seems natural, if we recall that assuming the truth of $PP(Q)$ amounts to assuming the soundness of Q. Apart from throwing a new light on the notion of erotetic implication, we strongly believe that this result may have significant consequences for the construction of erotetic search scenarios (mentioned in the first chapter), but this topic definitely goes beyond the scope of this book.

Another observation makes use of the notion of a *safe question of the language* $\mathcal{L}^?_{\mathsf{CPL}}$.

Definition 5.7 (*Safe and Risky Question of* $\mathcal{L}^?_{\mathsf{CPL}}$) A question Q of $\mathcal{L}^?_{\mathsf{CPL}}$ is *safe* iff $\varnothing \Vdash_{\mathsf{CPL}} dQ$. A question which is not safe is called *risky*.

Let us observe that in the case of safe questions the first condition defining our erotetic implications—the one warranting the transmission of soundness and truth into soundness—is quite trivially satisfied. Therefore:

Corollary 5.2 *Let* Q *and* Q^* *stand for safe questions of* $\mathcal{L}^?_{\mathsf{CPL}}$. *Then:*

1. $\mathsf{Im}(Q, \varnothing, Q \odot Q^*)$, *and*
2. $\mathsf{Im}(Q \odot Q^*, \varnothing, Q^*)$.

Proof Since Q^* is safe, $\varnothing \Vdash_{\mathsf{CPL}} dQ^*$. Trivially, for each $A \in dQ$, $\varnothing \cup \{A\} \Vdash_{\mathsf{CPL}} dQ^*$, which makes Theorem 5.1 applicable. Hence it follows that $\mathsf{Im}(Q, \varnothing, Q \odot Q^*)$ and $\mathsf{Im}(Q \odot Q^*, \{PP(Q)\}, Q^*)$. Therefore clause 1 is true. Now we show that also $\mathsf{Im}(Q \odot Q^*, \varnothing, Q^*)$.

By Definition 5.7, for every valuation v, the set dQ contains a d-wff true under v. dQ is a finite set and thus $PP(Q)$ is a well defined CPL-valid formula. Since, by Theorem 5.1, it holds that $\mathsf{Im}(Q \odot Q^*, \{PP(Q)\}, Q^*)$, by the definition of erotetic implication (the first clause): for each $B \in d(Q \odot Q^*)$, $\{PP(Q), B\} \Vdash_{\mathsf{CPL}} dQ^*$. But since $PP(Q)$ is true under every valuation, it yields that $\{B\} \Vdash_{\mathsf{CPL}} dQ^*$, which proves the first clause of Definition 5.2 for $\mathsf{Im}(Q \odot Q^*, \varnothing, Q^*)$.

The second clause of Definition 5.2 holds by an analogous argument. Briefly, since $\mathsf{Im}(Q \odot Q^*, \{PP(Q)\}, Q^*)$, for each $C \in d(Q^*)$, $\{PP(Q), C\} \Vdash_{\mathsf{CPL}} Y$ for some nonempty $Y \subset d(Q \odot Q^*)$. But $PP(Q)$ is CPL-valid, hence $\{C\} \Vdash_{\mathsf{CPL}} Y$, as required. □

If we take the set of declarative premises X to be empty, then the relation of erotetic implication $\mathsf{Im}(Q, \varnothing, Q^*)$ between questions on the basis of the empty set of declarative premises becomes a binary relation. Let us make this observation more formal. First, let us recall the following notion (see Definition 3.5 in Sect. 3.3): if R is a binary relation, then a directed R-chain is a finite sequence of elements such that $w_i R w_{i+1}$ holds between consecutive elements of the sequence. Let $\mathcal{Q}_{sf}$ stand for the *set of all safe questions of the language* $\mathcal{L}^?_{\mathsf{CPL}}$.

Definition 5.8 Let

$$\mathsf{Im}_{sf} \subseteq \mathcal{Q}_{sf} \times \mathcal{Q}_{sf}$$

be such that it holds between safe questions Q and Q^* of the language $\mathcal{L}^?_{\mathsf{CPL}}$ if $\mathsf{Im}(Q, \varnothing, Q^*)$; symbolically,

$$\mathsf{Im}_{sf}(Q, Q^*) \quad \text{iff} \quad \mathsf{Im}(Q, \varnothing, Q^*).$$

Then, by Im^{tr}_{sf}, we mean the *transitive closure of* Im_{sf}, that is, $\mathsf{Im}^{tr}_{sf}(Q, Q^*)$ iff there exists a directed Im_{sf}-$[Q, Q^*]$-chain of questions of $\mathcal{L}^?_{\mathsf{CPL}}$.

Now, by Corollary 5.2:

Corollary 5.3 $\mathsf{Im}^{tr}_{sf} = \mathcal{Q}_{sf} \times \mathcal{Q}_{sf}$.

Proof Let $Q, Q^* \in \mathcal{Q}_{sf}$. The sequence $\langle Q, Q \odot Q^*, Q^* \rangle$ is then the required directed chain. □

In other words, Im^{tr}_{sf} is the full relation in the set $\mathcal{Q}_{sf}$ of all safe questions of language $\mathcal{L}^?_{\mathsf{CPL}}$.

How should we interpret this result? It demonstrates that, as long as only safe questions are considered, the transitive closure of erotetic implication holds between any two questions. In a sense, this result in the field of the logic of questions parallels the fact that, when only valid declarative formulas are considered, entailment becomes a 'trivial' notion, since the relation holds between any two such formulas. From this perspective, the 'trivialization' of the transitive closure of erotetic implication among safe questions is not surprising; it merely shows that theories of questions focusing exclusively on safe questions may be overly restrictive. Any entailment relation between questions defined within such a framework risks collapsing into a relation that holds universally between all questions.

Conversely, it is important to emphasize that while entailment among valid formulas may not be particularly interesting, a *specific* transition from one valid formula to another within a proof system *is* significant, precisely because non-valid formulas exist. This highlights the necessity for proof systems of erotetic implication—systems such as the erotetic calculi presented in this book.

However, when risky questions come into the picture, the results presented in this section reveal another intriguing aspect. According to Theorem 5.1, if two questions fail to satisfy the second clause of Definition 5.2 but meet the first, then there is always a way to insert a 'connector' (an interpolant) between them to maintain the flow of erotetic implication.

Originally, the concepts of weak implication and falsificationist implication were introduced because the original relation was deemed too strong in a logical sense. Just as deductive reasoning serves as a normative benchmark in the realm of reasoning with declaratives—yet fails to adequately address some forms of non-normative reasoning, such as inductive or probabilistic reasoning—erotetic implication acts as a normative measure for reasoning with questions, while simultaneously being unable to capture certain interesting instances of erotetic reasoning. However, the results presented in this section indicate that weak implication and falsificationist implication do not possess greater explanatory power than erotetic implication, as every instance of 'weaker' implication can be captured by successive erotetic implications. In fact, in terms of explanatory power, it appears to be the opposite. Reasoning based on weak or falsificationist implications turns out to be an *enthymematic erotetic argument*. In the realm of declaratives, we demonstrate the validity of an enthymeme by revealing a 'hidden' premise, usually topically relevant to the conclusion. The same applies to weak and falsificationist implications: we can justify the correctness of a reasoning based on one of these relations by uncovering hidden erotetic premises and applying erotetic implication. The hidden premises are the assumption of the form $PP(Q)$, which amounts to assuming soundness of Q in a specific context, along with the interpolant question of the form '$Q \odot Q^*$'.

5.3 Erotetic Implication: Languages $\mathcal{L}^{?}_{\vdash \mathsf{L}}$

In Sect. 2.2 we have described the syntax of language $\mathcal{L}^{?}_{\vdash \mathsf{L}}$, but we have not presented the notion of an answer to a question. Let us recall that questions of this language are of the form:

$$?(S_1 \vdash T_1, \ldots, S_n \vdash T_n) \tag{5.6}$$

where $\langle S_1 \vdash T_1, \ldots, S_n \vdash T_n \rangle$ is a finite and non-empty sequence of sequents of $\mathcal{L}^{?}_{\vdash \mathsf{L}}$. As put originally in [21, 26], by *affirmative answer to question* (5.6) we mean an expression of the form:

$$S_1 \vdash T_1 \;\&\; (S_2 \vdash T_2 \;\&\ldots\&\; (S_{n-1} \vdash T_{n-1} \;\&\; S_n \vdash T_n)\ldots) \tag{5.7}$$

and by *negative answer to question* (5.6) we mean an expression of the form:

$$ng(S_1 \vdash T_1 \;\&\; (S_2 \vdash T_2 \;\&\ldots\&\; (S_{n-1} \vdash T_{n-1} \;\&\; S_n \vdash T_n)\ldots)) \tag{5.8}$$

We will use $af\,Q$ and $ng\,Q$ for the answers of the form (5.7) and (5.8), respectively. Thus all questions of $\mathcal{L}^{?}_{\vdash \mathsf{L}}$ are polar questions whose sets of direct answers are of the forms $\{af\,Q, ng\,Q\}$.

Definition 5.2 of erotetic implication in $\mathcal{L}^{?}_{\mathsf{CPL}}$ is based upon the semantic notion of Boolean valuation; in the case of $\mathcal{L}^{?}_{\vdash \mathsf{L}}$, however, we need something more general. The notions introduced below are central tools of the so-called *Minimal Erotetic Semantics* (MiES, for short), a very general framework for semantic analysis of both declaratives and questions developed by Andrzej Wiśniewski.[4] The primary notion here is that of a partition of a language (the notion comes from [10]).

Definition 5.9 (*Partition of Language* $\mathcal{L}^{?}_{\vdash \mathsf{L}}$) Let $\mathsf{D}_{\mathcal{L}^{?}_{\vdash \mathsf{L}}}$ be the set of d-wffs of the language $\mathcal{L}^{?}_{\vdash \mathsf{L}}$. By a *partition of* $\mathsf{D}_{\mathcal{L}^{?}_{\vdash \mathsf{L}}}$ (or a *partition of the language* $\mathcal{L}^{?}_{\vdash \mathsf{L}}$) we mean an ordered pair $\mathsf{P} = \langle \mathsf{T}_\mathsf{P}, \mathsf{U}_\mathsf{P} \rangle$ such that $\mathsf{T}_\mathsf{P} \cup \mathsf{U}_\mathsf{P} = \mathsf{D}_{\mathcal{L}^{?}_{\vdash \mathsf{L}}}$ and $\mathsf{T}_\mathsf{P} \cap \mathsf{U}_\mathsf{P} = \varnothing$.

Nothing more is assumed about T_P (the set of 'truths' in a partition) and U_P (the set of 'untruths' in a partition), except for the two conditions.

In the case of complex languages with questions, such as $\mathcal{L}^{?}_{\vdash \mathsf{L}}$, the counterpart of the semantic notion of Boolean valuation is that of *admissible partition*. Definition 5.10 presents the notion for the case of the language $\mathcal{L}^{?}_{\vdash \mathsf{CPL}}$. It is a straightforward generalization of the notion given by Andrzej Wiśniewski (see [22, pp. 28–29]), which works for all the three erotetic calculi for CPL presented in this book.

---

[4] As Wiśniewski himself admits (see [25, Sect. 2]), some ideas come from Belnap [1] and Shoesmith and Smiley [10], but MiES goes far beyond them. For the details of MiES, see [22] or [23].

Definition 5.10 (*Admissible Partition of* $\mathcal{L}^?_{\vdash\mathsf{CPL}}$) Let $\mathsf{P} = \langle \mathsf{T_P}, \mathsf{U_P} \rangle$ be a partition of the language $\mathcal{L}^?_{\vdash\mathsf{CPL}}$, and let $\mathfrak{t}$ and $\mathfrak{u}$ stand for arbitrary d-wffs of this language. We say that P is *admissible* iff the following conditions hold:

1. '$\mathfrak{t}\&\mathfrak{u}$' $\in \mathsf{T_P}$ iff both $\mathfrak{t} \in \mathsf{T_P}$ and $\mathfrak{u} \in \mathsf{T_P}$;
2. $\mathfrak{t} \in \mathsf{T_P}$ iff '$\mathbf{n}\mathbf{g}\mathfrak{t}$' $\notin \mathsf{T_P}$;
3. '$S \vdash T(\alpha)$' $\in \mathsf{T_P}$ iff both '$S \vdash T(\alpha_0)$' $\in \mathsf{T_P}$ and '$S \vdash T(\alpha_1)$' $\in \mathsf{T_P}$;
4. '$\vdash T(\beta)$' $\in \mathsf{T_P}$ iff '$\vdash T(\beta_0, \beta_1)$' $\in \mathsf{T_P}$;
5. '$S(\_) \vdash \beta$' $\in \mathsf{T_P}$ iff for $i \in \{0, 1\}$, '$S(-\beta_i) \vdash \beta_{1-i}$' $\in \mathsf{T_P}$;
6. '$S \vdash T(\kappa)$' $\in \mathsf{T_P}$ iff '$S \vdash T(\kappa_0)$' $\in \mathsf{T_P}$;
7. '$S(\alpha) \vdash T$' $\in \mathsf{T_P}$ iff '$S(\alpha_0, \alpha_1) \vdash T$' $\in \mathsf{T_P}$;
8. '$S(\beta) \vdash T$' $\in \mathsf{T_P}$ iff both '$S(\beta_0) \vdash T$' $\in \mathsf{T_P}$ and '$S(\beta_1) \vdash T$' $\in \mathsf{T_P}$;
9. '$S(\kappa) \vdash T$' $\in \mathsf{T_P}$ iff '$S(\kappa_0) \vdash T$' $\in \mathsf{T_P}$.

Let us recall that, in the case of CPL, κ is restricted to formulas of the form '$\neg\neg A$'. It is also worth recalling that the sequents are presented in a general form, but S and T are subject to constraints described in Sect. 2.2 of Chap. 2.

The notion of admissible partition for the case of erotetic calculus $\mathbf{E}^{\mathsf{PQ}}$ has been presented in [26]. The calculus needs both-sided, single succedent sequents, thus the notion presented in [26] concerns only sequents of this category. However, the language $\mathcal{L}^?_{\vdash\mathsf{FOL}}$ defined in this book has also right-sided and left-sided sequents, thus, here, we again present a more general version of the notion of admissible partition.

Definition 5.11 (*Admissible Partition of* $\mathcal{L}^?_{\vdash\mathsf{FOL}}$) Let $\mathsf{P} = \langle \mathsf{T_P}, \mathsf{U_P} \rangle$ be a partition of the language $\mathcal{L}^?_{\vdash\mathsf{FOL}}$. We say that P is *admissible* iff P satisfies clauses 1–9 from Definition 5.10, and, moreover, the following conditions:

10. For each closed term t of the language $\mathcal{L}_{\mathsf{FOL}}$ that does not occur in a sequent '$S \vdash T(\gamma)$',
    '$S \vdash T(\gamma)$' $\in \mathsf{T_P}$ iff '$S \vdash T(\gamma(t))$' $\in \mathsf{T_P}$.
11. For each closed term t that does not occur in a sequent '$S(\delta) \vdash T$',
    '$S(\delta) \vdash T$' $\in \mathsf{T_P}$ iff '$S(\delta(t)) \vdash T$' $\in \mathsf{T_P}$.
12. Finally, for each closed term t of the language $\mathcal{L}_{\mathsf{FOL}}$,

    12.1. '$S(\gamma) \vdash$' $\in \mathsf{T_P}$ iff '$S(\gamma, \gamma(t)) \vdash$' $\in \mathsf{T_P}$,
    12.2. '$S(\gamma) \vdash C$' $\in \mathsf{T_P}$ iff '$S(\gamma, \gamma(t)) \vdash C$' $\in \mathsf{T_P}$,
    12.3. '$\vdash S(\delta)$' $\in \mathsf{T_P}$ iff '$\vdash S(\delta, \delta(t))$' $\in \mathsf{T_P}$,
    12.4. '$S(\_) \vdash \delta$' $\in \mathsf{T_P}$ iff '$S(-\delta) \vdash \delta(t)$' $\in \mathsf{T_P}$.

Let us observe that clause 10 of Definition 5.11 is equivalent to the conjunction of the following two clauses:

10.1 If '$S \vdash T(\gamma)$' $\in \mathsf{T_P}$, then for each closed term t that does not occur in the sequent '$S \vdash T(\gamma)$', '$S \vdash T(\gamma(t))$' $\in \mathsf{T_P}$.
10.2 If for some closed term t that does not occur in the sequent '$S \vdash T(\gamma)$', '$S \vdash T(\gamma(t))$' $\in \mathsf{T_P}$, then '$S \vdash T(\gamma)$' $\in \mathsf{T_P}$.

Clause 11 of Definition 5.11 can be analyzed in the same manner.

Finally, we are in a position to define the notion of entailment between d-wffs of the languages $\mathcal{L}^?_{\vdash\mathsf{CPL}}$ and $\mathcal{L}^?_{\vdash\mathsf{FOL}}$, and the notion of erotetic implication between questions of these languages. The next definitions follow the pattern from [22, Sect. 3.1].

Definition 5.12 (*Entailment in* $\mathcal{L}^?_{\vdash\mathsf{L}}$) Let $\mathsf{L} \in \{\mathsf{CPL}, \mathsf{FOL}\}$. Suppose that $\mathfrak{X}$ is a set of d-wffs of the language $\mathcal{L}^?_{\vdash\mathsf{L}}$ and t is a single d-wff of this language. We say that the set $\mathfrak{X}$ *entails the formula* t *in the language* $\mathcal{L}^?_{\vdash\mathsf{L}}$, symbolically:

$$\mathfrak{X} \models_{\mathcal{L}^?_{\vdash\mathsf{L}}} \mathrm{t}$$

iff there is no admissible partition P of $\mathcal{L}^?_{\vdash\mathsf{L}}$ such that $\mathfrak{X} \subseteq \mathsf{T_P}$ and $\mathrm{t} \notin \mathsf{T_P}$.

As before, dQ stands for the set of direct answers to Q.

Definition 5.13 (*Erotetic Implication in* $\mathcal{L}^?_{\vdash\mathsf{L}}$) Let $\mathsf{L} \in \{\mathsf{CPL}, \mathsf{FOL}\}$. Suppose that Q and Q^* are questions of $\mathcal{L}^?_{\vdash\mathsf{L}}$ and that $\mathfrak{X}$ is a set of d-wffs of $\mathcal{L}^?_{\vdash\mathsf{L}}$. We say that Q *implies* Q^* *on the basis of the set* $\mathfrak{X}$ *of d-wffs*, symbolically, $\mathsf{Im_L}(Q, \mathfrak{X}, Q^*)$, iff for each admissible partition P of the language $\mathcal{L}^?_{\vdash\mathsf{L}}$ the following holds:

1. For each $\mathrm{t} \in dQ$: if $\mathfrak{X} \cup \{\mathrm{t}\} \subseteq \mathsf{T_P}$, then $dQ^* \cap \mathsf{T_P} \neq \varnothing$; and
2. For each $\mathrm{u} \in dQ^*$: there is a non-empty, proper subset $\mathfrak{Y}$ of dQ such that if $\mathfrak{X} \cup \{\mathrm{u}\} \subseteq \mathsf{T_P}$, then $\mathfrak{Y} \cap \mathsf{T_P} \neq \varnothing$.

In the case of questions of $\mathcal{L}^?_{\vdash\mathsf{L}}$, dQ is always a two-element set. For this reason, the second clause of Definition 5.13 may take the following simplified form.

2*. *For each* $\mathrm{u} \in dQ^*$, *there is a single answer* $\mathrm{t} \in dQ$ *such that if* $\mathfrak{X} \cup \{\mathrm{u}\} \subseteq \mathsf{T_P}$, *then* $\mathrm{t} \in \mathsf{T_P}$.

One may think, however, of more sophisticated cases of erotetic implication for which the more general clause is more appropriate (see [22]).

The next theorem states that if a question Q^* results from a question Q by a rule of $\mathbb{E}$, then the affirmative answer to one of these questions entails, in the sense of Definition 5.12, the affirmative answer to the second question, and the same holds with respect to the negative answers to these two questions. This type of relation is a special case of *equipollence of questions* defined by Tadeusz Kubiński.[5] It follows naturally that in this situation the question Q^* is erotetically implied by the question Q, in the sense given by Definition 5.13. In other words, clause 3 of the following theorem expresses the fact that the rules of $\mathbb{E}$ are semantically correct in terms of MiES semantics. Finally, let us also note that erotetic implication with empty set of declarative premises is called *pure erotetic implication* (see [22, Sect. 7.4] for more information).

Theorem 5.2 *Let* $\mathbb{E} \in \{\mathbb{E}^{\mathbf{Q}}, \mathbb{E}^{\mathbf{RQ}}, \mathbb{E}^{\mathbf{LQ}}, \mathbb{E}^{\mathbf{P}}, \mathbb{E}^{\mathbf{RP}}, \mathbb{E}^{\mathbf{LP}}\}$. *If a question* Q^* *results from a question* Q *by a rule of* $\mathbb{E}$, *then*

[5] See Sect. 1.3.2 in Chap. 1.

1. $\{af\,Q\} \vDash_{\mathcal{L}^?_{\vdash\mathsf{L}}} af\,Q^*$ *and* $\{af\,Q^*\} \vDash_{\mathcal{L}^?_{\vdash\mathsf{L}}} af\,Q$,
2. $\{ng\,Q\} \vDash_{\mathcal{L}^?_{\vdash\mathsf{L}}} ng\,Q^*$ *and* $\{ng\,Q^*\} \vDash_{\mathcal{L}^?_{\vdash\mathsf{L}}} ng\,Q$,
3. $\mathsf{Im}_\mathsf{L}(Q, \varnothing, Q^*)$.

Proof The proof is by cases, we analyze only that of R_α of $\mathbb{E}^{\mathbf{P}}$, L_γ of $\mathbb{E}^{\mathbf{LQ}}$ and R_γ of $\mathbb{E}^{\mathbf{RQ}}$.

In the first case, Q and Q^* are of the forms:

$$Q = ?(\Phi(S \vdash \alpha)) \qquad Q^* = ?(\Phi(S \vdash \alpha_0\,;\; S \vdash \alpha_1))$$

Let $\mathsf{P} = \langle \mathsf{T}_\mathsf{P}, \mathsf{U}_\mathsf{P} \rangle$ be an arbitrary admissible partition of $\mathcal{L}^?_{\vdash\mathsf{FOL}}$. The reasoning is straightforward. If $af\,Q \in \mathsf{T}_\mathsf{P}$, then, by clause 1 of Definition 5.10, also '$S \vdash \alpha$' $\in \mathsf{T}_\mathsf{P}$ and, by clause 3 of the same Definition, $\{S \vdash \alpha_0,\, S \vdash \alpha_1\} \subseteq \mathsf{T}_\mathsf{P}$. But then again, by clause 1, $af\,Q^* \in \mathsf{T}_\mathsf{P}$. If $ng\,Q \in \mathsf{T}_\mathsf{P}$, then we make use of clauses 1 and 2 of Definition 5.10, which yield together that at least one constituent of question Q is not in T_P. If the constituent is '$S \vdash \alpha$', then at least one of '$S \vdash \alpha_0$', '$S \vdash \alpha_1$' is not in T_P, which yields that $ng\,Q^* \in \mathsf{T}_\mathsf{P}$. Altogether this shows that clause 1 of the definition of erotetic implication (see Definition 5.13) is satisfied. If we recall clause 2*, then it is easy to see that the rest of the proof comes to showing that 'if $af\,Q^* \in \mathsf{T}_\mathsf{P}$, then $af\,Q \in \mathsf{T}_\mathsf{P}$' and 'if $ng\,Q^* \in \mathsf{T}_\mathsf{P}$, then $ng\,Q \in \mathsf{T}_\mathsf{P}$', which is analogous to what has been already shown.

Now suppose that Q^* results from Q by L_γ of $\mathbb{E}^{\mathbf{LQ}}$. Then the questions are of the forms:

$$Q = ?(\Phi(S(\gamma) \vdash)) \qquad Q^* = ?(\Phi(S(\gamma, \gamma(t)) \vdash))$$

where t is a closed term of $\mathcal{L}_{\mathsf{FOL}}$. By clause 12.1 of Definition 5.11, for each such closed term, '$S(\gamma) \vdash$' $\in \mathsf{T}_\mathsf{P}$ iff '$S(\gamma, \gamma(t)) \vdash$' $\in \mathsf{T}_\mathsf{P}$. Together with clause 1 of Definition 5.10 this proves that $af\,Q \in \mathsf{T}_\mathsf{P}$ iff $af\,Q^* \in \mathsf{T}_\mathsf{P}$, and together with clause 2—that $ng\,Q \in \mathsf{T}_\mathsf{P}$ iff $ng\,Q^* \in \mathsf{T}_\mathsf{P}$, which is exactly what we need for erotetic implication to hold.

For the case like R_γ of $\mathbb{E}^{\mathbf{RQ}}$ the reasoning is analogous. We omit the part involving clauses 1 and 2 of Definition 5.10, and use clause 10 of Definition 5.11 taking S to be empty. The questions are of the forms:

$$Q = ?(\Phi(\vdash T(\gamma))) \qquad Q^* = ?(\Phi(\vdash T(\gamma(t))))$$

where t is a closed term that does not occur in sequent '$\vdash T(\gamma)$'. Suppose that $af\,Q \in \mathsf{T}_\mathsf{P}$, it follows then that '$\vdash T(\gamma)$' $\in \mathsf{T}_\mathsf{P}$. Then also, by clause 10 (see clause 10.1), '$\vdash T(\gamma(t))$' $\in \mathsf{T}_\mathsf{P}$. Now suppose that $af\,Q^* \in \mathsf{T}_\mathsf{P}$. Then it follows that '$\vdash T(\gamma(t))$' $\in \mathsf{T}_\mathsf{P}$, and, since t does not occur in '$\vdash T(\gamma)$', clause 10 (see clause 10.2) applies to t, thus '$\vdash T(\gamma)$' $\in \mathsf{T}_\mathsf{P}$. The reasoning is analogous for the negative answers. Suppose that $ng\,Q \in \mathsf{T}_\mathsf{P}$ and assume that $\vdash T(\gamma) \notin \mathsf{T}_\mathsf{P}$. Then also $\vdash T(\gamma(t)) \notin \mathsf{T}_\mathsf{P}$, since clause 10 holds for every closed term t that does not occur in the relevant sequent (see the transposition of clause 10.2). Thus also $ng\,Q^* \in \mathsf{T}_\mathsf{P}$. Now assume that $ngQ^* \in \mathsf{T_P}$, and additionally that $\vdash T(\gamma(t)) \notin \mathsf{T_P}$, where t does not occur in $\vdash T(\gamma)$. Since clause 10 holds for every such closed term (transposition of 10.1), it follows that $\vdash T(\gamma) \notin \mathsf{T_P}$, and therefore $ngQ \in \mathsf{T_P}$. □

5.4 Erotetic Implication: The Modal Case

As in the classical case, the erotetic part of the language $\mathcal{M}^?_\vdash$ may be supplied with a semantics to the effect that the notion of erotetic implication can be defined. This subsection is partially based on Sect. 2.4 of [6], but the definitions are altered, mainly due to the differences in the erotetic calculi described in the previous chapter.

In this section L represents the 15 basic modal propositional logics characterized in Chap. 3. Until the end of this section, we will also use letters $\mathfrak{t}, \mathfrak{u}$ for d-wffs of the language $\mathcal{M}^?_\vdash$ and $\mathfrak{X}, \mathfrak{Y}$ for sets of d-wffs of this language. We define the notion of a partition of language $\mathcal{M}^?_\vdash$ just as before (*cf.* Definition 5.9), and we connect the notion of the admissibility of a partition with the underlying logic L. For instance, a K-admissible partition of the language $\mathcal{M}^?_\vdash$ is defined as follows:

Definition 5.14 (K-*admissible Partition of* $\mathcal{M}^?_\vdash$) A partition $\mathsf{P} = \langle \mathsf{T_P}, \mathsf{U_P} \rangle$ of the language $\mathcal{M}^?_\vdash$ is K-*admissible* iff the following conditions are satisfied:

1. '$\mathfrak{t}\&\mathfrak{u}$' $\in \mathsf{T_P}$ iff both $\mathfrak{t} \in \mathsf{T_P}$ and $\mathfrak{u} \in \mathsf{T_P}$;
2. $\mathfrak{t} \in \mathsf{T_P}$ iff '$ng\mathfrak{t}$' $\notin \mathsf{T_P}$;
3. '$S \vdash \Gamma(\alpha^{\varsigma})$' $\in \mathsf{T_P}$ iff both '$S \vdash \Gamma(\alpha_0^{\varsigma})$' $\in \mathsf{T_P}$ and '$S \vdash \Gamma(\alpha_1^{\varsigma})$' $\in \mathsf{T_P}$;
4. '$S \vdash \Gamma(\beta^{\varsigma})$' $\in \mathsf{T_P}$ iff '$S \vdash \Gamma(\beta_0^{\varsigma}, \beta_1^{\varsigma})$' $\in \mathsf{T_P}$;
5. '$S \vdash \Gamma(\neg\neg A^{\varsigma})$' $\in \mathsf{T_P}$ iff '$S \vdash \Gamma(A^{\varsigma})$' $\in \mathsf{T_P}$;
6. For each numeral j that is not an element of $W(S \vdash \Gamma(\nu^{\varsigma/i}))$, '$S \vdash \Gamma(\nu^{\varsigma/i})$' $\in \mathsf{T_P}$ iff '$S \vdash \Gamma(\nu_0^{\varsigma/i,j})$' $\in \mathsf{T_P}$;
7. For each numeral j such that $\langle i, j \rangle \in R(S \vdash \Gamma(\pi^{\varsigma/i}))$, '$S \vdash \Gamma(\pi^{\varsigma/i})$' $\in \mathsf{T_P}$ iff '$S \vdash \Gamma(\pi^{\varsigma/i}, \pi_0^{j})$' $\in \mathsf{T_P}$;
8. For each numeral $i \in W(S(A) \vdash \Gamma(\_))$, '$S(A) \vdash \Gamma(\_)$' $\in \mathsf{T_P}$ iff '$S(A) \vdash \Gamma(-A^i)$' $\in \mathsf{T_P}$.

For further use, let us recall that by $R(\varphi)^{\mathsf{L}}$ we mean the respective closure of relation $R(\varphi)$ that is specific to L (see Tables 3.4 and 3.5 in Sec. 3.3).

Definition 5.15 (L-*admissible Partition of* $\mathcal{M}^?_\vdash$) A partition $\mathsf{P} = \langle \mathsf{T_P}, \mathsf{U_P} \rangle$ of the language $\mathcal{M}^?_\vdash$ is L-*admissible* iff it satisfies clauses 1–6 and 8 of Definition 5.14, clause 7*:

7*. For each numeral j such that $\langle i, j \rangle \in R(S \vdash \Gamma(\pi^{\varsigma/i}))^{\mathsf{L}}$, '$S \vdash \Gamma(\pi^{\varsigma/i})$' $\in \mathsf{T_P}$ iff '$S \vdash \Gamma(\pi^{\varsigma/i}, \pi_0^{j})$' $\in \mathsf{T_P}$

and for $\mathsf{L} \in \{\mathsf{D}, \mathsf{DB}, \mathsf{D4}, \mathsf{D5}, \mathsf{D45}\}$ also clause D:

D. For each j that is not an element of $W(S \vdash \Gamma(\pi^{\varsigma/i}))$, '$S \vdash \Gamma(\pi^{\varsigma/i})$' $\in \mathsf{T_P}$ iff '$S \vdash \Gamma(\pi^{\varsigma/i}, \pi_0^{\varsigma/i,j})$' $\in \mathsf{T_P}$.

As in the case of classical logic, the notion of L-admissible partition allows to express both the entailment between d-wffs of $\mathcal{M}^?_\vdash$ and the erotetic implication between questions of this language. The definitions follow strictly those presented above.

Definition 5.16 (*Entailment in* $\mathcal{M}^?_\vdash$ *Relative to* L) Let $\mathfrak{X}$ be a set of d-wffs of the language $\mathcal{M}^?_\vdash$ and let t stand for a single d-wff of this language. We say that the set $\mathfrak{X}$ *entails the formula* t *in* $\mathcal{M}^?_\vdash$ *relatively to* L, symbolically:

$$\mathfrak{X} \models_{\mathcal{M}^?_{\vdash\mathsf{L}}} \mathsf{t}$$

iff there is no L-admissible partition P of the language $\mathcal{M}^?_\vdash$ such that $\mathfrak{X} \subseteq \mathsf{T_P}$ and $\mathsf{t} \notin \mathsf{T_P}$.

As previously, the notion of erotetic implication may be defined in a general form, as in Definition 5.17.

Definition 5.17 (*Erotetic Implication in* $\mathcal{M}^?_\vdash$ *Relative to* L) Let Q and Q^* be questions of $\mathcal{M}^?_\vdash$ and let $\mathfrak{X}$ be a set of d-wffs of $\mathcal{M}^?_\vdash$. We say that the question Q *implies in* $\mathcal{M}^?_\vdash$ *relatively to* L *the question* Q^* *on the basis of the set* $\mathfrak{X}$ (symbolically: $\mathsf{Im_L}(Q, \mathfrak{X}, Q^*)$) iff for each L-admissible partition P of the language $\mathcal{M}^?_\vdash$,

1. For each $\mathsf{t} \in dQ$, if $\mathfrak{X} \cup \{\mathsf{t}\} \subseteq \mathsf{T_P}$, then $dQ^* \cap \mathsf{T_P} \neq \varnothing$; and
2. For each $\mathsf{u} \in dQ^*$, there is a non-empty, proper subset $\mathfrak{Y}$ of dQ such that if $\mathfrak{X} \cup \{\mathsf{u}\} \subseteq \mathsf{T_P}$, then $\mathfrak{Y} \cap \mathsf{T_P} \neq \varnothing$.

The following theorem is an analogue of Theorem 5.2; clause 3 expresses the fact that the rules of $\mathbb{E}^\mathsf{L}$ capture appropriate—depending on L—cases of erotetic implication, hence the rules are semantically correct in terms of the Minimal Erotetic Semantics constructed for the modal case.

Theorem 5.3 *Let* Q *and* Q^* *be questions of* $\mathcal{M}^?_\vdash$. *If* Q^* *results from* Q *by a rule* **r** *of* $\mathbb{E}^\mathsf{L}$, *then*

1. $\{af\,Q\} \models_{\mathcal{M}^?_{\vdash\mathsf{L}}} af\,Q^*$ *and* $\{af\,Q^*\} \models_{\mathcal{M}^?_{\vdash\mathsf{L}}} af\,Q$,
2. $\{ng\,Q\} \models_{\mathcal{M}^?_{\vdash\mathsf{L}}} ng\,Q^*$ *and* $\{ng\,Q^*\} \models_{\mathcal{M}^?_{\vdash\mathsf{L}}} ng\,Q$,
3. $\mathsf{Im_L}(Q, \varnothing, Q^*)$.

Proof Let us consider the case of $\mathbf{r} = R_\pi$. Questions Q and Q^* are of the following forms:

$$Q = ?(\Phi(S \vdash \Gamma(\pi^{\varsigma/i}))) \qquad Q^* = ?(\Phi(S \vdash \Gamma(\pi^{\varsigma/i}, \pi_0{}^j)))$$

Suppose that $\mathsf{P} = \langle \mathsf{T_P}, \mathsf{U_P} \rangle$ is an L-admissible partition of the language $\mathcal{M}^?_\vdash$. Assume that $af\,Q \in \mathsf{T_P}$. Then, in particular, '$S \vdash \Gamma(\pi^{\varsigma/i})$' $\in \mathsf{T_P}$. Since the rule is applied in calculus $\mathbb{E}^\mathsf{L}$, the proviso of applicability of R_π must be satisfied, thus $\langle i, j \rangle \in R(S \vdash \Gamma(\pi^{\varsigma/i}))^\mathsf{L}$. By clause 7* of Definition 5.15, '$S \vdash \Gamma(\pi^{\varsigma/i}, \pi_0{}^j)$' $\in \mathsf{T_P}$, hence also $af\,Q^* \in \mathsf{T_P}$. The converse implication—if $af\,Q^* \in \mathsf{T_P}$, then $af\,Q \in \mathsf{T_P}$—follows by the same clause 7*.

We reason analogously for the negative answers. Numeral j is such that $\langle i, j\rangle \in R(S \vdash \Gamma(\pi^{\varsigma/i}))^{\mathsf{L}}$, therefore, by clause *7**, '$S \vdash \Gamma(\pi^{\varsigma/i})$' $\notin \mathsf{T_P}$ iff '$S \vdash \Gamma(\pi^{\varsigma/i}, \pi_0{}^{j})$' $\notin \mathsf{T_P}$. This is enough to show that $ngQ \in \mathsf{T_P}$ iff $ngQ^* \in \mathsf{T_P}$, we skip the remaining details.

The reasoning is analogous for the cases of $\mathbf{r} \in \{R_\nu, R_{\pi D}\}$. We make use of clause *6* of Definition 5.14 and clause *D* of Definition 5.15, respectively, and of the observation that if $j = \max(W(\phi)) + 1$ (where ϕ is $S \vdash \Gamma(\nu^{\varsigma/i})$ or $S \vdash \Gamma(\pi^{\varsigma/i})$, respectively), then it is not an element of ϕ.

All the remaining cases are straightforward. □

5.5 Erotetic Implication: The Intuitionistic Case

In this section, we repeat the above scheme to supply the erotetic part of the language $\mathcal{L}^?_{\vdash\mathsf{INT}}$ with a semantics. First, let us recall that by $R(\phi)^{tr}$ we mean the transitive closure of a relation $R(\phi)$, and by $R(\phi)^{ref,tr}$ we mean the reflexive and transitive closure of $R(\phi)$.

Definition 5.18 (INT-*admissible Partition of* $\mathcal{L}^?_{\vdash\mathsf{INT}}$) A partition $\mathsf{P} = \langle \mathsf{T_P}, \mathsf{U_P}\rangle$ of the language $\mathcal{L}^?_{\vdash\mathsf{INT}}$ is INT-*admissible* iff the following conditions are satisfied:

1. '$\mathsf{t}\&\mathsf{u}$' $\in \mathsf{T_P}$ iff both $\mathsf{t} \in \mathsf{T_P}$ and $\mathsf{u} \in \mathsf{T_P}$;
2. $\mathsf{t} \in \mathsf{T_P}$ iff '$ng\mathsf{t}$' $\notin \mathsf{T_P}$;
3. For each numeral j such that $\langle i, j\rangle \in R(\Gamma(p_n{}^{\varsigma/i}) \vdash \Delta)^{tr}$, where p_n is an arbitrary propositional variable,
   '$\Gamma(p_n{}^{\varsigma/i}) \vdash \Delta$' $\in \mathsf{T_P}$ iff '$\Gamma(p_n{}^{\varsigma/i}, p_n{}^{j}) \vdash \Delta$' $\in \mathsf{T_P}$;
4. For each j such that $\langle i, j\rangle \in R(\Gamma(\sim A^{\varsigma/i}) \vdash \Delta(\_))^{ref,tr}$,
   '$\Gamma(\sim A^{\varsigma/i}) \vdash \Delta(\_)$' $\in \mathsf{T_P}$ iff '$\Gamma(\sim A^{\varsigma/i}) \vdash \Delta(A^{j})$' $\in \mathsf{T_P}$;
5. For each $j \notin W(\Gamma(\_) \vdash \Delta(\sim A^{\varsigma/i}))$,
   '$\Gamma(\_) \vdash \Delta(\sim A^{\varsigma/i})$' $\in \mathsf{T_P}$ iff '$\Gamma(A^{\varsigma/i,j}) \vdash \Delta(\_)$' $\in \mathsf{T_P}$;
6. For each j such that $\langle i, j\rangle \in R(\Gamma(A \supset B^{\varsigma/i})(\_) \vdash \Delta(\_))^{ref,tr}$,
   '$\Gamma(A \supset B^{\varsigma/i})(\_) \vdash \Delta(\_)$' $\in \mathsf{T_P}$ iff both
   '$\Gamma(A \supset B^{\varsigma/i})(\_) \vdash \Delta(A^{j})$' $\in \mathsf{T_P}$ and '$\Gamma(A \supset B^{\varsigma/i})(B^{j}) \vdash \Delta(\_)$' $\in \mathsf{T_P}$;
7. For each $j \notin W(\Gamma(\_) \vdash \Delta(A \supset B^{\varsigma/i}))$,
   '$\Gamma(\_) \vdash \Delta(A \supset B^{\varsigma/i})$' $\in \mathsf{T_P}$ iff '$\Gamma(A^{\varsigma/i,j}) \vdash \Delta(B^{\varsigma/i,j})$' $\in \mathsf{T_P}$;
8. '$\Gamma(A \wedge B^{\varsigma/i}) \vdash \Delta$' $\in \mathsf{T_P}$ iff '$\Gamma(A^{\varsigma/i}, B^{\varsigma/i}) \vdash \Delta$' $\in \mathsf{T_P}$;
9. '$\Gamma \vdash \Delta(A \vee B^{\varsigma/i})$' $\in \mathsf{T_P}$ iff '$\Gamma \vdash \Delta(A^{\varsigma/i}, B^{\varsigma/i})$' $\in \mathsf{T_P}$;
10. '$\Gamma \vdash \Delta(A \wedge B^{\varsigma/i})$' $\in \mathsf{T_P}$ iff both '$\Gamma \vdash \Delta(A^{\varsigma/i})$' $\in \mathsf{T_P}$ and
    '$\Gamma \vdash \Delta(B^{\varsigma/i})$' $\in \mathsf{T_P}$;
11. '$\Gamma(A \vee B^{\varsigma/i}) \vdash \Delta$' $\in \mathsf{T_P}$ iff both '$\Gamma(A^{\varsigma/i}) \vdash \Delta$' $\in \mathsf{T_P}$ and
    '$\Gamma(B^{\varsigma/i}) \vdash \Delta$' $\in \mathsf{T_P}$.

Definition 5.19 (*Entailment in* $\mathcal{L}^?_{\vdash\mathsf{INT}}$) Let $\mathfrak{X}$ be a set of d-wffs of the language $\mathcal{L}^?_{\vdash\mathsf{INT}}$ and let t stand for a single d-wff of this language. We say that $\mathfrak{X}$ *entails the formula* t *in* $\mathcal{L}^?_{\vdash\mathsf{INT}}$, symbolically:

$$\mathfrak{X} \models_{\mathcal{L}^?_{\vdash \mathsf{INT}}} \mathfrak{t}$$

iff there is no INT-admissible partition P of the language $\mathcal{L}^?_{\vdash \mathsf{INT}}$ such that $\mathfrak{X} \subseteq \mathsf{T_P}$ and $\mathfrak{t} \notin \mathsf{T_P}$.

Definition 5.20 (*Erotetic Implication in* $\mathcal{L}^?_{\vdash \mathsf{INT}}$) Let Q and Q^* stand for questions of $\mathcal{L}^?_{\vdash \mathsf{INT}}$ and let $\mathfrak{X}$ be a set of d-wffs of this language. We say that Q *implies in* $\mathcal{L}^?_{\vdash \mathsf{INT}}$ Q^* on the basis of the set $\mathfrak{X}$ *of d-wffs* (symbolically: $\mathsf{Im_{INT}}(Q, \mathfrak{X}, Q^*)$) iff for each INT-admissible partition P of the language $\mathcal{L}^?_{\vdash \mathsf{INT}}$,

1. For each $\mathfrak{t} \in dQ$, if $\mathfrak{X} \cup \{\mathfrak{t}\} \subseteq \mathsf{T_P}$, then $dQ^* \cap \mathsf{T_P} \neq \varnothing$; and
2. For each $\mathfrak{u} \in dQ^*$, there is a non-empty, proper subset $\mathfrak{Y}$ of dQ such that if $\mathfrak{X} \cup \{\mathfrak{u}\} \subseteq \mathsf{T_P}$, then $\mathfrak{Y} \cap \mathsf{T_P} \neq \varnothing$.

The following theorem is true, we skip the proof.

Theorem 5.4 *Let Q and Q^* be questions of $\mathcal{L}^?_{\vdash \mathsf{INT}}$. If Q^* results from Q by a rule* **r** *of* $\mathbb{E}^{\mathsf{INT}}$, *then*

1. $\{af\,Q\} \models_{\mathcal{L}^?_{\vdash \mathsf{INT}}} af\,Q^*$ *and* $\{af\,Q^*\} \models_{\mathcal{L}^?_{\vdash \mathsf{INT}}} af\,Q$,
2. $\{ng\,Q\} \models_{\mathcal{L}^?_{\vdash \mathsf{INT}}} ng\,Q^*$ *and* $\{ng\,Q^*\} \models_{\mathcal{L}^?_{\vdash \mathsf{INT}}} ng\,Q$,
3. $\mathsf{Im_{INT}}(Q, \varnothing, Q^*)$.

Theorem 5.4 expresses the fact that the rules of $\mathbb{E}^{\mathsf{INT}}$ are correct in terms of the Minimal Erotetic Semantics constructed for the intuitionistic case.

5.6 Erotetic Calculi and Multiple-Conclusion Relations

In this section, we return to classical logic.

When we study questions concerning derivability, a natural issue arises regarding the type of the considered derivability relation. Typically, derivability is a concept defined relative to a specific proof system for a given logic. Examining the examples provided by Wiśniewski (e.g., in the introductions to [21, 26]) suggests that this notion often pertains to either an axiomatic framework or a natural deduction system for the underlying logic.

As discussed in Sect. 2.2, the sequents of the language $\mathcal{L}^?_{\vdash \mathsf{L}}$ have natural semantic interpretations. Both-sided, single succedent sequents may be interpreted as statements concerning semantic entailment in L; left-sided sequents may represent statements concerning semantically defined inconsistency in L; and right-sided sequents '$\vdash S$' can be understood, in the simplest terms, as assertions regarding the L-validity of the disjunction of the terms of S. This sets the stage for generalizations.

5.6.1 Semantic Interpretation

In the works by Wiśniewski, the relation of multiple-conclusion entailment is used in interpreting sequents. We have introduced the notion for the case of CPL (see Sect. 5.1). Let us now formulate the definition for FOL.

Definition 5.21 (*Multiple-Conclusion Entailment in* FOL) Let X and Y stand for arbitrary sets of formulas of the language $\mathcal{L}_{\mathsf{FOL}}$. We say that the set X *multiple-conclusion entails the set* Y *in* FOL, symbolically, $X \Vdash_{\mathsf{FOL}} Y$, iff:

- For every FOL-interpretation $\mathcal{I}$ of $\mathcal{L}_{\mathsf{FOL}}$, if each element of X is true under $\mathcal{I}$, then there is at least one element of Y true under $\mathcal{I}$.

Let $\mathsf{L} \in \{\mathsf{CPL}, \mathsf{FOL}\}$. Introducing the notion of multiple-conclusion entailment on the level of the underlying logic L yields the opportunity to interpret every sequent of the language $\mathcal{L}^{?}_{\vdash\mathsf{L}}$ as referring to multiple-conclusion entailment in L. At the moment, we do not use partitions, but refer to the underlying semantics.[6]

We will need one more notion—that of a *safeset*, introduced by Wiśniewski in the paper [24, p. 327].

Definition 5.22 (*Safeset in* L) A set X of formulas of $\mathcal{L}_{\mathsf{L}}$ is called a *safeset in* L iff for each L-interpretation $\mathcal{I}$ of $\mathcal{L}_{\mathsf{L}}$, there is a formula in X which is true under $\mathcal{I}$.

Example 5.1 For any sentence A of $\mathcal{L}_{\mathsf{L}}$, the set $\{A, -A\}$ is both semantically inconsistent in L and a safeset in L.

The following holds, the proof is straightforward and we skip it.

Corollary 5.4 *Let X and Y stand for sets of formulas of the language $\mathcal{L}_{\mathsf{L}}$.*

1. *X is semantically inconsistent in* L, *that is, there is no* L*-model of X, iff $X \Vdash_{\mathsf{L}} \varnothing$.*
2. *X entails a formula A in* L *iff $X \Vdash_{\mathsf{L}} \{A\}$.*
3. *Y is a safeset in* L *iff $\varnothing \Vdash_{\mathsf{L}} Y$.*

Thus, as we can see, the relation of multiple-conclusion entailment in L generalizes all the three properties/relations that underlie the semantic interpretation of sequents. Consequently, the rules transforming expressions which involve sequents, as questions of our languages do, can be interpreted as driving as towards a solution of problems concerning the multiple-conclusion entailment relation.

There is, however, an unnatural difference between the self-imposing syntactic interpretation of both-sided and left-sided sequents and hence the rules dealing with them, and no obvious interpretation of right-sided sequents in this aspect.

The next section is devoted to developing a syntactical relation which generalizes both the derivability in the underlying logic and syntactically defined inconsistency, and at the same time it encompasses the cases of right-sided sequents. In other words, we shall define a proof-theoretical counterpart of the semantic relation of multiple-conclusion entailment.

---

[6] In this respect, we do not follow Wiśniewski strictly, *cf.* [22, Sect. 3.2].

5.6.2 Proof-Theoretical Interpretation

The claim of this section is the following: each sequent of the languages designed for classical logic can be interpreted proof-theoretically (that is, by reference to a relation of derivability) in terms of a general relation of multiple-conclusion consequence, which will be defined in this section. Roughly speaking, the meaning of this relation is as follows: Y 'follows from' X iff Y presents a space of possibilities such that whatever can be derived from the entire Y **by cases** can be derived directly from X. In other words, Y presents the cases to consider in proofs by cases from X. This is precisely the situation encountered within natural deduction systems when applying the rule for eliminating disjunction.

This section is largely motivated by the account of consequence relations presented in the book [10]. Andrzej Wiśniewski developed many generalizations inspired by similar ideas in his papers [24, 25], but the specific problems addressed and the solutions developed in [24, 25] and here are distinct.

As mentioned earlier, the intuitive meaning of the turnstile symbol '$\vdash$', as an element of the object-level language $\mathcal{L}^?_{\vdash \mathsf{L}}$, is as referring to *derivability* in L.

Until the end of this chapter, $\mathsf{L} \in \{\mathsf{CPL}, \mathsf{FOL}\}$. Let $FORM_\mathsf{L}$ stand for the set of all formulas of the language $\mathcal{L}_\mathsf{L}$. Suppose that the following relation:

$$\vdash_\mathsf{L} \subseteq 2^{FORM_\mathsf{L}} \times FORM_\mathsf{L}$$

is a consequence relation generated by a proof system which is sound and complete with respect to $\models_\mathsf{L}$, that is:

$$X \vdash_\mathsf{L} A \quad \text{iff} \quad X \models_\mathsf{L} A.$$

Let us stress that this time '$\vdash_\mathsf{L}$' is not an object-level symbol. We think of '$\vdash_\mathsf{L}$' as referring to proof-theoretical *derivability* or *deducibility*, but do not assume anything more specific about the relation, except for compactness and the following three properties (see [10, p. 15]):

Overlap	If $A \in X$, then $X \vdash_\mathsf{L} A$.
Dilution	If $X^* \vdash_\mathsf{L} A$ and $X^* \subseteq X$, then $X \vdash_\mathsf{L} A$.
Cut for Sets	If $X \cup Z \vdash_\mathsf{L} A$ and $X \vdash_\mathsf{L} C$ for each $C \in Z$, then $X \vdash_\mathsf{L} A$.

The first property is also called *reflexivity* in the literature, the second one—*monotonicity*, and the third one expresses a form of *transitivity* of the relation.[7]

[7] The reader may compare the discussed properties with the standard conditions by Tarski. For a given language $\mathcal{L}$ and its set of formulas $FORM$, $C \subseteq 2^{FORM} \times 2^{FORM}$ is a *consequence operation in* $\mathcal{L}$ iff for each $X \subseteq FORM$:
$X \subseteq C(X)$—reflexivity,
If $X \subseteq Y$, Then $C(X) \subseteq C(Y)$—monotonicity,
$C(C(X)) \subseteq C(X)$—idempotence.

The last property may be expressed in a more familiar way:

Transitivity If $X \vdash_{\mathsf{L}} F$ and $\{F\} \vdash_{\mathsf{L}} A$, then $X \vdash_{\mathsf{L}} A$.

The following holds:

Corollary 5.5 *If* $\vdash_{\mathsf{L}} \subseteq 2^{FORM_{\mathsf{L}}} \times FORM_{\mathsf{L}}$ *is a relation satisfying* Dilution *and* Cut for Sets, *then it owns* Transitivity *as well.*

Proof Suppose that $X \vdash_{\mathsf{L}} F$ and $\{F\} \vdash_{\mathsf{L}} A$. By Dilution, $X \cup \{F\} \vdash_{\mathsf{L}} A$. Since $X \vdash_{\mathsf{L}} F$ (it holds for every formula from $\{F\}$), by Cut for Sets also $X \vdash_{\mathsf{L}} A$. □

We will write '$\vdash_{\mathsf{L}} A$' instead of '$\varnothing \vdash_{\mathsf{L}} A$'. A natural way to express the fact that A is a $\vdash_{\mathsf{L}}$-consequence of the empty set is to say that *A is a thesis of* L. As we wrote, we also assume that '$\vdash_{\mathsf{L}}$' is compact:

Compactness $X \vdash_{\mathsf{L}} A$ iff for some finite subset $Y \subseteq X$, $Y \vdash_{\mathsf{L}} A$.

The assumption is quite natural when the consequence relation is thought of as generated by one of, lets say, standard proof systems, where *proof* is usually a finite structure (sequence, tree, etc.) of formulas or other objects, like sequents. Assuming that $\vdash_{\mathsf{L}}$ is compact, it is also trivially true that:

Corollary 5.6 *If X is non-empty, then $X \vdash_{\mathsf{L}} A$ iff for some finite and non-empty $Y \subseteq X$, $Y \vdash_{\mathsf{L}} A$.*

The reason for taking non-emptiness into consideration becomes clear in the sequel. Now let us introduce the following auxiliary concept:

Definition 5.23 (*By-Cases-Consequence in* L) Let A stand for a formula of the language $\mathcal{L}_{\mathsf{L}}$ and let X be a set of formulas of this language. We say that A is a *by-cases-consequence of X in* L, symbolically,

$$X \vdash_{\mathsf{L}}^{bc} A$$

if for every finite and non-empty $Y \subseteq X$, $Y \vdash_{\mathsf{L}} A$.

Definition 5.23 states that independently of our choice of the (finite number of) premises from X, we will always arrive at A. As we can see from Corollary 5.6 and Definition 5.23, in the case of non-empty sets, the definition generalizes a property which in the case of '$\vdash_{\mathsf{L}}$' is expressed by an existential quantification. The relation of by-cases-consequence in L is stronger than $\vdash_{\mathsf{L}}$ in the following sense:

Corollary 5.7 *Let $X \neq \varnothing$. If $X \vdash_{\mathsf{L}}^{bc} A$, then $X \vdash_{\mathsf{L}} A$.*

C satisfying these three conditions is typically called *consequence operation in Tarski's sense*. See [13–15], compare [28, Chap. 1].

Proof Since X is non-empty, there exists a finite and non-empty subset Y of X, and, by assumption, $Y \vdash_{\mathsf{L}} A$. Then, by Dilution, $X \vdash_{\mathsf{L}} A$. □

Imagine that X is a search space, or a 'knowledge base,' however, not necessarily a consistent one. Suppose that we have some doubts regarding X—we have a reason to believe that some of the assumptions are wrong, but we do not know which are the problematic ones. Then we may have reasonable doubts as to a conclusion that follows from this or another choice of premises taken from X. If, however, the conclusion follows from every such (finite) choice of premises, then we can feel much more certain about it. The notion of by-cases-consequence expresses this kind of comfortable situation, when we can derive a conclusion with high level of certainty, although the premises are highly uncertain. In the margin, observe that this kind of certainty requires Dilution (i.e., monotonicity).

As a side effect of the universal quantification in Definition 5.23, we have:

Corollary 5.8 $\varnothing \vdash^{bc}_{\mathsf{L}} A$ *for any formula* A *of the language* $\mathcal{L}_{\mathsf{L}}$.

which may be interpreted as an expression of the fact that when there are no premises for reasoning by cases, then we are as helpless as in the face of inconsistency. Also the following holds:

Corollary 5.9 *If* A *is a thesis of* L, *then* $X \vdash^{bc}_{\mathsf{L}} A$ *for every set* X.

Proof Once again, it is worth to indicate that the result needs Dilution. For if A is a thesis of L, then $\varnothing \vdash_{\mathsf{L}} A$. Assume that X is arbitrary. Then, by the Dilution property, for each finite and non-empty subset Y of X, $Y \vdash_{\mathsf{L}} A$. Thus $X \vdash^{bc}_{\mathsf{L}} A$ for every set X. □

Let us now note that:

Corollary 5.10 $X \vdash^{bc}_{\mathsf{L}} A$ *iff for each singleton* $\{B\} \subseteq X$, $\{B\} \vdash_{\mathsf{L}} A$.

Proof The first implication follows from the very Definition 5.23.

For the converse implication, assume that $Y \subseteq X$ is finite and nonempty. Since it is nonempty, there is $B \in Y$. By assumption, $\{B\} \vdash_{\mathsf{L}} A$. By Dilution, $Y \vdash_{\mathsf{L}} A$. Since Y was arbitrary, we have shown that $X \vdash^{bc}_{\mathsf{L}} A$. □

Example 5.2 Suppose that $\vdash_{\mathsf{CPL}}$ is the consequence relation generated by the axiomatic account of CPL introduced by Hilbert and Bernays.[8] Thus, $X \vdash_{\mathsf{CPL}} A$ iff there exists a derivation of the formula A from X governed by the rules of the axiomatic system. We shall not prove this, but it is well-known that the relation satisfies Overlap, Dilution, and Cut for Sets. Formula 'p' of the language $\mathcal{L}_{\mathsf{CPL}}$ is a by-cases-consequence of the set: $\{p \wedge q, p, p \vee p\}$, and also of the set: $\{$'$p \wedge B$' $: B \in \mathrm{VAR}\}$, symbolically:

$$\{p \wedge q, p, p \vee p\} \vdash^{bc}_{\mathsf{CPL}} p$$

[8] Presented in [4].

$$\{`p \wedge B' : B \in VAR\} \vdash^{bc}_{\mathsf{CPL}} p$$

Conversely, $\{p, q\} \vdash_{\mathsf{CPL}} p$, but $\{p, q\} \nvdash^{bc}_{\mathsf{CPL}} p$, because $\{q\} \nvdash_{\mathsf{CPL}} p$.

Example 5.3 Let $\vdash_{\mathsf{FOL}}$ be the consequence relation generated by the axiomatic system Q_1 used by Smullyan.[9] Once again, $X \vdash_{\mathsf{FOL}} A$ iff there exists a derivation of the formula A from X governed by the rules of Q_1. As previously—the relation owns the properties of Overlap, Dilution, and Cut for Sets. Then, formula '$\exists x P(x)$' is a by-cases consequence in FOL of the set $\{P(a_i) : a_i$ is an individual constant of $\mathcal{L}_{\mathsf{FOL}}\}$.

Let us also note that:

Corollary 5.11 *Suppose that* $\vdash_{\mathsf{L}}$ *satisfies* Dilution, *and that for any finite set* $Y = \{A_1, \ldots, A_n\}$ *of formulas of* $\mathcal{L}_{\mathsf{L}}$, $\vdash_{\mathsf{L}}$ *respects the following behavior of the generalized disjunction connective,* $\bigvee$,

- $\{A_i\} \vdash_{\mathsf{L}} \bigvee Y$, $i = 1, \ldots, n$.

Then for each non-empty subset $Y^* \subseteq Y$, *it holds that* $Y^* \vdash_{\mathsf{L}} \bigvee Y$. *Moreover, it follows that* $Y \vdash^{bc}_{\mathsf{L}} \bigvee Y$.

Let us observe that the relation '$\vdash^{bc}_{\mathsf{L}}$' is reflexive in the following sense:

Corollary 5.12 *For each formula A of the language* $\mathcal{L}_{\mathsf{L}}$, $\{A\} \vdash^{bc}_{\mathsf{L}} A$.

However, this relation is not reflexive in the more general sense expressed by the property called 'Overlap'. Moreover, it is non-monotonic:

Example 5.4 Let $\vdash_{\mathsf{CPL}}$ be defined as in Example 5.2. We have: $\{p\} \vdash^{bc}_{\mathsf{CPL}} p$, but $\{p, q\} \nvdash^{bc}_{\mathsf{CPL}} p$. Hence, for $\mathsf{L} = \mathsf{CPL}$, the relation $\vdash^{bc}_{\mathsf{L}}$ does not satisfy *Dilution*.

Similarly:

Example 5.5 Let $\vdash_{\mathsf{FOL}}$ be defined as in Example 5.3. We have:
$\{\forall x P(x)\} \vdash^{bc}_{\mathsf{FOL}} \forall x P(x)$, but $\{\forall x P(x), \forall x Q(x)\} \nvdash^{bc}_{\mathsf{FOL}} \forall x P(x)$.

The above examples show that the relation of by-cases-consequence in L *is not* a consequence relation in the sense explicated by Shoesmith and Smiley. The relation is not reflexive in the sense defined by 'Overlap' (it is, however, reflexive in the most straightforward sense given by Corollary 5.12). What is more, the relation does not own the property of monotonicity expressed by 'Dilution', but, interestingly, it does own a property which is *dual* to Dilution with respect to set inclusion, namely:

Corollary 5.13 *If* $X \vdash^{bc}_{\mathsf{L}} A$ *and* $Y \subseteq X$, *then* $Y \vdash^{bc}_{\mathsf{L}} A$.

As to the property of Cut for Sets, if we take it literally as defined above, then it holds trivially by Corollary 5.13.

Corollary 5.14 *If* $X \cup Z \vdash^{bc}_{\mathsf{L}} A$ *and* $X \vdash_{\mathsf{L}} C$ *for each* $C \in Z$, *then* $X \vdash^{bc}_{\mathsf{L}} A$.

[9] See [11, p. 81].

Proof Let $Y \subseteq X$ be finite. Y is then a subset of $X \cup Z$, hence $X \vdash_{\mathsf{L}}^{bc} A$, by assumption. □

However, the following version, that may be conceived as dual to Cut for Sets, seems more interesting. More specifically, clause 1 of the following corollary expresses a restricted form of Dilution for $\vdash_{\mathsf{L}}^{bc}$, and clause 2 can be regarded as expressing transitivity of $\vdash_{\mathsf{L}}^{bc}$.

Corollary 5.15 *If $X \vdash_{\mathsf{L}}^{bc} C$ for each $C \in Z$, and $Z \vdash_{\mathsf{L}}^{bc} A$, then*

1. *$X \cup Z \vdash_{\mathsf{L}}^{bc} A$, and*
2. *$X \vdash_{\mathsf{L}}^{bc} A$.*

Proof If $X \vdash_{\mathsf{L}}^{bc} C$, then, by Corollary 5.10, (i) for each $B \in X$, $\{B\} \vdash_{\mathsf{L}} C$. Similarly, if $Z \vdash_{\mathsf{L}}^{bc} A$, then (ii) for each $C \in Z$, $\{C\} \vdash_{\mathsf{L}} A$. Corollary 5.5, together with (i) and (ii), yields that (iii) $\{B\} \vdash_{\mathsf{L}} A$ for each $B \in X$. (ii) and (iii) together show that for each $D \in X \cup Z$, $\{D\} \vdash_{\mathsf{L}}^{bc} A$, and Corollary 5.10 entails $X \cup Z \vdash_{\mathsf{L}}^{bc} A$.

Clause 2 follows from clause 1 by Corollary 5.13. □

Before we continue, let us prove one more result which also explains the meaning of our by-cases-consequence. Theorem 5.5 states that *if* the notion of *derivability* explicated by '$\vdash_{\mathsf{L}}$' presupposes that the *premises* of a *derivation* behave like joined with conjunction, *and* the implication connective is such that the deduction theorem holds, *then* the notion of *derivability by cases* explicated by '$\vdash_{\mathsf{L}}^{bc}$' presupposes *disjunctive* interpretation of the set of premises. Once again, an association with the rule for disjunction elimination present in many natural deduction systems seems inevitable. However, what may seem interesting is also the fact that the theorem and its proof highlight what needs to be assumed about the language and its connectives in order to derive the result. In particular, the assumptions concerning the behavior of conjunction and disjunction in the context of '$\vdash_{\mathsf{L}}$' coincide with those expressed by Shoesmith and Smiley in their book [10].

Theorem 5.5 *Suppose that the consequence relation $\vdash_{\mathsf{L}}$ is defined as above, that is, it satisfies, int.al.,* Dilution *and* Cut for Sets. *Suppose also that the behavior of the connectives $\wedge, \vee, \rightarrow$ in the context of $\vdash_{\mathsf{L}}$ is such that:*

(*a*) *For any formulas A, B of $\mathcal{L}_{\mathsf{L}}$, $\{A\} \vdash_{\mathsf{L}} A \vee B$ and $\{B\} \vdash_{\mathsf{L}} A \vee B$,*
(*b*) *For any formulas A, B, C of $\mathcal{L}_{\mathsf{L}}$, if $\{A\} \vdash_{\mathsf{L}} C$ and $\{B\} \vdash_{\mathsf{L}} C$, then $\{A \vee B\} \vdash_{\mathsf{L}} C$,*
(*c*) *For arbitrary sentences $B_1, \ldots, B_n$ of the language, and an arbitrary formula $A : \{B_1, \ldots, B_n\} \vdash_{\mathsf{L}} A$ iff $\vdash_{\mathsf{L}} (B_1 \wedge \ldots \wedge B_n) \rightarrow A$.*

Let X be a finite, non-empty set of sentences of the language $\mathcal{L}_{\mathsf{L}}$ and let A be a formula of $\mathcal{L}_{\mathsf{L}}$. Then:

$$X \vdash_{\mathsf{L}}^{bc} A \text{ iff } \vdash_{\mathsf{L}} \bigvee X \rightarrow A.$$

Proof Let $X = \{B_1, \ldots, B_n\}$ be a finite, non-empty set of sentences of language $\mathcal{L}_{\mathsf{L}}$ and let A be a formula of $\mathcal{L}_{\mathsf{L}}$.

If $X \vdash^{bc}_{\mathsf{L}} A$, then also $\{B_i\} \vdash_{\mathsf{L}} A$ for each $B_i \in X$, by Corollary 5.10. By (b) applied $n - 1$ times, $\{\bigvee X\} \vdash_{\mathsf{L}} A$. '$\bigvee X$' is a sentence of $\mathcal{L}_{\mathsf{L}}$, and hence by (c), $\vdash_{\mathsf{L}} \bigvee X \to A$.

For the second implication, assume that $\vdash_{\mathsf{L}} \bigvee X \to A$. Then, by (c), $\{\bigvee X\} \vdash_{\mathsf{L}} A$. Let $B_i \in X$ be arbitrary. By (a), we have both $\{B_i\} \vdash_{\mathsf{L}} B_i \vee B_{i+1}$ (for $1 \leqslant i \leqslant n - 1$) and $\{B_i\} \vdash_{\mathsf{L}} B_{i-1} \vee B_i$ (for $2 \leqslant i \leqslant n$). Referring to (a) an appropriate number of times, and by Transitivity of $\vdash_{\mathsf{L}}$, we arrive at $\{B_i\} \vdash_{\mathsf{L}} \bigvee X$ (as to Transitivity, by assumption, Dilution and Cut for Sets hold, hence Transitivity holds as well by Corollary 5.5). Once again, by Transitivity, $\{B_i\} \vdash_{\mathsf{L}} A$. But B_i was an arbitrary element of X, therefore, by Corollary 5.10, $X \vdash^{bc}_{\mathsf{L}} A$. □

Finally, let us define the following:

Definition 5.24 (*Multiple-Conclusion Consequence*) Let X and Y be sets of formulas of $\mathcal{L}_{\mathsf{L}}$, and let $\vdash_{\mathsf{L}}$ stand for a relation defined as above, that is, $\vdash_{\mathsf{L}} \subseteq 2^{FORM_{\mathsf{L}}} \times FORM_{\mathsf{L}}$ and $\vdash_{\mathsf{L}}$ satisfies *Overlap*, *Dilution* and *Cut for Sets*. We will say that Y is a *multiple-conclusion consequence of X in* L, symbolically,

$$X \Vdash_{\mathsf{L}} Y$$

iff for every formula A of $\mathcal{L}_{\mathsf{L}}$: if $Y \vdash^{bc}_{\mathsf{L}} A$, then $X \vdash_{\mathsf{L}} A$.

Let us recall the intuitions we have described at the beginning of this section. We aim at a description of a multiple-conclusion relation between sets of formulas, X and Y, which is such that Y presents an overall space of possibilities for reasoning by cases from X. Definition 5.24 says that $X \Vdash_{\mathsf{L}} Y$ holds iff every formula derivable from Y by cases is derivable from X. The next theorem states that, as far as finite sets of formulas are considered, if the relations $\vdash_{\mathsf{L}}$ and $\models_{\mathsf{L}}$ coincide, then the relations $\Vdash_{\mathsf{L}}$ and $\Vvdash_{\mathsf{L}}$ coincide as well.

Theorem 5.6 (Adequacy of $\Vdash_{\mathsf{L}}$ wrt $\Vvdash_{\mathsf{L}}$) *Let X stand for an arbitrary set of formulas of $\mathcal{L}_{\mathsf{L}}$ and let Y be a finite set of sentences of $\mathcal{L}_{\mathsf{L}}$. Assume also that $\vdash_{\mathsf{L}}$ is sound and complete with respect to $\models_{\mathsf{L}}$. Then $X \Vdash_{\mathsf{L}} Y$ iff $X \Vvdash_{\mathsf{L}} Y$.*

Proof Suppose that L = FOL. We actually show the proof only for this case, as the one for CPL may be obtained from this. (Let us recall that for L = CPL 'sentence' means 'formula'.)

Assume that $X \not\Vvdash_{\mathsf{L}} Y$. Then there is a FOL-interpretation $\mathcal{I}$ which is a model of X but is not a model of any element of Y. Since the elements are sentences and Y is finite, it then holds that '$\bigvee Y$' is not true under $\mathcal{I}$. For this reason, X does not entail in L formula '$\bigvee Y$', i.e. $X \not\models_{\mathsf{L}} \bigvee Y$. Therefore also $X \nvdash_{\mathsf{L}} \bigvee Y$. On the other hand, $Y \vdash^{bc}_{\mathsf{L}} \bigvee Y$ by Corollary 5.11. Sentence '$\bigvee Y$' is thus an example of a formula that is a by-cases consequence of Y, yet is not a consequence of X. By Definition 5.24, it follows that $X \nVdash_{\mathsf{L}} Y$.

For the converse direction, assume that $X \nVdash_{\mathsf{L}} Y$. Then there is a formula, A, such that $Y \vdash^{bc}_{\mathsf{L}} A$ and $X \nvdash_{\mathsf{L}} A$. Thus also $X \not\models_{\mathsf{L}} A$, that is, there is an L-model $\mathcal{I}$

of set X which is not a model of A. Since $Y \vdash^{bc}_{\mathsf{L}} A$, by Corollary 5.10, for each $B \in Y$, $\{B\} \vdash_{\mathsf{L}} A$, which yields also $\{B\} \models_{\mathsf{L}} A$. Since $\mathcal{I} \not\models A$, also $\mathcal{I} \not\models B$. Since it holds for every $B \in Y$, by Definition 5.21 it follows that $X \not\Vdash\!\models_{\mathsf{L}} Y$. □

Now, we establish the potential proof-theoretical interpretation of the three types of sequents considered in this book. For this purpose, let us start with a straightforward corollary from Theorem 5.6 and Corollary 5.4.

Corollary 5.16 *Let X be an arbitrary set of formulas of $\mathcal{L}_{\mathsf{L}}$, let Y be a finite set of sentences of $\mathcal{L}_{\mathsf{L}}$, and let A be a single sentence of this language. Then:*

1. *$X \Vdash_{\mathsf{L}} \varnothing$ iff $X \Vdash\!\models_{\mathsf{L}} \varnothing$ iff X is semantically* L*-inconsistent,*
2. *$X \Vdash_{\mathsf{L}} \{A\}$ iff $X \Vdash\!\models_{\mathsf{L}} \{A\}$ iff $X \models_{\mathsf{L}} A$,*
3. *$\varnothing \Vdash_{\mathsf{L}} Y$ iff $\varnothing \Vdash\!\models_{\mathsf{L}} Y$ iff Y is a safeset in* L.

By the considerations presented in this section, we finally arrive at:

Corollary 5.17 *The following conditions are equivalent for each closed sequent '$S \vdash T$' of the language $\mathcal{L}^{?}_{\vdash\mathsf{L}}$.*

1. *$S \vdash T$ has a Socratic proof in $\mathbb{E} \in \{\mathbb{E}^{\mathbf{Q}}, \mathbb{E}^{\mathbf{RQ}}, \mathbb{E}^{\mathbf{LQ}}, \mathbb{E}^{\mathbf{P}}, \mathbb{E}^{\mathbf{RP}}, \mathbb{E}^{\mathbf{LP}}\}$.*
2. $\mathsf{set}(S) \Vdash\!\models_{\mathsf{L}} \mathsf{set}(T)$.
3. $\mathsf{set}(S) \Vdash_{\mathsf{L}} \mathsf{set}(T)$.

Proof By Corollary 2.21, a closed sequent $S \vdash T$ has a Socratic proof in $\mathbb{E}$ iff the sequent is correct under every L-interpretation of language $\mathcal{L}_{\mathsf{L}}$, that is, for each such L-interpretation $\mathcal{I}$, either there is a sentence in S false under $\mathcal{I}$ or there is a sentence in T true under $\mathcal{I}$. It follows that clause 1 holds iff clause 2 holds. Moreover, by Theorem 5.6, clause 2 holds iff clause 3 holds. □

Let us establish some more coincidences.

Definition 5.25 (*Syntactic Inconsistency in* L) Let X be a set of formulas of language $\mathcal{L}_{\mathsf{L}}$. We say that X is *syntactically inconsistent in* L iff $X \vdash_{\mathsf{L}} A$ for every formula A of $\mathcal{L}_{\mathsf{L}}$. We say that a formula B of $\mathcal{L}_{\mathsf{L}}$ is *syntactically inconsistent in* L iff the singleton set $\{B\}$ is syntactically inconsistent in L.

Corollary 5.18 *$X \Vdash_{\mathsf{L}} \varnothing$ iff X is syntactically inconsistent in* L.

Proof By Corollary 5.8, $\varnothing \vdash^{bc}_{\mathsf{L}} A$ for any formula A. Thus if $X \Vdash_{\mathsf{L}} \varnothing$, then by Definition 5.24, $X \vdash_{\mathsf{L}} A$ for any A, and hence X is syntactically inconsistent in L.

For the converse direction, let us observe that if $X \vdash_{\mathsf{L}} A$ for any A, then the implication defining the multiple-conclusion consequence must be satisfied for the case of $X \Vdash_{\mathsf{L}} \varnothing$. □

Corollary 5.19 *$\varnothing \Vdash_{\mathsf{L}} Y$ iff for every A such that $Y \vdash^{bc}_{\mathsf{L}} A$, A is a thesis of* L.

Proof Directly from Definition 5.24: $\varnothing \Vdash_{\mathsf{L}} Y$ iff for every formula A of $\mathcal{L}_{\mathsf{L}}$, if $Y \vdash^{bc}_{\mathsf{L}} A$, then $\varnothing \vdash_{\mathsf{L}} A$, that is, A is a thesis of L. □

In other words, Corollary 5.19 states that $\varnothing \Vdash_{\mathsf{L}} Y$, or, semantically: Y is a safeset in L, iff the only formulas that are by-cases-consequences (in L) of Y are the theses of L.

Corollary 5.20 *If* $\vdash_{\mathsf{L}}$ *satisfies Dilution, then* $X \vdash_{\mathsf{L}} A$ *iff* $X \Vdash_{\mathsf{L}} \{A\}$.

Proof Assume that $X \vdash_{\mathsf{L}} A$ and that $\{A\} \vdash^{bc}_{\mathsf{L}} B$. Then $\{A\} \vdash_{\mathsf{L}} B$, by Corollary 5.7, and also $X \vdash_{\mathsf{L}} B$ by Dilution. Thus if $X \vdash_{\mathsf{L}} A$, then $X \Vdash_{\mathsf{L}} \{A\}$. For the other direction observe that $\{A\} \vdash^{bc}_{\mathsf{L}} A$ (Corollary 5.12). By Definition 5.24, $X \vdash_{\mathsf{L}} A$. □

We have shown in this section that erotetic calculi, no matter what kind of sequents are used, can be interpreted as transforming questions concerning the relation of multiple-conclusion entailment (semantic version), or the relation of multiple-conclusion consequence (syntactic, i.e., proof-theoretical interpretation).

An open question is to provide a full interpretation of modal and intuitionistic logics in these lines.

References

1. Belnap, N. D., & Steel, T. P. (1976). *The logic of questions and answers*. Yale University Press.
2. Chlebowski, S., & Leszczyńska-Jasion, D. (2015). Dual erotetic calculi and the minimal LFI. *Studia Logica, 103*, 1245–1278. https://doi.org/10.1007/s11225-015-9617-0
3. Grobler, A. (2006). *Metodologia nauk*. Znak & Aureus.
4. Hilbert, D., & Bernays, P. (1968). *Grundlagen der mathematik I* (2nd ed.). Berlin: Springer.
5. Leszczyńska, D. (2004). Socratic proofs for some normal modal propositional logics. *Logique et Analyse, 47*(185–188), 259–285.
6. Leszczyńska, D. (2007). The method of Socratic proofs for normal modal propositional logics. Adam Mickiewicz University Press.
7. Leszczyńska-Jasion, D. (2008). The method of Socratic proofs for modal propositional logics: K5, S4.2, S4.3, S4M, S4F, S4R and G. *Studia Logica, 89*(3), 371–405.
8. Leszczyńska-Jasion, D. (2009). A loop-free decision procedure for modal propositional logics K4, S4 and S5. *Journal of Philosophical Logic, 38*(2), 151–177.
9. Negri, S., & von Plato, J. (2001). *Structural proof theory*. Cambridge University Press.
10. Shoesmith, D. J., & Smiley, T. J. (1978). *Multiple-conclusion logic*. Cambridge University Press.
11. Smullyan, R. M. (1968). *First-Order Logic*. Berlin: Springer.
12. Szczepiński, R. (2018). *The method of Socratic proofs for first degree entailment* [Unpublished manuscript received from the author in March 2018].
13. Tarski, A. (1930a). Fundamentale Begriffe der Methodologie der Deductiven Wissenschaften [English translation in (Tarski:1956)]. *Monatshefte für Mathematik und Physik, 37*, 361–404.
14. Tarski, A. (1930b). Über einige fundamentalen Begriffe der Metamathematik [English translation in (Tarski:1956)]. *Comptes Rendus des séances de la Société des Sciences et les Lettres de Varsovie, 23*, 22–29.
15. Tarski, A. (1956). *Logic, Semantics, Metamathematics. Papers from 1923 to 1938* (J. Woodger, Trans.). Clarendon Press.
16. Urbański, M., Paluszkiewicz, K., & Urbańska, J. (2016). From real data to formal models: An analysis of solutions to erotetic reasoning test task. In F. Paglieri, L. Bonelli, & S. Felletti (Eds.), *The psychology of argument: Cognitive approaches to argumentation and persuasion* (pp. 31–46). College Publications.

17. Wiśniewski, A. (1990). Implied questions. *Manuscrito, 13*(2), 23–38.
18. Wiśniewski, A. (1994). Erotetic implications. *Journal of Philosophical Logic, 23*, 173–195.
19. Wiśniewski, A. (1995). *The posing of questions: Logical foundations of erotetic inferences*. Kluwer Academic Publishers.
20. Wiśniewski, A. (2001). Questions and inferences. *Logique et Analyse, 173–175*, 5–43.
21. Wiśniewski, A. (2004). Socratic proofs. *Journal of Philosophical Logic, 33*(3), 299–326.
22. Wiśniewski, A. (2013). *Questions, inferences, and scenarios*. College Publications.
23. Wiśniewski, A. (2015). Semantics of questions. In S. Lappin & C. Fox (Eds.), *The handbook of contemporary semantic theory* (2nd ed., pp. 273–313). Wiley-Blackwell.
24. Wiśniewski, A. (2017). Generalized entailments. *Logic and Logical Philosophy, 26*, 321–356. https://doi.org/10.12775/LLP.2017.014
25. Wiśniewski, A. (2018). Deduction and reduction theorems for inferential erotetic logic. *Studia Logica, 106*, 295–309.
26. Wiśniewski, A., & Shangin, V. (2006). Socratic proofs for quantifiers. *Journal of Philosophical Logic, 35*(2), 147–178. https://doi.org/10.1007/s10992-005-9000-0
27. Wiśniewski, A., Vanackere, G., & Leszczyńska, D. (2005). Socratic proofs and paraconsistency: a case study. *Studia Logica, 80*(2–3), 433–468.
28. Wójcicki, R. (1988). *Theory of logical calculi: Basic theory of consequence operations* (softcover reprint of the hardcover 1st edn. by Kluwer Academic Publishers). Springer–Science+Business Media.

Chapter 6
From Erotetic Calculi to Proofs. The Classical Case

In this chapter, we deal with constructing proofs within sequent systems, building on the groundwork laid in prior chapters. Here, our focus turns to translating Socratic transformations into derivations within a Gentzen-style sequent calculus. The discussions in this and the next chapter highlight essential aspects of proof construction in these two proof systems, including comparative analysis and methods for simulating one system within another. Specifically, we aim to achieve:

- a deeper understanding of the interrelations between different proof methods;
- structured algorithms for translation, primarily presented in pseudocode;
- a novel algorithm for constructing proofs in implicit sequent systems for modal and intuitionistic logics without relying on backtracking—transforming proofs derived in explicit systems into proofs in implicit systems.

These objectives lay the groundwork for advancing the study of proof methods in the logic of questions, opening pathways to more efficient and comprehensive proof search strategies.

6.1 Socratic Trees

In the previous chapters, we have seen that the structure of a sequent is a rich medium of expression in proof theory. Moreover, the structure of seqsequents seems to be highly convenient for formulating proof-search algorithms—see Sect. 2.7. However, for the purpose of translations between an erotetic calculus and a sequent calculus, a tree structure is necessary. In this section, we 'extract' the structure of a tree from the structure of a Socratic transformation to use it further in the translation procedures.

We adopt a set-theoretic approach to the notion of tree. The reader can consult Appendix A if necessary.

D. Leszczyńska-Jasion, *The Method of Socratic Proofs*, Trends in Logic 64,
https://doi.org/10.1007/978-3-031-82451-7_6

The content of this section is based on Sect. 2 of [15]. The aim of [15] was to define a translation of a Socratic transformation into a tree, called a *Socratic tree*, whose nodes are *annotated sequents*. The authors of [15] considered only classical logic. Here, we present the construction described in [15] and generalize it to all erotetic calculi considered in the previous chapters. Therefore, in this section $\mathbb{E} \in \{\mathbb{E}^{\mathbf{Q}}, \mathbb{E}^{\mathbf{P}}, \mathbb{E}^{\mathbf{LQ}}, \mathbb{E}^{\mathbf{LP}}, \mathbb{E}^{\mathbf{RQ}}, \mathbb{E}^{\mathbf{RP}}, \mathbb{E}^{\mathsf{K}}, \mathbb{E}^{\mathsf{D}}, \mathbb{E}^{\mathsf{T}}, \mathbb{E}^{\mathsf{KB}}, \mathbb{E}^{\mathsf{DB}}, \mathbb{E}^{\mathsf{B}}, \mathbb{E}^{\mathsf{K4}}, \mathbb{E}^{\mathsf{D4}}, \mathbb{E}^{\mathsf{S4}}, \mathbb{E}^{\mathsf{KB4}}, \mathbb{E}^{\mathsf{K5}}, \mathbb{E}^{\mathsf{D5}}, \mathbb{E}^{\mathsf{K45}}, \mathbb{E}^{\mathsf{D45}}, \mathbb{E}^{\mathsf{S5}}, \mathbb{E}^{\mathsf{INT}}\}$.

In [15], the notion of an *annotated sequent* was introduced. A sequent is annotated to retain information about its position within a Socratic transformation. More specifically, if $\mathbf{s} = \langle s_1, s_2, \ldots \rangle$ is a Socratic transformation, then an ordered triple $\langle n, i, \phi \rangle$, where the sequent ϕ is the i-th constituent of the n-th question s_n of $\mathbf{s}$, is called an *annotated sequent of the Socratic transformation* $\mathbf{s}$. We will use '$\mathbf{s}_i^n$' for the i-th constituent of the n-th question of $\mathbf{s}$. It is important to note that a sequent indicated by the inscription '$\mathbf{s}_i^n$' may occur in many places of the Socratic transformation $\mathbf{s}$, but the sequent together with its coordinates in $\mathbf{s}$ is a unique entity with respect to $\mathbf{s}$.

An annotated sequent '$\langle n, i, \phi \rangle$' will sometimes be denoted by ϕ_i^n, just to shorten the strings of symbols. We will also use $\mathfrak{a}$, $\mathfrak{b}$, possibly with subscripts, to represent annotated sequents.

Recall that we say a sequent ψ *results in* $\mathbf{s}$ *from* sequent ϕ *by erotetic rule* $\mathbf{r}$ if, for some term s_n of $\mathbf{s}$, the pair s_n/s_{n+1} is an instance of $\mathbf{r}$, ϕ is the sequent-premise of s_n, and ψ is the (or one of the two) sequent-conclusion(s) of s_{n+1}. Additionally, we will say that an *annotated sequent* ϕ_i^n *is active in* s_n if the i-th constituent of question s_n is the sequent-premise in this question. Similarly, we will say that an *annotated sequent* ϕ_k^{n+1} *results from* annotated sequent ψ_i^n if the k-th constituent of question s_{n+1} results from the i-th constituent of question s_n.

The rules of erotetic calculi operate on questions based on seqsequents. However, each rule acts upon only one constituent of a question, leaving the remaining sequents unchanged. From the perspective of the logic of questions, this is a desirable effect, since an application of an erotetic rule reformulates the problem expressed by the question-premise. Moreover, this construction of rules ensures that they are semantically invertible, which implies, on the one hand, that the relation of erotetic implication holds between the question-premise and the question-conclusion (as we saw in the previous chapter) and, on the other hand, that the erotetic system is *confluent*. In a confluent system, whenever one begins with a provable formula, there are no 'bad moves' in the derivation construction that would lead to a 'dead end' where no proof can be found.[1] Confluency is a desirable property, particularly favorable for implementation.

[1] See [11, p. 121], 'A tableau calculus is *proof confluent*, if from every tableau for an unsatisfiable set of sentences, a closed tableau can be constructed.'

From the present perspective, however, the unchanged sequents 'left' in the resulting question are redundant. Thus, we will aim to omit them from the tree (cf. Definition 6.1). First, we define a relation $P_{\mathbf{s}}$ that links two annotated sequents of Socratic transformation $\mathbf{s}$ whenever one of these annotated sequents is active and the other is its sequent-conclusion. Moreover, if after an application of a rule a sequent is rewritten in the next question (that is, the rule has been applied with respect to some other constituent), then the relation $P_{\mathbf{s}}$ links the annotated sequent to its annotated repetition (cf. Definition 6.2). Next, we will consider the transitive closure of the relation $P_{\mathbf{s}}$. Finally, to eliminate redundant sequent repetitions, we will restrict the transitive closure of $P_{\mathbf{s}}$ to the set $X_{\mathbf{s}}$ (cf. Definition 6.3).

Suppose that $\mathbf{s}$ is an arbitrary Socratic transformation of a question of the form '$?(\phi)$' in $\mathbb{E}$. For the purposes of this section, it is important that the question is based on a single sequent. To ensure precision, let us recall that sequent ϕ has various forms depending on the underlying logic and erotetic calculus. Specifically:

- For $\mathbb{E} \in \{\mathbb{E}^{\mathbf{Q}}, \mathbb{E}^{\mathbf{P}}, \mathbb{E}^{\mathbf{LQ}}, \mathbb{E}^{\mathbf{LP}}, \mathbb{E}^{\mathbf{RQ}}, \mathbb{E}^{\mathbf{RP}}\}$, ϕ takes the form $S \vdash T$, where S and T are suitably restricted.
- For $\mathbb{E}$ in modal logics, ϕ is of the form $S \vdash \Gamma$, where Γ contains indexed formulas.
- For $\mathbb{E} = \mathbb{E}^{\mathsf{INT}}$, ϕ takes the form $\Gamma \vdash \Delta$, with indexed formulas on both sides of the sequent.

However, as we will see, these syntactic differences in syntax are inessential to the results discussed in this section.

Definition 6.1 specifies the set of nodes for the 'target' Socratic tree—the nodes are the annotated sequents relevant for further translations.

Definition 6.1 Let $\mathbf{s}$ be an a Socratic transformation of a question of the form '$?(\phi)$' in $\mathbb{E}$. By the set of relevant annotated sequents of $\mathbf{s}$ we mean the smallest set, $X_{\mathbf{s}}$, of annotated sequents of $\mathbf{s}$ such that $\phi_1^1 \in X_{\mathbf{s}}$ and, for each $n \geqslant 1$, if ψ_i^n is the active annotated sequent in question s_n, then:

1. if s_{n+1} results from s_n by a non-branching rule, then $\chi_i^{n+1} \in X_{\mathbf{s}}$,
2. if s_{n+1} results from s_n by a branching rule, then $(\omega_1)_i^{n+1} \in X_{\mathbf{s}}$ and $(\omega_2)_{i+1}^{n+1} \in X_{\mathbf{s}}$.

According to Definition 6.1, the set $X_{\mathbf{s}}$ contains the annotated sequent ϕ_1^1, which takes the role of the origin of the tree, and the annotated sequents-conclusions of the questions of $\mathbf{s}$. The following example illustrates the idea (Example 6.1 also comes from [15]).

Example 6.1 Here is a Socratic transformation $\mathbf{s}\star$ of question $?((p \vee p) \vee (q \vee q) \vdash p \vee q)$ in $\mathbb{E}^{\mathbf{P}}$; $\mathbf{s}\star$ is a Socratic proof of sequent $(p \vee p) \vee (q \vee q) \vdash p \vee q$ in $\mathbb{E}^{\mathbf{P}}$. The elements of $X_{\mathbf{s}\star}$, that is, the relevant annotated sequents of $\mathbf{s}\star$, are framed.

$$?(\boxed{(p \vee p) \vee (q \vee q) \vdash p \vee q})$$
$$\overline{\qquad\qquad\qquad\qquad\qquad\qquad} \; L_\beta$$
$$?(\boxed{p \vee p \vdash p \vee q} \; ; \; \boxed{q \vee q \vdash p \vee q})$$
$$\overline{\qquad\qquad\qquad\qquad\qquad\qquad} \; L_\beta$$
$$?(\boxed{p \vdash p \vee q} \; ; \; \boxed{p \vdash p \vee q} \; ; \; q \vee q \vdash p \vee q)$$
$$\overline{\qquad\qquad\qquad\qquad\qquad\qquad} \; R_\beta$$
$$?(\boxed{p, \neg p \vdash q} \; ; \; p \vdash p \vee q \; ; \; q \vee q \vdash p \vee q)$$
$$\overline{\qquad\qquad\qquad\qquad\qquad\qquad} \; R_\beta$$
$$?(p, \neg p \vdash q \; ; \; \boxed{p, \neg p \vdash q} \; ; \; q \vee q \vdash p \vee q)$$
$$\overline{\qquad\qquad\qquad\qquad\qquad\qquad} \; L_\beta$$
$$?(p, \neg p \vdash q \; ; \; p, \neg p \vdash q \; ; \; \boxed{q \vdash p \vee q} \; ; \; \boxed{q \vdash p \vee q})$$
$$\overline{\qquad\qquad\qquad\qquad\qquad\qquad} \; R_\beta$$
$$?(p, \neg p \vdash q \; ; \; p, \neg p \vdash q \; ; \; \boxed{q, \neg p \vdash q} \; ; \; q \vdash p \vee q)$$
$$\overline{\qquad\qquad\qquad\qquad\qquad\qquad} \; R_\beta$$
$$?(p, \neg p \vdash q \; ; \; p, \neg p \vdash q \; ; \; q, \neg p \vdash q \; ; \; \boxed{q, \neg p \vdash q})$$

For a definition of a relation $R_{\mathbf{s}}$ on the set $X_{\mathbf{s}}$, we need a certain auxiliary concept. Again, let $\mathbf{s}$ stand for a Socratic transformation of a question based on a single sequent.

Definition 6.2 Let $Y_{\mathbf{s}}$ be the set of all annotated sequents of $\mathbf{s}$. By $P_{\mathbf{s}}$ we mean the smallest binary relation on $Y_{\mathbf{s}}$ such that for each $n \geqslant 1$:

1. If $s_{n+1} = ?(\psi_1; \ldots; \psi_j)$ results from $s_n = \;?(\phi_1; \ldots; \phi_j)$ by a non-branching rule, then for each $i \leqslant j$: $\langle (\phi_i)^n_i, \; (\psi_i)^{n+1}_i \rangle \in P_{\mathbf{s}}$.
2. If $s_{n+1} = ?(\psi_1; \ldots; \psi_{j+1})$ results from $s_n = ?(\phi_1; \ldots; \phi_j)$ by a branching rule and the k-th constituent of s_n is active in s_n, then:

   (a) for each $i \leqslant k$, $\langle (\phi_i)^n_i, \; (\psi_i)^{n+1}_i \rangle \in P_{\mathbf{s}}$,
   (b) for each $i : k \leqslant i \leqslant j$, $\langle (\phi_i)^n_i, (\psi_{i+1})^{n+1}_{i+1} \rangle \in P_{\mathbf{s}}$.

It follows from Definition 6.2 that if $\langle \phi^n_i, \psi^m_j \rangle \in P_{\mathbf{s}}$, then either ϕ and ψ is the same, differently annotated, sequent or ϕ is active in some question of $\mathbf{s}$ and ψ results from ϕ.

Once again, we illustrate the definitions with Socratic transformation $\mathbf{s}\star$ (see Fig. 6.1). The lines display the $P_{\mathbf{s}\star}$ relation. The relevant annotated sequents of $\mathbf{s}\star$ are framed.

Finally, let us introduce:

Definition 6.3 Let $P^{tr}_{\mathbf{s}}$ be the transitive closure of $P_{\mathbf{s}}$. Then by $R_{\mathbf{s}}$ we mean the relation $P^{tr}_{\mathbf{s}}$ restricted to $X_{\mathbf{s}}$, that is $\langle \mathfrak{a}, \mathfrak{b} \rangle \in R_{\mathbf{s}}$ iff both $\langle \mathfrak{a}, \mathfrak{b} \rangle \in P^{tr}_{\mathbf{s}}$ and $\mathfrak{a}, \mathfrak{b} \in X_{\mathbf{s}}$.

If we apply Definition 6.3 to our previous example, we will receive the structure presented in Fig. 6.2. Only the elements of $X_{\mathbf{s}\star}$ are displayed, the lines represent relation $R_{\mathbf{s}\star}$.

By $\mathbf{Tr}(\mathbf{s})$ we will refer to the structure $\langle X_{\mathbf{s}}, R_{\mathbf{s}} \rangle$. It is easy to see that the structure is a tree with origin ϕ^1_1. A proof of this fact for the case of both-sided erotetic calculi, for $\mathsf{L} = \mathsf{CPL}, \mathsf{FOL}$ may be found in [15]. We skip the other cases. After the authors

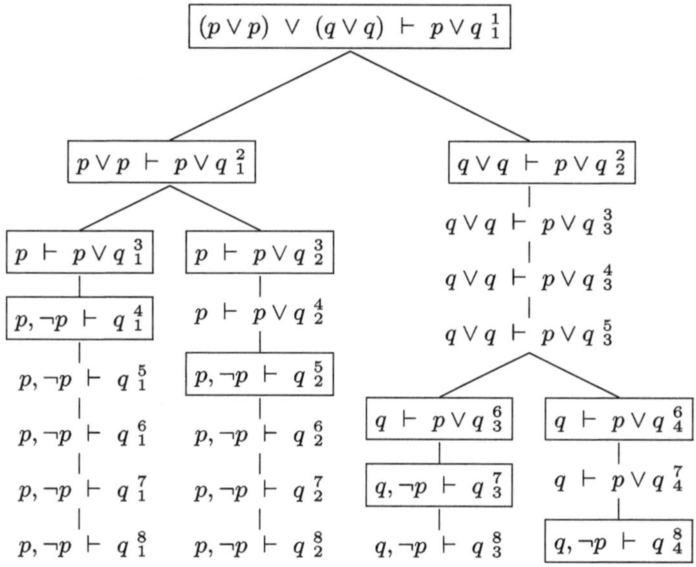

Fig. 6.1 Illustration of Definition 6.2

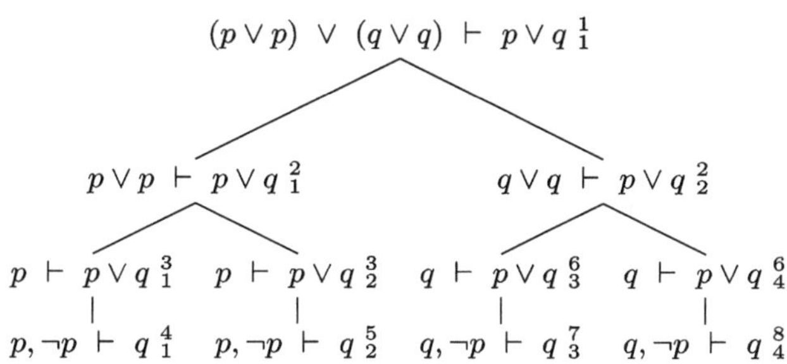

Fig. 6.2 Illustration of Definition 6.3

of [15], we will call the structure **Tr**(**s**) the *Socratic tree determined by a Socratic transformation* **s** of a question based on a single sequent. Definitions 6.1, 6.2 and 6.3 give us a recipe how to construct **Tr**(**s**) for a given Socratic transformation **s**. In [15], also an algorithm of construction of a Socratic tree is given.

Let us also observe that the relation of immediate $R_{\mathbf{s}}$-successor coincides with the erotetic rules in the following sense[2]:

Corollary 6.1 *Let* **s** *be a Socratic transformation in* $\mathbb{E}$ *and let* $\mathbf{Tr}(\mathbf{s}) = \langle X_{\mathbf{s}}, R_{\mathbf{s}} \rangle$ *be the Socratic tree determined by* **s**. *If an annotated sequent* $\langle m, k, \psi \rangle$ *of* **Tr**(**s**) *is an immediate* $R_{\mathbf{s}}$*-successor of an annotated sequent* $\langle n, i, \phi \rangle$ *of* **Tr**(**s**), *then the sequent* ψ *results in* **s** *from the sequent* ϕ *by a rule of* $\mathbb{E}$.

[2] See also [15, Corollary 2, p. 970].

Proof By Definitions 6.1, 6.2 and 6.3: for each $\langle \mathfrak{a}, \mathfrak{b} \rangle \in R_{\mathbf{s}}$, $\langle \mathfrak{a}, \mathfrak{b} \rangle \in P_{\mathbf{s}}$ and $\mathfrak{a}, \mathfrak{b} \in X_{\mathbf{s}}$. By Definition 6.2, annotated sequents $\mathfrak{a}$ and $\mathfrak{b}$ either represent the same sequent (different annotations) or a sequent-premise and a sequent-conclusion of a rule of $\mathbb{E}$. Since both $\mathfrak{a}, \mathfrak{b} \in X_{\mathbf{s}}$, it is the second option. □

In [15], the construction of Socratic tree is described for the case of Socratic transformations obtained by the rules of $\mathbf{E}^*$ (the propositional part of $\mathbf{E}^{\mathbf{PQ}}$[3]) or $\mathbf{E}^{\mathbf{PQ}}$, but there is nothing in the construction that depends on this version of erotetic calculus. More specifically, the construction of Socratic trees depends on the notion of a sequent-conclusion and the relation between sequents generated by the erotetic rules, and not on 'both-sidedness' of sequents. Therefore, our considerations on Socratic trees can be applied to Socratic transformations generated in the other erotetic calculi considered here, also those for modal and intuitionistic logics.

Theorem 6.1 *Any finite Socratic transformation of a question of the form $?(\phi)$ in $\mathbb{E}$ may be transformed into a Socratic tree with triple $\langle 1, 1, \phi \rangle$ in the root.*

Proof The procedure of picking up the relevant annotated sequents and constructing relation $R_{\mathbf{s}}$ between them is indicated in Definitions 6.1–6.3. □

Let us introduce some further examples.

Example 6.2 Here is a Socratic proof of sequent:

$$\vdash (\exists x P(x) \to \forall x(Q(x) \to R(x))) \to (\exists x(P(x) \wedge Q(x)) \to \exists x R(x))$$

in calculus $\mathbb{E}^{\mathbf{Q}}$. To increase the readability of the example, we put A for formula '$\exists x P(x)$', and B for '$\forall x(Q(x) \to R(x))$'. Starting with the 8th question we also put ϕ for sequent '$\neg A, \neg P(a), P(a), Q(a) \vdash \exists x R(x)$', and ψ for sequent '$B, \neg Q(a), P(a), Q(a) \vdash \exists x R(x)$'.

$$\frac{1.\ ?(\vdash (A \to B) \to (\exists x(P(x) \wedge Q(x)) \to \exists x R(x)))}{2.\ ?(A \to B \vdash \exists x(P(x) \wedge Q(x)) \to \exists x R(x))}\ R_\beta$$

$$\frac{2.\ ?(A \to B \vdash \exists x(P(x) \wedge Q(x)) \to \exists x R(x))}{3.\ ?(A \to B, \exists x(P(x) \wedge Q(x)) \vdash \exists x R(x))}\ R_\beta$$

$$\frac{3.\ ?(A \to B, \exists x(P(x) \wedge Q(x)) \vdash \exists x R(x))}{4.\ ?(A \to B, P(a) \wedge Q(a) \vdash \exists x R(x))}\ L_\delta$$

$$\frac{4.\ ?(A \to B, P(a) \wedge Q(a) \vdash \exists x R(x))}{5.\ ?(A \to B, P(a), Q(a) \vdash \exists x R(x))}\ L_\alpha$$

$$\frac{5.\ ?(A \to B, P(a), Q(a) \vdash \exists x R(x))}{6.\ ?(\neg A, P(a), Q(a) \vdash \exists x R(x)\ ;\ B, P(a), Q(a) \vdash \exists x R(x))}\ L_\beta$$

$$\frac{6.\ ?(\neg A, P(a), Q(a) \vdash \exists x R(x)\ ;\ B, P(a), Q(a) \vdash \exists x R(x))}{7.\ ?(\neg A, \neg P(a), P(a), Q(a) \vdash \exists x R(x)\ ;\ B, P(a), Q(a) \vdash \exists x R(x))}\ L_\gamma$$

$$\frac{7.\ ?(\neg A, \neg P(a), P(a), Q(a) \vdash \exists x R(x)\ ;\ B, P(a), Q(a) \vdash \exists x R(x))}{8.\ ?(\phi\ ;\ B, Q(a) \to R(a), P(a), Q(a) \vdash \exists x R(x))}\ L_\gamma$$

$$\frac{8.\ ?(\phi\ ;\ B, Q(a) \to R(a), P(a), Q(a) \vdash \exists x R(x))}{9.\ ?(\phi\ ;\ B, \neg Q(a), P(a), Q(a) \vdash \exists x R(x)\ ;\ B, R(a), P(a), Q(a) \vdash \exists x R(x))}\ L_\beta$$

$$\frac{9.\ ?(\phi\ ;\ B, \neg Q(a), P(a), Q(a) \vdash \exists x R(x)\ ;\ B, R(a), P(a), Q(a) \vdash \exists x R(x))}{10.\ ?(\phi\ ;\ \psi\ ;\ B, R(a), P(a), Q(a), \neg\exists x R(x) \vdash R(a))}\ R_\delta$$

[3] See also footnote 7 in Chap. 2 for the relation between $\mathbf{E}^{\mathbf{PQ}}$ and $\mathbb{E}$.

And here is the Socratic tree determined by this Socratic proof:

$$\vdash (A \to B) \to (\exists x(P(x) \wedge Q(x)) \to \exists x R(x)) \ {}^{1}_{1}$$

$$|$$

$$A \to B \vdash \exists x(P(x) \wedge Q(x)) \to \exists x R(x) \ {}^{2}_{1}$$

$$|$$

$$A \to B, \exists x(P(x) \wedge Q(x)) \vdash \exists x R(x) \ {}^{3}_{1}$$

$$|$$

$$A \to B, P(a) \wedge Q(a) \vdash \exists x R(x) \ {}^{4}_{1}$$

$$|$$

$$A \to B, P(a), Q(a) \vdash \exists x R(x) \ {}^{5}_{1}$$

Left branch:

$$\neg A, P(a), Q(a) \vdash \exists x R(x) \ {}^{6}_{1}$$

$$|$$

$$\phi \ {}^{7}_{1}$$

Right branch:

$$B, P(a), Q(a) \vdash \exists x R(x) \ {}^{6}_{2}$$

$$|$$

$$B, Q(a) \to R(a), P(a), Q(a) \vdash \exists x R(x) \ {}^{8}_{2}$$

$$\psi \ {}^{9}_{2} \qquad \chi^{10}_{3}$$

where $\chi = B, R(a), P(a), Q(a), \neg\exists x R(x) \vdash R(a)$.

Let us observe that the Socratic tree displayed above bears a resemblance to a sequent calculus proof.

Example 6.3 Here is a Socratic proof of the following left-sided sequent:

$$\neg(\forall x(P(x) \wedge Q(x)) \to \forall x P(x) \wedge \exists x Q(x)) \vdash$$

in calculus $\mathbb{E}^{\mathbf{LQ}}$. Below, A stands for '$\forall x(P(x) \wedge Q(x))$'.

$$1.\ ?(\ \neg(\forall x(P(x) \wedge Q(x)) \to \forall x P(x) \wedge \exists x Q(x))\ \vdash)$$

$$\overline{}\ L_\alpha$$

$$2.\ ?(\ \forall x(P(x) \wedge Q(x)), \neg(\forall x P(x) \wedge \exists x Q(x)) \vdash)$$

$$\overline{}\ L_\beta$$

$$3.\ ?(\ A, \neg\forall x P(x) \vdash;\ A, \neg\exists x Q(x) \vdash)$$

$$\overline{}\ L_\delta$$

$$4.\ ?(\ A, \neg P(a) \vdash;\ A, \neg\exists x Q(x) \vdash)$$

$$\overline{}\ L_\gamma$$

$$5.\ ?(\ A, P(a) \wedge Q(a), \neg P(a) \vdash;\ A, \neg\exists x Q(x) \vdash)$$

$$\overline{}\ L_\alpha$$

$$6.\ ?(\ A, P(a), Q(a), \neg P(a) \vdash;\ A, \neg\exists x Q(x) \vdash)$$

$$\overline{}\ L_\gamma$$

$$7.\ ?(\ A, P(a), Q(a), \neg P(a) \vdash;\ A, P(a) \wedge Q(a), \neg\exists x Q(x) \vdash)$$

$$\overline{}\ L_\alpha$$

$$8.\ ?(\ A, P(a), Q(a), \neg P(a) \vdash;\ A, P(a), Q(a), \neg\exists x Q(x) \vdash)$$

$$\overline{}\ L_\gamma$$

$$9.\ ?(\ A, P(a), Q(a), \neg P(a) \vdash;\ A, P(a), Q(a), \neg\exists x Q(x), \neg Q(a) \vdash)$$

Here is the Socratic tree determined by this Socratic proof.

$$\neg(\forall x(P(x) \wedge Q(x)) \rightarrow \forall x P(x) \wedge \exists x Q(x)) \vdash {}^{1}_{1}$$

$$\forall x(P(x) \wedge Q(x)), \neg(\forall x P(x) \wedge \exists x Q(x)) \vdash {}^{2}_{1}$$

Left branch:

$$A, \neg\forall x P(x) \vdash {}^{3}_{1}$$

$$A, \neg P(a) \vdash {}^{4}_{1}$$

$$A, P(a) \wedge Q(a), \neg P(a) \vdash {}^{5}_{1}$$

$$A, P(a), Q(a), \neg P(a) \vdash {}^{6}_{1}$$

Right branch:

$$A, \neg\exists x Q(x) \vdash {}^{3}_{2}$$

$$A, P(a) \wedge Q(a), \neg\exists x Q(x) \vdash {}^{7}_{2}$$

$$A, P(a), Q(a), \neg\exists x Q(x) \vdash {}^{8}_{2}$$

$$A, P(a), Q(a), \neg\exists x Q(x), \neg Q(a) \vdash {}^{9}_{2}$$

The above construction resembles a Hintikka-style tableau.

Example 6.4 Here is a Socratic proof of sequent $\vdash (\Box(p \rightarrow q) \rightarrow (\Box p \rightarrow \Box q))^0$ in calculus $\mathbb{E}^{\mathsf{K}}$ (see also Example 3.2). Let us recall that this time numeral 0 in the upper index is a part of indexed formulas. A stands for formula '$\neg\Box(p \rightarrow q)$' and B for '$\neg\Box p$'.

$$\frac{\dfrac{\dfrac{\dfrac{\dfrac{\dfrac{?(\vdash (\Box(p \rightarrow q) \rightarrow (\Box p \rightarrow \Box q))^0\ )}{?(\vdash (\neg\Box(p \rightarrow q))^0, (\Box p \rightarrow \Box q)^0\ )}\ R_\beta}{?(\vdash (\neg\Box(p \rightarrow q))^0, (\neg\Box p)^0, (\Box q)^0\ )}\ R_\beta}{?(\vdash (\neg\Box(p \rightarrow q))^0, (\neg\Box p)^0, (q)^{0,1}\ )}\ R_\nu}{?(\vdash A^0, (\neg(p \rightarrow q))^1, (\neg\Box p)^0, (q)^{0,1}\ )}\ R_\pi}{?(\vdash A^0, (\neg(p \rightarrow q))^1, B^0, (\neg p)^1, (q)^{0,1}\ )}\ R_\pi}{?(\vdash A^0, (p)^1, B^0, (\neg p)^1, (q)^{0,1}\ ;\ \vdash A^0, (\neg q)^1, B^0, (\neg p)^1, (q)^{0,1}\ )}\ R_\alpha$$

Figure 6.3 presents the Socratic tree determined by this Socratic proof. In order to increase the readability of the example, we do not display the annotations. Let us observe that the Socratic tree bears resemblance to a proof in a sequent system for modal logics.

Example 6.5 A Socratic proof of sequent $?(\vdash (\Box p \rightarrow \Box\Box\Diamond p)^0)$ in calculus $\mathbb{E}^{\mathsf{D5}}$. This time, we aim at a resemblance with Priest's analytic tableaux for modal logics (see [18, Chaps. 2 and 3]). When R_ν is applied, the information concerning accessibility relation is written in the form of a relational formula 'iRj' placed after $(\nu_0)^j$, instead of extending the index in '$(\nu_0)^{i,j}$'; similarly when $R_{\pi D}$ is applied.

$$\vdash (\Box(p \to q) \to (\Box p \to \Box q))^0$$

$$\vdash (\neg\Box(p \to q))^0, (\Box p \to \Box q)^0$$

$$\vdash (\neg\Box(p \to q))^0, (\neg\Box p)^0, (\Box q)^0$$

$$\vdash (\neg\Box(p \to q))^0, (\neg\Box p)^0, (q)^{0,1}$$

$$\vdash A^0, (\neg(p \to q))^1, (\neg\Box p)^0, (q)^{0,1}$$

$$\vdash A^0, (\neg(p \to q))^1, B^0, (\neg p)^1, (q)^{0,1}$$

$$\vdash A^0, (p)^1, B^0, (\neg p)^1, (q)^{0,1} \qquad \vdash A^0, (\neg q)^1, B^0, (\neg p)^1, (q)^{0,1}$$

Fig. 6.3 Socratic tree determined by the Socratic proof presented in Example 6.4

$$\cfrac{\cfrac{\cfrac{\cfrac{\cfrac{?(\vdash (\Box p \to \Box\Box\Diamond p)^0)}{?(\vdash (\neg\Box p)^0, (\Box\Box\Diamond p)^0)}\, R_\beta}{?(\vdash (\neg\Box p)^0, (\Box\Diamond p)^1, 0R1)}\, R_\nu}{?(\vdash (\neg\Box p)^0, (\Diamond p)^2, 1R2, 0R1)}\, R_\nu}{?(\vdash (\neg\Box p)^0, (\neg p)^3, 0R3, (\Diamond p)^2, 1R2, 0R1)}\, R_{\pi D}}{?(\vdash (\neg\Box p)^0, (\neg p)^3, 0R3, (\Diamond p)^2, (p)^3, 1R2, 0R1)}\, R_\pi$$

The last application of R_π is permitted since $\langle 2, 3\rangle$ belongs to the Euclidean closure of $\{\langle 0, 1\rangle, \langle 1, 2\rangle, \langle 0, 3\rangle\}$, as the following diagrams illustrate. To the left: the diagram of relation $R = \{\langle 0, 1\rangle, \langle 1, 2\rangle, \langle 0, 3\rangle\}$; to the right: the diagram of the Euclidean closure of R.

Let us also observe that if the proviso of applicability of R_ν is formulated in terms of chains, then the rule may be applied, since there are the following directed $R(\varphi)$-chains (where φ stands for '$\vdash (\neg\Box p)^0, (\neg p)^3, 0R3, (\Diamond p)^2, 1R2, 0R3$'): directed $R(\varphi)$-[0, 2]-chain $\langle 0, 1, 2\rangle$ and directed $R(\varphi)$-[0, 3]-chain $\langle 0, 3\rangle$.

Here is the Socratic tree determined by the last Socratic proof.

$$
\begin{array}{c}
\vdash (\Box p \rightarrow \Box\Box\Diamond p)^0 \\
| \\
\vdash (\neg\Box p)^0, (\Box\Box\Diamond p)^0 \\
| \\
\vdash (\neg\Box p)^0, (\Box\Diamond p)^1, 0R1 \\
| \\
\vdash (\neg\Box p)^0, (\Diamond p)^2, 1R2, 0R1 \\
| \\
\vdash (\neg\Box p)^0, (\neg p)^5, 0R3, (\Diamond p)^2, 1R2, 0R1 \\
| \\
\vdash (\neg\Box p)^0, (\neg p)^3, 0R3, (\Diamond p)^2, p^3, 1R2, 0R1
\end{array}
$$

The last example comes from intuitionistic logic.

Example 6.6 Here is a Socratic proof of sequent $\vdash\ ((\sim p \wedge \sim q) \supset \sim (p \vee q))^0$ in $\mathbb{E}^{\mathsf{INT}}$.

$$
\begin{array}{cl}
?(\vdash\ ((\sim p \wedge \sim q) \supset \sim (p \vee q))^0) & \\
\hline
?((\sim p \wedge \sim q)^{0,1} \vdash (\sim (p \vee q))^{0,1}) & R_{\supset} \\
\hline
?((\sim p)^{0,1}, (\sim q)^{0,1} \vdash (\sim (p \vee q))^{0,1}) & L_{\wedge} \\
\hline
?((\sim p)^{0,1}, (\sim q)^{0,1}, (p \vee q)^{0,1,2} \vdash) & R_{\sim} \\
\hline
?((\sim p)^{0,1}, (\sim q)^{0,1}, p^{0,1,2} \vdash\ ;\ (\sim p)^{0,1}, (\sim q)^{0,1}, q^{0,1,2} \vdash) & L_{\vee} \\
\hline
?((\sim p)^{0,1}, (\sim q)^{0,1}, p^{0,1,2} \vdash p^2\ ;\ (\sim p)^{0,1}, (\sim q)^{0,1}, q^{0,1,2} \vdash) & L_{\sim} \\
\hline
?((\sim p)^{0,1}, (\sim q)^{0,1}, p^{0,1,2} \vdash p^2\ ;\ (\sim p)^{0,1}, (\sim q)^{0,1}, q^{0,1,2} \vdash q^2) & L_{\sim}
\end{array}
$$

The Socratic tree determined by this Socratic proof should be compared with Priest's analytic tableaux for intuitionistic logic [18, Chap. 6], where the occurrence of an indexed formula $A^{s/i}$ in the antecedent/succedent of a sequent is translated as $A, +i$, respectively, $A, -i$.

$$
\begin{array}{c}
\vdash\ ((\sim p \wedge \sim q) \supset \sim (p \vee q))^0 \\
| \\
(\sim p \wedge \sim q)^{0,1} \vdash (\sim (p \vee q))^{0,1} \\
| \\
(\sim p)^{0,1}, (\sim q)^{0,1} \vdash (\sim (p \vee q))^{0,1} \\
| \\
(\sim p)^{0,1}, (\sim q)^{0,1}, (p \vee q)^{0,1,2} \vdash \\
\swarrow \qquad\qquad \searrow \\
\begin{array}{cc}
(\sim p)^{0,1}, (\sim q)^{0,1}, p^{0,1,2} \vdash & (\sim p)^{0,1}, (\sim q)^{0,1}, q^{0,1,2} \vdash \\
| & | \\
(\sim p)^{0,1}, (\sim q)^{0,1}, p^{0,1,2} \vdash p^2 & (\sim p)^{0,1}, (\sim q)^{0,1}, q^{0,1,2} \vdash q^2
\end{array}
\end{array}
$$

In the next section, we exploit some of the indicated resemblances between proof systems. It is worth to stress at this point that displaying and formalizing the resemblances is quite easy when the method of Socratic proofs is used. We have already stressed that this is due to the fact that the structures of seqsequents, on which questions are based, are very rich semantically, as they retain all the semantic information after each application of an erotetic rule. It makes the method of Socratic proofs suitable for proof-search in various proof systems.

6.2 Translation Into Sequent Systems for Classical Logic

The content of this section is partially based on the paper [15].

In [15], the authors have shown how Socratic proofs in $\mathbf{E}^{\mathbf{PQ}}$ determine proofs in Gentzen-style sequent calculus called $\mathbf{G}^{\mathbf{PQ}}$. Given soundness and completeness of the original $\mathbf{E}^{\mathbf{PQ}}$ (proved in [24]), soundness and completeness of $\mathbf{G}^{\mathbf{PQ}}$ is proved with respect to $\mathbf{E}^{\mathbf{PQ}}$ *via* the procedure 'translating' a Socratic proof into a proof of sequent in $\mathbf{G}^{\mathbf{PQ}}$.

Here, we adopt the same strategy with respect to proving soundness and completeness. Let us recall that the original erotetic calculus $\mathbf{E}^{\mathbf{PQ}}$ have rules for quantifiers expressed without the use of γ-, δ-notation. Since we have chosen to work with a variant of $\mathbf{E}^{\mathbf{PQ}}$ expressed in the γ-, δ-notation, we need to introduce an analogous variant of sequent system. We will call this system '$\mathbb{G}^{\mathbf{Q}}$' (the one introduced in [15] was called '$\mathbf{G}^{\mathbf{PQ}}$' and the one examined in [13]: '$\mathbb{G}^{\mathbf{PQ}}$'). The proof of soundness of the sequent calculus $\mathbb{G}^{\mathbf{Q}}$ follows the proof of soundness of $\mathbb{E}^{\mathbf{Q}}$, and completeness of the sequent calculus is entailed by the existence of the translating procedure.

Let us recall that the rules of $\mathbb{E}^{\mathbf{Q}}$ operate on questions based on both-sided, single succedent sequents of the form (2.6), that is:

$$S \vdash A$$

where S is a finite, possibly empty sequence of formulas of the language $\mathcal{L}_{\mathsf{FOL}}$, and A is a single formula of this language. Since in the proof-theoretical context the formulas occurring in a sequent are never sentential functions, we shall further assume that the elements of S are sentences, and so is A.

Similarly, the rules of calculus $\mathbb{G}^{\mathbf{Q}}$ operate on sequents of the form (2.6). Axioms of $\mathbb{G}^{\mathbf{Q}}$ are sequents falling under one of the schemes:

(a) $S(B) \vdash B$
(b) $S(B)(-B) \vdash C$

specified in Definition 2.3 (of a Socratic proof in $\mathbb{E}$, see Chap. 2).

The rules of calculus $\mathbb{G}^{\mathbf{Q}}$ are displayed in Table 6.1.

Let us observe that there are no primary structural rules in $\mathbb{G}^{\mathbf{Q}}$. In particular, there is no weakening, hence the general form of axioms. The rules are context-shearing (multiplicative). As we shall see, the rules are also semantically invertible, hence the calculus is confluent.

For the sake of the semantics of sequents, let us recall that a sequent of the form '$S \vdash A$' is called *correct under* FOL-*interpretation* $\mathcal{I}$ if either there is a sentence in

Table 6.1 Rules of calculus $\mathbb{G}^{\mathbf{Q}}$

$\dfrac{S(\alpha_0, \alpha_1) \vdash C}{S(\alpha) \vdash C}\ L_\alpha^{\mathbb{G}}$	$\dfrac{S \vdash \alpha_0 \quad S \vdash \alpha_1}{S \vdash \alpha}\ R_\alpha^{\mathbb{G}}$
$\dfrac{S(\beta_0) \vdash C \quad S(\beta_1) \vdash C}{S(\beta) \vdash C}\ L_\beta^{\mathbb{G}}$	$\dfrac{S(-\beta_i) \vdash \beta_{1-i}}{S(\_) \vdash \beta}\ R_\beta^{\mathbb{G}}$
$\dfrac{S(\kappa_0) \vdash C}{S(\kappa) \vdash C}\ L_\kappa^{\mathbb{G}}$	$\dfrac{S \vdash \kappa_0}{S \vdash \kappa}\ R_\kappa^{\mathbb{G}}$
$\dfrac{S(\gamma, \gamma(t)) \vdash C}{S(\gamma) \vdash C}\ L_\gamma^{\mathbb{G}}$ t is any closed term	$\dfrac{S \vdash \gamma(a_i)}{S \vdash \gamma}\ R_\gamma^{\mathbb{G}}$ a_i is an individual constant which does not occur in $S \vdash \gamma$
$\dfrac{S(\delta(a_i)) \vdash C}{S(\delta) \vdash C}\ L_\delta^{\mathbb{G}}$ a_i is an individual constant which does not occur in $S(\delta) \vdash C$	$\dfrac{S(-\delta) \vdash \delta(t)}{S(\_) \vdash \delta}\ R_\delta^{\mathbb{G}}$ t is any closed term

S false under $\mathcal{I}$ or A is true under $\mathcal{I}$. The following is easy to see (compare Corollary 2.5 in Chap. 2).

Corollary 6.2 *Each axiom of* $\mathbb{G}^{\mathbf{Q}}$ *is correct under every* FOL*-interpretation of language* $\mathcal{L}_{\mathsf{FOL}}$.

The proofs of the next three lemmas follow the proofs of Lemma 2.1 and 2.2 presented in Sect. 2.4 of Chap. 2; we skip these proofs.

Lemma 6.1 (Correctness and Invertibility of Rules, 1) *Let* **r** *stand for one of the following one-premise rules of* $\mathbb{G}^{\mathbf{Q}}$: $L_\alpha^{\mathbb{G}}$, $R_\beta^{\mathbb{G}}$, $L_\kappa^{\mathbb{G}}$, $R_\kappa^{\mathbb{G}}$, $L_\gamma^{\mathbb{G}}$, $R_\delta^{\mathbb{G}}$; *thus* **r** *is any of the one-premise rules except for* $R_\gamma^{\mathbb{G}}$ *and* $L_\delta^{\mathbb{G}}$. *Then for each* FOL*-interpretation* $\mathcal{I}$ *of the language* $\mathcal{L}_{\mathsf{FOL}}$, *a sequent falling under the schema of the premise of* **r** *is correct under* $\mathcal{I}$ *iff the respective sequent falling under the schema of the conclusion of* **r** *is correct under* $\mathcal{I}$.

Lemma 6.2 (Correctness and Invertibility of Rules, 2) *Let* **r** *be a two-premises rule of* $\mathbb{G}^{\mathbf{Q}}$, *that is,* $\mathbf{r} \in \{R_\alpha^{\mathbb{G}}, L_\beta^{\mathbb{G}}\}$. *Then, for each* FOL*-interpretation* $\mathcal{I}$ *of the language* $\mathcal{L}_{\mathsf{FOL}}$, *sequents falling under the schema of the premises of* **r** *are both correct under* $\mathcal{I}$ *iff the respective sequent falling under the schema of the conclusion of* **r** *is correct under* $\mathcal{I}$.

Lemma 6.3 (Correctness and Invertibility of Rules, 3) *Let* $\mathbf{r} \in \{R_\gamma^{\mathbb{G}}, L_\delta^{\mathbb{G}}\}$. *Then:*

1. *if* $\mathcal{I}$ *is a* FOL*-interpretation of* $\mathcal{L}_{\mathsf{FOL}}$ *such that a sequent falling under the schema of the premise of* **r** *is correct under* $\mathcal{I}$, *then the respective sequent falling under the schema of the conclusion of* **r** *is correct under* $\mathcal{I}$ *as well; and moreover*

2. *if a sequent falling under the conclusion schema of* **r** *is correct under* ***every*** FOL*-interpretation of* $\mathcal{L}_{\mathsf{FOL}}$*, then the respective sequent-premise of* **r** *is correct under* ***every*** FOL*-interpretation of* $\mathcal{L}_{\mathsf{FOL}}$.

In accordance with traditional accounts (see, for example, [23, p. 20]), by a *proof of sequent '*$S \vdash A$*' in* $\mathbb{G}^{\mathbf{Q}}$ we mean a finite tree labeled with sequents, regulated by the rules of $\mathbb{G}^{\mathbf{Q}}$, with the root labeled with sequent '$S \vdash A$', and the leaves labeled with axioms of the calculus. What we mean by 'regulation' is that each node-label is connected with the label(s) of the immediate successor(s) node(s) (if there are any) according to one of the rules of the calculus, that is, the node-label falls under the schema of the conclusion of the rule, whereas the immediate successor(s) label(s) falls (fall) under the schema of its premise(s).

Lemmas 6.1, 6.2 and 6.3 entail:

Theorem 6.2 (Soundness of $\mathbb{G}^{\mathbf{Q}}$ wrt Semantics of Sequents) *If a sequent is provable in calculus* $\mathbb{G}^{\mathbf{Q}}$*, then it is correct under every* FOL*-interpretation.*

Proof By Corollary 6.2, Lemmas 6.1–6.3 and induction on some measure of size of a proof of a sequent—e.g. the height of the proof. □

Theorem 6.3 (Soundness of $\mathbb{G}^{\mathbf{Q}}$ wrt Semantics of $\mathcal{L}_{\mathsf{FOL}}$) *If sequent of the form '*$\vdash A$*' is provable in calculus* $\mathbb{G}^{\mathbf{Q}}$*, then* A *is* FOL*-valid.*

Proof Follows from the previous theorem. □

Now, we show that after a certain reformulation of its structure, a Socratic tree **Tr(s)** $= \langle X_{\mathbf{s}}, R_{\mathbf{s}} \rangle$ produces a proof in $\mathbb{G}^{\mathbf{Q}}$. In [15], the authors have used the structure of annotated sequents and assumed that if a node of a Socratic tree is an object of the form $\langle n, i, \phi \rangle$, then the pair $\langle n, i \rangle$ constitutes a node of the parallel Gentzen-style tree, and the sequent ϕ constitutes its label.[4]

Thus suppose that $\langle X_{\mathbf{s}}, R_{\mathbf{s}} \rangle$ is the Socratic tree determined by a Socratic transformation **s**. Then we put: $X_{\mathbf{sG}}$—the set of pairs of numerals which are parts of the annotated sequents in $X_{\mathbf{s}}$, $R_{\mathbf{sG}}$—the relation extracted from relation $R_{\mathbf{s}}$ in a similar way; formally:

Definition 6.4 Let $\langle X_{\mathbf{s}}, R_{\mathbf{s}} \rangle$ be the Socratic tree determined by a Socratic transformation **s**. Then set $X_{\mathbf{sG}}$ and relation $R_{\mathbf{sG}}$ on this set are defined as follows:

1. $\langle n, i \rangle \in X_{\mathbf{sG}}$ iff for some sequent ϕ, $\langle n, i, \phi \rangle \in X_{\mathbf{s}}$,
2. $\langle \langle n_1, i_1 \rangle, \langle n_2, i_2 \rangle \rangle \in R_{\mathbf{sG}}$ iff for some sequents ϕ_1, ϕ_2, $\langle \langle n_1, i_1, \phi_1 \rangle, \langle n_2, i_2, \phi_2 \rangle \rangle \in R_{\mathbf{s}}$.

Obviously, the structure $\langle X_{\mathbf{sG}}, R_{\mathbf{sG}} \rangle$ is a tree, since $\langle X_{\mathbf{s}}, R_{\mathbf{s}} \rangle$ is a tree. Moreover, the tree is labeled with labeling function $\eta_{\mathbf{s}}$ such that if $\langle i, j, \phi \rangle \in X_{\mathbf{s}}$, then sequent ϕ is the label $\eta_{\mathbf{s}}(\langle i, j \rangle)$ of node $\langle i, j \rangle$. A triple $\langle X_{\mathbf{sG}}, R_{\mathbf{sG}}, \eta_{\mathbf{s}} \rangle$, that is, the tree $\langle X_{\mathbf{sG}}, R_{\mathbf{sG}} \rangle$ together with its labeling function $\eta_{\mathbf{s}}$, will be called the *labeled tree determined by a Socratic transformation* **s**.

[4] In [15], the authors have also shown how this feature of the construction of a tree may be used in order to simplify the algorithm of tree construction. It is also a good idea to compare the construction of trees used in [15] with that in [12].

Theorem 6.4 *If $\langle X_{\mathbf{s}}, R_{\mathbf{s}}\rangle$ is the Socratic tree determined by a Socratic proof of a sequent '$S \vdash A$' in $\mathbb{E}^{\mathbf{Q}}$, then the labeled tree $\langle X_{\mathbf{sG}}, R_{\mathbf{sG}}, \eta_{\mathbf{s}}\rangle$ is a proof of the sequent '$S \vdash A$' in $\mathbb{G}^{\mathbf{Q}}$.*

Proof Let $\mathbf{s}$ be a Socratic proof of '$S \vdash A$' in $\mathbb{E}^{\mathbf{Q}}$, and let $\langle X_{\mathbf{s}}, R_{\mathbf{s}}\rangle$ stand for the Socratic tree determined by $\mathbf{s}$.

The root of $\langle X_{\mathbf{s}}, R_{\mathbf{s}}\rangle$ is the triple $\langle 1, 1, S \vdash A\rangle$, hence the sequent '$S \vdash A$' labels the root $\langle 1, 1\rangle$ of tree $\langle X_{\mathbf{sG}}, R_{\mathbf{sG}}, \eta_{\mathbf{s}}\rangle$.

By Corollary 6.1, if $\langle m, k, \psi\rangle$ is an immediate $R_{\mathbf{s}}$-successor of $\langle n, i, \phi\rangle$, then ψ results from ϕ by a rule of $\mathbb{E}^{\mathbf{Q}}$. This yields that ψ is a premise and ϕ a conclusion of a rule of $\mathbb{G}^{\mathbf{Q}}$. Hence the labeled tree $\langle X_{\mathbf{sG}}, R_{\mathbf{sG}}, \eta_{\mathbf{s}}\rangle$ is regulated by the rules of $\mathbb{G}^{\mathbf{Q}}$.

Last but not least, $\mathbf{s}$ is a Socratic proof, hence each leaf of $\langle X_{\mathbf{s}}, R_{\mathbf{s}}\rangle$ is a basic sequent. It follows that the leaves of $\langle X_{\mathbf{sG}}, R_{\mathbf{sG}}, \eta_{\mathbf{s}}\rangle$ are labeled with axioms of $\mathbb{G}^{\mathbf{Q}}$. □

Theorem 6.4, together with Theorems 2.5 and 2.6 (completeness of $\mathbb{E}$) entail:

Theorem 6.5 (Completeness of $\mathbb{G}^{\mathbf{Q}}$) *If a sequent '$S \vdash A$' is correct under every* FOL*-interpretation of $\mathcal{L}_{\mathsf{FOL}}$, then there exists a proof of the sequent in $\mathbb{G}^{\mathbf{Q}}$. Moreover, if A is a* FOL*-valid formula, then there exists a proof of the sequent '$\vdash A$' in $\mathbb{G}^{\mathbf{Q}}$.*

If $\mathbf{s}$ is a Socratic proof of a sequent '$S \vdash A$' in $\mathbb{E}^{\mathbf{Q}}$, then the labeled tree $\langle X_{\mathbf{sG}}, R_{\mathbf{sG}}, \eta_{\mathbf{s}}\rangle$ will be called the *proof of $S \vdash A$ in $\mathbb{G}^{\mathbf{Q}}$ based upon* $\mathbf{s}$. We will denote it by $\mathbf{G}(\mathbf{s})$.

Example 6.7 Here is the proof of sequent $(p \vee p) \vee (q \vee q) \vdash p \vee q$ based upon Socratic transformation from Example 6.1.

$$\dfrac{\dfrac{\dfrac{p, \neg p \vdash q}{p \vdash p \vee q}R^{\mathbb{G}}_{\beta} \quad \dfrac{p, \neg p \vdash q}{p \vdash p \vee q}R^{\mathbb{G}}_{\beta}}{p \vee p \vdash p \vee q}L^{\mathbb{G}}_{\beta} \qquad \dfrac{\dfrac{q, \neg p \vdash q}{q \vdash p \vee q}R^{\mathbb{G}}_{\beta} \quad \dfrac{q, \neg p \vdash q}{q \vdash p \vee q}R^{\mathbb{G}}_{\beta}}{q \vee q \vdash p \vee q}L^{\mathbb{G}}_{\beta}}{(p \vee p) \vee (q \vee q) \vdash p \vee q}L^{\mathbb{G}}_{\beta}$$

Example 6.8 Here is the proof of sequent

$$\vdash (\exists x P(x) \to \forall x(Q(x) \to R(x))) \to (\exists x(P(x) \wedge Q(x)) \to \exists x R(x))$$

based upon Socratic transformation from Example 6.2. We put:

$A = \exists x P(x)$, $B = \forall x(Q(x) \to R(x))$

$\phi = \neg A, \neg P(a), P(a), Q(a) \vdash \exists x R(x)$, $\psi = B, \neg Q(a), P(a), Q(a) \vdash \exists x R(x)$,

$\chi = B, R(a), P(a), Q(a), \neg\exists x R(x) \vdash R(a)$.

$$
\dfrac{\dfrac{\dfrac{\dfrac{\dfrac{\dfrac{\phi}{\neg A, P(a), Q(a) \vdash \exists x R(x)}\, L^{\mathbb{G}}_{\delta} \qquad \dfrac{\dfrac{\psi \quad \dfrac{\chi}{B, R(a), P(a), Q(a) \vdash \exists x R(x)}\, R^{\mathbb{G}}_{\delta}}{B, Q(a) \to R(a), P(a), Q(a) \vdash \exists x R(x)}\, L^{\mathbb{G}}_{\beta}}{B, P(a), Q(a) \vdash \exists x R(x)}\, L^{\mathbb{G}}_{\gamma}}{A \to B, P(a), Q(a) \vdash \exists x R(x)}\, L^{\mathbb{G}}_{\beta}}{A \to B, P(a) \wedge Q(a) \vdash \exists x R(x)}\, L^{\mathbb{G}}_{\alpha}}{A \to B, \exists x(P(x) \wedge Q(x)) \vdash \exists x R(x)}\, L^{\mathbb{G}}_{\delta}}{A \to B \vdash \exists x(P(x) \wedge Q(x)) \to \exists x R(x)}\, R^{\mathbb{G}}_{\beta}}{\vdash (A \to B) \to (\exists x(P(x) \wedge Q(x)) \to \exists x R(x))}\, R^{\mathbb{G}}_{\beta}
$$

6.3 A Note on Dual Erotetic Calculi for CPL and Distributive Deductive Systems

In this section, we briefly describe how the possibility of using different sequents may be utilized to improve the process of proof-search. This time, by 'different sequents' we do not mean the diversity of left-, both- and right-sided sequents, but the so-called *reversed sequents*, dual in nature to the 'canonical sequents' used so far in this book.

We will consider CPL only and we will focus on right-sided sequents. The choice of 'sidedness' is inessential to the result presented here.

We will concentrate upon an erotetic language containing the symbol '$\dashv$' instead of '$\vdash$'. The language is labeled with '$\mathcal{L}^{?}_{\dashv\mathsf{CPL}}$', where $\mathcal{L}_{\mathsf{CPL}}$ is the language of the underlying classical propositional logic. The vocabulary of $\mathcal{L}^{?}_{\dashv\mathsf{CPL}}$ includes the vocabulary of $\mathcal{L}_{\mathsf{CPL}}$, the comma, the semicolon and the following signs: ?, $\dashv$, *ng* ($\mathcal{L}^{?}_{\dashv\mathsf{CPL}}$-negation), $\sqcup$ ($\mathcal{L}^{?}_{\dashv\mathsf{CPL}}$-disjunction). As before, the language $\mathcal{L}^{?}_{\dashv\mathsf{CPL}}$ has two categories of well-formed expressions: declarative well-formed formulas (d-wffs for short) and erotetic formulas, that is, questions. As we said, we take right-sided sequents only, thus an *atomic d-wff* of $\mathcal{L}^{?}_{\dashv\mathsf{CPL}}$ is an expression of the form:

$$\dashv S \tag{6.1}$$

where $S = \langle A_1, \ldots, A_n \rangle$ is a finite (possibly empty) sequence of formulas of $\mathcal{L}_{\mathsf{CPL}}$. Expressions of the above form are called *reversed sequents* or simply *sequents of* $\mathcal{L}^{?}_{\dashv\mathsf{CPL}}$. If S is empty (which is permitted), then we display the right-sided reversed sequent (6.1) as follows:

$$\dashv \varnothing \tag{6.2}$$

Compound d-wffs of $\mathcal{L}^{?}_{\dashv\mathsf{CPL}}$ are built from atomic d-wffs of $\mathcal{L}^{?}_{\dashv\mathsf{CPL}}$ by means of $\sqcup$ and *ng*. Questions of $\mathcal{L}^{?}_{\dashv\mathsf{CPL}}$ are of the following form:

$$?(\Phi) \tag{6.3}$$

where Φ is a finite and non-empty sequence of reversed sequents of $\mathcal{L}^?_{\dashv\mathsf{CPL}}$.

The general idea is that questions of the language $\mathcal{L}^?_{\dashv\mathsf{CPL}}$ are questions concerning *refutability*, semantically understood as *falsifiablity in a model*. In the case of CPL, a question of the form:

$$?(\dashv A^1_1, \ldots, A^1_{n_1} \ ; \ \ldots \ ; \ \dashv A^k_1, \ldots, A^k_{n_k}) \tag{6.4}$$

can be read: 'Is it the case that at least one of formulas $A^1_1, \ldots, A^1_{n_1}$ is false (refuted) under a valuation V **and at the same time**...**and at the same time** at least one of formulas $A^k_1, \ldots, A^k_{n_k}$ is false under the same valuation V'. To put it in other words, the question asks if it is the case that V **does not satisfy** any of the sets $\{A^1_1, \ldots, A^1_{n_1}\}$, ..., $\{A^k_1, \ldots, A^k_{n_k}\}$. Consequently, a question of the form:

$$?(\dashv A)$$

asks whether A is false under (some) V, that is, whether A is refutable.

With a mechanism involving refutability, erotetic calculi with reversed sequents *seem* to refer to the tradition of refutation deductive systems. However, this is not an accurate comparison, as we will explain. Let us mention at this point that the idea of refutation systems goes back to Łukasiewicz, who introduced the very notion of *rejection* in [16] (see also [21, 27]). The *refutability*, as captured by erotetic calculi with reversed sequents, is to be understood as proving falsifiability or proving satisfiability of the converse in the sense known from the resolution systems. As the authors of [5] put it: 'The difference between the notion of refutability considered here and the one known from the tradition of refutation systems is clear in this context, as one can have p and $\neg p$ refuted in a refutation system (both are non-theses), but one cannot have them both falsified under the same valuation. In other words, the rules of refutation systems preserve non-validity, whereas the rules presented in this paper preserve—as we shall see—falsifiability under a valuation.'

The notion of reversed sequent has been introduced in the paper [5], where the language with reversed sequents was used to simulate resolution inside an erotetic calculus. The erotetic resolution systems in [5] are presented for CPL, the propositional parts of paraconsistent logics CLuN[5] and CLuNs, and the minimal Logic of Formal Inconsistency called mbC. As previously mentioned, [3] contains a full exposition of both canonical and dual erotetic calculi for the Classical Logic. Some applications of dual erotetic calculi have been studied also in [4].

Since we have used the term 'dual', let us also mention another line of research focusing on dual consequence operation, see [25, 26, 28]. Examining in detail the relation between different types of consequence operations on the one hand and derivability as determined by various erotetic calculi on the other, as well as an examination of the very concept of duality in deductive systems, is an extremely

[5] In [1] Diderik Batens introduced paraconsistent propositional logic named 'PI'; the name 'CLuN' appeared later, but is nowadays common. For more information on the mentioned logics, see e.g. our [5], and the references indicated there.

Table 6.2 Rules of $\mathbb{E}^{\mathsf{CPL}}_{res}$

$$\frac{?(\Phi(\dashv S(\beta)))}{?(\Phi(\dashv S(\beta_1);\ \dashv S(\beta_2)))}\ R_\beta$$

$$\frac{?(\Phi(\dashv S(\alpha)))}{?(\Phi(\dashv S(\alpha_1,\ \alpha_2)))}\ R_\alpha \qquad \frac{?(\Phi(\dashv S(\neg\neg A)))}{?(\Phi(\dashv S(A)))}\ R_{\neg\neg}$$

$$\frac{?(\Phi(\dashv S(A))(\dashv T(-A)))}{?(\dashv \underline{S}'\underline{T};\ \Phi(\_)(\_);\ \dashv S(A);\ \dashv T(-A))}\ R_{res}$$

interesting enterprise for future research. It seems that we have barely touched this issue in [5].

The rules of dual erotetic calculus $\mathbb{E}^{\mathsf{CPL}}_{res}$ for CPL, based on right-sided reversed sequents, are displayed in Table 6.2. For simplicity, we use the same symbols for the rules as we did before. The resolution rule R_{res} may be applied provided that there are some complementary formulas, A and $-A$, in the displayed sequents-premises.

The sequences $\underline{S}$ and $\underline{T}$ ($\underline{U}$ and $\underline{V}$) are obtained from sequences S and T (U and V, respectively) by deleting all occurrences of formulas of the form A ($-A$, respectively), if there are any. Thus, in the case of the resolution rule, sequences of formulas are forced to 'behave' like ordinary sets.

Let us stress that we did not restrict the applications of R_{res} to clauses—the elements of sequences S, T, U and V may be compound, as well as the 'cut-formulas' A and $-A$. Thus, $\mathbb{E}^{\mathsf{CPL}}_{res}$ constitutes an account of *non-clausal resolution*. (See also [10] for non-clausal resolution.)

Rewriting the premise sequents in the conclusion of rule R_{res} is to warrant semantic invertibility of the rule. The consequence of this fact is that in the case of implementation of the method no backtracking is needed in applying the resolution rule. Obviously, in this account the computational cost is in searching through the growing sequences of sequents for the premises of R_{res}.

When the erotetic rules are applied to the initial question, one arrives at a sequence of questions called *Socratic transformation*. The notion is defined as in the canonical case (see Definition 2.2 in Sect. 2.3). Calculus $\mathbb{E}^{\mathsf{CPL}}_{res}$ is a resolution system, thus instead of the notion of a proof we need the notion of a refutation. This is given below.

Definition 6.5 (Socratic Refutation) Let $\dashv A$ be a reversed sequent. A *Socratic refutation of the sequent* $\dashv A$ *in* $\mathbb{E}^{\mathsf{CPL}}_{res}$ is a finite Socratic transformation $\mathbf{s} = \langle Q_1, \ldots, Q_n\rangle$ of the question $Q_1 = ?(\dashv A)$ in $\mathbb{E}^{\mathsf{CPL}}_{res}$, such that the last question, Q_n, of $\mathbf{s}$ has the following form:

(#) $?(\dashv \varnothing;\ \Omega)$

where Ω is a finite sequence of sequents.

If there is a Socratic refutation of a sequent, $\dashv A$, in $\mathbb{E}^{\mathsf{CPL}}_{res}$, then we say that $\dashv A$ is *refutable* in $\mathbb{E}^{\mathsf{CPL}}_{res}$. Let us observe that we *refute reversed sequents* not *the formula concerned by the sequent*. By refuting the reversed sequent we answer to the negative to the question about refutability of A. Thus a Socratic refutation of '$\dashv A$' proves A, just as a refutation of '$\neg A$' in a standard resolution system proves A.

Example 6.9 An example of a Socratic refutation of the reversed sequent $\dashv (p \to q) \to (\neg q \to \neg p)$ in $\mathbb{E}^{\mathsf{CPL}}_{res}$.

$$
\begin{array}{cl}
?(\dashv (p \to q) \to (\neg q \to \neg p)) & \\
\hline
?(\dashv \neg(p \to q) \; ; \; \dashv \neg q \to \neg p) & R_\beta \\
\hline
?(\dashv p, \neg q \; ; \; \dashv \neg q \to \neg p) & R_\alpha \\
\hline
?(\dashv p, \neg q \; ; \; \dashv \neg\neg q \; ; \; \dashv \neg p) & R_\beta \\
\hline
?(\dashv p, \neg q \; ; \; \dashv q \; ; \; \dashv \neg p) & R_{\neg\neg} \\
\hline
?(\dashv p \; ; \; \dashv \neg p \; ; \; \dashv p, \neg q \; ; \; \dashv q) & R_{res} \\
\hline
?(\dashv \varnothing \; ; \; \dashv p, \neg q \; ; \; \dashv q \; ; \; \dashv p \; ; \; \dashv \neg p) & R_{res}
\end{array}
$$

In [3], it is proved that the systems of erotetic resolution presented in this paper, including the one discussed here, are sound and complete with respect to the respective semantics.

Let us observe the dual behavior of αs and βs in the context of the erotetic rules. In the right-sided setting of reversed sequents, α-formulas 'behave' like β-formulas do on the right side of canonical sequents, and conversely—βs take the roles of αs. One can say that the same dual shift is observed inside the canonical calculi with both-sided sequents, where αs on the right side of the turnstile 'behave' like βs on the left side, and conversely. It would be misleading, however, to take the two phenomena as features of the same relation of duality. There are at least two relations of duality worth study. One is located at the level of different sides of the turnstile and explains the similarities in 'behavior' of αs and βs on the two sides. Semantically, it is the relation between truth (the left side) and false (the right side). In proof-theoretical terms, the first relation of duality explains both the differences and the similarities between the method of analytic tableaux (the left side of the turnstile) and, for example, the method of diagrams of formulas introduced by Rasiowa and Sikorski (the right side).[6]

But one can also examine relation of duality located at the level of proof-procedures; it is the second relation of duality, the one characterising canonical and dual erotetic calculi, and it also may be expressed in proof-theoretical and in semantic sense.

Proof-theoretically, the difference between the two types of calculi is in the nature of the closing conditions: these may be arrived at through a kind of decomposition of formulas and inspection of complementary formulas (canonical calculi), or through

[6] The method of diagrams of formulas was introduced in [19, 20] and further developed for many logics. It found various important applications, int.al, in computer science. The reader may find more references in [17] or [14]. In [14] the authors develop some new Rasiowa-Sikorski systems and also indicate their connection with the method of Socratic proofs.

a decomposition and resolution (dual calculi). Semantically, this relation of duality can be expressed as follows. Suppose S is a finite sequence of formulas of a given formal language, and V is a valuation function defined for the language and with values in $\{0, 1\}$. Then we may define two semantic, dual to each other, properties: the first property consists in S having at least one term true under V, whereas the second one consists in S having at least one term false under V. The erotetic rules of canonical calculi preserve the first property, and the dual erotetic calculi preserve the second property.

To sum up, there are at least two relations of duality worth study. One is located at the level of two sides of a sequent. The second relation is at the level of proof-procedures; it is what we do with the sequents and how we arrive at the conclusion that a proof has been obtained. Finally, it is worth to stress that on the second level of duality the cut rule known from sequent calculi is 'canonical' and it expresses the Principle of Bivalence, whereas the resolution rule is its dual and it expresses the Principle of Non-contradiction. The previous parts of this book explore, in a way, the similarities between proof systems determined by the first relation of duality. However, we may take a look at the merits that possibly follow from the syntactic opportunity of joining canonical and dual sequents in one (meta) proof-format.

It is well-known that the same formulas may have very long tableau-proofs and short resolution refutations, or conversely—see [2, 6–9], more references may be found in [6]. The same difference may be observed at the level of conjunctive and disjunctive normal forms of a formula. It is enough to imagine what happens when we insist to present formula (6.5) in conjunctive normal form (CNF, for short), or formula (6.6) in disjunctive normal form (DNF, for short) by using a standard algorithm with the laws of distributivity.

$$\begin{aligned} A = & (p \wedge q \wedge r) \vee (p \wedge q \wedge \neg r) \vee (p \wedge \neg q \wedge r) \vee (p \wedge \neg q \wedge \neg r) \vee \\ \vee \; & (\neg p \wedge q \wedge r) \vee (\neg p \wedge q \wedge \neg r) \vee (\neg p \wedge \neg q \wedge r) \vee (\neg p \wedge \neg q \wedge \neg r) \end{aligned} \tag{6.5}$$

$$\begin{aligned} B = & (p \vee q \vee r) \wedge (p \vee q \vee \neg r) \wedge (p \vee \neg q \vee r) \wedge (p \vee \neg q \vee \neg r) \wedge \\ \wedge \; & (\neg p \vee q \vee r) \wedge (\neg p \vee q \vee \neg r) \wedge (\neg p \vee \neg q \vee r) \wedge (\neg p \vee \neg q \vee \neg r) \end{aligned} \tag{6.6}$$

In the case of the above formulas, the features which make (6.5) unsuitable for converting it to CNF, and (6.6) unsuitable for converting it to DNF, are clearly visible; it is the presence of relatively many occurrences of '$\wedge$' in the scope of '$\vee$' (and conversely, for (6.6)). But in a general case, deciding which form (CNF or DNF) is easier to derive—or in other words, which is shorter—is far from obvious. Here is a proposal of a syntactic characterisation of the features that make the difference.

Definition 6.6 Let $FORM$ be the set of formulas of $\mathcal{L}_{\mathsf{CPL}}$. Functions

$$\alpha c : FORM \longrightarrow \mathbb{N} \quad \beta c : FORM \longrightarrow \mathbb{N}$$

called: α-complexity and β-complexity, respectively, are defined recursively as follows:

1. $\alpha c(p_i) = \alpha c(\neg p_i) = 1$	1. $\beta c(p_i) = \beta c(\neg p_i) = 1$
2. $\alpha c(\neg\neg A) = \alpha c(A)$	2. $\beta c(\neg\neg A) = \beta c(A)$
3. $\alpha c(\alpha) = \alpha c(\alpha_1) + \alpha c(\alpha_2)$	3. $\beta c(\alpha) = \beta c(\alpha_1) \cdot \beta c(\alpha_2)$
4. $\alpha c(\beta) = \alpha c(\beta_1) \cdot \alpha c(\beta_2)$	4. $\beta c(\beta) = \beta c(\beta_1) + \beta c(\beta_2)$

By and large, the α-complexity measures the number of branches of an α-branching tree, and the β-complexity measures the number of branches of a β-branching tree. Before we decide to search for a proof of a formula in one or another proof system, the measures may be calculated in order to estimate the complexity of the search.

Suppose, for example, that we have to decide validity of (6.5) and (6.6) either by analytic tableaux or by resolution. In terms of complexity, it is exactly the choice between a canonical and a dual erotetic calculus. We have:

$$\alpha c(\neg A) = \alpha c(\neg(p \wedge q \wedge r)) + \alpha c(\neg(p \wedge q \wedge \neg r)) + \ldots +$$

$$\alpha c(\neg(\neg p \wedge \neg q \wedge \neg r)) = 1 \times 8 = 8$$

and

$$\alpha c(\neg B) = \alpha c(\neg(p \vee q \vee r)) \cdot \alpha c(\neg(p \vee q \vee \neg r)) \cdot \ldots \cdot$$

$$\alpha c(\neg(\neg p \vee \neg q \vee \neg r)) = 3^8 = 6561$$

It follows that in a complete resolution derivation for $\neg A$ we would obtain 8 clauses after deriving CNF of $\neg A$ (one may derive the empty clause after 7 applications of the resolution rule). In the case of formula B, we would have to derive 6561 clauses before applying the resolution rule. It is not the case, however, that testing validity of formula B is a hopeless task. For we have:

$$\beta c(\neg A) = 6561 \qquad \beta c(\neg B) = 8$$

A Smullyan-type analytic tableau for $\neg B$ has at most 8 branches, each of them may be completed immediately, each is open. On the other hand, a complete analytic tableau for $\neg A$ has 6561 branches. It is possible to construct a smaller tableau—the reader may check that the minimal one has 33 branches; but in general, the lower bound of the number of branches of this kind of tableau is $n!$, where n is the number of distinct variables in a formula (see [6–8]). Therefore, it seems reasonable to treat problems akin to (6.5) with resolution, while problems akin to (6.6) should be dealt with by analytic tableaux.

Obviously, we are aware of the diagnosis of this situation which is derived by the authors of [6–8]—it is the absence of cut that leads to the computational collapse, therefore analytic tableau system **KE**, equipped with a form of non-eliminable analytic cut, is proposed by D'Agostino and Mondadori as a remedy. For the same reason the resolution system works well, where the standard tableaux grow too fast—the rule of resolution is, as we have observed, a dual of the rule of cut. But in the present context, we would like to draw another solution, independent of the use of any form of cut.

At the start, the general diagnosis is this: before you treat a problem with analytic tableaux or resolution, calculate the α- and β-complexity of the problem. The same diagnosis may be applied with respect to different variants of erotetic calculi—if the α-complexity of a formula is smaller than its β-complexity, use the canonical version of erotetic calculi, otherwise use the dual version. But suppose now that the formula we have to examine is not simply (6.5) or (6.6), but the conjunction $A \wedge B$. No matter if we decide to calculate this by analytic tableaux or by resolution—in both cases we will have to deal with expansion of complexity on one of the conjuncts. But we know that it is better to calculate validity of A by rules which branch on α-formulas, and to use β-branching rules in order to calculate the validity of B. So why not join the two formats to check the conjuncts by different sets of rules?

The observations expressed above gave rise to the idea of *Distributive Deductive Systems*. A Distributive Deductive System (DDS, for short) consists of two layers: the first one contains various proof systems, called *modules* of DDS, and the second contains tools to decide which module suits which formal problem. Functions αc, βc are the main citizens of the second layer (though certainly not the only possible ones, see [22] where the authors examine a number of possible functions to generate tableaux optimal in size).

Figure 6.4 presents an illustration of this idea. '$B \setminus p \vee q \vee r$' stands for formula B, displayed as (6.6), without conjunct '$p \vee q \vee r$', that is:

$$B = (p \vee q \vee \neg r) \wedge (p \vee \neg q \vee r) \wedge (p \vee \neg q \vee \neg r) \wedge (\neg p \vee q \vee r) \wedge$$

$$\wedge (\neg p \vee q \vee \neg r) \wedge (\neg p \vee \neg q \vee r) \wedge (\neg p \vee \neg q \vee \neg r),$$

similarly for $A \setminus p \wedge q \wedge r$. '$\mathcal{D}$' refers to the following derivation.

$$\cfrac{\cfrac{\cfrac{\vdash B}{\vdash p \vee q \vee r \,;\, \vdash B \setminus p \vee q \vee r}\,R^c_\alpha}{\vdash p, q \vee r \,;\, \vdash B \setminus p \vee q \vee r}\,R^c_\beta}{\vdash p, q, r \,;\, \vdash B \setminus p \vee q \vee r}\,R^c_\beta$$

In the description of the rules in Fig. 6.4, we have used 'c' in the upper index of the name of a rule for 'canonical', and 'd' for 'dual'.

Since $\alpha c(A \wedge B) < \beta c(A \wedge B)$, we start the transformation presented in Fig. 6.4 with an application of a rule for conjunction taken from the canonical erotetic calculus (the propositional part of $\mathbb{E}^{\mathbf{RPQ}}$). Then we obtain a sequent with formula A, where

$$\dfrac{\vdash A \wedge B}{\vdash A \;\; ; \;\; \vdash B}\, R^{c}_{\alpha}$$

Left branch (under $\vdash A$), right branch: $\mathcal{D}$

$$\dfrac{\vdash A}{\dashv A}\; c \mapsto d$$

$$\dfrac{\dashv A}{\dashv p \wedge q \wedge r \; ; \; \dashv A \setminus p \wedge q \wedge r}\, R^{d}_{\beta}$$

$$\dfrac{}{\dashv p \wedge q \wedge r \; ; \; \dashv p \wedge q \wedge \neg r \; ; \; \dashv A \setminus 1, 2}\, R^{d}_{\beta}$$

$$\dfrac{}{\dashv 1 \; ; \dashv 2 \; ; \dashv 3 \; ; \dashv A \setminus 1, 2, 3}\, R^{d}_{\beta}$$

$$\dfrac{}{\dashv 1 \; ; \dashv 2 \; ; \dashv 3 \; ; \dashv 4 \; ; \dashv A \setminus 1, 2, 3, 4}\, R^{d}_{\beta}$$

$$\dfrac{}{\dashv 1 \; ; \dashv 2 \; ; \dashv 3 \; ; \dashv 4 \; ; \dashv 5 \; ; \dashv A \setminus 1, 2, 3, 4, 5}\, R^{d}_{\beta}$$

$$\dfrac{}{\dashv 1 \; ; \dashv 2 \; ; \dashv 3 \; ; \dashv 4 \; ; \dashv 5 \; ; \dashv 6 \; ; \dashv A \setminus 1, 2, 3, 4, 5, 6}\, R^{d}_{\beta}$$

$$\dfrac{}{\dashv 1 \; ; \dashv 2 \; ; \dashv 3 \; ; \dashv 4 \; ; \dashv 5 \; ; \dashv 6 \; ; \dashv 7 \; ; \dashv 8}\, R^{d}_{\beta}$$

$$\dfrac{}{\dashv p, q \wedge r \; ; \dashv 2 \; ; \dashv 3 \; ; \dashv 4 \; ; \dashv 5 \; ; \dashv 6 \; ; \dashv 7 \; ; \dashv 8}\, R^{d}_{\alpha}$$

$$\vdots$$

$$\dfrac{\dashv p, q, r \; ; \dashv p, q, \neg r \; ; \dashv p, \neg q, r \; ; \dashv p, \neg q, \neg r \; ; \dashv 5' - \dashv 8'}{\dashv p, q \; ; \dashv p, \neg q, r \; ; \dashv p, \neg q, \neg r \; ; \dashv 5' - \dashv 8' \; ; W_1}\, R_{res}$$

$$\dfrac{}{\dashv p, \neg q \; ; \dashv p, q \; ; \dashv 5' - \dashv 8' \; ; W_2}\, R_{res}$$

$$\dfrac{}{\dashv p \; ; \dashv 5' - \dashv 8' \; ; W_3}\, R_{res}$$

$$\vdots$$

$$\dashv \varnothing \; ; W_7$$

Fig. 6.4 Transformation in DDS: illustration

αc of A is much bigger than βc, thus the sequent is 'switched' into the reversed one (this is indicated by the inscription '$c \mapsto d$') and then the rules of $\mathbb{E}^{\mathbf{RPQ}}_{dual}$ are applied. On the other hand, since αc of B is smaller than its βc, we continue to apply the canonical rules on this side of the derivation. Observe also that the derivation is not a proof of anything—it does not end with a success in the 'B-part', which is fine, as formula '$A \wedge B$' is not valid.

We believe that the ideas presented in this section deserve further examination.

References

1. Batens, D. (1980). Paraconsistent extensional propositional logics. *Logique et Analyse, 90–91*, 195–234.
2. Boolos, G. (1984). Don't eliminate cut. *Journal of Philosophical Logic, 13*(4), 373–378.
3. Chlebowski, S. (2018). *Canonical and dual erotetic calculi for first-order logic* (Doctoral dissertation). Adam Mickiewicz University.
4. Chlebowski, S., Gajda, A., & Urbański, M. (2022). An abductive question-answer system for the minimal logic of formal inconsistency mbC. *Studia Logica, 110*, 479–509.

5. Chlebowski, S., & Leszczyńska-Jasion, D. (2015). Dual erotetic calculi and the minimal LFI. *Studia Logica, 103,* 1245–1278. https://doi.org/10.1007/s11225-015-9617-0
6. D'Agostino, M. (1990). *Investigations into the complexity of some propositional calculi: Technical monograph.* Oxford University Computing Laboratory, Programming Research Group.
7. D'Agostino, M. (1992). Are tableaux an improvement on truth-tables? Cut-free proofs and bivalence. *Journal of Logic, Language and Computation, 1*, 235–252.
8. D'Agostino, M., & Mondadori, M. (1994). The taming of the cut: Classical refutations with analytic cut. *Journal of Logic and Computation, 4*(3), 285–319.
9. Dunham, B., & Wang, H. (1976). Towards feasible solutions of the tautology problem. *Annals of Mathematical Logic, 10*(2), 117–154.
10. Fitting, M. (1990). *First-order logic and automated theorem proving*. Berlin: Springer.
11. Hählne, R. (2001). Tableaux and related methods. In A. Robinson & A. Voronkov (Eds.), *Handbook of automated reasoning* (pp. 101–175). Elsevier Science Publishers.
12. Kaye, R. (2008). *The mathematics of logic*. Cambridge University Press.
13. Leszczyńska-Jasion, D. (2018). *From questions to proofs: Between the logic of questions and proof theory*. Faculty of Social Science Publishers, Adam Mickiewicz University.
14. Leszczyńska-Jasion, D., Ignaszak, M., & Chlebowski, S. (2018). Rasiowa-Sikorski deduction systems with the rule of cut: A case study. *Studia Logica*. https://doi.org/10.1007/s11225-018-9795-7
15. Leszczyńska-Jasion, D., Urbański, M., & Wiśniewski, A. (2013). Socratic trees. *Studia Logica, 101*(5), 959–986. https://doi.org/10.1007/s11225-012-9404-0
16. Łukasiewicz, J. (1921). Logika dwuwartościowa (two-valued logic). *Przeglad Filozoficzny, 23*, 189–205.
17. Orłowska, E., & Golińska-Pilarek, J. (2011). *Dual tableaux: Foundations, methodology, case studies*. Berlin: Springer.
18. Priest, G. (2008). *An introduction to non-classical logic: From if to is*. Cambridge University Press.
19. Rasiowa, H., & Sikorski, R. (1960). On the Gentzen theorem. *Fundamenta Mathematicae, 48*, 57–69.
20. Rasiowa, H., & Sikorski, R. (1963). *The mathematics of metamathematics*. Polish Academy of Sciences.
21. Skura, T. F. (2011). Refutation systems in propositional logic. In D. M. Gabbay & F. Guenthner (Eds.), *Handbook of philosophical logic.* (Vol. 16). Berlin: Springer.
22. Sochański, M., Leszczyńska-Jasion, D., Chlebowski, S., Tomczyk, A., & Jukiewicz, M. (2022). Synthetic tableaux: Minimal tableau search heuristics. In J. Blanchette, L. Kovács, & D. Pattinson (Eds.), *Automated reasoning: IJCAR 2022* (pp. 407–425). Cham: Springer. https://doi.org/10.1007/978-3-031-10769-6_25
23. Troelstra, A. S., & Schwichtenberg, H. (2000). *Basic proof theory* (2nd ed.). Cambridge University Press.
24. Wiśniewski, A., & Shangin, V. (2006). Socratic proofs for quantifiers. *Journal of Philosophical Logic, 35*(2), 147–178. https://doi.org/10.1007/s10992-005-9000-0
25. Wójcicki, R. (1973). Dual counterparts of consequences operations. *Bulletin of the Section of Logic, 2*(2), 54–57.
26. Wójcicki, R. (1988). *Theory of logical calculi: Basic theory of consequence operations* (softcover reprint of the hardcover 1st edition by Kluwer Academic Publishers). Springer–Science+Business Media.
27. Wybraniec-Skardowska, U. (2018). Rejection in Łukasiewicz's and Słupecki's sense. In U. Wybraniec-Skardowska & Á. Garrido (Eds.), *The Lvov-Warsaw School: Past and present* (pp. 575–597). Springer-Birkhäuser.
28. Wybraniec-Skardowska, U. (2023). Operator counterparts of types of reasoning. *Logica Universalis, 17*, 511–528.

Chapter 7
From Erotetic Calculi to Proofs. The Non-classical Cases

In this chapter, we examine the relationship between erotetic calculi for six basic modal logics and more standard sequent calculi for those logics. We will focus on basic modal logics that are neither symmetric nor Euclidean; hence, the six we consider are K, D, T, $\mathsf{K4}$, $\mathsf{D4}$, $\mathsf{S4}$. In terms of semantic intuitions, without the two mentioned properties, the information captured by modalities always travels in one direction—whether forward or backward depends on how we read derivations, but the direction is nevertheless fixed.

As we shall see, this directional transfer of information is similar in explicit and implicit systems for modal logics, allowing us to construct translations between such systems, as we do in this chapter. However, when at least one of the two additional properties (symmetry, Euclideanness) is included, semantic information flows in multiple directions, significantly complicating the picture. For this reason, extending the work presented here to the remaining basic modal logics remains an open problem to be addressed in the future.

After discussing the six basic modal logics, in the last section, we conclude with a sketch of the translation procedure for intuitionistic propositional logic, INT. In this case, the construction is, unsurprisingly, based on that for modal logic $\mathsf{S4}$.

As for the modal erotetic calculi, for reasons explained below, we have decided to work with a different version of these calculi, which is nearly identical to those presented in [8]. Thus, in this chapter, we use $\mathbb{E} \in \{\mathbb{E}^{\mathsf{K}\tau}, \mathbb{E}^{\mathsf{D}\tau}, \mathbb{E}^{\mathsf{T}\tau}, \mathbb{E}^{\mathsf{K4}\tau}, \mathbb{E}^{\mathsf{D4}\tau}, \mathbb{E}^{\mathsf{S4}\tau}, \mathbb{E}^{\mathsf{INT}}\}$.

We will not examine the soundness and completeness of the modified versions of the modal $\mathbb{E}$, as doing so would yield no new insights; it would largely involve repeating the arguments discussed in Chap. 3 and/or in [8].

D. Leszczyńska-Jasion, *The Method of Socratic Proofs*, Trends in Logic 64,
https://doi.org/10.1007/978-3-031-82451-7_7

7.1 Translation Into Sequent System for Modal Logic K

We will derive proofs of sequents with modal formulas in a similar manner for the cases of K, D, T, K4, D4, S4—that is, for all normal modal logics that are neither symmetric nor Euclidean. In this chapter, however, we must address semantic information expressed in the form of indices. Perhaps the most interesting aspect of the translations provided here is the relationship between purely syntactic entities, such as sequents that do not explicitly represent semantic information, and sequents enhanced with additional technical details, like indices and labels.

At the end of Chap. 3, following [4], we introduced a distinction between implicit and explicit proof systems for modal logics. As we observed, the method of Socratic proofs is an explicit proof method, as it uses indices to encode semantic information within the syntax. In this section, we demonstrate how to transform a Socratic proof of a sequent containing indexed formulas, constructed in an erotetic calculus, into a proof of a corresponding 'ordinary' sequent in an implicit sequent calculus—that is, a sequent calculus in which semantic information is not represented directly. Thus, the algorithms presented here describe the relationship between an explicit erotetic calculus and an implicit sequent calculus.

Another aspect of explicit and implicit systems worth comparing is the property of *purity* of rules. In [1], we find the following definition: a rule of a monotonic Gentzen-type system is called *pure* if, whenever $\Gamma \Rightarrow \Delta$ can be inferred by it from $\Gamma_i \Rightarrow \Delta_i$ ($i = 1, \ldots, n$), then $\Gamma, \Gamma' \Rightarrow \Delta, \Delta'$ can also be inferred by it from $\Gamma_i, \Gamma' \Rightarrow \Delta_i, \Delta'$ ($i = 1, \ldots, n$, Γ' and Δ' are arbitrary sets of formulas).[1] The rules of erotetic calculi for modal logics are pure, whereas the modal rules of implicit sequent calculi are not.

A further notable difference between these two types of proof systems, closely related to the above discussion, also motivates the study of translation algorithms: the property of *confluency*, which we defined in the previous chapter (see Sect. 6.1). Recall that confluency is a desirable property for implementation, as it prevents the necessity of backtracking by ensuring there are no 'bad moves' in the construction of a derivation. Since the modal rules of implicit sequent systems are not invertible, backtracking may be required to determine whether a failure in a backward proof-search is due to a 'bad move' or to unprovability of a sequent.

There are, however, numerous systems that use various types of labels and have invertible modal rules, thus possessing the property of confluency (we list several of these at the beginning of Chap. 3). Such systems may be useful for automated proof search. Our results demonstrate that proofs in an implicit sequent system for modal logics can be obtained through translation from an explicit proof format that is more conductive to proof search, eliminating the need for backtracking. Naturally, the computational effort occurs within the level of the explicit proof system; therefore, using the algorithms presented here together with an inefficient proof search algorithm in the explicit system does not provide a computational speedup. However,

---

[1] See also [9, Chap. 1], for additional important properties of sequent calculus rules.

we believe that examining the relations between different proof systems may bring us closer to developing new, efficient algorithms.

Below, we present the rules of the erotetic calculus $\mathbb{E}^{\mathsf{K}\tau}$ alongside the corresponding rules of Gentzen-style (implicit) sequent calculus $\mathbb{G}^{\mathsf{K}\tau}$. The Gentzen-style system appears in the right columns of Tables 7.1 and 7.2. The rules for classical operators are standard, while the rules for modal operators are based on those found in [3, Chap. 3] (see also [11, Sect. 1] for further comparison). The rules of the erotetic calculus $\mathbb{E}^{\mathsf{K}\tau}$ have been designed to align with the rules of $\mathbb{G}^{\mathsf{K}\tau}$. This alignment results in a two-sided, multisuccedent version of the Socratic proof method for modal logics, shown in the left columns of Tables 7.1 and 7.2. In both calculi, the antecedent and the succedent of a sequent may be empty, though not simultaneously.

Conventions: Language, Concatenation, Replacement

We need a separate language of sequents for the implicit systems. The language will be called $\mathcal{M}_{\vdash}^{imp}$. It is built upon $\mathcal{M}$, but contains additionally the turnstile symbol '$\vdash$' and the comma ','. Sequents of $\mathcal{M}_{\vdash}^{imp}$ are expressions of the form:

$$A_1, \ldots, A_n \vdash B_1, \ldots, B_k$$

where $\langle A_1, \ldots, A_n \rangle$ and $\langle B_1, \ldots, B_k \rangle$ represent finite sequences of formulas of $\mathcal{M}$. It may happen that $n = 0$ or $k = 0$, though not both at the same time. Below we will use S, T for finite, possibly empty sequences of formulas of $\mathcal{M}$. Letters Γ, Δ refer to finite, possibly empty, sequences of indexed formulas of $\mathcal{M}_{\vdash}^{?}$. Letters ϕ, ψ, χ will be used for sequents of the language $\mathcal{M}_{\vdash}^{?}$; we will not need additional metavariables for sequents of the language $\mathcal{M}_{\vdash}^{imp}$.

We will present most of the rules in accordance with notation conventions introduced previously in Sect. 2.3. Recall that by '$T(A)$' in the premise of rule $L_{\neg}^{\mathbb{G}}$ (see Table 7.1) we refer to a sequence of formulas that has A among its terms, where the position of A is arbitrary. Then, by '$T(\_)$' we mean the result of removing the one indicated occurrence of formula A. Moreover, if the expression '$S(\_)$' is used first in the premise of a rule—like in the case of $L_{\neg}^{\mathbb{G}}$—then '$S(\neg A)$' used in the conclusion refers to the result of adding $\neg A$ at the position indicated by '_'. These conventions also apply to sequences of indexed formulas of the language $\mathcal{M}_{\vdash}^{?}$.

Table 7.1 presents the non-branching rules for the classical connectives, Table 7.2 presents the branching rules for the classical connectives, and Tables 7.3, 7.5—the modal rules.

None of the calculi considered in this chapter contains the structural rule of Exchange, which allows to change the order of formulas in the antecedent or the succedent of a sequent. Therefore the rules are formulated in a way which allows to put the resulting formula-conclusion (that is, e.g., formula A^{ς} in the case of rule $L_{\neg}^{\mathbb{E}}$, and formula $\neg A$ in the case of rule $L_{\neg}^{\mathbb{G}}$) in an arbitrary place.

Let us observe that in the case of rules $L_{\wedge}^{\mathbb{G}}$ and $R_{\vee}^{\mathbb{G}}$, formula B immediately succeeds formula A in the premise. The branching rules of $\mathbb{G}^{\mathsf{K}\tau}$ have shared contexts. Let us also explain that, according to the introduced conventions, in the rule $R_{\wedge}^{\mathbb{G}}$ the

Table 7.1 Non-branching rules of $\mathbb{E}^{\mathsf{K}\tau}$ and $\mathbb{G}^{\mathsf{K}\tau}$

$$\frac{?(\Phi(\Gamma((\neg A)^{\varsigma}) \vdash \Delta(\_)))}{?(\Phi(\Gamma(\_) \vdash \Delta(A^{\varsigma})))}\ L^{\mathbb{E}}_{\neg} \qquad \frac{S(\_) \vdash T(A)}{S(\neg A) \vdash T(\_)}\ L^{\mathbb{G}}_{\neg}$$

$$\frac{?(\Phi(\Gamma(\_) \vdash \Delta((\neg A)^{\varsigma})))}{?(\Phi(\Gamma(A^{\varsigma}) \vdash \Delta(\_)))}\ R^{\mathbb{E}}_{\neg} \qquad \frac{S(A) \vdash T(\_)}{S(\_) \vdash T(\neg A)}\ R^{\mathbb{G}}_{\neg}$$

$$\frac{?(\Phi(\Gamma((A \wedge B)^{\varsigma}) \vdash \Delta))}{?(\Phi(\Gamma(A^{\varsigma}, B^{\varsigma}) \vdash \Delta))}\ L^{\mathbb{E}}_{\wedge} \qquad \frac{S(A, B) \vdash T}{S(A \wedge B) \vdash T}\ L^{\mathbb{G}}_{\wedge}$$

$$\frac{?(\Phi(\Gamma \vdash \Delta((A \vee B)^{\varsigma})))}{?(\Phi(\Gamma \vdash \Delta(A^{\varsigma}, B^{\varsigma})))}\ R^{\mathbb{E}}_{\vee} \qquad \frac{S \vdash T(A, B)}{S \vdash T(A \vee B)}\ R^{\mathbb{G}}_{\vee}$$

$$\frac{?(\Phi(\Gamma(\_) \vdash \Delta((A \to B)^{\varsigma})))}{?(\Phi(\Gamma(A^{\varsigma}) \vdash \Delta(B^{\varsigma})))}\ R^{\mathbb{E}}_{\to} \qquad \frac{S(A) \vdash T(B)}{S(\_) \vdash T(A \to B)}\ R^{\mathbb{G}}_{\to}$$

Table 7.2 Branching rules of $\mathbb{E}^{\mathsf{K}\tau}$ and $\mathbb{G}^{\mathsf{K}\tau}$

$$\frac{?(\Phi(\Gamma \vdash \Delta((A \wedge B)^{\varsigma})))}{?(\Phi(\Gamma \vdash \Delta(A^{\varsigma})\ ;\ \Gamma \vdash \Delta(B^{\varsigma})))}\ R^{\mathbb{E}}_{\wedge} \qquad \frac{S \vdash T(A) \quad S \vdash T(B)}{S \vdash T(A \wedge B)}\ R^{\mathbb{G}}_{\wedge}$$

$$\frac{?(\Phi(\Gamma((A \vee B)^{\varsigma}) \vdash \Delta))}{?(\Phi(\Gamma(A^{\varsigma}) \vdash \Delta\ ;\ \Gamma(B^{\varsigma}) \vdash \Delta))}\ L^{\mathbb{E}}_{\vee} \qquad \frac{S(A) \vdash T \quad S(B) \vdash T}{S(A \vee B) \vdash T}\ L^{\mathbb{G}}_{\vee}$$

$$\frac{?(\Phi(\Gamma((A \to B)^{\varsigma}) \vdash \Delta(\_)))}{?(\Phi(\Gamma(\_) \vdash \Delta(A^{\varsigma})\ ;\ \Gamma(B^{\varsigma}) \vdash \Delta(\_)))}\ L^{\mathbb{E}}_{\to}$$

$$\frac{S(\_) \vdash T(A) \quad S(B) \vdash T}{S(A \to B) \vdash T(\_)}\ L^{\mathbb{G}}_{\to}$$

indicated occurrences of A and B in the sequences $T(A)$ and $T(B)$, respectively, occur in the same position. The situation is similar in the case of rule $L^{\mathbb{G}}_{\vee}$.

Finally, the rules for the modal operators, presented in Table 7.3, are non-branching. In the case of the rules of $\mathbb{E}^{\mathsf{K}\tau}$ 'creating a new world', that is, $L^{\mathbb{E}}_{\Diamond}$ and $R^{\mathbb{E}}_{\Box}$, j is the smallest numeral new with respect to the active constituent of the question-premise, that is:

$$j = \max(W(\phi)) + 1, \tag{7.1}$$

where ϕ is the sequent-premise of the rule. We have used the same restriction previously (see Sect. 3.3), but not in a mandatory form, as it was necessary to relax it for the aims of structural analysis. This time, however, (7.1) is mandatory.

Recall also that in the case of the rules 'settling the worlds', that is, $\mathbf{L}_{\Box}$, $\mathbf{R}_{\Diamond}$, numeral j is already present in the active constituent of the question-premise. The numeral must satisfy the relevant condition of the applicability of the rule. In the case of logic K, it means that j immediately succeeds i in some index of the active sequent.

We will use the notion of an *active formula* and an *active indexed formula* as previously. If the active indexed formula of $\mathbf{r}$ is of the form '$(A)^{\varsigma/i}$', then numeral

Table 7.3 Modal rules of $\mathbb{E}^{\mathsf{K}\tau}$

$\dfrac{?(\,\Phi\,(\,\Gamma\,(\,\Diamond A^{\varsigma/i})\vdash\Delta\,)\,)}{?(\,\Phi\,(\,\Gamma\,(\,A^{\varsigma/i,j}\,)\vdash\Delta\,)\,)}\,L^{\mathbb{E}}_{\Diamond}$	$\dfrac{?(\,\Phi\,(\,\Gamma\vdash\Delta\,(\,\Diamond A^{\varsigma/i}\,)\,)\,)}{?(\,\Phi\,(\,\Gamma\vdash\Delta\,(\,\Diamond A^{\varsigma/i},\,A^{j}\,)\,)\,)}\,R^{\mathbb{E}}_{\Diamond}$
$\dfrac{?(\,\Phi\,(\,\Gamma\,(\,\Box A^{\varsigma/i}\,)\vdash\Delta\,)\,)}{?(\,\Phi\,(\,\Gamma\,(\,\Box A^{\varsigma/i},\,A^{j}\,)\vdash\Delta\,)\,)}\,L^{\mathbb{E}}_{\Box}$	$\dfrac{?(\,\Phi\,(\,\Gamma\vdash\Delta\,(\,\Box A^{\varsigma/i}\,)\,)\,)}{?(\,\Phi\,(\,\Gamma\vdash\Delta\,(\,A^{\varsigma/i,j}\,)\,)\,)}\,R^{\mathbb{E}}_{\Box}$

Table 7.4 Modal rules of sequent calculi by Melvin Fitting

$\dfrac{U\sharp, A \longrightarrow V\flat}{U, \Diamond A \longrightarrow V}\,\mathbf{L}_{\Diamond}$	$\dfrac{U\sharp \longrightarrow V\flat, A}{U \longrightarrow V, \Box A}\,\mathbf{R}_{\Box}$

where $U\sharp = \{B : \Box B \in U\}$, $V\flat = \{B : \Diamond B \in V\}$

i will be called the *active numeral* of $\mathbf{r}$. Further, next to the notion of a constituent-conclusion (sequent-conclusion), we will sometimes use the notion of an indexed *formula-conclusion of a rule* $\mathbf{r}$, meaning the indexed formula (or formulas) whose schema is distinguished in the sequent-conclusion of $\mathbf{r}$.

We will also need the notion of a *principal numeral*. The principal numeral of a rule $\mathbf{r}$ is the last one in the index of the indexed formula-conclusion—if there is exactly one indexed formula-conclusion. In the case of rules $R^{\mathbb{E}}_{\Diamond}$ and $L^{\mathbb{E}}_{\Box}$, there are two formulas-conclusions with different numerals at the end of their indices: $\Diamond A^{\varsigma/i}$, A^{j}, or $\Box A^{\varsigma/i}$, A^{j}, respectively. In this situation, we take j to be the principal numeral.[2]

Calculus $\mathbb{G}^{\mathsf{K}\tau}$ used here has been obtained from the sequent calculus considered in [3] by dropping the unified notation, so we first recall the original account by Melvin Fitting before we continue (see [3, p. 90]). In [3] the author uses '$\longrightarrow$' for the sequent sign; sequents have finite sets of formulas on the left and right side of the sign. Finally, the ν, π-notation is used, as in Chap. 3 (see Table 7.4).

Though not obvious at the first glance, the above account of the modal rules incorporates weakening, since U, V in the conclusions may contain some extra formulas. Below we illustrate this phenomenon, see Example 7.3.

Table 7.5 presents the modal rules of calculus $\mathbb{G}^{\mathsf{K}\tau}$. If $S = \langle A_1, \ldots, A_n\rangle$ is a finite sequence of formulas of $\mathcal{M}$, then by '$\Box S$' and $\Diamond S$' we mean:

$$\Box S = \langle \Box A_1, \ldots, \Box A_n\rangle$$

$$\Diamond S = \langle \Diamond A_1, \ldots, \Diamond A_n\rangle$$

[2] In the context of sequent calculi, a formula-conclusion is usually called a *principal formula*. However, since the rules of erotetic calculi are reversed with respect to the rules of sequent calculi, using this terminology would be confusing.

Table 7.5 Modal rules and weakening in $\mathbb{G}^{\mathsf{K}\tau}$

$\dfrac{S_1 \,'\, A \,'\, S_2 \vdash T}{\Box S_1 \,'\, \Diamond A \,'\, \Box S_2 \vdash \Diamond T}\, L^{\mathbb{G}}_{\Diamond}$	$\dfrac{S \vdash T_1 \,'\, A \,'\, T_2}{\Box S \vdash \Diamond T_1 \,'\, \Box A \,'\, \Diamond T_2}\, R^{\mathbb{G}}_{\Box}$
$\dfrac{S \vdash T}{S^* \vdash T}\, L^{\mathbb{G}}_{W}$ where S^* is a proper supersequence of S	$\dfrac{S \vdash T}{S \vdash T^*}\, R^{\mathbb{G}}_{W}$ where T^* is a proper supersequence of T

We find it much more convenient to present the modal rules of $\mathbb{G}^{\mathsf{K}\tau}$ in a different convention than the one utilized so far. Namely, we use '′' as a sign of concatenation. The inscription

$$S_1 \,'\, A \,'\, S_2$$

in the antecedent of the premise of rule $L^{\mathbb{G}}_{\Diamond}$ refers to the concatenation of sequence S_1 with sequence $\langle A \rangle$ and sequence S_2. The same rule $L^{\mathbb{G}}_{\Diamond}$ in the replacement-convention would look as follows:

$$\frac{S\,(\,A\,) \vdash T}{\Box S\,(\,\Diamond A\,) \vdash \Diamond T}\, L^{\mathbb{G}}_{\Diamond}$$

Since weakening is not incorporated in the modal rules of $\mathbb{G}^{\mathsf{K}\tau}$, we also state it explicitly. The notion of (proper) supersequence of a sequence is understood as in Definition 2.9.

The axioms of calculus $\mathbb{G}^{\mathsf{K}\tau}$ are sequents of the form '$A \vdash A$', exclusively, where A is a formula of language $\mathcal{M}$. It is common to formulate the axioms of a sequent calculus as a no-premise rule Ax, as below. Let us also observe that the no-premise rule Ax_{+W} (to the right) is derivable in $\mathbb{G}^{\mathsf{K}\tau}$ by Ax, $L^{\mathbb{G}}_{W}$ and $R^{\mathbb{G}}_{W}$.

$$\frac{}{A \vdash A}\, Ax \qquad \frac{}{S(A) \vdash T(A)}\, Ax_{+W}$$

Calculus $\mathbb{G}^{\mathsf{K}\tau}$ is the set of rules:

$$\{L^{\mathbb{G}}_{\neg}, R^{\mathbb{G}}_{\neg}, L^{\mathbb{G}}_{\wedge}, R^{\mathbb{G}}_{\vee}, R^{\mathbb{G}}_{\rightarrow}, R^{\mathbb{G}}_{\wedge}, L^{\mathbb{G}}_{\vee}, L^{\mathbb{G}}_{\rightarrow}, L^{\mathbb{G}}_{\Diamond}, R^{\mathbb{G}}_{\Box}, L^{\mathbb{G}}_{W}, R^{\mathbb{G}}_{W}, Ax\}.$$

The notion of proof in calculus $\mathbb{G}^{\mathsf{K}\tau}$ is defined in a standard manner:

Definition 7.1 Let A be a formula of $\mathcal{M}$. A *proof of the sequent* $\vdash A$ *in calculus* $\mathbb{G}^{\mathsf{K}\tau}$ is a finite tree labeled with sequents of the language $\mathcal{M}^{imp}_{\vdash}$, regulated by the rules of calculus $\mathbb{G}^{\mathsf{K}\tau}$, with the root labeled with the sequent $\vdash A$, and the leaves labeled with axioms of the calculus.

On the other hand, calculus $\mathbb{E}^{\mathsf{K}\tau}$ is the following set of rules:

$$\{L^{\mathbb{E}}_{\neg}, R^{\mathbb{E}}_{\neg}, L^{\mathbb{E}}_{\wedge}, R^{\mathbb{E}}_{\vee}, R^{\mathbb{E}}_{\rightarrow}, R^{\mathbb{E}}_{\wedge}, L^{\mathbb{E}}_{\vee}, L^{\mathbb{E}}_{\rightarrow}, L^{\mathbb{E}}_{\Diamond}, R^{\mathbb{E}}_{\Diamond}, L^{\mathbb{E}}_{\Box}, R^{\mathbb{E}}_{\Box}\}.$$

Now we introduce the following definition:

Definition 7.2 A *Socratic transformation of a question* Q *in* $\mathbb{E}^{\mathsf{K}\tau}$ is a sequence $\mathbf{s} = \langle Q_1, Q_2, \ldots \rangle$ of questions of the language $\mathcal{M}^?_{\vdash}$ such that $Q_1 = Q$, and for each $n > 1$, the question Q_n results from Q_{n-1} by the application of one of the rules of $\mathbb{E}^{\mathsf{K}\tau}$.

Definition 7.3 Let $\vdash (A)^0$ be a sequent of the language $\mathcal{M}^?_{\vdash}$. A *Socratic proof of* $\vdash (A)^0$ *in* $\mathbb{E}^{\mathsf{K}\tau}$ is a finite Socratic transformation $\mathbf{s}$ of the question $?(\vdash (A)^0)$ in $\mathbb{E}^{\mathsf{K}\tau}$ such that each constituent of the last question of $\mathbf{s}$ is of the following form:

(i) $\Gamma(B^{\varsigma_1/i}) \vdash \Delta(B^{\varsigma_2/i})$.

Example 7.1 Here is an example of a Socratic proof in $\mathbb{E}^{\mathsf{K}\tau}$. A stands for '$\Box(p \to q)$'. (The numbers of lines are for further reference).

$$1.\ ?(\vdash (\Box(p \to q) \to (\Box p \to \Box q))^0) \quad R^{\mathbb{E}}_{\to}$$
$$2.\ ?((A)^0 \vdash (\Box p \to \Box q)^0) \quad R^{\mathbb{E}}_{\to}$$
$$3.\ ?((A)^0, (\Box p)^0 \vdash (\Box q)^0) \quad R^{\mathbb{E}}_{\Box}$$
$$4.\ ?((A)^0, (\Box p)^0 \vdash (q)^{0,1}) \quad L^{\mathbb{E}}_{\Box}$$
$$5.\ ?((A)^0, (\Box p)^0, (p)^1 \vdash (q)^{0,1}) \quad L^{\mathbb{E}}_{\Box}$$
$$6.\ ?((A)^0, (p \to q)^1, (\Box p)^0, (p)^1 \vdash (q)^{0,1}) \quad R^{\mathbb{E}}_{\to}$$
$$7.\ ?((A)^0, (\Box p)^0, (p)^1 \vdash (q)^{0,1}, (p)^1\ ;\ (A)^0, (q)^1, (\Box p)^0, (p)^1 \vdash (q)^{0,1})$$

Example 7.2 A proof of the sequent $\vdash \Box(p \to q) \to (\Box p \to \Box q)$ in $\mathbb{G}^{\mathsf{K}\tau}$.

$$\dfrac{\dfrac{\dfrac{\dfrac{\dfrac{\dfrac{p \vdash p}{p \vdash q, p}\,R^{\mathbb{G}}_{W} \qquad \dfrac{q \vdash q}{q, p \vdash q}\,L^{\mathbb{G}}_{W}}{p \to q, p \vdash q}\,L^{\mathbb{G}}_{\to}}{\Box(p \to q), \Box p \vdash \Box q}\,R^{\mathbb{G}}_{\Box}}{\Box(p \to q) \vdash \Box p \to \Box q}\,R^{\mathbb{G}}_{\to}}{\vdash \Box(p \to q) \to (\Box p \to \Box q)}\,R^{\mathbb{G}}_{\to}}{}$$

In the above example, one application of the modal rule $R^{\mathbb{G}}_{\Box}$ of $\mathbb{G}^{\mathsf{K}\tau}$ corresponds to a sequence of applications of the modal rules of $\mathbb{E}^{\mathsf{K}\tau}$ in lines 4–6 of Example 7.1. Semantically speaking, one application of $R^{\mathbb{G}}_{\Box}$ of $\mathbb{G}^{\mathsf{K}\tau}$ describes the process of creating a new world and settling the world by means of the available information. In the case of calculus $\mathbb{E}^{\mathsf{K}\tau}$, the same process is described by the three applications of modal rules. However, in order to see the correspondence clearly, the modal rules of $\mathbb{E}^{\mathsf{K}\tau}$ must be applied in an appropriate order. Therefore in the sequel we introduce into calculus $\mathbb{E}^{\mathsf{K}\tau}$ a derived rule which gathers a whole sequence of applications of the modal rules.

Another observation is that since indexed sequents carry more semantic information than 'ordinary' sequents, we will need to 'skip' part of this information—namely, we will skip these indexed formulas which are rewritten in Socratic transformations in order to retain semantic invertibility of the rules.

Another example presents proofs of sequent '$\vdash \Box(p \to p) \vee q$' in calculus $\mathbb{G}^{\mathsf{K}\tau}$ (the leftmost column), in Fitting's sequent calculus (the middle), and a proof of sequent '$\vdash (\Box(p \to p) \vee q)^0$' in calculus $\mathbb{E}^{\mathsf{K}\tau}$ (the right column). The example illustrates the need of weakening in calculus $\mathbb{G}^{\mathsf{K}\tau}$, where the rule is used to add information—encoded in formula 'q'—which is unnecessary for the derivation to become a proof. The remaining two proofs show how the same problem is undertaken in the other calculi.

Example 7.3

calculus$\mathbb{G}^{\mathsf{K}\tau}$

$$\cfrac{\cfrac{\cfrac{\cfrac{p \vdash p}{\vdash p \to p}\, R^{\mathbb{G}}_{\to}}{\vdash \Box(p \to p)}\, R^{\mathbb{G}}_{\Box}}{\vdash \Box(p \to p), q}\, R^{\mathbb{G}}_{W}}{\vdash \Box(p \to p) \vee q}\, R^{\mathbb{G}}_{\vee}$$

Fitting's account

$$\cfrac{\cfrac{\cfrac{p \longrightarrow p}{\longrightarrow p \supset p}}{\longrightarrow \Box(p \supset p), q}\, \mathbf{R}_{\Box}}{\longrightarrow \Box(p \supset p) \vee q}$$

calculus$\mathbb{E}^{\mathsf{K}\tau}$

$$\begin{array}{cl} ?(\vdash (\Box(p \to p) \vee q)^0) & \\ \hline ?(\vdash (\Box(p \to p))^0, (q)^0) & R^{\mathbb{E}}_{\vee} \\ \hline ?(\vdash (p \to p)^{0,1}, (q)^0) & R^{\mathbb{E}}_{\Box} \\ \hline ?((p)^{0,1} \vdash (p)^{0,1}, (q)^0) & R^{\mathbb{E}}_{\to} \end{array}$$

In Fitting's account '$\longrightarrow$' is used for the sequent sign and '$\supset$' for implication. In the case of the application of $\mathbf{R}_{\Box}$, for $V = \{\Box(p \supset p), q\}$ we have $V\flat = \{p \supset p\}$, thus the application of $\mathbf{R}_{\Box}$ hides weakening.

It seems like weakening is *the* structural rule characterising modal logics. It is present in the implicit formulations of sequent systems for modal logics, even though sometimes it is hidden, like we have seen. There are cut-free calculi for modal logics, but not weakening-free; reflexivity can be weakened to atoms (i.e. for example axioms only of the form $p_i \vdash p_i$), but weakening cannot. (See [5, 11].)

Analogously to the first-order case, we introduce a generalization of the modal rules. The rules presented in Table 7.6 are derivable rules in $\mathbb{E}^{\mathsf{K}\tau}$. The idea is that when a new world accessible from i is created (numeral j) all the properly modalised formulas that live in i are activated so they can send their suitable subformulas to settle the new world.

Algorithm 7.1 defines the result of operation $\sharp^i_j$ performed on a finite sequence Γ of indexed formulas. The result will be written as '$\Gamma\sharp^i_j$'. In the first line we use '$\varnothing$' for the empty sequence.

Similarly for the right side of the turnstile—Algorithm 7.2 defines the result, '$\Delta\flat^i_j$', of operation $\flat^i_j$ performed on a finite sequence of indexed formulas Δ. (On the margin, the complexity of Algorithms 7.1 and 7.2 depends linearly on the length, that is, the number of terms, of sequence Γ/sequence Δ, respectively.)

The two operations, $\sharp^i_j$ and $\flat^i_j$, are used in the definition of rules presented in Table 7.6.

By '$\mathbb{E}^{\mathsf{K}\tau} \cup \{L^{\mathbb{E}}_{\Diamond\Box}, R^{\mathbb{E}}_{\Box\Diamond}\}$' we will refer to the erotetic calculus resulting by adding the rules $L^{\mathbb{E}}_{\Diamond\Box}$, $R^{\mathbb{E}}_{\Box\Diamond}$ to $\mathbb{E}^{\mathsf{K}\tau}$. The fact that the rules are derivable in $\mathbb{E}^{\mathsf{K}\tau}$ can be proved as in the first-order case (see Lemma 2.13 in Chap. 2).

Algorithm 7.1: operation $\sharp^i_j$

Data: finite sequence $\Gamma = \langle (A_1)^{\varsigma_1}, \ldots, (A_m)^{\varsigma_m} \rangle$ of indexed formulas
Result: finite sequence $\Gamma\sharp^i_j$ of indexed formulas

```
Γ♯^i_j ← ∅
for k = 1 to m do
    if k-th term of Γ is of the form (□B)^{ς/i} then
        Γ♯^i_j ← Γ♯^i_j ' ⟨(□B)^{ς/i}, (B)^j⟩
    else
        Γ♯^i_j ← Γ♯^i_j ' ⟨(A_k)^{ς_k}⟩
    end
end
return Γ♯^i_j
```

Algorithm 7.2: operation $\flat^i_j$

Data: finite sequence $\Delta = \langle (A_1)^{\varsigma_1}, \ldots, (A_m)^{\varsigma_m} \rangle$ of indexed formulas
Result: finite sequence $\Delta\flat^i_j$ of indexed formulas

```
Δ♭^i_j ← ∅
for k = 1 to m do
    if k-th term of Δ is of the form (◇B)^{ς/i} then
        Δ♭^i_j ← Δ♭^i_j ' ⟨(◇B)^{ς/i}, (B)^j⟩
    else
        Δ♭^i_j ← Δ♭^i_j ' ⟨(A_k)^{ς_k}⟩
    end
end
return Δ♭^i_j
```

Table 7.6 General modal (erotetic) rules

$$\frac{?(\ \Phi\ (\ \Gamma_1\ '\ (\Diamond A)^{\varsigma/i}\ '\ \Gamma_2\ \vdash\ \Delta\ )\ )}{?(\ \Phi\ (\ \Gamma_1\sharp^i_j\ '\ (A^{\varsigma/i,j})\ '\ \Gamma_2\sharp^i_j\ \vdash\ \Delta\flat^i_j\ )\ )}\ L^{\mathbb{E}}_{\Diamond\Box}$$

where $j = max\{W(\Gamma_1\ '\ (\Diamond A)^{\varsigma/i}\ '\ \Gamma_2 \vdash \Delta)\} + 1$

$$\frac{?(\ \Phi\ (\ \Gamma\ \vdash \Delta_1\ '\ (\Box A)^{\varsigma/i}\ '\ \Delta_2\ )\ )}{?(\ \Phi\ (\ \Gamma\sharp^i_j\ \vdash \Delta_1\flat^i_j\ '\ A^{\varsigma/i,j}\ '\ \Delta_2\flat^i_j\ )\ )}\ R^{\mathbb{E}}_{\Box\Diamond}$$

where $j = max\{W(\Gamma \vdash \Delta_1\ '\ (\Box A)^{\varsigma/i}\ '\ \Delta_2)\} + 1$

If an indexed formula of the form $(A)^{\varsigma/i}$ occurs in a sequence Γ, then we will say that *formula A lives in i in Γ*.

Corollary 7.1 *Rules $L^{\mathbb{E}}_{\Diamond\Box}$ and $R^{\mathbb{E}}_{\Box\Diamond}$ are derivable in $\mathbb{E}^{\mathsf{K}\tau}$.*

Proof In order to simulate the application of $L^{\mathbb{E}}_{\Diamond\Box}$ to a question with respect to sequent $\Gamma_1{}' (\Diamond A)^{\varsigma/i}{}' \Gamma_2 \vdash \Delta$, the application of rule $L^{\mathbb{E}}_{\Diamond}$ is followed by an application of rule $L^{\mathbb{E}}_{\Box}$ for each formula '$\Box B$' such that the formula lives in i in Γ_1, then by an application of the same rule for each '$\Box C$' that lives in i in Γ_2, and then also by an application of rule $R^{\mathbb{E}}_{\Diamond}$ for each formula '$\Diamond D$' that lives in i in Δ.

Analogously for the other rule. □

If we now go back to Example 7.1, we can modify the proof by applying one of the general modal rules as follows (as before, 'A' stands for '$\Box(p \to q)$'):

Example 7.4 A proof of the sequent $\vdash (\Box(p \to q) \to (\Box p \to \Box q))^0$ in $\mathbb{E}^{\mathsf{K}\tau} \cup \{L^{\mathbb{E}}_{\Diamond\Box}, R^{\mathbb{E}}_{\Box\Diamond}\}$

$$\dfrac{\dfrac{\dfrac{\dfrac{?(\vdash (\Box(p \to q) \to (\Box p \to \Box q))^0)}{?((A)^0 \vdash (\Box p \to \Box q)^0)}\, R^{\mathbb{E}}_{\to}}{?((A)^0, (\Box p)^0 \vdash (\Box q)^0)}\, R^{\mathbb{E}}_{\to}}{?((A)^0, (p \to q)^1, (\Box p)^0, (p)^1 \vdash (q)^{0,1})}\, R^{\mathbb{E}}_{\Box\Diamond}}{?((A)^0, (\Box p)^0, (p)^1 \vdash (q)^{0,1}, (p)^1 \; ; \; (A)^0, (q)^1, (\Box p)^0, (p)^1 \vdash (q)^{0,1})}\, R^{\mathbb{E}}_{\to}$$

We proceed to the description of the translation procedure.

Stage 1: Regular Socratic Transformation

An indexed formula occurring in a sequent $\Gamma \vdash \Delta$ will be called *atomic in* $\Gamma \vdash \Delta$, if it is either an indexed propositional variable, an indexed formula of the form $(\Box A)^{\varsigma}$ occurring in Γ, or an indexed formula of the form $(\Diamond A)^{\varsigma}$ occurring in Δ. We will say that sequent $\Gamma \vdash \Delta$ is *atomic* if every member of Γ and Δ is atomic in it. Moreover, we will use the notion of a basic sequent, as we did in Sect. 2.7; thus by a *basic sequent* we will mean a sequent satisfying condition (i) specified in Definition 7.3.

All Socratic transformations considered until the end of Sect. 7.1 will be constructed according to Algorithm 7.3. We will call them *regular Socratic transformations*. When a regular Socratic transformation is constructed, the rules are applied always with respect to the leftmost indexed formula of the appropriate form that occurs in a sequent. The non-branching rules for the classical connectives are prior to the branching rules (see Procedure CLASSICAL).

The algorithm uses three procedures defined below:

MOVE-RIGHT, CLASSICAL, MODAL.

The procedures are defined separately in order to make the algorithm easy to comprehend. The Procedure MOVE-RIGHT is called when the current sequent is a basic one but it is not the last constituent of the current question. In the presentation we use the notation 'right-context =[ψ|Tail]' which means that 'right-context' is a sequence of sequents whose first element is ϕ, and Tail is the sequence of the remaining sequents. The notation refers to Prolog standards for lists. Let us also recall that in the case of sequences of sequents we use ';' as the sign of concatenation.

Algorithm 7.3: regular Socratic transformation in $\mathbb{E}^{\mathsf{K}\tau} \cup \{L^{\mathbb{E}}_{\Diamond\Box}, R^{\mathbb{E}}_{\Box\Diamond}\}$

Data: question $?(\vdash (A)^0)$
Result: regular Socratic transformation of question $?(\vdash (A)^0)$

$n \leftarrow 1$; /* n counts the number of constituents */
$i \leftarrow 1$; /* i indicates the current constituent */
$\mathbf{s} \leftarrow \langle ?(\vdash (A)^0) \rangle$; /* **s** stores the Socratic transformation */
$\Phi \leftarrow \varnothing$; /* the left context */
$\Psi \leftarrow \varnothing$; /* the right context */
$\phi \leftarrow \vdash (A)^0$; /* the current constituent */
while *ϕ is not atomic* **do**
  CLASSICAL($?(\Phi; \phi; \Psi), n, \mathbf{s}, Q, n, \mathbf{s}$)
  if *ϕ is not a basic sequent* **then**
    MODAL($?(\Phi; \phi; \Psi), \mathbf{s}, Q, \mathbf{s}$)
  end
  while *ϕ is a basic sequent* **do**
    if $n = i$ **then**
      return s *is a Socratic proof of the sequent in* $\mathbb{E}^{\mathsf{K}\tau} \cup \{L^{\mathbb{E}}_{\Diamond\Box}, R^{\mathbb{E}}_{\Box\Diamond}\}$.
    else
      MOVE-RIGHT($\Phi, \Psi, \phi, i, \Phi, \Psi, \phi, i$)
    end
  end
end
return *The sequent is not provable in* $\mathbb{E}^{\mathsf{K}\tau} \cup \{L^{\mathbb{E}}_{\Diamond\Box}, R^{\mathbb{E}}_{\Box\Diamond}\}$.

Procedure MOVE-RIGHT(left-context, right-context, sequent, numeral, Φ, Ψ, ϕ, i)

Data: left-context, right-context =[ψ|Tail], sequent, numeral

$\Phi \leftarrow \Phi$; $\langle$ sequent $\rangle$
$\Psi \leftarrow$ Tail
$\phi \leftarrow \psi$
$i \leftarrow$ numeral $+1$

If we take $|A|$ to be the length of formula A understood as the number of occurrences of connectives in A, then we may set the length of indexed formulas, finite sequences of such formulas and sequents as follows:

$$|A^{\varsigma}| := |A|$$

$$|A_1^{\varsigma_1}, \ldots, A_n^{\varsigma_n}| := \sum_{i=1}^{n} |A_i^{\varsigma_i}|$$

$$|\Gamma \vdash \Delta| := |\Gamma| + |\Delta|$$

Using these measures one can say that the cost of executing procedure CLASSICAL on an active sequent ϕ is $\mathcal{O}(n \log n)$, where $n = |\phi|$, provided that the applications of rules are counted as basic operations.[3] Under the same assumptions the cost of calling procedure MODAL is linear with respect to n.

Procedure CLASSICAL(question,numeral,transformation,Q, n, **s**)

Data: `question` $= ?(\Phi(\phi))$, `numeral`, `transformation`, $\Gamma = \langle (A_1)^{\varsigma_1}, \ldots, (A_m)^{\varsigma_m} \rangle$ is the sequence of all indexed formulas occurring in ϕ

1 $Q \leftarrow$ `question`
2 $n \leftarrow$ `numeral;` `/* counts the number of constituents */`
3 **s** $\leftarrow$ `transformation`
4 $j \leftarrow 1$
5 **while** $j \leqslant m$ **do**
6   **if** $(A_j)^{\varsigma_j}$ *is a formula-premise of a classical non-branching rule* **r then**
7     $\Delta \leftarrow$ the sequence of the indexed formula-conclusion(s) of **r**
8     $\Gamma \leftarrow \langle (A_1)^{\varsigma_1}, \ldots, (A_{j-1})^{\varsigma_{j-1}} \rangle \,'\, \Delta \,'\, \langle (A_{j+1})^{\varsigma_{j+1}}, \ldots, (A_m)^{\varsigma_m} \rangle$
9     $\psi \leftarrow$ sequent-conclusion of **r**
10     $Q \leftarrow ?(\Phi(\psi))$
11   **s** $\leftarrow$ **s** ; $\langle Q \rangle$
12   **else**
13   **if** $(A_j)^{\varsigma_j}$ *is a formula-premise of a classical branching rule* **r then**
14       $\Delta \leftarrow$ 1-term sequence with the left ind. formula-conclusion of **r**
15       $\Gamma \leftarrow \langle (A_1)^{\varsigma_1}, \ldots, (A_{j-1})^{\varsigma_{j-1}} \rangle \,'\, \Delta \,'\, \langle (A_{j+1})^{\varsigma_{j+1}}, \ldots, (A_m)^{\varsigma_m} \rangle$
16       $\psi_1, \psi_2 \leftarrow$ the left and right sequent-conclusions of **r**
17       $Q \leftarrow ?(\Phi(\psi_1, \psi_2))$
18   **s** $\leftarrow$ **s** ; $\langle Q \rangle$
19       $n \leftarrow n + 1$
20   **else**
21       $j \leftarrow j + 1$
22   **end**
23   **end**
24 **end**

In Procedures CLASSICAL and MODAL we start with forming a sequence Γ of all occurrences of indexed formulas in sequent ϕ. Δ in line 7 of Procedure CLASSICAL is a one-term or a two-term sequence—depending on the rule. In the case of Δ in line 14, the sequence has one term. Sequence Γ is a part of the control condition in the while-loop. If a classical rule is applied, the sequence is updated, so in the next round classical rules can be applied to the previously introduced indexded formula-conclusions. In the case of Procedure MODAL, only the initial sequence Γ is important, as the procedure aims at *at most one* application of a general modal rule—if it is applied, the assignment in line no. 8 of the procedure ends the execution of the while-loop.

[3] This is a very rough analysis and it says nothing about complexity of the method with respect to the number of different variables used in the problem to be solved, that is, sequent. It also does not take into account the cost of searching for the active formula. The estimation $\mathcal{O}(n \log n)$ is derived under the assumption that all rules applied are branching, and each application divides 'problem' to two equal 'subproblems', that is the number of occurrences of connectives is divided by two.

After going through Procedure CLASSICAL in Algorithm 7.3, the 'current' sequent of the last question contains only indexed variables and/or indexed modal formulas. After that a check of the closing condition (basic sequent, line 9 of Algorithm 7.3) is performed. Procedure MODAL is called provided that the active sequent is not a basic one already. The check in line 9 of Algorithm 7.3 warrants that if the closing condition is met in a world i (that is, there is a formula and its negation with indices ending with numeral i), then this fact is recognised before another world can be created. In turn, if after going through Procedure CLASSICAL the current sequent is a basic one, and it is the last constituent, then **s** is a Socratic proof. If the sequent is a basic one, but not the last one, then we move to the next constituent to the right (line 16 of Algorithm 7.3). This time the check condition is in the while-loop, in order to find the right-most constituent of the last question which is not basic.

Procedure MODAL(question,transformation,Q, **s**)

Data: question = $?(\Phi(\phi))$, `transformation`, $\Gamma = \langle (A_1)^{\varsigma_1}, \ldots, (A_m)^{\varsigma_m} \rangle$ is the sequence of all indexed formulas occurring in ϕ

s $\leftarrow$ `transformation`
$Q \leftarrow ?(\Phi(\phi))$
$j \leftarrow 1$
while $j \leqslant m$ **do**
  if $(A_j)^{\varsigma_j}$ *is a formula-premise of* **r** $\in \{L^{\mathbb{E}}_{\Diamond\Box}, R^{\mathbb{E}}_{\Box\Diamond}\}$ **then**
    $Q \leftarrow$ question-conclusion of **r**
    $s \leftarrow s\,;\ \langle Q \rangle$
    $j \leftarrow m + 1$;   `/* ends the execution of the while-loop */`
  else
    $j \leftarrow j + 1$
  end
end

The notions of *Socratic transformation in calculus* $\mathbb{E}^{\mathsf{K}\tau} \cup \{L^{\mathbb{E}}_{\Diamond\Box}, R^{\mathbb{E}}_{\Box\Diamond}\}$ and *Socratic proof in calculus* $\mathbb{E}^{\mathsf{K}\tau} \cup \{L^{\mathbb{E}}_{\Diamond\Box}, R^{\mathbb{E}}_{\Box\Diamond}\}$ are understood as in the case of calculus $\mathbb{E}^{\mathsf{K}\tau}$. A Socratic proof of a sequent which is a regular Socratic transformation of the relevant question is also called *regular Socratic proof*. Finally, observe that since a regular Socratic transformation of a question is given by a deterministic algorithm, it is a unique object.

Stage 2: Simplified Socratic Trees

Regular Socratic transformations obtained at the previous stage may be further transformed into Socratic trees according to the pattern described previously in Sect. 6.1. We assume that the Socratic tree is constructed together with a regular Socratic transformation, by marking the sequents-conclusions. In fact, the algorithm of construction of regular Socratic transformations can be easily transformed into an algorithm of construction of the Socratic trees determined by the regular Socratic transformation.

Before we continue, let us note some important properties of the Socratic tree determined by a regular Socratic transformation of a question of the form $?(\vdash (A)^0)$.

First of all, when the rules are applied according to Algorithm 7.3, the applications follow certain pattern:

1. Classical rules are applied with respect to formulas living in world 0 (as we start with $?(\vdash (A)^0)$) in the antecedent or the succedent of a sequent, until there are no classical connectives to be eliminated (see the construction of the while-loop in lines 5-24 of Procedure CLASSICAL). If a branching rule is applied, the right sequent-conclusion is left to be analyzed later.
2. A general modal rule is applied once, if possible (line 10 of Algorithm 7.3 and line 8 of Procedure MODAL): world 1 accessible to 0 is created and settled.
3. Once again, classical connectives are eliminated in formulas living in world 1 (on the left or the right side of the turnstile), and only in 1, since those in 0 have been analyzed already.

⋮

n. A general modal rule is applied once: world j, accessible to i, is created and settled (if possible).

$n+1$. Classical connectives are eliminated in formulas living in world j (left or right of the turnstile), and only j, since the classical connectives of formulas living in the other worlds have been already eliminated.

⋮

The above pattern warrants that the following corollary is true.

Corollary 7.2 *If* **s** *is a regular Socratic transformation and* ϕ *is a constituent-conclusion of a question of* **s**, *then the principal numeral of* ϕ *is the largest one in this sequent. Moreover, if* ϕ *is a constituent-premise of a classical rule, then the numeral active in* ϕ *is the largest in* ϕ.

Let us observe that applying rules $L_{\Diamond}^{\mathbb{E}}$ and $R_{\Box}^{\mathbb{E}}$ under proviso (7.1) is necessary for Corollary 7.2 to hold. Let us also emphasize that the construction of regular Socratic transformations is justified by the fact that we do not analyze symmetric and/or Euclidean logics—when a new world is created, settled, and the 'settlers' are analyzed by the classical rules, we know that the world is settled once for good.

On the other hand, the applications of modal rules need not be 'linear' with respect to the numerals, since there may be more than two modal formulas in a world, say i. It may happen that the leftmost such formula is analyzed and it gives birth to a chain of worlds. But then the transformation does not end with a success, so we analyze another modal formula in i. In other words, when ψ results from ϕ by a modal rule, the active numeral in ϕ is not necessarily the largest one in the sequent, but the principal numeral of ψ is the largest one in ψ.

Before we continue, let us recall that together with the notion of a Socratic tree we have introduced the symbol '$R_{\mathbf{s}}$', which refers to the relation in the Socratic tree $\mathbf{Tr}(\mathbf{s})$ determined by **s**. If we consider Socratic trees determined by regular Socratic transformations, then we may observe what follows:

Corollary 7.3 *Let* $\mathbf{Tr}(\mathbf{s})$ *stand for a Socratic tree determined by a regular Socratic transformation* $\mathbf{s}$ *of a questions of the form* $?(\vdash (A)^0)$. *If a node of* $\mathbf{Tr}(\mathbf{s})$*, labeled with* ψ*, results in* $\mathbf{s}$ *by a classical rule, and its immediate* $R_{\mathbf{s}}$*-successor, labeled with* ϕ*, results in* $\mathbf{s}$ *by a classical rule, then the principal numeral in* ψ *and* ϕ *is the same numeral.*

Proof As we have emphasized in the above pattern of the applications of rules in a regular Socratic transformation: when we start with a formula living in 0, we apply the rules for classical connectives to the consecutive sequents as long as the rules are applicable; and once we are finished with 0, and a new world is introduced, we never come back to 0. The situation is analogous when one of the general modal rules is applied, and a new world is created, settled, and the classical rules are applied with respect to the formulas living in the new world. □

In other words, Corollary 7.3 states that in Socratic trees determined by regular Socratic transformations, if two consecutive (linked by $R_{\mathbf{s}}$) nodes result in $\mathbf{s}$ by classical rules, then they have the same principal numeral. By the same arguments also:

Corollary 7.4 *Let* $\mathbf{Tr}(\mathbf{s})$ *stand for a Socratic tree determined by a regular Socratic transformation* $\mathbf{s}$ *of a questions of the form* $?(\vdash (A)^0)$. *If a node of* $\mathbf{Tr}(\mathbf{s})$ *labeled with* ψ *results in* $\mathbf{s}$ *by a modal rule, where the principal numeral in* ψ *is* j*, and its immediate* $R_{\mathbf{s}}$*-successor labeled with* ϕ *results in* $\mathbf{s}$ *by a classical rule, then the principal numeral in* ϕ *is also* j.

As we shall see, however, the transition from regular Socratic transformations to Socratic trees is not enough for our present purpose, since Socratic transformations may contain way too much information that must be now precisely cut off. Before we present the solution to this problem, let us first illustrate the point with examples.

If we compare the Socratic proof in Example 7.4 with the target proof in calculus $\mathbb{G}^{\mathsf{K}\tau}$ (see Example 7.2), then we can see that each of the 'ordinary' sequents of language $\mathcal{M}_{\vdash}^{imp}$ represents the settlers of exactly one world, whereas the sequents of language $\mathcal{M}_{\vdash}^{?}$ keep all the information about the various worlds (there are two worlds in this example) together with the information concerning the relation between the worlds. Therefore the next step should result in cleaning away the redundant information. This will be done by a function τ which takes a sequence of indexed formulas and simplifies the sequence by leaving only these formulas that live in a specified world. In the definition we use, as previously, '$\varnothing$' for the empty sequence.

Definition 7.4 Let $i \in \mathbb{N}$.

$$\tau(i, \varnothing) = \varnothing$$

$$\tau(i, \langle A_1^{\varsigma_1}, \ldots, A_k^{\varsigma_k}\rangle) = \begin{cases} \langle A_1\rangle\ '\ \tau(i, \langle A_2^{\varsigma_2}, \ldots, A_k^{\varsigma_k}\rangle) & \text{if } \varsigma_1 \text{ ends with } i, \\ \tau(i, \langle A_2^{\varsigma_2}, \ldots, A_k^{\varsigma_k}\rangle) & \text{otherwise.} \end{cases}$$

Next, we extend the definition of τ to act on sequents.

Definition 7.5 Let $i \in \mathbb{N}$ and let $\Gamma \vdash \Delta$ be a sequent of the language $\mathcal{M}^?_{\vdash}$. Then:

$$\tau(i, \Gamma \vdash \Delta) = \tau(i, \Gamma) \vdash \tau(i, \Delta).$$

Let us observe that the result $\tau(i, \Gamma \vdash \Delta)$ is either a sequent of language $\mathcal{M}^{imp}_{\vdash}$, or—if there are no formulas living in i—the sequent with empty antecedent and empty succedent. The empty sequent is not a properly built expression of language $\mathcal{M}^?_{\vdash}$ (nor of language $\mathcal{M}^{imp}_{\vdash}$), but it will never occur in the derivations considered below. We need it only for the purpose of this definition.

We use function τ to modify sequents in Example 7.4. First, we generate the Socratic tree on the basis of this example; A stands for $\Box(p \to q)$:

$$\vdash (\Box(p \to q) \to (\Box p \to \Box q))^0$$

$$(A)^0 \vdash (\Box p \to \Box q)^0$$

$$(A)^0, (\Box p)^0 \vdash (\Box q)^0$$

$$(A)^0, (p \to q)^1, (\Box p)^0, (p)^1 \vdash (q)^{0,1}$$

$$(A)^0, (\Box p)^0, (p)^1 \vdash (q)^{0,1}, (p)^1 \qquad (A)^0, (q)^1, (\Box p)^0, (p)^1 \vdash (q)^{0,1}$$

next, we treat the sequents with function τ, and turn the tree upside down (or rather—turn it back into the right position):

$$\dfrac{\dfrac{\dfrac{\dfrac{\tau(1, (A)^0, (\Box p)^0, (p)^1 \vdash (q)^{0,1}, (p)^1) \quad \tau(1, (A)^0, (q)^1, (\Box p)^0, (p)^1 \vdash (q)^{0,1})}{\tau(1, (A)^0, (p \to q)^1, (\Box p)^0, (p)^1 \vdash (q)^{0,1})}}{\tau(0, (A)^0, (\Box p)^0 \vdash (\Box q)^0)}}{\tau(0, (A)^0 \vdash (\Box p \to \Box q)^0)}}{\tau(0, \vdash (\Box(p \to q) \to (\Box p \to \Box q))^0)}$$

arriving at the tree below, which is almost the proof presented in Example 7.2, but without the initial applications of weakening to axioms:

$$\dfrac{\dfrac{\dfrac{\dfrac{p \vdash q, p \quad q, p \vdash q}{p \to q, p \vdash q}\, L^{\mathbb{G}}_{\to}}{\Box(p \to q), \Box p \vdash \Box q}\, R^{\mathbb{G}}_{\Box}}{\Box(p \to q) \vdash \Box p \to \Box q}\, R^{\mathbb{G}}_{\to}}{\vdash \Box(p \to q) \to (\Box p \to \Box q)}\, R^{\mathbb{G}}_{\to}$$

As we can see, function τ changes its first argument, which is numeral 0, to numeral 1. The change is driven by the application of the modal rule, and we can feel intuitively that this is the right step here. This observation regarding the choice of numeral—the first argument of τ—will be generalised and stated formally as the use of another function called $\tau_{\mathbf{s}}$, where **s** indicates the whole Socratic transformation (see Definition 7.6). Before we introduce the definition, let us recall that according to Corollary 6.1, if sequent ϕ is an immediate $R_{\mathbf{s}}$-successor of sequent ψ in a Socratic tree, then ψ is the active sequent of an erotetic rule and ϕ is its sequent-conclusion (possibly one of the two). Thus with regard to every sequent of a Socratic tree, except for the one in the root, we can say that the sequent is a sequent-conclusion of an erotetic rule applied in **s**, and there is always exactly one rule satisfying this claim.

The next definition introduces a *relabelling function*, since it acts on the sequents labelling the nodes of $\mathbf{Tr}(\mathbf{s})$ and assigns them sequents of language $\mathcal{M}_{\vdash}^{imp}$; that is to say, the function replaces one label of a node with another. The relabelling function works as follows: for a numeral i and a sequent ϕ (with indexed formulas), $\tau(i, \phi)$ is a sequent obtained from ϕ by, first, deleting each indexed formula whose index ends with a numeral other than i, and, second, deleting all the indices of the remaining formulas.

Definition 7.6 Let **s** be a regular Socratic transformation of a question of the form '$?(\vdash (A)^0)$' and let $\mathbf{Tr}(\mathbf{s})$ be the Socratic tree determined by **s**. Then $\tau_{\mathbf{s}}$ is a relabelling function defined recursively as follows:

1. If ϕ labels the root of $\mathbf{Tr}(\mathbf{s})$, then $\tau_{\mathbf{s}}(\phi) = \tau(0, \phi)$.
2. If ϕ labels a node, ψ labels the immediate $R_{\mathbf{s}}$-predecessor of the node, ϕ results from ψ in **s** by a classical rule, and i is the numeral such that $\tau_{\mathbf{s}}(\psi) = \tau(i, \psi)$, then $\tau_{\mathbf{s}}(\phi) = \tau(i, \phi)$. **(conservative extension)**
3. If ϕ is a label of a node that results in **s** from its immediate $R_{\mathbf{s}}$-predecessor by a modal rule, then $\tau_{\mathbf{s}}(\phi) = \tau(j, \phi)$, where j is the principal numeral in ϕ. **(radical change)**

If $\mathbf{Tr}(\mathbf{s}) = \langle X_{\mathbf{s}}, R_{\mathbf{s}} \rangle$ is a Socratic tree, then the structure $\langle X_{\mathbf{s}}, R_{\mathbf{s}}, \tau_{\mathbf{s}} \rangle$ is a labeled tree. $\tau_{\mathbf{s}}$ is a two-argument function; if the second argument, ϕ, is given, then the first argument will be called $\tau_{\mathbf{s}}$ *-numeral for* ϕ. The choice of $\tau_{\mathbf{s}}$-numerals by function $\tau_{\mathbf{s}}$ has the following nice property:

Corollary 7.5 *Let* **s** *be a regular Socratic transformation of a question of the form* $?(\vdash (A)^0)$ *and let* $\tau_{\mathbf{s}}$ *be defined as in Definition 7.6. Then:*

1. *if* ϕ *is a sequent-conclusion of a classical rule, and it is the only immediate* $R_{\mathbf{s}}$*-successor of* ψ*, then*
$$\frac{\tau_{\mathbf{s}}(\phi)}{\tau_{\mathbf{s}}(\psi)}$$
is an instance of one of the non-branching classical rules of $\mathbb{G}^{\mathsf{K}\tau}$.
2. *if* ϕ_1 *is one of the two immediate* $R_{\mathbf{s}}$*-successors of* ψ*, and the other is* ϕ_2*, then at least one of the following schemes:*
$$\frac{\tau_{\mathbf{s}}(\phi_1) \quad \tau_{\mathbf{s}}(\phi_2)}{\tau_{\mathbf{s}}(\psi)} \quad or \quad \frac{\tau_{\mathbf{s}}(\phi_2) \quad \tau_{\mathbf{s}}(\phi_1)}{\tau_{\mathbf{s}}(\psi)}$$

is an instance of one of the branching classical rules of $\mathbb{G}^{\mathsf{K}\tau}$.

Proof We consider only one rule as an example.

Suppose that sequent ψ occurs in the tree (more precisely, that there is $(m, k, \psi) \in X_{\mathbf{s}}$) and it has exactly two immediate $R_{\mathbf{s}}$-successors ϕ_1 and ϕ_2 (more precisely, (m_1, k_1, ϕ_1) and (m_2, k_2, ϕ_2) for some m_1, k_1, m_2, k_2). Since there are no modal branching rules in $\mathbb{E}^{\mathsf{K}\tau} \cup \{L^{\mathbb{E}}_{\Diamond\Box}, R^{\mathbb{E}}_{\Box\Diamond}\}$, ψ is the active sequent of a classical rule $\mathbf{r}$ of $\mathbb{E}^{\mathsf{K}\tau}$ and ϕ_1, ϕ_2 are its sequents-conclusions. Suppose further that the rule is $\mathbf{r} = L^{\mathbb{E}}_{\vee}$. We have

$$\psi = \Gamma_1 \,'\, (A \vee B)^{\varsigma/i} \,'\, \Gamma_2 \vdash \Delta$$

and one of the two:

$$\phi_1 = \Gamma_1 \,'\, (A)^{\varsigma/i} \,'\, \Gamma_2 \vdash \Delta \qquad \phi_2 = \Gamma_1 \,'\, (B)^{\varsigma/i} \,'\, \Gamma_2 \vdash \Delta$$

or

$$\phi_2 = \Gamma_1 \,'\, (A)^{\varsigma/i} \,'\, \Gamma_2 \vdash \Delta \qquad \phi_1 = \Gamma_1 \,'\, (B)^{\varsigma/i} \,'\, \Gamma_2 \vdash \Delta$$

(displayed with the use of the concatenation sign). We assume the first possibility, the second one is analogous.

Obviously, the form of $\tau_{\mathbf{s}}(\psi)$, $\tau_{\mathbf{s}}(\phi_1)$ and $\tau_{\mathbf{s}}(\phi_2)$ depends on their $\tau_{\mathbf{s}}$-numerals. Definition 7.6 warrants that the $\tau_{\mathbf{s}}$-numeral for ϕ_1 and for ϕ_2 is identical to the $\tau_{\mathbf{s}}$-numeral for ψ (see ***conservative extension***), but we do not know the $\tau_{\mathbf{s}}$-numeral for ψ. In order to make the proof more transparent, let us make an assumption which we will justify below, namely that for ψ we have:

$$\tau_{\mathbf{s}}(\psi) = \tau(i, \psi) \tag{7.2}$$

that is, we assume that the $\tau_{\mathbf{s}}$-numeral for ψ is the same numeral i that is *the active numeral* of the application of $L^{\mathbb{E}}_{\vee}$. *If* this is the case, then

$$\tau_{\mathbf{s}}(\psi) = \tau(i, \Gamma_1) \,'\, A \vee B \,'\, \tau(i, \Gamma_2) \vdash \tau(i, \Delta)$$

where $\tau(i, \Gamma_1)$, $\tau(i, \Gamma_2)$ and $\tau(i, \Delta)$ are possibly empty sequences of formulas of $\mathcal{M}$. For simplicity, let $\tau(i, \Gamma_1) = S_1$, $\tau(i, \Gamma_2) = S_2$ and $\tau(i, \Delta) = T$. Then under the assumption (7.2) also:

$$\tau_{\mathbf{s}}(\phi_1) = S_1 \,'\, A \,'\, S_2 \vdash T \qquad \tau_{\mathbf{s}}(\phi_2) = S_1 \,'\, B \,'\, S_2 \vdash T$$

which was to be proved, since then:

$$\frac{\tau_{\mathbf{s}}(\phi_1) \quad \tau_{\mathbf{s}}(\phi_2)}{\tau_{\mathbf{s}}(\psi)}$$

is an instance of $L^{\mathbb{G}}_{\vee}$.

Now we go back to the justification of (7.2). We prove it by induction with respect to the structure of **Tr(s)** which is—let us recall—determined by a *regular* Socratic transformation. More specifically, the induction is with respect to the level of node ψ in **Tr(s)**. (If necessary, see Appendix A for the notion of a level of a node in a tree.)

Base step: suppose that ψ is at level 0, then ψ is in the root and, by Definition 7.6, $\tau_{\mathbf{s}}(\psi) = \tau(0, \psi)$. On the other hand, 0 must be the active numeral in ψ, as there are no other numerals in ψ.

Inductive step: we assume that the inductive hypothesis is satisfied for the nodes above ψ and that ϕ results from ψ by a classical rule (non-branching or branching). Then we need to consider the previous application of a rule, that is, the application of a rule $\mathbf{r}*$ that ψ resulted from. If the rule was classical, then by the inductive hypothesis, the fact that **s** is regular and by Corollary 7.3, it follows that the two applications of classical rules are in the same active world. If $\mathbf{r}*$ was modal, then the $\tau_{\mathbf{s}}$-numeral for ψ is the new one in ψ, say j, and by Definition 7.6, $\tau_{\mathbf{s}}(\psi) = \tau(j, \psi)$. Then, by Corollary 7.4, the classical rule applied immediately after that must act upon a formula living in j.

The reasoning is the same for the other rules. □

The above Corollary states that in the case of the applications of the classical rules the choice of the numeral made by function $\tau_{\mathbf{s}}$ is the proper one, that is, the result of $\tau_{\mathbf{s}}$-translation of a sequent-premise and its sequent(s)-conclusion(s) is an instance of a rule of $\mathbb{G}^{\mathsf{K}\tau}$. Unfortunately, in the general case the choice of the numeral by the use of function $\tau_{\mathbf{s}}$ does not guarantee that the resulting tree is regulated by the rules of $\mathbb{G}^{\mathsf{K}\tau}$. The following example illustrates the difficulty.

Example 7.5 The regular Socratic proof of the sequent $\vdash ((\Box p \vee \Box q) \vee \Diamond\neg q)^0$ in calculus $\mathbb{E}^{\mathsf{K}\tau} \cup \{L^{\mathbb{E}}_{\Diamond\Box}, R^{\mathbb{E}}_{\Box\Diamond}\}$:

$$
\begin{array}{ll}
?(\vdash ((\Box p \vee \Box q) \vee \Diamond\neg q)^0\,) & \\ \hline
?(\vdash (\Box p \vee \Box q)^0, (\Diamond\neg q)^0\,) & R^{\mathbb{E}}_{\vee} \\ \hline
?(\vdash (\Box p)^0, (\Box q)^0, (\Diamond\neg q)^0\,) & R^{\mathbb{E}}_{\vee} \\ \hline
?(\vdash (p)^{0,1}, (\Box q)^0, (\Diamond\neg q)^0, (\neg q)^1\,) & R^{\mathbb{E}}_{\Box\Diamond} \\ \hline
?(\,(q)^1 \vdash (p)^{0,1}, (\Box q)^0, (\Diamond\neg q)^0\,) & R^{\mathbb{R}}_{\neg} \\ \hline
?(\,(q)^1 \vdash (p)^{0,1}, (q)^{0,2}, (\Diamond\neg q)^0, (\neg q)^2\,) & R^{\mathbb{E}}_{\Box\Diamond} \\ \hline
?(\,(q)^1, (q)^2 \vdash (p)^{0,1}, (q)^{0,2}, (\Diamond\neg q)^0\,) & R^{\mathbb{R}}_{\neg}
\end{array}
$$

Suppose that we turn the Socratic proof into a Socratic tree, turn it upside down and act on the sequents with function τ, changing the numeral (the first argument of the function) every time a modal rule is applied:

$$
\begin{array}{c}
\tau(2, (q)^1, (q)^2 \vdash (p)^{0,1}, (q)^{0,2}, (\Diamond\neg q)^0) \\ \hline
\tau(2, (q)^1 \vdash (p)^{0,1}, (q)^{0,2}, (\Diamond\neg q)^0, (\neg q)^2) \\ \hline
\tau(1, (q)^1 \vdash (p)^{0,1}, (\Box q)^0, (\Diamond\neg q)^0) \\ \hline
\tau(1, \vdash (p)^{0,1}, (\Box q)^0, (\Diamond\neg q)^0, (\neg q)^1) \\ \hline
\tau(0, \vdash (\Box p)^0, (\Box q)^0, (\Diamond\neg q)^0) \\ \hline
\tau(0, \vdash (\Box p \vee \Box q)^0, (\Diamond\neg q)^0) \\ \hline
\tau(0, \vdash ((\Box p \vee \Box q) \vee \Diamond\neg q)^0)
\end{array}
\qquad
\begin{array}{c}
q \vdash q \\ \hline
\vdash q, \neg q \\ \hline
q \vdash p \\ \hline
\vdash p, \neg q \\ \hline
\vdash \Box p, \Box q, \Diamond\neg q \\ \hline
\vdash \Box p \vee \Box q, \Diamond\neg q \\ \hline
\vdash (\Box p \vee \Box q) \vee \Diamond\neg q
\end{array}
$$

The resulting tree (to the right) is not a proof in $\mathbb{G}^{\mathsf{K}\tau}$. If the second sequent (from the top) was followed by the fifth one, we could just add an application of the proper rule of weakening and it would be fine, but the two sequents in the middle spoil this picture. We can see the reason of this failure. The choice of formula $\Box p$ to be acted upon (in the third question of Example 7.5) was wrong. The mistake is corrected in the Socratic transformation on the further stage of its construction due to confluency of the erotetic system; it is possible because all the information is still here, in every question of the transformation. But in the ordinary sequent calculus this mistake can be corrected only by means of backtracking, which leads to a new proof—one, in which the mistake is not committed. (Let us recall that ordinary, i.e. implicit, sequent calculi for modal logics are not confluent.)

The situation gets even more complicated if the wrong choice accidentally leads to a partial success, as in the example below.

Example 7.6 Here is the regular Socratic proof of the sequent '$\vdash (\Box(p \wedge r) \vee (\Box p \to \Box(q \to p)))^0$' in calculus $\mathbb{E}^{\mathsf{K}\tau} \cup \{L^{\mathbb{E}}_{\Diamond\Box}, R^{\mathbb{E}}_{\Box\Diamond}\}$, '$\chi$' stands for the sequent '$(\Box p)^0, (p)^1 \vdash (p)^{0,1}, (\Box(q \to p))^0$'.

$$
\begin{array}{cl}
?(\vdash (\Box(p \wedge r) \vee (\Box p \to \Box(q \to p)))^0) & \\ \hline
?(\vdash (\Box(p \wedge r))^0, (\Box p \to \Box(q \to p))^0) & R^{\mathbb{E}}_{\vee} \\ \hline
?((\Box p)^0 \vdash (\Box(p \wedge r))^0, (\Box(q \to p))^0) & R^{\mathbb{E}}_{\to} \\ \hline
?((\Box p)^0, (p)^1 \vdash (p \wedge r)^{0,1}, (\Box(q \to p))^0) & R^{\mathbb{E}}_{\Box\Diamond} \\ \hline
?(\chi \,;\, (\Box p)^0, (p)^1 \vdash (r)^{0,1}, (\Box(q \to p))^0) & R^{\mathbb{E}}_{\wedge} \\ \hline
?(\chi \,;\, (\Box p)^0, (p)^2, (p)^1 \vdash (r)^{0,1}, (q \to p)^{0,2}) & R^{\mathbb{E}}_{\Box\Diamond} \\ \hline
?(\chi \,;\, (\Box p)^0, (p)^2, (p)^1, (q)^{0,2} \vdash (r)^{0,1}, (p)^{0,2}) & R^{\mathbb{E}}_{\to}
\end{array}
$$

Figure 7.1 presents the Socratic tree determined by this Socratic proof treated with function τ.

Turning this tree into a proof in calculus $\mathbb{G}^{\mathsf{K}\tau}$ requires (apart from weakening in the leaf) cutting off the three boxed nodes: that is, the whole left branch and a part of the right branch. The challenge now is to settle precisely what must be cut off and how to do it. The idea is the following.

As long as we consider non-symmetric, non-Euclidean normal modal logics, if a formula is a validity of the logic, then in the construction of the counter-model a chain (linear order) of worlds suffices to prove it: it is the chain leading from the world in

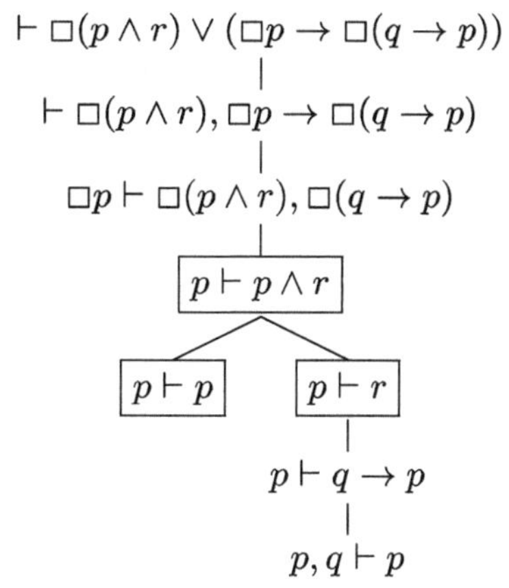

Fig. 7.1 The Socratic tree determined by the Socratic proof from Example 7.6

which a given formula was assumed to be false to the world in which a contradiction appears. In other words, when constructing a potential counter-model for these logics the semantic information is always carried from a world to its R-successor, never to its R-predecessor. For this reason, linear samples of counter-models are sufficient to show that a formula has no counter-model. A sequence of numerals representing such a 'linear sample' (a chain of worlds) in a Socratic transformation will be called a *world line*.[4]

More specifically, a *world line of a branch* of a Socratic tree is a sequence of numerals leading from 0 to a numeral i such that there is a formula that lives in i both in the antecedent and in the succedent of the sequent labelling the leaf of the branch. Let us observe that if a Socratic tree is determined by a regular Socratic transformation, then each branch of the tree has exactly one world line; it follows from the fact that a regular Socratic transformation ends as soon as a contradiction appears—compare our remarks above concerning the check of the closing condition. Formally,

Definition 7.7 Let **T**(**s**) stand for a Socratic tree determined by a regular Socratic transformation **s** of a question of the form $?(\vdash (A)^0)$, such that **s** is a Socratic proof of $\vdash (A)^0$. Let ϕ be the label of the leaf of a branch $\mathcal{B}$ of **T**(**s**). The *world line of branch* $\mathcal{B}$ is a maximal finite sequence $\langle i_1, \ldots, i_n \rangle$ of numerals such that:

1. the last numeral, i_n, is such that for some formula A, $(A)^{\varsigma_1/i_n}$ occurs on the left side and $(A)^{\varsigma_2/i_n}$ occurs on the right side of the turnstile,
2. for each i_k, where $k > 1$, i_{k-1} is the numeral that immediately precedes i_k in an index of some indexed formula in ϕ.

[4] This way of thinking of countermodel constructions is present from the very beginning of the so-called Kripke semantics: 'the semantical theory would lose no generality if only tree models were admitted' [7]. As Kripke points out, after Hintikka, in the cited article, there are formulas not valid in S4, that have no finite *tree* countermodels. But still, we consider provable sequents, containing provable formulas, and for provable formulas it holds true that a linear construction is sufficient.

Let us note that due to the properties of regular Socratic transformations, we could have defined i_n in clause *1* as the principal numeral in the leaf. Let us also note that the condition expressed in *2* is unambiguous, as long as ϕ occurs in a Socratic transformation—in a Socratic transformation a numeral occurring in an index of a formula in ϕ cannot have two different immediate predecessors in some indices in ϕ; it follows form the fact that indices are extended only by *new* numerals. Finally, maximality of the defined sequence warrants that it starts with numeral 0.

Let us recall the following notion. If R is a relation defined on X and $Y \subseteq X$, then by $R\mid_Y$ we mean the restriction of R to Y (and similarly for functions).

We define the following operation of deleting a node from a tree.[5]

Definition 7.8 If $\langle X, R\rangle$ is a tree and $\mathbf{n} \in X$ is a node of this tree, then *deleting* **n** *from* $\langle X, R\rangle$, symbolically $\langle X, R\rangle \setminus \mathbf{n}$, is an operation performed on the tree that results in the following structure: $\langle X \setminus \{\mathbf{n}\}, R\mid_{X\setminus\{\mathbf{n}\}}\rangle$. Thus

$$\langle X, R\rangle \setminus \mathbf{n} \ :=^{df} \ \langle X \setminus \{\mathbf{n}\}, R\mid_{X\setminus\{\mathbf{n}\}}\rangle.$$

It is easy to see that:

Corollary 7.6 *If* $\langle X, R\rangle$ *is a tree and* $\mathbf{n} \in X$ *is not a root of the tree, then the structure* $\langle X, R\rangle \setminus \mathbf{n}$ *is also a tree.*

Proof It is enough to observe that the operation of restriction does not spoil such properties as reflexivity, transitivity, antisymmetry and linearity. We consider only one case: that of transitivity. Thus let $\langle X, R\rangle$ be a tree and suppose that x, y, z are elements of $\langle X, R\rangle \setminus \mathbf{n}$ such that relation $R\mid_{X\setminus\{\mathbf{n}\}}$ holds between x and y, between y and z, but not between x and z. Then also xRy and yRz, and as R is transitive, xRz. If the pair $\langle x, z\rangle$ belongs to R but not to $R\mid_{X\setminus\{\mathbf{n}\}}$, then one of x, z must be the deleted node **n**. However, if $x = \mathbf{n}$, then the pair $\langle x, y\rangle$ is not in $R\mid_{X\setminus\{\mathbf{n}\}}$, and if $z = \mathbf{n}$, then the pair $\langle y, z\rangle$ is not in $R\mid_{X\setminus\{\mathbf{n}\}}$. A contradiction.

If the deleted node is not the root, then the result must be a tree. Observe, however, that if the root of a tree has exactly one immediate successor, then the result of deleting the root is also a tree. Then the immediate successor becomes the root of the new tree. □

If $\mathbf{Tr} = \langle X, R, \eta\rangle$ is a labeled tree, then by $\mathbf{Tr} \setminus \mathbf{n}$ we mean the tree $\langle X, R\rangle \setminus \mathbf{n}$ together with the restricted labelling function $\eta\mid_{X\setminus\{\mathbf{n}\}}$.

Example 7.7 $\langle X, R\rangle$ is such that $X = \{1, 2, 3, 4, 5, 6\}$ and R is the transitive closure of $\{\langle 1, 2\rangle, \langle 2, 3\rangle, \langle 3, 4\rangle, \langle 3, 5\rangle, \langle 5, 6\rangle\}$. Below we present the tree (a), and the result of deleting nodes 3–5 from $\langle X, R\rangle$ ((b)–(d)).

[5] There are well-known definitions of the operation of deletion of a node from a tree which did not, however, suit our needs, therefore we have introduced our own definition. For example, the operation of deletion of a node from a binary search tree produces different result than our operation when a branching point is deleted. Compare [2, pp. 295–296].

(a) $\langle X, R\rangle$ (b) $\langle X, R\rangle \setminus 3$ (c) $\langle X, R\rangle \setminus 4$ (d) $\langle X, R\rangle \setminus 5$

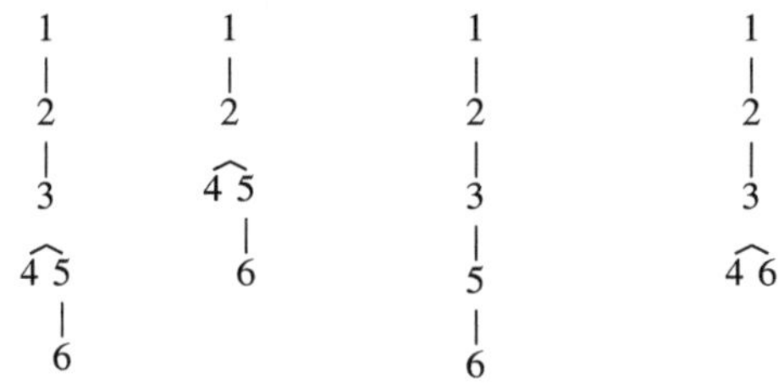

Stage 3: Cutting off the 'Dry Branches'

The idea is the following. We take the rightmost branch of a Socratic tree and find its world line. Starting with the leaf, we cut off the nodes labeled with sequents whose principal numeral is from outside the world line. If a deleted node is a branching point, then the whole (left or right) branch must be cut off. After reaching the root, we go to the branch on the left. The idea is stated formally in the form of Procedure PRUNE. The Socratic tree introduced as data must be determined by a regular Socratic *proof*, otherwise the assignment in line 5 may fail, since an open branch has no world line.

When the procedure PRUNE is executed, it calls the recursive procedure REMOVE which deletes a whole subtree of the tree, if necessary. This happens when we arrive at a node whose principal numeral is not in the appropriate world line, thus the node should be removed, but the node has a sibling, which means that it has resulted from the previous node by an application of a branching rule. The branching point will be analyzed and removed, if necessary, during another iteration of the while-loop, but the whole subtree starting with the sibling must be removed first, otherwise we will loose the information about its redundancy.

Now let us go back to Examples 7.5 and 7.6.

Example 7.8 The result of acting with Procedure PRUNE on the Socratic tree determined by the regular Socratic proof of the sequent $\vdash ((\Box p \vee \Box q) \vee \Diamond\neg q)^0$ (see Example 7.5). The tree has one branch and the world line of its leaf is $\langle 0, 2\rangle$, hence the procedure PRUNE removes the nodes with the principal numeral 1. For clarity, right to each sequent we add its principal numeral in square brackets. Further, the function $\tau_{\mathbf{s}}$ acts on the nodes, and the final result is displayed in the left down corner of Fig. 7.2.

As we have seen in Example 7.6 above, even after pruning we may need to modify the leaves by applications of $L_W^{\mathbb{G}}$ and/or $R_W^{\mathbb{G}}$. From the above example we can see another case when an application of weakening may be needed—between the premise and the conclusion of the general modal rule. The second sequent (from the top) of the pruned tree is:

$$(q)^1 \vdash (p)^{0,1}, (q)^{0,2}, (\Diamond\neg q)^0, (\neg q)^2$$

its principal numeral is 2, and it is translated by function $\tau_{\mathbf{s}}$ into '$\vdash q, \neg q$'. The next node in the pruned Socratic tree is labeled with:

Procedure PRUNE(Socratic tree,s)

```
Data: Socratic tree determined by a regular Socratic proof of a sequent of the form
      ⊢ (A)^0
Result: Socratic tree s pruned
s ← Socratic tree
b ← the rightmost branch of s
repeat
    n ← the leaf of b
    wl ← the world line of b
    while n is not the root of s do
        if the principal numeral of n is in wl then
            n ← the immediate predecessor of n
        else
            if n has a sibling then
                k ← the sibling of n
                REMOVE(s,k,s)
                m ← the immediate predecessor of n
                s ← s \ n
                n ← m
            else
                m ← the immediate predecessor of n
                s ← s \ n
                n ← m
            end
        end
    end
    if b is not the leftmost branch of s then
        b ← the branch which is next to the left of b
    end
until b is the leftmost branch of s;
```

$$\vdash (\Box p)^0, (\Box q)^0, (\Diamond\neg q)^0$$

and its principal numeral is 0. The result after translation by $\tau_{\mathbf{s}}$ is '$\vdash \Box p, \Box q, \Diamond\neg q$'. Unfortunately, the following

$$\frac{\vdash q, \neg q}{\vdash \Box p, \Box q, \Diamond\neg q}$$

is not an instance of a rule of $\mathbb{G}^{\mathsf{K}\tau}$, but the conclusion can be obtained from the premise by an application of $R^{\mathbb{G}}_{\Box}$ *and* $R^{\mathbb{G}}_{W}$. The boxed sequent is the missing one to make the tree a proof in $\mathbb{G}^{\mathsf{K}\tau}$.

In order to add applications of weakening, every case of application of a modal rule needs to be analyzed in the following way. Let ϕ stand for the sequent-conclusion and ψ for the sequent-premise of an application of erotetic rule $L^{\mathbb{E}}_{\Diamond\Box}$ or $R^{\mathbb{E}}_{\Box\Diamond}$. Suppose that $\tau_{\mathbf{s}}(\phi) = S' \vdash T'$ and $\tau_{\mathbf{s}}(\psi) = S \vdash T$ are the respective nodes treated with function $\tau_{\mathbf{s}}$. Find the leftmost formula of the form '$\Diamond A$' in S such that A occurs in S'. If there is no such formula, then find the leftmost in T formula of the form '$\Box A$' such that A

Procedure REMOVE(tree,node,s)

```
Data: tree, node
Result: tree with the whole subtree starting with node removed
s ← tree
n ← node
if n is a leaf of s then
    s ← s \ n
    return s
else
    if n is a branching point then
        m1 ← the left immediate successor of n
        m2 ← the right immediate successor of n
        s ← s \ n
        REMOVE(s,m1,s)
        REMOVE(s,m2,s)
    else
        m ← the immediate successor of n
        s ← s \ n
        REMOVE(s,m,s)
    end
end
```

is in T'. Let $S^\star \vdash T^\star$ stand for the result (conclusion) of application of, respectively, $L_\Diamond^\mathbb{G}$ or $R_\Box^\mathbb{G}$ with respect to A. If $S' \vdash T'$ differs from $S^\star \vdash T^\star$, then add a node labeled with $S^\star \vdash T^\star$ between these labeled with $S \vdash T$ and $S' \vdash T'$. If both: S' differs from $S^\star$ and T' differs from $T^\star$, then add also a node labeled with $S' \vdash T^\star$ between these labeled with $S^\star \vdash T^\star$ and $S' \vdash T'$. Then the result is:

$$\cfrac{\cfrac{\cfrac{S \vdash T}{S^\star \vdash T^\star}\,L_\Diamond^\mathbb{G}/R_\Box^\mathbb{G}}{S' \vdash T^\star}\,L_W^\mathbb{G}}{S' \vdash T'}\,R_W^\mathbb{G}$$

It will be convenient to shorten the above derivation into an application of a derived rule which will be called $L_{\Diamond+W}^\mathbb{G}$ or $R_{\Box+W}^\mathbb{G}$, depending on the initial use of $L_\Diamond^\mathbb{G}$ or $R_\Box^\mathbb{G}$. This operation reminds 'hiding' weakening, but let us emphasize that we have introduced weakening as a primary rule, and $L_{\Diamond+W}^\mathbb{G}$, $R_{\Box+W}^\mathbb{G}$ are derived ones.

$$\frac{S_1 \;'\; A \;'\; S_2 \vdash T}{S_1^* \;'\; \Diamond A \;'\; S_2^* \vdash T^*}\,L_{\Diamond+W}^\mathbb{G} \qquad \frac{S \vdash T_1 \;'\; A \;'\; T_2}{S^* \vdash T_1^* \;'\; \Box A \;'\; T_2^*}\,R_{\Box+W}^\mathbb{G}$$

where 'S_1^*' ('S_2^*', 'S^*') is a supersequence, not necessarily proper, of '$\Box S_1$' ('$\Box S_2$', '$\Box S$', respectively), and 'T^*' ('T_1^*', 'T_2^*') is a supersequence, not necessarily proper, of '$\Diamond T$' ('$\Diamond T_1$', '$\Diamond T_2$', respectively).

We 'weaken' the leaves in a similar manner. Let us recall that each leaf results from a basic sequent ϕ with indexed formulas, which is such that the $\tau_\mathbf{s}$-numeral of

Socratic tree $\mathbf{Tr}(\mathbf{s})$:

$$(q)^1, (q)^2 \vdash (p)^{0,1}, (q)^{0,2}, (\Diamond\neg q)^0 \ [2]$$
$$(q)^1 \vdash (p)^{0,1}, (q)^{0,2}, (\Diamond\neg q)^0, (\neg q)^2 \ [2]$$
$$(q)^1 \vdash (p)^{0,1}, (\Box q)^0, (\Diamond\neg q)^0 \ [1]$$
$$\vdash (p)^{0,1}, (\Box q)^0, (\Diamond\neg q)^0, (\neg q)^1 \ [1]$$
$$\vdash (\Box p)^0, (\Box q)^0, (\Diamond\neg q)^0 \ [0]$$
$$\vdash (\Box p \vee \Box q)^0, (\Diamond\neg q)^0 \ [0]$$
$$\vdash ((\Box p \vee \Box q) \vee \Diamond\neg q)^0$$

$\Longrightarrow$

$\mathbf{Tr}(\mathbf{s})$ pruned

$$(q)^1, (q)^2 \vdash (p)^{0,1}, (q)^{0,2}, (\Diamond\neg q)^0 \ [2]$$
$$(q)^1 \vdash (p)^{0,1}, (q)^{0,2}, (\Diamond\neg q)^0, (\neg q)^2 \ [2]$$
$$\vdash (\Box p)^0, (\Box q)^0, (\Diamond\neg q)^0 \ [0]$$
$$\vdash (\Box p \vee \Box q)^0, (\Diamond\neg q)^0 \ [0]$$
$$\vdash ((\Box p \vee \Box q) \vee \Diamond\neg q)^0$$

$\Downarrow$

$\tau_{\mathbf{s}}$ acting on sequents:

$$\tau(2, (q)^1, (q)^2 \vdash (p)^{0,1}, (q)^{0,2}, (\Diamond\neg q)^0)$$
$$\tau(2, (q)^1 \vdash (p)^{0,1}, (q)^{0,2}, (\Diamond\neg q)^0, (\neg q)^2)$$
$$\tau(0, \vdash (\Box p)^0, (\Box q)^0, (\Diamond\neg q)^0)$$
$$\tau(0, \vdash (\Box p \vee \Box q)^0, (\Diamond\neg q)^0)$$
$$\tau(1, \vdash ((\Box p \vee \Box q) \vee \Diamond\neg q)^0)$$

$\Longleftarrow$

the final tree with the boxed node added

$$q \vdash q$$
$$\vdash q, \neg q$$
$$\boxed{\vdash \Box q, \Diamond\neg q}$$
$$\vdash \Box p, \Box q, \Diamond\neg q$$
$$\vdash \Box p \vee \Box q, \Diamond\neg q$$
$$\vdash (\Box p \vee \Box q) \vee \Diamond\neg q$$

Fig. 7.2 Example 7.8

the sequent is the one in which the closing condition was met. Therefore there is a formula occurring both on the left and on the right side of the turnstile in $\tau_{\mathbf{s}}(\phi)$. If there is more than one formula on the right, then we add an application of $R_W^{\mathbb{G}}$, and if there is more than one formula on the left, then we add an application of $L_W^{\mathbb{G}}$. However, if we are not very fond of 'clean' axioms, we may feel satisfied with sequents falling under the scheme of Ax_{+W}.

Example 7.9 The Socratic tree determined by the regular Socratic proof of the sequent $\vdash (\Box(p \wedge r) \vee (\Box p \rightarrow \Box(q \rightarrow p)))^0$ in calculus $\mathbb{E}^{\mathsf{K}\tau} \cup \{L_{\Diamond\Box}^{\mathbb{E}}, R_{\Box\Diamond}^{\mathbb{E}}\}$ (see Example 7.6).

$$\vdash (\Box(p \wedge r) \vee (\Box p \to \Box(q \to p)))^0$$
$$\vdash (\Box(p \wedge r))^0, (\Box p \to \Box(q \to p))^0 \; [0]$$
$$\vdash (\Box(p \wedge r))^0, (\Box p \to \Box(q \to p))^0 \; [0]$$
$$(\Box p)^0 \vdash (\Box(p \wedge r))^0, (\Box(q \to p))^0 \; [0]$$
$$(\Box p)^0, (p)^1 \vdash (p \wedge r)^{0,1}, (\Box(q \to p))^0 \; [1]$$

$$(\Box p)^0, (p)^1 \vdash (p)^{0,1}, (\Box(q \to p))^0 \; [1] \qquad (\Box p)^0, (p)^1 \vdash (r)^{0,1}, (\Box(q \to p))^0 \; [1]$$
$$(\Box p)^0, (p)^2, (p)^1 \vdash (r)^{0,1}, (q \to p)^{0,2} \; [2]$$
$$(\Box p)^0, (p)^2, (p)^1, (q)^{0,2} \vdash (r)^{0,1}, (p)^{0,2} \; [2]$$

The world line of the left branch is $\langle 0, 1\rangle$, and that of the right branch is $\langle 0, 2\rangle$. The analysis by procedure PRUNE starts with the right branch and removes the nodes with 1 as the principal numeral, leaving the following tree (to the left), transformed by function $\tau_{\mathbf{s}}$ (to the right), which further, supplemented with the applications of weakening, becomes a proof in $\mathbb{G}^{\mathsf{K}\tau}$ (below).

$$(\Box p)^0, (p)^2, (p)^1, (q)^{0,2} \vdash (r)^{0,1}, (p)^{0,2} \; [2] \Longrightarrow \qquad p, q \vdash p$$
$$(\Box p)^0, (p)^2, (p)^1 \vdash (r)^{0,1}, (q \to p)^{0,2} \; [2] \qquad p \vdash q \to p$$
$$(\Box p)^0 \vdash (\Box(p \wedge r))^0, (\Box(q \to p))^0 \; [0] \qquad \Box p \vdash \Box(p \wedge r), \Box(q \to p)$$
$$\vdash (\Box(p \wedge r))^0, (\Box p \to \Box(q \to p))^0 \; [0] \qquad \vdash \Box(p \wedge r), \Box p \to \Box(q \to p)$$
$$\vdash (\Box(p \wedge r))^0, (\Box p \to \Box(q \to p))^0 \; [0] \qquad \vdash \Box(p \wedge r), \Box p \to \Box(q \to p)$$
$$\vdash (\Box(p \wedge r) \vee (\Box p \to \Box(q \to p)))^0 \qquad \vdash \Box(p \wedge r) \vee (\Box p \to \Box(q \to p))$$

$$\Downarrow$$

$$\dfrac{\dfrac{\dfrac{\dfrac{\dfrac{\dfrac{p \vdash p}{p, q \vdash p}\, L_W^{\mathbb{G}}}{p \vdash q \to p}\, R_{\to}^{\mathbb{G}}}{\Box p \vdash \Box(q \to p)}\, R_{\Box}^{\mathbb{G}}}{\Box p \vdash \Box(p \wedge r), \Box(q \to p)}\, R_W^{\mathbb{G}}}{\vdash \Box(p \wedge r), \Box p \to \Box(q \to p)}\, R_{\to}^{\mathbb{G}}}{\vdash \Box(p \wedge r) \vee (\Box p \to \Box(q \to p))}\, R_{\vee}^{\mathbb{G}}$$

Let us observe that the fact that the procedure starts with the right branch is inessential for the final result. If we started with the left branch, the world line would be $\langle 0, 1\rangle$, no nodes would be removed as the left branch has no sequents with principle numerals from outside the world line. After reaching the root, we would move to the right branch, change the world line to $\langle 0, 2\rangle$, and cut off the nodes with 1 as active, which would lead, however, to cutting off the left branch.

The last task in this section is to prove the following theorem:

Theorem 7.1 *Let* **s** *be a regular Socratic proof of a sequent '$\vdash (A)^0$' in calculus* $\mathbb{E}^{\mathsf{K}\tau} \cup \{L^{\mathbb{E}}_{\Diamond\Box}, R^{\mathbb{E}}_{\Box\Diamond}\}$. *If* $\langle X, R\rangle$ *is a tree obtained from* **s** *by the procedure PRUNE, then* $\langle X, R, \tau_{\mathbf{s}} |_X\rangle$ *is a proof of the sequent '$\vdash A$' in calculus* $\mathbb{G}^{\mathsf{K}\tau} \cup \{Ax_{+W}, L^{\mathbb{G}}_{\Diamond+W}, R^{\mathbb{G}}_{\Box+W}\}$.

First, we prove the following lemma from which it follows immediately that a tree $\langle X, R, \tau_s |_X\rangle$ obtained as above is regulated by the rules of $\mathbb{G}^{\mathsf{K}\tau} \cup \{L^{\mathbb{G}}_{\Diamond+W}, R^{\mathbb{G}}_{\Box+W}\}$. Then we go back to the above theorem.

Lemma 7.1 *Let* **s** *denote a regular Socratic proof of the sequent '$\vdash (A)^0$' in calculus* $\mathbb{E}^{\mathsf{K}\tau} \cup \{L^{\mathbb{E}}_{\Diamond\Box}, R^{\mathbb{E}}_{\Box\Diamond}\}$, *and let* $\mathcal{T} = \langle X, R, \tau_{\mathbf{s}} |_X\rangle$ *be the tree obtained from* $\mathbf{Tr}(\mathbf{s})$ *by procedure PRUNE.*

1. *If a sequent ψ labels a node of $\mathcal{T}$ that immediately precedes exactly one node labeled by ϕ, and the two sequents have the same principal numeral, then the following:*
$$\frac{\tau_{\mathbf{s}}(\phi)}{\tau_{\mathbf{s}}(\psi)}$$
is an instance of a classical non-branching rule of $\mathbb{G}^{\mathsf{K}\tau}$.
2. *If a sequent ψ labels a node of $\mathcal{T}$ that immediately precedes two nodes labeled by ϕ_1 and ϕ_2, then the three sequents have the same principal numeral, and the following:*
$$\frac{\tau_{\mathbf{s}}(\phi_1) \quad \tau_{\mathbf{s}}(\phi_2)}{\tau_{\mathbf{s}}(\psi)}$$
is an instance of a classical branching rule of $\mathbb{G}^{\mathsf{K}\tau}$.
3. *If a sequent ψ labels a node of $\mathcal{T}$ that immediately precedes exactly one node labeled by ϕ, and the two sequents have different principal numerals, then the following:*
$$\frac{\tau_{\mathbf{s}}(\phi)}{\tau_{\mathbf{s}}(\psi)}$$
is an instance of one of the rules: $L^{\mathbb{G}}_{\Diamond}, R^{\mathbb{G}}_{\Box}, L^{\mathbb{G}}_{\Diamond+W}, R^{\mathbb{G}}_{\Box+W}$.

Proof The classical cases *1* and 2 seem to be a repetition of the content of Corollary 7.5. They *seem to be*, but now we deal with a tree after pruning by Procedure PRUNE, so we need to show that the pruning mechanism does not spoil the property expressed by Corollary 7.5.

Fortunately, it is easy to see that it does not. Let us recall Corollary 7.2; it states that the principal numeral in a sequent—which is a constituent-conclusion in a question of a regular Socratic transformation, and then becomes a label of a node of a Socratic tree determined by the transformation—is always the largest in the sequent. This entails that no nodes between these labeled by ψ and ϕ could have been deleted during pruning. The same pertains to case *2*.

The situation in *3* is different, since whole subtrees between ψ and ϕ could have been deleted. Let us recall a property of Socratic trees—by Corollary 6.1, sequent ϕ results from ψ by one of the erotetic rules used in the construction of **s**. Since the two sequents have different principal numerals, it must be $L^{\mathbb{E}}_{\Diamond\Box}$ or $R^{\mathbb{E}}_{\Box\Diamond}$.

Suppose that the modal rule is $L^{\mathbb{E}}_{\Diamond\Box}$, the case with $R^{\mathbb{E}}_{\Box\Diamond}$ is analogous. Now we will make an additional assumption to be proved later. Let i, j stand for the principal numerals of ψ, ϕ, respectively.

- **Additional Assumption**: The formula-premise of the application of $L^{\mathbb{E}}_{\Diamond\Box}$ is a formula, $\Diamond A$, living in i, that is, the principal numeral of ψ.

What happens then is that the sequents are of the forms:

$$\psi \quad = \quad \Gamma((\Diamond A)^{\varsigma/i}) \vdash \Delta$$

$$\phi \quad = \quad \Pi((A)^{\rho/i,j})\sharp^{i}_{j} \vdash \Theta\flat^{i}_{j}$$

Let us establish what we know about the relations between the contexts on the right: Δ and $\Theta\flat^{i}_{j}$. Recall that '$\Theta\flat^{i}_{j}$' is Θ 'enlarged' with 'settlers' that come from i to j. Every indexed formula of the form '$(\Diamond B)^{\varrho/i}$' occurring in Δ (if there was any) is present also in $\Theta\flat^{i}_{j}$, since these formulas are not eliminated in the course of a Socratic transformation. What is more, the converse is true: every indexed formula of this form that occurs in $\Theta\flat^{i}_{j}$ is also in Δ. Once again, this follows from the fact that our tree has been pruned: the nodes between ψ and ϕ (if there were any) had their principal numerals outside the world line of the branch, whereas numerals i, j belong to the world line. Hence no rules applied between ψ and ϕ could result in introducing new formulas living in i. The same pertains to the context of formula '$(\Diamond A)^{\varsigma/i}$' in '$\Gamma((\Diamond A)^{\varsigma/i})$' and the context of '$(A)^{\rho/i,j}$' in '$\Pi((A)^{\rho/i,j})\sharp^{i}_{j}$', that is, for every indexed formula of the form '$(\Box B)^{\varrho/i}$': the formula occurs in '$\Gamma((\Diamond A)^{\varsigma/i})$' iff it occurs in '$\Pi((A)^{\rho/i,j})\sharp^{i}_{j}$'. On the other hand, the following is also true for the same reason: for every indexed formula of the form '$(\Box B)^{\varrho/i}$', the formula occurs in Δ (on the right side) iff it occurs in '$\Theta\flat^{i}_{j}$'. And for the left side: for every indexed formula of the form '$(\Diamond B)^{\varrho/i}$' *except for* '$(\Diamond A)^{\varsigma/i}$': the formula occurs in '$\Gamma((\Diamond A)^{\varsigma/i})$' iff it occurs in '$\Pi((A)^{\rho/i,j})\sharp^{i}_{j}$'. The formulas of the form '$\Box B$', that were present at the right side, together with the additional (that is, except for '$\Diamond A$') formulas of the form '$\Diamond B$', that were on the left side, constitute the 'extra' context that must be added in calculus $\mathbb{G}^{\mathsf{K}\tau}$ by weakening.

Now we are almost done. Let us calculate $\tau_{\mathbf{s}}(\phi)$. The $\tau_{\mathbf{s}}$-numeral for ϕ is j, thus $\tau_{\mathbf{s}}(\phi)$ is a sequent of language $\mathcal{M}^{imp}_{\vdash}$ (no indices) with formulas that lived in j. Let us represent it as:

$$\tau_{\mathbf{s}}(\phi) \quad = \quad S_1\,'\,A\,'\,S_2 \vdash T$$

where $S_1\,'\,A\,'\,S_2$ is the sequence of all formulas that have been introduced to world j by operation $\sharp^i_j$ (among which A occurs), and T is the sequence of all formulas introduced to world j by operation $\flat^i_j$ (see Algorithms 7.1 and 7.2). On the other hand, $\tau_{\mathbf{s}}(\psi) = \tau(i, \psi)$, and the formulas were introduced to $\tau_{\mathbf{s}}(\phi)$ only by the elimination of suitable modal operators. If there were no additional indexed modal formulas of the form '$(\Box B)^{\varrho/i}$' in Δ nor '$(\Diamond B)^{\varrho/i}$' in '$\Gamma((\Diamond A)^{\varsigma/i})$', then we would have that:

$$\tau_{\mathbf{s}}(\psi) \;=\; \Box S_1\,'\,\Diamond A\,'\,\Box S_2 \vdash \Diamond T$$

and then $\tau_{\mathbf{s}}(\phi)/\tau_{\mathbf{s}}(\psi)$ would be an instance of $L^{\mathbb{G}}_{\Diamond}$. Otherwise, additional $\Diamond$-formula(s) occur on the left side of the turnstile and/or additional $\Box$-formula(s) occur on the right side of the turnstile. Then after the application of $L^{\mathbb{G}}_{\Diamond}$, another application of $L^{\mathbb{G}}_{W}$ and/or $R^{\mathbb{G}}_{W}$ produces sequent $\tau_{\mathbf{s}}(\psi)$, which was to be proved.

Now we go back to the additional assumption which assumes a nice coincidence consisting in the fact that the principal numeral i of sequent ψ must be at the same time the active numeral for sequent ϕ, even if ϕ results from some other sequent-premise, deleted in the course of pruning. The nice coincidence follows from the fact that the world line is a sequence of numerals that are linearly ordered by $<$. We justify this statement.

Let k stand for the numeral which was active in the application of $L^{\mathbb{E}}_{\Diamond\Box}$ that ϕ resulted from. Numeral k must be in the world line, for otherwise j would be outside the world line (see Definition 7.7). The sequents between ψ and ϕ had their principal numerals outside the world line, therefore k must be present in ψ. Suppose that $k < i$. Then the numerals in ϕ form a tree that has at least two leaves: j is one of the leaves, but i is another one. Since world lines are linearly ordered, the numerals cannot belong to the world line of the branch at the same time. Therefore $k \not< i$, which means that $k = i$, since By Corollary 7.2 numeral i is the largest in ψ. □

Now we may proceed to the proof of Theorem 7.1.

Proof of Theorem 7.1 Let $\mathbf{s}$ be a regular Socratic proof of sequent '$\vdash (A)^0$' in calculus $\mathbb{E}^{\mathsf{K}\tau} \cup \{L^{\mathbb{E}}_{\Diamond\Box}, R^{\mathbb{E}}_{\Box\Diamond}\}$, and assume that $\langle X, R, \tau_{\mathbf{s}}|_X\rangle$ is a tree obtained from $\mathbf{s}$ by the procedure PRUNE, together with the labelling function $\tau_{\mathbf{s}}$ restricted to X. Then

- $\langle X, R, \tau_{\mathbf{s}}|_X\rangle$ is a finite tree labeled with sequents of language $\mathcal{M}^{imp}_{\vdash}$ (by the definition of $\tau_{\mathbf{s}}$),
- by Lemma 7.1, the tree is regulated by the rules of calculus $\mathbb{G}^{\mathsf{K}\tau} \cup \{L^{\mathbb{G}}_{\Diamond+W}, R^{\mathbb{G}}_{\Box+W}\}$,
- the root of the tree is labeled with sequent $\tau_{\mathbf{s}}(\vdash (A)^0) = \tau(0, \vdash (A)^0)$, which is '$\vdash A$',
- each leaf of $\mathbf{Tr}(\mathbf{s})$ is labeled by a basic sequent, and for each such label ϕ, its $\tau_{\mathbf{s}}$-numeral is the one in which there is a formula on the two sides of the turnstile, therefore $\tau_{\mathbf{s}}(\phi)$ is an axiom Ax or a general axiom Ax_{+W}.

All this entails that $\langle X, R, \tau_{\mathbf{s}}|_X\rangle$ is a proof of sequent $\vdash A$ in calculus $\mathbb{G}^{\mathsf{K}\tau} \cup \{L^{\mathbb{G}}_{\Diamond+W}, R^{\mathbb{G}}_{\Box+W}, Ax_{+W}\}$. □

7.2 Translations for Non-Symmetric, Non-Euclidean Extensions of K

The erotetic calculi for the extensions of K inherit the rules of $\mathbb{E}^{\mathsf{K}\tau}$ and possibly obtain some additional rules indicated below. The algorithm for regular Socratic transformations constructed in an erotetic calculus will be based on Algorithm 7.3, but the Procedures CLASSICAL and/or MODAL may be altered. Moreover, the definition of relabelling function $\tau_{\mathbf{s}}$ will be modified in some cases.

The Gentzen-style sequent calculi for the extensions of K are based on the calculi presented in [3]. As before, they all have been obtained by dropping the unified notation. Interestingly, we have found the calculi presented here convergent with implicit Hintikka-style tableau systems presented in [4]; their convergence follows from the fact that the tableau systems examined by Rajeev Goré where also inspired by [3] (and also by [6, 10], as the author admitted). Soundness and completeness of the sequent calculi presented here may be proved by using the techniques used in [4], or by reference to arguments used in [3], and we will not examine these issues here.

Calculus $\mathbb{G}^{\mathsf{D}\tau}$ consists of the primary and the derived rules of $\mathbb{G}^{\mathsf{K}\tau}$, and additionally the rule:

$$\frac{S \vdash T}{\Box S \vdash \Diamond T}\; R^{\mathbb{G}}_{\mathsf{D}}$$

where, let us recall, $\Box S$ is a sequence of formulas such that $\Box A$ is i-th term of $\Box S$ iff A is i-th term of sequence S; in turn, $\Diamond A$ is i-th term of $\Diamond T$ iff A is i-th term of T. As previously, it is convenient to add a derived rule summarising the possible combination of applications of $R^{\mathbb{G}}_{\mathsf{D}}$, $L^{\mathbb{G}}_{W}$ and $R^{\mathbb{G}}_{W}$. The rule is called '$R^{\mathbb{G}}_{\mathsf{D}+W}$' and has the following scheme:

$$\frac{S \vdash T}{S^{*} \vdash T^{*}}\; R^{\mathbb{G}}_{\mathsf{D}+W}$$

where S^{*} is a proper supersequence of $\Box S$ and T^{*} is a proper supersequence of $\Diamond T$.

Calculus $\mathbb{G}^{\mathsf{T}\tau}$ is the set of rules $\mathbb{G}^{\mathsf{K}\tau}$ (the primary and the derived ones) and the following:

$$\frac{S(A) \vdash T}{S(\Box A) \vdash T}\; L^{\mathbb{G}}_{ref} \qquad\qquad \frac{S \vdash T(A)}{S \vdash T(\Diamond A)}\; R^{\mathbb{G}}_{ref}$$

It is the only modal rule added to obtain an extension of K that does not need to be 'spoiled' with weakening.

Calculus $\mathbb{G}^{\mathsf{K4}\tau}$ contains the primary and the derived rules of $\mathbb{G}^{\mathsf{K}\tau}$, except for the modal rules $L^{\mathbb{G}}_{\Diamond}$, $R^{\mathbb{G}}_{\Box}$ which must be replaced with $L^{\mathbb{G}}_{tr}$, $R^{\mathbb{G}}_{tr}$. The use of operations $\sharp$ and $\flat$ is inspired by Fitting's account but the operations' definitions are more complicated due to the fact that we deal with sequences, not sets (see Algorithms 7.4 and 7.5). Moreover, this account does not hide weakening as Fitting's account does: we still

Algorithm 7.4: Operation $\sharp$

Data: n-term sequence $\Box S = \langle \Box A_1, \ldots, \Box A_n \rangle$ of formulas of $\mathcal{M}$, $n \geqslant 0$
Result: $2n$-term sequence $(\Box S)\sharp$ of formulas of $\mathcal{M}$

$(\Box S)\sharp \leftarrow \varnothing$
for $k = 1$ **to** n **do**
| $(\Box S)\sharp \leftarrow (\Box S)\sharp \,'\, \langle \Box A_k, A_k \rangle$
end
return $(\Box S)\sharp$

define the rule separately. The use of '$\Box S_i$' and '$\Diamond T_i$' in the conclusions of rules $L^{\mathbb{G}}_{tr}$, $R^{\mathbb{G}}_{tr}$ indicates that all formulas in the conclusion are properly 'modalised'.

Algorithm 7.5: Operation $\flat$

Data: n-th term sequence $\Diamond T = \langle \Diamond A_1, \ldots, \Diamond A_n \rangle$ of formulas of $\mathcal{M}$, $n \geqslant 0$
Result: $2n$ term sequence $(\Diamond T)\flat$ of formulas of $\mathcal{M}$

$(\Diamond T)\flat \leftarrow \varnothing$
for $k = 1$ **to** n **do**
| $(\Diamond T)\flat \leftarrow (\Diamond T)\flat \,'\, \langle \Diamond A_k, A_k \rangle$
end
return $(\Diamond T)\flat$

$$\frac{(\Box S_1)\sharp \,'\, A \,'\, (\Box S_2)\sharp \;\vdash\; (\Diamond T)\flat}{\Box S_1 \,'\, \Diamond A \,'\, \Box S_2 \;\vdash\; \Diamond T}\; L^{\mathbb{G}}_{tr}$$

$$\frac{(\Box S)\sharp \;\vdash\; (\Diamond T_1)\flat \,'\, A \,'\, (\Diamond T_2)\flat}{\Box S \;\vdash\; \Diamond T_1 \,'\, \Box A \,'\, \Diamond T_2}\; R^{\mathbb{G}}_{tr}$$

Calculus $\mathbb{G}^{\mathsf{D4}\tau}$ contains the rules of $\mathbb{G}^{\mathsf{K}\tau}$ for the classical connectives and for weakening, rules $L^{\mathbb{G}}_{tr}$, $R^{\mathbb{G}}_{tr}$ (for transitivity) and rule $R^{\mathbb{G}}_{\mathsf{D}tr}$ for seriality. We need a new one, since $R^{\mathbb{G}}_{\mathsf{D}}$ does not account for the transition of information which is specific to the transitive logics.

$$\frac{(\Box S)\sharp \;\vdash\; (\Diamond T)\flat}{\Box S \;\vdash\; \Diamond T}\; R^{\mathbb{G}}_{\mathsf{D}tr}$$

Finally, calculus $\mathbb{G}^{\mathsf{S4}\tau}$ consists of the rules of $\mathbb{G}^{\mathsf{K}\tau}$ for the classical connectives and for weakening, rules $L^{\mathbb{G}}_{tr}$, $R^{\mathbb{G}}_{tr}$ for transitivity, and $L^{\mathbb{G}}_{ref4}$ and $R^{\mathbb{G}}_{ref4}$, which express reflexivity in the presence of transitivity.

$$\frac{S(\Box A, A) \;\vdash\; T}{S(\Box A) \;\vdash\; T}\; L^{\mathbb{G}}_{ref4} \qquad\qquad \frac{S \;\vdash\; T(\Diamond A, A)}{S \;\vdash\; T(\Diamond A)}\; R^{\mathbb{G}}_{ref4}$$

In $L^{\mathbb{G}}_{ref4}$ two occurrences of formulas: '$\Box A$, A' in the premise are replaced with one occurrence: '$\Box A$' in the conclusion, analogously for $R^{\mathbb{G}}_{ref4}$.

Table 7.7 Calculi $\mathbb{G}^{\mathsf{L}\tau}$

Calculus	Primary rules	Derived rules
$\mathbb{G}^{\mathsf{K}\tau}$	$L^{\mathbb{G}}_{\neg}, R^{\mathbb{G}}_{\neg}, L^{\mathbb{G}}_{\wedge}, R^{\mathbb{G}}_{\vee}, R^{\mathbb{G}}_{\rightarrow}, R^{\mathbb{G}}_{\wedge}, L^{\mathbb{G}}_{\vee}, L^{\mathbb{G}}_{\rightarrow}, L^{\mathbb{G}}_{\Diamond}, R^{\mathbb{G}}_{\Box}, L^{\mathbb{G}}_{W}, R^{\mathbb{G}}_{W}, Ax$	$Ax_{+W}, L^{\mathbb{G}}_{\Diamond+W}, R^{\mathbb{G}}_{\Box+W}$
$\mathbb{G}^{\mathsf{D}\tau}$	all above and $R^{\mathbb{G}}_{\mathsf{D}}$	all above and rule $R^{\mathbb{G}}_{\mathsf{D}+W}$
$\mathbb{G}^{\mathsf{T}\tau}$	all primary rules of $\mathbb{G}^{\mathsf{K}\tau}$ and $L^{\mathbb{G}}_{ref}, R^{\mathbb{G}}_{ref}$	as in $\mathbb{G}^{\mathsf{K}\tau}$
$\mathbb{G}^{\mathsf{K4}\tau}$	$L^{\mathbb{G}}_{\neg}, R^{\mathbb{G}}_{\neg}, L^{\mathbb{G}}_{\wedge}, R^{\mathbb{G}}_{\vee}, R^{\mathbb{G}}_{\rightarrow}, R^{\mathbb{G}}_{\wedge}, L^{\mathbb{G}}_{\vee}, L^{\mathbb{G}}_{\rightarrow}, L^{\mathbb{G}}_{tr}, R^{\mathbb{G}}_{tr}, L^{\mathbb{G}}_{W}, R^{\mathbb{G}}_{W}, Ax$	$Ax_{+W}, L^{\mathbb{G}}_{tr+W}, R^{\mathbb{G}}_{tr+W}$
$\mathbb{G}^{\mathsf{D4}\tau}$	$L^{\mathbb{G}}_{\neg}, R^{\mathbb{G}}_{\neg}, L^{\mathbb{G}}_{\wedge}, R^{\mathbb{G}}_{\vee}, R^{\mathbb{G}}_{\rightarrow}, R^{\mathbb{G}}_{\wedge}, L^{\mathbb{G}}_{\vee}, L^{\mathbb{G}}_{\rightarrow}, R^{\mathbb{G}}_{\mathsf{D}tr}, L^{\mathbb{G}}_{tr}, R^{\mathbb{G}}_{tr}, L^{\mathbb{G}}_{W}, R^{\mathbb{G}}_{W}, Ax$	$Ax_{+W}, L^{\mathbb{G}}_{tr+W}, R^{\mathbb{G}}_{tr+W}, R^{\mathbb{G}}_{\mathsf{D}tr+W}$
$\mathbb{G}^{\mathsf{S4}\tau}$	$L^{\mathbb{G}}_{\neg}, R^{\mathbb{G}}_{\neg}, L^{\mathbb{G}}_{\wedge}, R^{\mathbb{G}}_{\vee}, R^{\mathbb{G}}_{\rightarrow}, R^{\mathbb{G}}_{\wedge}, L^{\mathbb{G}}_{\vee}, L^{\mathbb{G}}_{\rightarrow}, L^{\mathbb{G}}_{ref4}, R^{\mathbb{G}}_{ref4}, L^{\mathbb{G}}_{tr}, R^{\mathbb{G}}_{tr}, L^{\mathbb{G}}_{W}, R^{\mathbb{G}}_{W}, Ax$	$Ax_{+W}, L^{\mathbb{G}}_{tr+W}, R^{\mathbb{G}}_{tr+W}$

All rules of our Gentzen-style calculi for modal logics are gathered in Table 7.7. Below we continue with the description of erotetic calculi modified for the purpose of this chapter. The table summarising the descriptions of erotetic calculi is quite large, so we have placed it in Appendix B.

Translation Procedure for D

We start with logic D. In the original erotetic account [8], the serial logics are formalized by the addition of rule $R_{\pi\mathsf{D}}$. We first add analogues of it in the standard notation (rules $L^{\mathbb{E}}_{\mathsf{D}}$, $R^{\mathbb{E}}_{\mathsf{D}}$ below) and then derive a generalisation (as in the case of K).

$$\frac{?(\ \Phi\ (\ \Gamma\ (\ (\Box A)^{\varsigma/i}\ )\ \vdash\ \Delta\ )\ )}{?(\ \Phi\ (\ \Gamma\ (\ (\Box A)^{\varsigma/i},\ A^{\,i,j}\ )\ \vdash\ \Delta\ )\ )}\ L^{\mathbb{E}}_{\mathsf{D}}$$

$$\frac{?(\ \Phi\ (\ \Gamma\ \vdash\ \Delta\ (\ (\Diamond A)^{\varsigma/i}\ )\ )\ )}{?(\ \Phi\ (\ \Gamma\ \vdash\ \Delta\ (\ (\Diamond A)^{\varsigma/i},\ A^{\,i,j}\ )\ )\ )}\ R^{\mathbb{E}}_{\mathsf{D}}$$

Proviso: j must be new with respect to the sequent-premise. This is enough for the rule to be sound, but as usually we will put more restrictions on applicability of the rule. First, j is not arbitrary but must represent the smallest number not present in the sequent-premise. Second, the rule will be applied only when there is no other way to create the new world accessible from i.

Calculus $\mathbb{E}^{\mathsf{D}\tau}$ contains the rules of $\mathbb{E}^{\mathsf{K}\tau}$ together with $L^{\mathbb{E}}_{\Diamond\Box}$, $R^{\mathbb{E}}_{\Box\Diamond}$ and $L^{\mathbb{E}}_{\mathsf{D}}$, $R^{\mathbb{E}}_{\mathsf{D}}$. The generalisation of the erotetic rules for seriality is straightforward. Numeral i is active in the sequent-premise of rule $R^{\mathbb{E}}_{\mathsf{D}g}$, numeral j is new with respect to the active sequent.

$$\frac{?(\Phi\ ;\ \Gamma \vdash \Delta\ ;\ \Psi)}{?(\Phi\ ;\ \Gamma\sharp\mathsf{D}^i_j \vdash \Delta\flat\mathsf{D}^i_j\ ;\ \Psi)}\ R^{\mathbb{E}}_{\mathsf{D}g}$$

Algorithm 7.6: Operation $\sharp\mathsf{D}^i_j$

Data: finite sequence $\Gamma = \langle (A_1)^{\varsigma_1}, \ldots, (A_m)^{\varsigma_m} \rangle$ of indexed formulas
Result: finite sequence $\Gamma\sharp\mathsf{D}^i_j$ of indexed formulas

1 $\Gamma\sharp\mathsf{D}^i_j \leftarrow \varnothing$
2 **for** $k = 1$ **to** m **do**
3 | **if** *k-th term of* Γ *is of the form* $(\Box B)^{\varsigma/i}$ **then**
4 | | $\Gamma\sharp\mathsf{D}^i_j \leftarrow \Gamma\sharp\mathsf{D}^i_j\ '\ \langle (\Box B)^{\varsigma/i}, (B)^{i,j} \rangle$
5 | **else**
6 | | $\Gamma\sharp\mathsf{D}^i_j \leftarrow \Gamma\sharp\mathsf{D}^i_j\ '\ \langle (A_k)^{\varsigma_k} \rangle$
7 | **end**
8 **end**
9 **return** $\Gamma\sharp\mathsf{D}^i_j$

Operations $\sharp\mathsf{D}^i_j$ and $\flat\mathsf{D}^i_j$ are slightly modified versions of operations $\sharp^i_j$, $\flat^i_j$, respectively, defined by Algorithms 7.6 and 7.7 (compare with Algorithms 7.1 and 7.2).

The only difference between Algorithms 7.1 and 7.6 is that in the assignment in line 2 there is '$(F_k)^j$' in the former and '$(F_k)^{i,j}$' in the latter. The point is to make sure that we have added to the resulting sequent the information that i sees j. The definition of $\flat^i_j$ is modified in the same way (compare Algorithms 7.2 and 7.7).

Algorithm 7.7: Operation $\flat\mathsf{D}^i_j$

Data: finite sequence $\Delta = \langle (A_1)^{\varsigma_1}, \ldots, (A_m)^{\varsigma_m} \rangle$ of indexed formulas
Result: finite sequence $\Delta\flat\mathsf{D}^i_j$ of indexed formulas

1 $\Delta\flat^i_j \leftarrow \varnothing$
2 **for** $k = 1$ **to** m **do**
3 | **if** *k-th term of* Δ *is of the form* $(\Diamond B)^{\varsigma/i}$ **then**
4 | | $\Delta\flat\mathsf{D}^i_j \leftarrow \Delta\flat\mathsf{D}^i_j\ '\ \langle (\Diamond B)^{\varsigma/i}, (B)^{i,j} \rangle$
5 | **else**
6 | | $\Delta\flat\mathsf{D}^i_j \leftarrow \Delta\flat\mathsf{D}^i_j\ '\ \langle (A_k)^{\varsigma_k} \rangle$
7 | **end**
8 **end**
9 **return** $\Delta\flat\mathsf{D}^i_j$

Let us also observe that there is no need to apply the rule if there is an indexed formula of the form '$(\Diamond A)^{\varsigma/i}$' on the left, or of the form '$(\Box A)^{\varsigma/i}$' on the right, side of the turnstile. Similarly, there is no sense to apply the rule if there is no formula of the form '$(\Box A)^{\varsigma/i}$' on the left and no formula of the form '$(\Diamond A)^{\varsigma/i}$' on the right. If in the sequent-premise of an application of $R^{\mathbb{E}}_{\mathsf{D}g}$ there is at least one expression '$(\Box A)^{\varsigma/i}$' on the left side or '$(\Diamond A)^{\varsigma/i}$' on the right side, then we call this application of the rule *non-trivial*.

Corollary 7.7 *Every non-trivial application of* $R^{\mathbb{E}}_{\mathsf{D}g}$ *is derivable in* $\mathbb{E}^{\mathsf{D}\tau}$.

Proof As in the previous similar cases: we apply m_1 times rule $L^{\mathbb{E}}_{\mathsf{D}}$ with respect to every '$\Box A_k$', $1 \leqslant k \leqslant m_1$ such that '$\Box A_k$' lives in i in Γ, and then we apply rule $R^{\mathbb{E}}_{\mathsf{D}}$ m_2 times with respect to every '$\Diamond A_k$', $1 \leqslant k \leqslant m_2$ such that '$\Diamond A_k$' lives in i in Δ. □

Regular Socratic transformations in $\mathbb{E}^{\mathsf{D}\tau} \cup \{L^{\mathbb{E}}_{\Diamond\Box}, R^{\mathbb{E}}_{\Box\Diamond}, L^{\mathbb{E}}_{\mathsf{D}}, R^{\mathbb{E}}_{\mathsf{D}}\}$ are defined according to Algorithm 7.3, but with a modified procedure of applications of the modal rules—we want the rule for seriality to be applied when the general modal rules are not applicable. If numeral i has no immediate successor in the indices of indexed formulas of a sequent ϕ, then we say that i is a *dead end in* ϕ. If an indexed formula of the form '$A^{\varsigma/i}$' occurs in ϕ, then we also say that A *is in a dead end in* ϕ. See Procedure MODAL-D.

Procedure MODAL-D(question,transformation,Q, **s**)

Data: question $=?(\Phi(\phi))$, `transformation`, $\Gamma = \langle (A_1)^{\varsigma_1}, \ldots, (A_m)^{\varsigma_m} \rangle$ is the sequence of all indexed formulas occurring in ϕ

s $\leftarrow$ `transformation`
$Q \leftarrow ?(\Phi(\phi))$
$j \leftarrow 1$
$i \leftarrow 0$; `/* used to decide whether to use seriality or not */`
while $j \leqslant m$ **do**
  if $(A_j)^{\varsigma_j}$ *is a formula-premise of* **r** $\in \{L^{\mathbb{E}}_{\Diamond\Box}, R^{\mathbb{E}}_{\Box\Diamond}\}$ **then**
    $Q \leftarrow$ question-conclusion of **r**
    s $\leftarrow$ **s** ; $\langle Q \rangle$
    $i \leftarrow 1$
    $j \leftarrow m + 1$; `/* ends the execution of the while-loop */`
  else
    $j \leftarrow j + 1$
  end
end
if $i = 0$ **then**
  if *there is a formula of the form '$\Box A$' in a dead end on the left side, or a formula of the form '$\Diamond A$' in a dead end on the right side* **then**
    apply $R^{\mathbb{E}}_{\mathsf{D}g}$ with the numeral representing the dead end as active
    $Q \leftarrow$ question-conclusion of **r**
    s $\leftarrow$ **s** ; $\langle Q \rangle$
  end
end

The next example illustrates the result of applying the tools defined for logic D.

Example 7.10 A regular Socratic proof of the sequent '$\vdash (\Box p \to \Diamond p)^0$' in calculus $\mathbb{E}^{\mathsf{D}\tau} \cup \{L^{\mathbb{E}}_{\Diamond\Box}, R^{\mathbb{E}}_{\Box\Diamond}, R^{\mathbb{E}}_{\mathsf{D}g}\}$ translated into a proof of the sequent '$\vdash \Box p \to \Diamond p$' in calculus $\mathbb{G}^{\mathsf{D}\tau} \cup \{Ax_{+W}, L^{\mathbb{G}}_{\Diamond+W}, R^{\mathbb{G}}_{\Box+W}, R^{\mathbb{G}}_{\mathsf{D}+W}\}$. (In fact, in the case of this example the translation goes from $\mathbb{E}^{\mathsf{D}\tau} \cup \{R^{\mathbb{E}}_{\mathsf{D}g}\}$ to $\mathbb{G}^{\mathsf{D}\tau}$, but, in general, all the derived rules may be needed—see Theorem 7.2.) The Socratic tree has one branch with world line $\langle 0, 1\rangle$.

$$\cfrac{\cfrac{?(\vdash (\Box p \to \Diamond p)^0)}{?((\Box p)^0 \vdash (\Diamond p)^0)}\; R^{\mathbb{E}}_{\to}}{?((\Box p)^0, p^{0,1} \vdash (\Diamond p)^0, p^{0,1})}\; R^{\mathbb{E}}_{\mathsf{D}g}$$

$$\Downarrow$$

$$\cfrac{\cfrac{\tau(1, (\Box p)^0, p^{0,1} \vdash (\Diamond p)^0, p^{0,1})}{\tau(0, (\Box p)^0 \vdash (\Diamond p)^0)}}{\tau(0, \vdash (\Box p \to \Diamond p)^0)} \quad \Longrightarrow \quad \cfrac{\cfrac{p \vdash p}{\Box p \vdash \Diamond p}\; R^{\mathbb{G}}_{\mathsf{D}}}{\vdash \Box p \to \Diamond p}\; R^{\mathbb{G}}_{\to}$$

The following theorem is true:

Theorem 7.2 *Let* **s** *be a regular Socratic proof of the sequent '$\vdash (A)^0$' in calculus $\mathbb{E}^{\mathsf{D}\tau} \cup \{L^{\mathbb{E}}_{\Diamond\Box}, R^{\mathbb{E}}_{\Box\Diamond}, R^{\mathbb{E}}_{\mathsf{D}g}\}$. If $\langle X, R\rangle$ is a tree obtained from* **s** *by the procedure PRUNE, then $\langle X, R, \tau_{\mathbf{s}} |_X\rangle$ is a proof of the sequent '$\vdash A$' in calculus $\mathbb{G}^{\mathsf{D}\tau} \cup \{Ax_{+W}, L^{\mathbb{G}}_{\Diamond+W}, R^{\mathbb{G}}_{\Box+W}, R^{\mathbb{G}}_{\mathsf{D}+W}\}$.*

For the purpose of proving Theorem 7.2, we need to analyze the content of Lemma 7.1 and adjust it to the present case. It is easy to see that nothing changes in the analysis of applications of the classical rules, nor in the justification of **Additional Assumption**. What is specific to D is the change in the situation described in clause *3* of Lemma 7.1:

Lemma 7.2 *Let* **s** *stand for a regular Socratic proof of the sequent '$\vdash (A)^0$' in calculus $\mathbb{E}^{\mathsf{D}\tau} \cup \{L^{\mathbb{E}}_{\Diamond\Box}, R^{\mathbb{E}}_{\Box\Diamond}, R^{\mathbb{E}}_{\mathsf{D}g}\}$, and let $\mathcal{T} = \langle X, R, \tau_{\mathbf{s}} |_X\rangle$ be the tree obtained from* **Tr**(**s**) *by Procedure PRUNE.*

1. *and 2 as in Lemma 7.1.*
3. *If a sequent ψ labels a node of $\mathcal{T}$ that immediately precedes exactly one node labeled by ϕ, and the two sequents have different principal numerals, then the following:*

$$\frac{\tau_{\mathbf{s}}(\phi)}{\tau_{\mathbf{s}}(\psi)}$$

is an instance of one of the rules: $L^{\mathbb{G}}_{\Diamond}$, $R^{\mathbb{G}}_{\Box}$, $L^{\mathbb{G}}_{\Diamond+W}$, $R^{\mathbb{G}}_{\Box+W}$, $R^{\mathbb{G}}_{\mathsf{D}}$ or $R^{\mathbb{G}}_{\mathsf{D}+W}$.

Proof We reason as in the proof of Lemma 7.1: the fact that **s** is regular and that $\mathcal{T}$ has been pruned yields that ϕ results in **Tr**(**s**) from its predecessor by one of the modal rules: $L^{\mathbb{E}}_{\Diamond\Box}$, $R^{\mathbb{E}}_{\Box\Diamond}$, $R^{\mathbb{E}}_{\mathsf{D}g}$. If it is $L^{\mathbb{E}}_{\Diamond\Box}$ or $R^{\mathbb{E}}_{\Box\Diamond}$, then we show that $\tau_{\mathbf{s}}(\phi)/\tau_{\mathbf{s}}(\psi)$ is an instance of one of $L^{\mathbb{G}}_{\Diamond}$, $R^{\mathbb{G}}_{\Box}$, $L^{\mathbb{G}}_{\Diamond+W}$, $R^{\mathbb{G}}_{\Box+W}$. If it is $R^{\mathbb{E}}_{\mathsf{D}g}$, then in an analogous way we show that $\tau_{\mathbf{s}}(\phi)/\tau_{\mathbf{s}}(\psi)$ is an instance of $R^{\mathbb{G}}_{\mathsf{D}}$ or $R^{\mathbb{G}}_{\mathsf{D}+W}$. □

Translation Procedure for T

Calculus $\mathbb{E}^{\mathsf{T}\tau}$ contains the rules of $\mathbb{E}^{\mathsf{K}\tau}$ and the two rules displayed below, which capture reflexivity. The derived rules of $\mathbb{E}^{\mathsf{K}\tau}$: $L^{\mathbb{E}}_{\Diamond\Box}$ and $R^{\mathbb{E}}_{\Box\Diamond}$, are also derived rules of $\mathbb{E}^{\mathsf{T}\tau}$.

$$\frac{?(\ \Phi\ (\ \Gamma\ (\ (\Box A)^{\varsigma/i}\ )\ \vdash\ \Delta\ )\ )}{?(\ \Phi\ (\ \Gamma\ (\ (\Box A)^{\varsigma/i},\ A^{i}\ )\ \vdash\ \Delta\ )\ )}\ L^{\mathbb{E}}_{ref}$$

$$\frac{?(\ \Phi\ (\ \Gamma\ \vdash\ \Delta\ (\ (\Diamond A)^{\varsigma/i}\ )\ )\ )}{?(\ \Phi\ (\ \Gamma\ \vdash\ \Delta\ (\ (\Diamond A)^{\varsigma/i},\ A^{i}\ )\ )\ )}\ R^{\mathbb{E}}_{ref}$$

In the original account [8] there was one scheme to capture the rules corresponding to $L^{\mathbb{E}}_{\Box}$, $R^{\mathbb{E}}_{\Diamond}$ and to the above two, with an appropriate provisos of its applicability. But in the present context it is advisable to set apart the cases of $i \neq j$ and $i = j$.

As we shall see, reflexive logics may require labelling some formulas with additional information; this will happen during the generation of a regular Socratic transformation. The additional information will be placed in the bottom index of formulas. We assume that when the construction of a regular Socratic transformation starts, there are no labels in the bottom indices; labels are added during the execution of the algorithm. If the label is added to a formula A, then we say that this occurrence of the formula in a Socratic transformation *has a bottom label*. Then the occurrence takes the following form: 'A_{ref}'. If there is no bottom label, then we say simply that the formula is unlabeled.

Regular Socratic transformations in calculus $\mathbb{E}^{\mathsf{T}\tau}$ are defined in line with the general scheme of Algorithm 7.3, but Procedure CLASSICAL is replaced with Procedure SETTLE-T. The latter includes the applications of rules $L^{\mathbb{E}}_{ref}$ and $R^{\mathbb{E}}_{ref}$. We used the following abbreviations for simplicity: the 'ϕ-sequence' in the Data description refers to the sequence of all indexed formulas occurring in ϕ, symbol '$\circledast$' in line 21 is '$\Box$', if the indexed formula in question occurs in the antecedent (then $\mathbf{r}$ in line 24 is $L^{\mathbb{E}}_{ref}$), and it is '$\Diamond$', if the formula occurs in the succedent (then $\mathbf{r} = R^{\mathbb{E}}_{ref}$).

Let us emphasize again that the information concerning the bottom index is not defined in the rule, but it is used in the algorithm applying the rules. The respective assignment is present in line 22 of Procedure SETTLE-T. The bottom labels play two roles. They prevent redundant repetitions of applications of the reflexivity-rules $L^{\mathbb{E}}_{ref}$, $R^{\mathbb{E}}_{ref}$ (line 21), but they are also used on the level of 'decoding' indexed transformations into purely syntactic ones. The bottom label is a hint that the sequents after translation must fit the rules $L^{\mathbb{G}}_{ref}$, $R^{\mathbb{G}}_{ref}$ of $\mathbb{G}^{\mathsf{T}\tau}$.

Function τ will be replaced with τ_{T} (Definition 7.9). Formulas of the form '$\Box A$', even if in the appropriate world, will be neglected by the function if they have the bottom label.

Procedure SETTLE-T(question,numeral,transformation,Q, n, **s**)

```
Data: question = ?(Φ(ϕ)), numeral, transformation, Γ = ⟨(A_1)^{ς_1}, ..., (A_m)^{ς_m}⟩
      is the ϕ-sequence
Q ← question
n ← numeral;                      /* counts the number of constituents */
s ← transformation
j ← 1
while j ⩽ m do
    if (A_j)^{ς_j} is a formula-premise of a classical non-branching rule r then
        Δ ← the sequence of the indexed formula-conclusions of r
        Γ ← ⟨(A_1)^{ς_1}, ..., (A_{j-1})^{ς_{j-1}}⟩ ′ Δ ′ ⟨(A_{j+1})^{ς_{j+1}}, ..., (A_m)^{ς_m}⟩
        ψ ← sequent-conclusion of r
        Q ← ?(Φ(ψ))
        s ← s ; ⟨Q⟩
    else
        if (A_j)^{ς_j} is a formula-premise of a classical branching rule r then
            Δ ← sequence with the left indexed formula-conclusion of r
            Γ ← ⟨(A_1)^{ς_1}, ..., (A_{j-1})^{ς_{j-1}}⟩ ′ Δ ′ ⟨(A_{j+1})^{ς_{j+1}}, ..., (A_m)^{ς_m}⟩
            ψ_1, ψ_2 ← the left and right sequent-conclusions of r
            Q ← ?(Φ(ψ_1, ψ_2))
            s ← s ; ⟨Q⟩
            n ← n + 1
        else
            if j-th term of Γ is of the form (⊛B)^{ς/i} and is unlabeled then
                ⊛B ← ⊛B_{ref} ;              /* bottom label added */
                Γ ← ⟨(A_1)^{ς_1}..(A_{j-1})^{ς_{j-1}}, (⊛B)^{ς/i}, B^i, (A_{j+1})^{ς_{j+1}}..(A_m)^{ς_m}⟩
                ψ ← sequent-conclusion of r ∈ {L^E_{ref}, R^E_{ref}}
                Q ← ?(Φ(ψ))
                s ← s ; ⟨Q⟩
                j ← j + 1
            else
                j ← j + 1
            end
        end
    end
end
```

Definition 7.9 Let $i \in \mathbb{N}$.

$$\tau_{\mathsf{T}}(i, \varnothing) = \varnothing$$

$$\tau_{\mathsf{T}}(i, \langle A_1^{\varsigma_1}, \ldots, A_k^{\varsigma_k}\rangle) = \begin{cases} \langle A_1\rangle \,'\, \tau_{\mathsf{T}}(i, \langle A_2^{\varsigma_2}, \ldots, A_k^{\varsigma_k}\rangle) & \text{if } \varsigma_1 \text{ ends with } i \\ & \text{and } A_1 \text{ is unlabeled} \\ \tau_{\mathsf{T}}(i, \langle A_2^{\varsigma_2}, \ldots, A_k^{\varsigma_k}\rangle) & \text{otherwise} \end{cases}$$

Similarly, Definition 7.6 of relabelling function $\tau_{\mathbf{s}}$ must be altered—we will use function $\tau_{\mathsf{T}\mathbf{s}}$ instead of $\tau_{\mathbf{s}}$.

Definition 7.10 Let $\mathbf{s}$ be a regular Socratic transformation of a question of the form '$?(\vdash (A)^0)$' in $\mathbb{E}^{\mathsf{T}\tau} \cup \{L^{\mathbb{E}}_{\Diamond\Box}, R^{\mathbb{E}}_{\Box\Diamond}\}$, and let $\mathbf{Tr}(\mathbf{s})$ be the Socratic tree determined by $\mathbf{s}$. Then $\tau_{\mathsf{T}\mathbf{s}}$ is a relabelling function defined recursively as follows:

1. If ϕ labels the root of $\mathbf{Tr}(\mathbf{s})$, then $\tau_{\mathsf{T}\mathbf{s}}(\phi) = \tau_{\mathsf{T}}(0, \phi)$.
2. If ϕ labels a node, ψ labels the immediate $R_{\mathbf{s}}$-predecessor of this node, ϕ results from ψ in $\mathbf{s}$ by a classical rule, and i is the numeral such that $\tau_{\mathsf{T}\mathbf{s}}(\psi) = \tau_{\mathsf{T}}(i, \psi)$, then $\tau_{\mathsf{T}\mathbf{s}}(\phi) = \tau_{\mathsf{T}}(i, \phi)$. **(conservative extension)**
3. If ϕ labels a node, ψ labels the immediate $R_{\mathbf{s}}$-predecessor of this node, ϕ results from ψ in $\mathbf{s}$ by $L^{\mathbb{E}}_{ref}$ or $R^{\mathbb{E}}_{ref}$, and i is the numeral such that $\tau_{\mathsf{T}\mathbf{s}}(\psi) = \tau_{\mathsf{T}}(i, \psi)$, then $\tau_{\mathsf{T}\mathbf{s}}(\phi) = \tau_{\mathsf{T}}(i, \phi)$. **(conservative extension reflectively continued)**
4. If ϕ labels a node and ϕ results in $\mathbf{s}$ from its immediate $R_{\mathbf{s}}$-predecessor by a modal rule other than $L^{\mathbb{E}}_{ref}$, $R^{\mathbb{E}}_{ref}$, then $\tau_{\mathsf{T}\mathbf{s}}(\phi) = \tau_{\mathsf{T}}(j, \phi)$, where j is the principal numeral in ϕ. **(radical change)**

As a matter of fact, clause 3, **(conservative extension reflectively continued)** is a special case of 4, **(radical change)** (believe me, it happens!), as the principal numeral in the case of application of $L^{\mathbb{E}}_{ref}$, $R^{\mathbb{E}}_{ref}$ is i.

Example 7.11 The following example shows how the regular Socratic proof of the sequent '$\vdash (\Box p \to p)^0$' (the specific axiom of T) in $\mathbb{E}^{\mathsf{T}\tau} \cup \{L^{\mathbb{E}}_{\Diamond\Box}, R^{\mathbb{E}}_{\Box\Diamond}\}$ is translated into a proof of '$\vdash \Box p \to p$' in $\mathbb{G}^{\mathsf{T}\tau} \cup \{Ax_{+W}, L^{\mathbb{G}}_{\Diamond+W}, R^{\mathbb{G}}_{\Box+W}\}$. The resulting Socratic proof has only one branch, and the world line has only one element: 0. We can see how, by the use of the bottom label, one occurrence of '$\Box p$' is saved in this transformation and the other lost, as required.

$$\cfrac{\cfrac{?(\vdash (\Box p \to p)^0)}{?((\Box p)^0 \vdash p^0)}\, R^{\mathbb{E}}_{\to}}{?((\Box p)^0_{ref}, p^0 \vdash p^0)}\, L^{\mathbb{E}}_{ref}$$

$$\Downarrow$$

$$\cfrac{\cfrac{\tau_{\mathsf{T}}(0, (\Box p)^0_{ref}, p^0 \vdash p^0)}{\tau_{\mathsf{T}}(0, (\Box p)^0 \vdash p^0)}}{\tau_{\mathsf{T}}(0, \vdash (\Box p \to p)^0)} \quad \Longrightarrow \quad \cfrac{\cfrac{p \vdash p}{\Box p \vdash p}\, L^{\mathbb{G}}_{ref}}{\vdash \Box p \to p}\, R^{\mathbb{G}}_{\to}$$

As previously, we establish the following:

Theorem 7.3 *Let* $\mathbf{s}$ *be a regular Socratic proof of the sequent* '$\vdash (A)^0$' *in calculus* $\mathbb{E}^{\mathsf{T}\tau} \cup \{L^{\mathbb{E}}_{\Diamond\Box}, R^{\mathbb{E}}_{\Box\Diamond}\}$. *If* $\langle X, R\rangle$ *is a tree obtained from* $\mathbf{s}$ *by the Procedure PRUNE, then* $\langle X, R, \tau_{\mathsf{T}\mathbf{s}}|_X\rangle$ *is a proof of* '$\vdash A$' *in calculus* $\mathbb{G}^{\mathsf{T}\tau} \cup \{Ax_{+W}, L^{\mathbb{G}}_{\Diamond+W}, R^{\mathbb{G}}_{\Box+W}\}$.

In the present case, clause *3* of Lemma 7.1 remains the same, but we need to modify clause *1*.

Lemma 7.3 *Let* **s** *stand for a regular Socratic proof of the sequent '$\vdash (A)^{0}$' in calculus $\mathbb{E}^{\mathsf{T}\tau} \cup \{L^{\mathbb{E}}_{\Diamond\Box}, R^{\mathbb{E}}_{\Box\Diamond}\}$, and let $\mathcal{T} = \langle X, R, \tau_{\mathsf{Ts}}\,|_X\rangle$ be the tree obtained from* **Tr(s)** *by procedure PRUNE.*

1. *If a sequent ψ labels a node of $\mathcal{T}$ that immediately precedes exactly one node labeled by ϕ, and the two sequents have the same principal numeral, then the following:*

$$\frac{\tau_{\mathsf{Ts}}(\phi)}{\tau_{\mathsf{Ts}}(\psi)}$$

is either an instance of a classical non-branching rule of $\mathbb{G}^{\mathsf{T}\tau}$ or an instance of one of $L^{\mathbb{G}}_{ref}$, $R^{\mathbb{G}}_{ref}$.

2. *and 3 as in Lemma 7.1.*

Proof If ϕ results in **Tr(s)** from its predecessor by one of the modal rules: $L^{\mathbb{E}}_{ref}$, $R^{\mathbb{E}}_{ref}$, then the active occurrence of modal formula '$(\Box A)^{\varsigma}$' (respectively, '$(\Diamond A)^{\varsigma}$') has no bottom label yet; otherwise the rule would not have been applied—see line 21 of Procedure SETTLE-T. Therefore the occurrence of '$\Box A$' (of '$\Diamond A$') is not lost under τ_{T} in sequent $\tau_{\mathsf{Ts}}(\psi)$, but it is lost in $\tau_{\mathsf{Ts}}(\phi)$, where it already has the bottom index. For this reason $\tau_{\mathsf{Ts}}(\phi)/\tau_{\mathsf{Ts}}(\psi)$ is an instance of $L^{\mathbb{G}}_{ref}$ ($R^{\mathbb{G}}_{ref}$, respectively). □

Translation Procedure for K4

It is time to deal with the transitive logics. In the original erotetic calculi the difference between K and K4 is in the proviso of applicability of R_{π}. Let us recall: A^{j} may be introduced (from world i) for any pair $\langle i, j\rangle$ that belongs to the transitive closure of the set of pairs of numerals in the indices. The rules of $\mathbb{E}^{\mathsf{K4}\tau}$ are the classical rules of $\mathbb{E}^{\mathsf{K}\tau}$, modal rules $L^{\mathbb{E}}_{\Diamond}$ and $R^{\mathbb{E}}_{\Box}$ (as in the case of $\mathbb{E}^{\mathsf{K}\tau}$), and rules $L^{\mathbb{E}}_{\Box}$, $R^{\mathbb{E}}_{\Diamond}$ with the suitable proviso. In order to generalise the modal rules we need to modify (again) the definitions of operations $\sharp^{i}_{j}$ and $\flat^{i}_{j}$—see Algorithms 7.8 and 7.9.

Algorithm 7.8: Operation $\sharp_j$ for transitive logics

Data: finite sequence $\Gamma = \langle (A_1)^{\varsigma_1}, \ldots, (A_m)^{\varsigma_m}\rangle$ of indexed formulas,
set $R(\phi)$ of pairs of numerals
Result: finite sequence $\Gamma\sharp_j$ of indexed formulas

$\Gamma\sharp_j \leftarrow \varnothing$
for $k = 1$ **to** m **do**
  if *k-th term of Γ is of the form $(\Box B)^{\varsigma/i}$ where $\langle i, j\rangle \in R(\phi)^{tr}$* **then**
    $\Gamma\sharp_j \leftarrow \Gamma\sharp_j\,'\,\langle(\Box B)^{\varsigma/i}, (B)^{j}\rangle$
  else
    $\Gamma\sharp_j \leftarrow \Gamma\sharp_j\,'\,\langle(A_k)^{\varsigma_k}\rangle$
  end
end
return $\Gamma\sharp_j$

Here is the generalisation of modal rules for the case of K4.

Algorithm 7.9: Operation $\flat_j$ for transitive logics

Data: finite sequence $\Delta = \langle (A_1)^{\varsigma_1}, \ldots, (A_m)^{\varsigma_m} \rangle$ of indexed formulas, set $R(\phi)$ of pairs of numerals
Result: finite sequence $\Delta\flat_j$ of indexed formulas

```
Δ♭_j ← ∅
for k = 1 to m do
    if k-th term of Δ is of the form (◇B)^{ς/i} where ⟨i, j⟩ ∈ R(φ)^tr then
        Δ♭_j ← Δ♭_j ' ⟨(◇B)^{ς/i}, (B)^j⟩
    else
        Δ♭_j ← Δ♭_j ' ⟨(A_k)^{ς_k}⟩
    end
end
return Δ♭_j
```

$$\frac{?(\ \Phi\ (\ \Gamma_1\ '\ (\Diamond A)^{\varsigma/i}\ '\ \Gamma_2\ \vdash\ \Delta\ )\ )}{?(\ \Phi\ (\ \Gamma_1\sharp_j\ '\ A^{\,\varsigma/i,j}\ '\ \Gamma_2\sharp_j\ \vdash\ \Delta\flat_j\ )\ )}\ L^{\mathbb{E}}_{tr}$$

$$\frac{?(\ \Phi\ (\ \Gamma\ \vdash\ \Delta_1\ '\ (\Box A)^{\varsigma/i}\ '\ \Delta_2\ )\ )}{?(\ \Phi\ (\ \Gamma\sharp_j\ \vdash\ \Delta_1\flat_j\ '\ A^{\,\varsigma/i,j}\ '\ \Delta_2\flat_j\ )\ )}\ R^{\mathbb{E}}_{tr}$$

This time we state without proof:

Corollary 7.8 *Rules $L^{\mathbb{E}}_{tr}$ and $R^{\mathbb{E}}_{tr}$ are derivable in $\mathbb{E}^{\mathsf{K4}\tau}$.*

For the sake of defining regular Socratic transformations in $\mathbb{E}^{\mathsf{K4}\tau}$ we modify only Procedure MODAL: in line 5 we use $\mathbf{r} \in \{L^{\mathbb{E}}_{tr}, R^{\mathbb{E}}_{tr}\}$ instead of $\mathbf{r} \in \{L^{\mathbb{E}}_{\Diamond\Box}, R^{\mathbb{E}}_{\Box\Diamond}\}$—see Sect. 7.1. This modification will be called 'MODAL-K4'.

The main difference between the explicit and implicit (purely syntactical) proof methods for transitive modal logics is that in the latter accounts the information about transitivity is transferred by rewriting formulas of the form '$\Box A$' on the left ('$\Diamond A$' on the right). In explicit formats the same information is encoded in the additional indices/labels. We need to modify function τ to the effect of expressing the information about transitivity in the syntactical way. Unfortunately, the function must act in a different way on the left and right side (see Definition 7.11).

Definition 7.11 Let $i \in \mathbb{N}$ and let ϕ be a sequent of $\mathcal{M}^{?}_{\vdash}$. We define functions τ_{4L} and τ_{4R} as follows:

$$\tau_{4L}(\phi, i, \varnothing) = \varnothing$$

$$\tau_{4L}(\phi, i, \langle A_1^{\varsigma_1}, \ldots, A_k^{\varsigma_k} \rangle) =$$

$$= \begin{cases} \langle A_1 \rangle\ '\ \tau_{4L}(\phi, i, \langle A_2^{\varsigma_2}, \ldots, A_k^{\varsigma_k} \rangle) & \text{if } \varsigma_1 \text{ ends with } i, \textbf{ or } A_1 = \Box B \text{ and} \\ & \varsigma_1 \text{ ends with } k \text{ such that } \langle k, i \rangle \in R^{tr}(\phi) \\ \tau_{4L}(\phi, i, \langle A_2^{\varsigma_2}, \ldots, A_k^{\varsigma_k} \rangle) & \text{otherwise} \end{cases}$$

and

$$\tau_{4R}(\phi, i, \varnothing) = \varnothing$$

$$\tau_{4R}(\phi, i, \langle A_1^{\varsigma_1}, \ldots, A_k^{\varsigma_k}\rangle) =$$

$$= \begin{cases} \langle A_1\rangle \,'\, \tau_{4R}(\phi, i, \langle A_2^{\varsigma_2}, \ldots, A_k^{\varsigma_k}\rangle) & \text{if } \varsigma_1 \text{ ends with } i, \textbf{ or } A_1 = \Diamond B \text{ and} \\ & \varsigma_1 \text{ ends with } k \text{ such that } \langle k, i\rangle \in R^{tr}(\phi) \\ \tau_{4R}(\phi, i, \langle A_2^{\varsigma_2}, \ldots, A_k^{\varsigma_k}\rangle) & \text{otherwise} \end{cases}$$

We join the two functions in the following definition:

Definition 7.12 Let $i \in \mathbb{N}$ and let $\phi = \Gamma \vdash \Delta$ be a sequent of $\mathcal{M}_{\vdash}^{?}$. Then:

$$\tau_4(i, \Gamma \vdash \Delta) = \tau_{4L}(\phi, i, \Gamma) \vdash \tau_{4R}(\phi, i, \Delta)$$

Let us present:

Example 7.12 A regular Socratic proof of '$\vdash (\Box p \to \Box\Box p)^0$' in $\mathbb{E}^{\mathsf{K4}\tau} \cup \{L_{tr}^{\mathbb{E}}, R_{tr}^{\mathbb{E}}\}$ translated into a proof of '$\vdash \Box p \to \Box\Box p$' in calculus $\mathbb{G}^{\mathsf{K4}\tau} \cup \{Ax_{+W}, L_{tr+W}^{\mathbb{G}}, R_{tr+W}^{\mathbb{G}}\}$. The Socratic tree has one branch with world line $\langle 0, 1, 2\rangle$.

$$\dfrac{\dfrac{\dfrac{?(\vdash (\Box p \to \Box\Box p)^0)}{?((\Box p)^0 \vdash (\Box\Box p)^0)}\, R_{\to}^{\mathbb{E}}}{?((\Box p)^0, p^1 \vdash (\Box p)^{0,1})}\, R_{tr}^{\mathbb{E}}}{?((\Box p)^0, p^2, p^1 \vdash p^{0,1,2})}\, R_{tr}^{\mathbb{E}} \quad \Longrightarrow \quad \dfrac{\dfrac{\dfrac{\tau_4(2, (\Box p)^0, p^2, p^1 \vdash p^{0,1,2})}{\tau_4(1, (\Box p)^0, p^1 \vdash (\Box p)^{0,1})}}{\tau_4(0, (\Box p)^0 \vdash (\Box\Box p)^0)}}{\tau_4(0, \vdash (\Box p \to \Box\Box p)^0)}$$

$$\Downarrow$$

$$\dfrac{\dfrac{\dfrac{\Box p, p \vdash p}{\Box p, p \vdash \Box p}}{\Box p \vdash \Box\Box p}}{\vdash \Box p \to \Box\Box p}$$

Similarly as before, a modification *modulo* weakening is needed in order to obtain a proof in $\mathbb{G}^{\mathsf{K4}\tau}$, though here the situation is less transparent. The transition:

$$\dfrac{\Box p, p \vdash \Box p}{\Box p \vdash \Box\Box p}$$

(in the middle of the proof of $\vdash \Box p \to \Box\Box p$) is an instance of $R_{tr}^{\mathbb{G}}$:

$$\dfrac{(\Box S)\sharp \ \vdash \ (\Diamond T_1)\flat \,'\, A \,'\, (\Diamond T_2)\flat}{\Box S \ \vdash \ \Diamond T_1 \,'\, \Box A \,'\, \Diamond T_2}\, R_{tr}^{\mathbb{G}}$$

with $A = \Box p, \Box A = \Box\Box p, \Box S = \langle \Box p\rangle$ in the conclusion and $(\Box S)\sharp = \langle \Box p, p\rangle$ in the premise. The top transition:

$$\frac{\Box p,\ p \vdash p}{\Box p,\ p \vdash \Box p}$$

has weakening hidden, since from $\Box p, p \vdash p$ we can derive $\Box p \vdash \Box p$ by $R^{\mathbb{G}}_{tr}$ ($A = p$, $\Box A = \Box p$, $(\Box S)\sharp = \langle \Box p, p\rangle$, $\Box S = \langle \Box p\rangle$), and then we can move to $\Box p, p \vdash \Box p$ by weakening. The whole proof in $\mathbb{G}^{\mathsf{K4}\tau}$ should go as follows:

$$\cfrac{\cfrac{\cfrac{\cfrac{\cfrac{p \vdash p}{\Box p,\ p \vdash p}\ L^{\mathbb{G}}_{W}}{\Box p \vdash \Box p}\ R^{\mathbb{G}}_{tr}}{\Box p,\ p \vdash \Box p}\ L^{\mathbb{G}}_{W}}{\Box p \vdash \Box\Box p}\ R^{\mathbb{G}}_{tr}}{\vdash \Box p \to \Box\Box p}\ R^{\mathbb{G}}_{\to}$$

We can see clearly that the first steps are redundant (we could start with $\Box p \vdash \Box p$) but redundancy seems inevitable in such translations. To sum up, the above example indicates that, as previously, we need to add weakening to the modal rules.

$$\frac{(\Box S_1)\sharp\ '\ A\ '\ (\Box S_2)\sharp\ \vdash\ (\Diamond T)\flat}{S_1^{*}\ '\ \Diamond A\ '\ S_2^{*}\ \vdash\ T^{*}}\ L^{\mathbb{G}}_{tr+W}$$

$$\frac{(\Box S)\sharp\ \vdash\ (\Diamond T_1)\flat\ '\ A\ '\ (\Diamond T_2)\flat}{S^{*}\ \vdash\ T_1^{*}\ '\ \Box A\ '\ T_2^{*}}\ R^{\mathbb{G}}_{tr+W}$$

where in $L^{\mathbb{G}}_{tr+W}$: 'S_1^*' is a supersequence of '$\Box S_1$', 'S_2^*' is a supersequence of '$\Box S_2$', and 'T^*' is a supersequence of '$\Diamond T$', and at least in one case the supersequence is proper; similarly, in $R^{\mathbb{G}}_{tr+W}$: 'S^*' is a supersequence of '$\Box S$', 'T_1^*' is a supersequence of '$\Diamond T_1$', and 'T_2^*' is a supersequence of '$\Diamond T_2$', and at least in one case the supersequence is proper.

Function $\tau_{4\mathbf{s}}$ is defined as function $\tau_{\mathbf{s}}$ in Definition 7.6 but with τ_4 playing the role of τ.

Definition 7.13 Let **s** be a regular Socratic transformation of a question of the form '$?(\vdash (A)^0)$' in $\mathbb{E}^{\mathsf{K4}\tau} \cup \{L^{\mathbb{E}}_{tr}, R^{\mathbb{E}}_{tr}\}$, and let $\mathbf{Tr}(\mathbf{s})$ be the Socratic tree determined by **s**. Then $\tau_{4\mathbf{s}}$ is a relabelling function defined recursively as follows:

1. If ϕ labels the root of $\mathbf{Tr}(\mathbf{s})$, then $\tau_{4\mathbf{s}}(\phi) = \tau_4(0, \phi)$.
2. If ϕ labels a node, ψ labels the immediate $R_{\mathbf{s}}$-predecessor of this node, ϕ results from ψ in **s** by a classical rule, and i is the numeral such that $\tau_{4\mathbf{s}}(\psi) = \tau_4(i, \psi)$, then $\tau_{4\mathbf{s}}(\phi) = \tau_4(i, \phi)$. **(conservative extension)**
3. If ϕ is a label of a node that results in **s** from its immediate $R_{\mathbf{s}}$-predecessor by a modal rule, then $\tau_{4\mathbf{s}}(\phi) = \tau_4(j, \phi)$, where j is the principal numeral in ϕ. **(radical change)**

For simplicity, in the sequel we will say that *numeral i can see numeral j in a sequent ϕ* iff the pair $\langle i, j\rangle$ belongs to the transitive closure of the set of pairs $R(\phi)$. Lemma 7.1 is modified in clause *3* as follows.

Lemma 7.4 *Let* **s** *stand for a regular Socratic proof of* '$\vdash (A)^0$' *in calculus* $\mathbb{E}^{\mathsf{K4}\tau} \cup \{L^{\mathbb{E}}_{tr}, R^{\mathbb{E}}_{tr}\}$, *and let* $\mathcal{T} = \langle X, R, \tau_{4\mathbf{s}} \mid_X \rangle$ *be the tree obtained from* **Tr(s)** *by procedure PRUNE.*

1. *and 2 as in Lemma 7.1.*
3. *If a sequent* ψ *labels a node of* $\mathcal{T}$ *that immediately precedes exactly one node labeled by* ϕ, *and the two sequents have different principal numerals, then the following:*

$$\frac{\tau_{4\mathbf{s}}(\phi)}{\tau_{4\mathbf{s}}(\psi)}$$

is an instance of one of the rules: $L^{\mathbb{G}}_{tr}$, $R^{\mathbb{G}}_{tr}$, $L^{\mathbb{G}}_{tr+W}$, *or* $R^{\mathbb{G}}_{tr+W}$.

Proof As previously, the proof is almost a repetition of the proof of Lemma 7.1. We observe that **s** is regular and that $\mathcal{T}$ has been pruned, and for this reason the two sequents: ϕ and ψ share the same modalised formulas (modalised with '$\Box$' on the left side of the turnstile, and with '$\Diamond$' on the right side) that live in worlds which can see the principal j. If there are any other formulas in $\tau_{4\mathbf{s}}(\psi)$ that do not fit the scheme of conclusion, then $\tau_{4\mathbf{s}}(\phi)/\tau_{4\mathbf{s}}(\psi)$ is an instance of $L^{\mathbb{G}}_{tr+W}$ or $R^{\mathbb{G}}_{tr+W}$. Otherwise it is an instance of $L^{\mathbb{G}}_{tr}$ or $R^{\mathbb{G}}_{tr}$. $\Box$

Lemma 7.4 yields that the following theorem is true.

Theorem 7.4 *Let* **s** *be a regular Socratic proof of the sequent* '$\vdash (A)^0$' *in calculus* $\mathbb{E}^{\mathsf{K4}\tau} \cup \{L^{\mathbb{E}}_{tr}, R^{\mathbb{E}}_{tr}\}$. *If* $\langle X, R \rangle$ *is a tree obtained from* **s** *by the procedure PRUNE, then* $\langle X, R, \tau_{4\mathbf{s}} \mid_X \rangle$ *is a proof of* A *in calculus* $\mathbb{G}^{\mathsf{K4}\tau} \cup \{Ax_{+W}, L^{\mathbb{G}}_{tr+W}, R^{\mathbb{G}}_{tr+W}\}$.

Translation Procedure for D4

Erotetic calculus $\mathbb{E}^{\mathsf{D4}\tau}$ for logic D4 contains the rules for the classical connectives, modal rules $L^{\mathbb{E}}_{\Diamond}$, $R^{\mathbb{E}}_{\Box}$ of calculus $\mathbb{E}^{\mathsf{K}\tau}$, modal rules $L^{\mathbb{E}}_{\mathsf{D}}$ and $R^{\mathbb{E}}_{\mathsf{D}}$ for seriality, and rules $L^{\mathbb{E}}_{\Box}$, $R^{\mathbb{E}}_{\Diamond}$ with the proviso of applicability for transitive logics. As previously, we add the generalised versions. In the following rules, that capture seriality, numeral j is new with respect to the active sequent:

$$\frac{?(\ \Phi\ (\ \Gamma_1\ '\ (\Box A)^{\varsigma/i}\ '\ \Gamma_2 \vdash \Delta\ ))}{?(\ \Phi\ (\ \Gamma_1\sharp_j\ '\ (\Box A)^{\varsigma/i}\ '\ A^{i,j}\ '\ \Gamma_2\sharp_j \vdash \Delta\flat_j\ ))}\ L^{\mathbb{E}}_{\mathsf{D}g}$$

$$\frac{?(\ \Phi\ (\ \Gamma \vdash \Delta_1\ '\ (\Diamond A)^{\varsigma/i}\ '\ \Delta_2\ ))}{?(\ \Phi\ (\ \Gamma\sharp_j \vdash \Delta_1\flat_j\ '\ (\Diamond A)^{\varsigma/i}\ '\ A^{i,j}\ '\ \Delta_2\flat_j\ ))}\ R^{\mathbb{E}}_{\mathsf{D}g2}$$

where $\sharp_j$ and $\flat_j$ are defined as for the transitive logics (see Algorithm 7.8), so they gather all the information from the worlds that can see the new j. (Let us also recall that we have already introduced rule $R^{\mathbb{E}}_{\mathsf{D}g}$ for logic D, for this reason the above rule is called '$R^{\mathbb{E}}_{\mathsf{D}g2}$'.)

As previously, we state without a proof:

Corollary 7.9 *Rules* $L^{\mathbb{E}}_{\mathsf{D}g}$, $R^{\mathbb{E}}_{\mathsf{D}g2}$ *are derivable in* $\mathbb{E}^{\mathsf{D4}\tau}$.

Corollary 7.10 *Rules $L^{\mathbb{E}}_{tr}$ and $R^{\mathbb{E}}_{tr}$ are derivable in $\mathbb{E}^{\mathsf{D4}\tau}$.*

For the case of defining regular Socratic transformations constructed *via* the rules of $\mathbb{E}^{\mathsf{D4}\tau}$ we use Procedure MODAL-D, with the following modifications:

- line 6 of the procedure: '$\mathbf{r} \in \{L^{\mathbb{E}}_{tr}, R^{\mathbb{E}}_{tr}\}$' instead of '$\mathbf{r} \in \{L^{\mathbb{E}}_{\Diamond\Box}, R^{\mathbb{E}}_{\Box\Diamond}\}$',
- line 17 of the procedure: '$R^{\mathbb{E}}_{\mathsf{D}g2}$' instead of '$R^{\mathbb{E}}_{\mathsf{D}g}$'.

This modification of Procedure MODAL-D will be called 'MODAL-D4'. For translating sequent we use the relabelling function $\tau_{4\mathbf{s}}$ as defined for K4 (see Definition 7.13).

Example 7.13 The regular Socratic proof of '$\vdash (\Box p \to \Diamond\Diamond p)^0$' in $\mathbb{E}^{\mathsf{D4}\tau} \cup \{L^{\mathbb{E}}_{tr}, R^{\mathbb{E}}_{tr}, L^{\mathbb{E}}_{\mathsf{D}g}, R^{\mathbb{E}}_{\mathsf{D}g2}\}$ translated into a proof of '$\vdash \Box p \to \Diamond\Diamond p$' in $\mathbb{G}^{\mathsf{D4}\tau} \cup \{Ax_{+W}, L^{\mathbb{G}}_{tr+W}, R^{\mathbb{G}}_{tr+W}, R^{\mathbb{G}}_{\mathsf{D}tr+W}\}$.

$$\cfrac{\cfrac{\cfrac{?(\vdash (\Box p \to \Diamond\Diamond p)^0)}{?((\Box p)^0 \vdash (\Diamond\Diamond p)^0)}\ R^{\mathbb{E}}_{\to}}{?((\Box p)^0, p^{0,1} \vdash (\Diamond\Diamond p)^0, (\Diamond p)^1)}\ L^{\mathbb{E}}_{\mathsf{D}g}}{?((\Box p)^0, p^2, p^{0,1} \vdash (\Diamond\Diamond p)^0, (\Diamond p)^2, (\Diamond p)^1, p^{1,2})}\ R^{\mathbb{E}}_{\mathsf{D}g2}$$

$$\Downarrow$$

$$\cfrac{\cfrac{\cfrac{\tau_4(2, (\Box p)^0, p^2, p^{0,1} \vdash (\Diamond\Diamond p)^0, (\Diamond p)^2, (\Diamond p)^1, p^{1,2})}{\tau_4(1, (\Box p)^0, p^{0,1} \vdash (\Diamond\Diamond p)^0, (\Diamond p)^1)}}{\tau_4(0, (\Box p)^0 \vdash (\Diamond\Diamond p)^0)}}{\tau_4(0, \vdash (\Box p \to \Diamond\Diamond p)^0)}$$

$$\Downarrow$$

$$\cfrac{\cfrac{\cfrac{\Box p, p \vdash \Diamond\Diamond p, \Diamond p, \Diamond p, p}{\Box p, p \vdash \Diamond\Diamond p, \Diamond p}}{\Box p \vdash \Diamond\Diamond p}}{\vdash \Box p \to \Diamond\Diamond p}$$

We present some calculations to explain the above example. We indicate the first argument of functions τ_{4L}, τ_{4R} by 'ϕ', as the sequent is clear from the context. We have:

$\tau_{4L}(\phi, 1, \langle(\Box p)^0, p^{0,1}\rangle) = \langle\Box p, p\rangle$, because 0 can see 1 in the sequent, similarly: $\tau_{4R}(\phi, 1, \langle(\Diamond\Diamond p)^0, (\Diamond p)^1\rangle) = \langle\Diamond\Diamond p, \Diamond p\rangle$

$\tau_{4L}(\phi, 2, \langle(\Box p)^0, p^2, p^{0,1}\rangle) = \langle\Box p, p\rangle$, as 0 can see 2 in the sequent. The occurrence of 'p' in 2 is left, and the other is lost.

$\tau_{4R}(\phi, 2, \langle(\Diamond\Diamond p)^0, (\Diamond p)^2, (\Diamond p)^1, p^{1,2}\rangle) = \langle\Diamond\Diamond p, \Diamond p, \Diamond p, p\rangle$, since 0 and 1 can see 2.

The final tree is not a proof in $\mathbb{G}^{\mathsf{D4}\tau}$, but we can see how to 'fix it' by weakening.

$$\dfrac{\dfrac{\dfrac{\dfrac{\dfrac{\dfrac{p \vdash p}{\Box p, p \vdash p}\,L^{\mathbb{G}}_{W}}{\Box p, p \vdash \Diamond\Diamond p, \Diamond p, \Diamond p, p}\,R^{\mathbb{G}}_{W}}{\Box p \vdash \Diamond\Diamond p, \Diamond p}\,R^{\mathbb{G}}_{\mathsf{D}tr}}{\Box p, p \vdash \Diamond\Diamond p, \Diamond p}\,L^{\mathbb{G}}_{W}}{\Box p \vdash \Diamond\Diamond p}\,R^{\mathbb{G}}_{\mathsf{D}tr}}{\vdash \Box p \to \Diamond\Diamond p}\,R^{\mathbb{G}}_{\to}$$

Therefore the only missing element in the picture is the following derived rule of $\mathbb{G}^{\mathsf{D4}\tau}$:

$$\frac{(\Box S)\sharp \;\vdash\; (\Diamond T)\flat}{S^{*} \;\vdash\; T^{*}}\; R^{\mathbb{G}}_{\mathsf{D}tr+W}$$

where 'S^*' is a supersequence of '$\Box S$', 'T^*' is a supersequence of '$\Diamond T$', and at least one of the supersequences is proper.

We are ready to conclude:

Theorem 7.5 *Let* **s** *be a regular Socratic proof of the sequent* '$\vdash (A)^{0}$' *in calculus* $\mathbb{E}^{\mathsf{D4}\tau} \cup \{L^{\mathbb{E}}_{tr}, R^{\mathbb{E}}_{tr}, L^{\mathbb{E}}_{\mathsf{D}g}, R^{\mathbb{E}}_{\mathsf{D}g2}\}$. *If* $\langle X, R\rangle$ *is a tree obtained from* **s** *by the procedure PRUNE, then* $\langle X, R, \tau_{4\mathbf{s}}\,|_X\rangle$ *is a proof of* '$\vdash A$' *in calculus* $\mathbb{G}^{\mathsf{D4}\tau} \cup \{Ax_{+W}, L^{\mathbb{G}}_{tr+W}, R^{\mathbb{G}}_{tr+W}, R^{\mathbb{G}}_{\mathsf{D}tr+W}\}$.

Lemma 7.1 is modified in clause *3* as follows.

Lemma 7.5 *Let* **s** *stand for a regular Socratic transformation in calculus* $\mathbb{E}^{\mathsf{D4}\tau} \cup \{L^{\mathbb{E}}_{tr}, R^{\mathbb{E}}_{tr}, L^{\mathbb{E}}_{\mathsf{D}g}, R^{\mathbb{E}}_{\mathsf{D}g2}\}$, *and let* $\mathcal{T} = \langle X, R, \tau_{4\mathbf{s}}\,|_X\rangle$ *be the tree obtained from* **Tr(s)** *by procedure PRUNE.*

1. *and 2 as in Lemma 7.1.*
3. *If a sequent* ψ *labels a node of* $\mathcal{T}$ *that immediately precedes exactly one node labeled by* ϕ, *and the two sequents have different principal numerals, then the following:*

$$\frac{\tau_{4\mathbf{s}}(\phi)}{\tau_{4\mathbf{s}}(\psi)}$$

is an instance of one of: $L^{\mathbb{G}}_{tr}, R^{\mathbb{G}}_{tr}, L^{\mathbb{G}}_{tr+W}, R^{\mathbb{G}}_{tr+W}, R^{\mathbb{G}}_{\mathsf{D}tr}, R^{\mathbb{G}}_{\mathsf{D}tr+W}$.

Proof As before, the classical cases listed in Lemma 7.1, the justification of the fact that a modal rule has been applied 'between' ψ and ϕ, and the reasons for **Additional Assumption**—they all run as in Lemma 7.1. Further, if the modal rule was, e.g., $L^{\mathbb{E}}_{tr}$, then $\psi = \Gamma_1 \sharp_j\,'\,A^{\varsigma/i,j}\,'\,\Gamma_2 \sharp_j \vdash \Delta\flat_j$, and either $\phi = \Gamma_1\,'\,(\Diamond A)^{\varsigma/i}\,'\,\Gamma_2 \vdash \Delta$ or it has a proper supersequence of $\Gamma_1\,'\,(\Diamond A)^{\varsigma/i}\,'\,\Gamma_2$ on the left and/or a proper supersequence of Δ on the right. We compare the definitions of $\sharp_j$ and $\sharp$ (and so on for $\flat$) and arrive at the conclusion that $\tau_{4\mathbf{s}}(\phi)/\tau_{4\mathbf{s}}(\psi)$ is an instance of $L^{\mathbb{G}}_{tr}$ or $L^{\mathbb{G}}_{tr+W}$. Similarly for the other cases. □

Translation Procedure for S4

Calculus $\mathbb{E}^{\mathsf{S4}\tau}$ for S4 contains the rules for the classical connectives and $L_{\Diamond}^{\mathbb{E}}$, $R_{\Box}^{\mathbb{E}}$ of calculus $\mathbb{E}^{\mathsf{K}\tau}$, rules $L_{ref}^{\mathbb{E}}$ and $R_{ref}^{\mathbb{E}}$ for reflexivity and rules $L_{\Box}^{\mathbb{E}}$, $R_{\Diamond}^{\mathbb{E}}$ with the proviso of applicability for transitive logics. Next, we add—as derived rules—the generalised versions of transitivity rules, that is, $L_{tr}^{\mathbb{E}}$ and $R_{tr}^{\mathbb{E}}$. The proof of their derivability in $\mathbb{E}^{\mathsf{S4}\tau}$ is as before, we skip it.

Corollary 7.11 *Rules $L_{tr}^{\mathbb{E}}$ and $R_{tr}^{\mathbb{E}}$ are derivable in $\mathbb{E}^{\mathsf{S4}\tau}$.*

The notion of regular Socratic transformation is defined in line with Algorithm 7.3 with reference to Procedures SETTLE-T (instead of CLASSICAL) and MODAL-K4 (instead of MODAL). Let us recall that Procedure SETTLE-T makes use of bottom labels, marking a modal formula after an application of the rules for reflexivity. In the case of logic T, the bottom label played two roles: preventing from redundant applications of the reflexivity rules (see the condition expressed in line 21 of Procedure SETTLE-T), and distinguishing between the occurrences of modal formulas that must be saved, and these that must be lost under the translation function (see Definition 7.9). The situation is different in the case of S4. We use Procedure SETTLE-T, and we want the bottom labels to play their preventing role. On the other hand, the relevant occurrences of modal formulas will not disappear during the translation, since now they are needed to encode the property of transitivity in the syntactic manner.

For this reason, we assume that after generating a regular Socratic proof by Algorithm 7.3 using Procedure SETTLE-T instead of CLASSICAL, and Procedure MODAL-K4 instead of MODAL, all bottom labels (if there are any) are removed from the Socratic proof. The translation function to work for S4 will be τ_4, thus we assume that it takes as the second argument a sequent that contains no indexed formulas with bottom labels.

The final relabelling function, called $\tau_{\mathsf{S4}\mathbf{s}}$, is defined as in the case of T (function $\tau_{\mathsf{T}\mathbf{s}}$, Definition 7.10), but when the numerals are settled, then function τ_4 is used (Definition 7.12). Here is the whole definition:

Definition 7.14 Let **s** be a regular Socratic transformation of a question of the form '$?(\vdash (A)^0)$' in $\mathbb{E}^{\mathsf{S4}\tau} \cup \{L_{tr}^{\mathbb{E}}, R_{tr}^{\mathbb{E}}\}$ and let $\mathbf{Tr}(\mathbf{s})$ be the Socratic tree determined by **s**. Then $\tau_{\mathsf{S4}\mathbf{s}}$ is a relabelling function defined recursively as follows:

1. If ϕ labels the root of $\mathbf{Tr}(\mathbf{s})$, then $\tau_{\mathsf{S4}\mathbf{s}}(\phi) = \tau_4(0, \phi)$.
2. If ϕ labels a node, ψ labels the immediate $R_{\mathbf{s}}$-predecessor of the node, ϕ results from ψ in **s** by a classical rule or by one of $L_{ref4}^{\mathbb{E}}$, $R_{ref4}^{\mathbb{E}}$, and i is the numeral such that $\tau_{\mathsf{S4}\mathbf{s}}(\psi) = \tau_4(i, \psi)$, then $\tau_{\mathsf{S4}\mathbf{s}}(\phi) = \tau_4(i, \phi)$.
3. If ϕ labels a node and ϕ results in **s** from its immediate $R_{\mathbf{s}}$-predecessor by a modal rule other than $L_{ref}^{\mathbb{E}}$, $R_{ref}^{\mathbb{E}}$, then $\tau_{\mathsf{S4}\mathbf{s}}(\phi) = \tau_4(j, \phi)$, where j is the principal numeral in ϕ.

Example 7.14

$$
\begin{array}{cl}
?(\vdash (\Diamond\Box\Diamond p \rightarrow \Diamond p)^0) & \\ \hline
?((\Diamond\Box\Diamond p)^0 \vdash (\Diamond p)^0) & R^{\mathbb{E}}_{\rightarrow} \\ \hline
?((\Diamond\Box\Diamond p)^0 \vdash (\Diamond p)^0_{ref}, p^0) & R^{\mathbb{E}}_{ref} \\ \hline
?((\Box\Diamond p)^{0,1} \vdash (\Diamond p)^0_{ref}, p^1, p^0) & L^{\mathbb{E}}_{tr} \\ \hline
?((\Box\Diamond p)^{0,1}_{ref}, (\Diamond p)^1 \vdash (\Diamond p)^0_{ref}, p^1, p^0) & L^{\mathbb{E}}_{ref} \\ \hline
?((\Box\Diamond p)^{0,1}_{ref}, (\Diamond p)^2, p^{1,2} \vdash (\Diamond p)^0_{ref}, p^2, p^1, p^0) & L^{\mathbb{E}}_{tr}
\end{array}
$$

Function $\tau_{\mathbf{S4s}}$ assigns numerals to the nodes of the above regular Socratic transformation **s** as follows (the bottom labels disappear):

$$
\begin{array}{c}
\tau_4(2, (\Box\Diamond p)^{0,1}, (\Diamond p)^2, p^{1,2} \vdash (\Diamond p)^0, p^2, p^1, p^0) \\ \hline
\tau_4(1, (\Box\Diamond p)^{0,1}, (\Diamond p)^1 \vdash (\Diamond p)^0, p^1, p^0) \\ \hline
\tau_4(1, (\Box\Diamond p)^{0,1} \vdash (\Diamond p)^0, p^1, p^0) \\ \hline
\tau_4(0, (\Diamond\Box\Diamond p)^0 \vdash (\Diamond p)^0, p^0) \\ \hline
\tau_4(0, (\Diamond\Box\Diamond p)^0 \vdash (\Diamond p)^0) \\ \hline
\tau_4(0, \vdash (\Diamond\Box\Diamond p \rightarrow \Diamond p)^0)
\end{array}
$$

Below to the left we have the resulting tree, to the right the tree supplemented with the missing intermediate nodes. We can see that the 'gaps' may be filled with weakening, so the tree (to the right) is constructed with the use of more general derived rules $L^{\mathbb{G}}_{tr+W}$, $R^{\mathbb{G}}_{tr+W}$.

$$
\begin{array}{c}
\Box\Diamond p, \Diamond p, p \vdash \Diamond p, p \\ \hline
\Box\Diamond p, \Diamond p \vdash \Diamond p, p \\ \hline
\Box\Diamond p \vdash \Diamond p, p \\ \hline
\Diamond\Box\Diamond p \vdash \Diamond p, p \\ \hline
\Diamond\Box\Diamond p \vdash \Diamond p \\ \hline
\vdash \Diamond\Box\Diamond p \rightarrow \Diamond p
\end{array}
\quad \Longrightarrow \quad
\begin{array}{cl}
\Box\Diamond p, \Diamond p, p \vdash \Diamond p, p & \\ \hline
\boxed{\Box\Diamond p, \Diamond p \vdash \Diamond p} & L^{\mathbb{G}}_{tr} \\ \hline
\Box\Diamond p, \Diamond p \vdash \Diamond p, p & R^{\mathbb{G}}_{W} \\ \hline
\Box\Diamond p \vdash \Diamond p, p & L^{\mathbb{G}}_{ref4} \\ \hline
\boxed{\Diamond\Box\Diamond p \vdash \Diamond p} & L^{\mathbb{G}}_{tr} \\ \hline
\Diamond\Box\Diamond p \vdash \Diamond p, p & R^{\mathbb{G}}_{W} \\ \hline
\Diamond\Box\Diamond p \vdash \Diamond p & R^{\mathbb{G}}_{ref4} \\ \hline
\vdash \Diamond\Box\Diamond p \rightarrow \Diamond p & R^{\mathbb{G}}_{\rightarrow}
\end{array}
$$

In the transition from the topmost sequent '$\Box\Diamond p, \Diamond p, p \vdash \Diamond p, p$' to sequent '$\Box\Diamond p, \Diamond p \vdash \Diamond p$', in the sequent-premise we have:

$(\Box S_1)\flat = \Box\Diamond p, \Diamond p$, $A = p$, $(\Diamond T)\flat = \Diamond p, p$,

and in the sequent-conclusion we have:

$\Box S_1 = \Box\Diamond p$, $\Diamond A = \Diamond p$, and $\Diamond T = \Diamond p$.

In the transition from '$\Box\Diamond p \vdash \Diamond p, p$' to '$\Diamond\Box\Diamond p \vdash \Diamond p$' (the middle application of $L^{\mathbb{G}}_{tr}$), we have: $A = \Box\Diamond p$ and $\Diamond A = \Diamond\Box\Diamond p$, and the succedent as in the topmost sequent and its successor.

For the last time in the modal realm, we prove the S4-counterpart of Lemma 7.1:

Lemma 7.6 *Let* **s** *stand for a regular Socratic proof of the sequent '$\vdash (A)^0$' in calculus $\mathbb{E}^{\mathsf{S4}\tau} \cup \{L^{\mathbb{E}}_{tr}, R^{\mathbb{E}}_{tr}\}$, and let $\mathcal{T} = \langle X, R, \tau_{\mathsf{S4s}} \mid_X \rangle$ be the tree obtained from* **Tr(s)** *by procedure PRUNE.*

1. *If a sequent ψ labels a node of $\mathcal{T}$ that immediately precedes exactly one node labeled by ϕ, and the two sequents have the same principal numeral, then the following:*
$$\frac{\tau_{\mathsf{S4s}}(\phi)}{\tau_{\mathsf{S4s}}(\psi)}$$
is either an instance of a classical non-branching rule of $\mathbb{G}^{\mathsf{T}\tau}$ or an instance of one of $L^{\mathbb{G}}_{ref4}$, $R^{\mathbb{G}}_{ref4}$.
2. *as in Lemma 7.1.*
3. *If a sequent ψ labels a node of $\mathcal{T}$ that immediately precedes exactly one node labeled by ϕ, and the two sequents have different principal numerals, then the following:*
$$\frac{\tau_{\mathsf{S4s}}(\phi)}{\tau_{\mathsf{S4s}}(\psi)}$$
is an instance of one of the rules: $L^{\mathbb{G}}_{tr}$, $R^{\mathbb{G}}_{tr}$, $L^{\mathbb{G}}_{tr+W}$, or $R^{\mathbb{G}}_{tr+W}$.

Proof As in the previous cases. □

Theorem 7.6 *Let* **s** *be a regular Socratic proof of the sequent '$\vdash (A)^0$' in calculus $\mathbb{E}^{\mathsf{S4}\tau} \cup \{L^{\mathbb{E}}_{tr}, R^{\mathbb{E}}_{tr}\}$. If $\langle X, R\rangle$ is a tree obtained from* **s** *by the procedure PRUNE, then $\langle X, R, \tau_{\mathsf{S4s}} \mid_X \rangle$ is a proof of '$\vdash A$' in calculus $\mathbb{G}^{\mathsf{S4}\tau} \cup \{Ax_{+W}, L^{\mathbb{G}}_{tr+W}, R^{\mathbb{G}}_{tr+W}\}$.*

7.3 Translation Into Sequent System for Intuitionistic Logic

In this section, we will use the erotetic calculus $\mathbb{E}^{\mathsf{INT}}$, introduced in Chapter 4, for the translation. We will start with a presentation of a sequent system for propositional INT.

Language and Rules of $\mathbb{G}^{\mathsf{INT}}$

First, we present a language of sequents for the implicit system. The language is built upon $\mathcal{L}_{\mathsf{INT}}$ and contains, in addition, '$\vdash$', and the comma ','. The language will be called $\mathcal{L}^{imp}_{\vdash}$. Sequents of $\mathcal{L}^{imp}_{\vdash}$ are expressions of the form

Table 7.8 Rules of $\mathbb{G}^{\mathsf{INT}}$

$$\frac{S(\sim A) \vdash T(A)}{S(\sim A) \vdash T(\_)}\ L^{\mathbb{G}}_{\sim} \qquad \frac{(S)^{\sharp\mathsf{INT}}\,'\, A \vdash T}{S \vdash \sim A}\ R^{\mathbb{G}}_{\sim}$$

$$\frac{S(A \supset B) \vdash T(A) \quad S(A \supset B, B) \vdash T(\_)}{S(A \supset B) \vdash T(\_)}\ L^{\mathbb{G}}_{\supset} \quad \frac{(S)^{\sharp\mathsf{INT}}\,'\, A \vdash T(B)}{S \vdash A \supset B}\ R^{\mathbb{G}}_{\supset}$$

$$\frac{S(A, B) \vdash T}{S(A \wedge B) \vdash T}\ L^{\mathbb{G}i}_{\wedge} \qquad \frac{S \vdash T(A) \quad S \vdash T(B)}{S \vdash T(A \wedge B)}\ R^{\mathbb{G}i}_{\wedge}$$

$$\frac{S(A) \vdash T \quad S(B) \vdash T}{S(A \vee B) \vdash T}\ L^{\mathbb{G}i}_{\vee} \quad \frac{S \vdash T(A, B)}{S \vdash T(A \vee B)}\ R^{\mathbb{G}i}_{\vee}$$

$$A_1, \ldots, A_n \vdash B_1, \ldots, B_k$$

where $\langle A_1, \ldots, A_n \rangle$, $\langle B_1, \ldots, B_k \rangle$ are finite sequences of formulas of $\mathcal{L}_{\mathsf{INT}}$. As in the modal case, it may be the case that $n = 0$ or $k = 0$, but not both. S, T will stand for finite, possibly empty, sequences of formulas of $\mathcal{L}_{\mathsf{INT}}$.

Table 7.8 presents the rules of sequent calculus $\mathbb{G}^{\mathsf{INT}}$. Axioms of $\mathbb{G}^{\mathsf{INT}}$ are sequents of the form

$$S(A) \vdash T(A). \qquad Ax + W$$

The operation $(\bullet)^{\sharp\mathsf{INT}}$ acts on sequences (in Chap. 4 we have defined in for sets) by 'deleting' every element of the sequence that is not a propositional variable or an 'intuitionistic necessity'—this is how we called formulas of $\mathcal{L}_{\mathsf{INT}}$ whose main connective is negation or implication. Below we define the operation formally. We use '$\varnothing$' for the empty sequence, and, for simplicity, write just '$=$' for 'is of the form'.

$$(\varnothing)^{\sharp\mathsf{INT}} = \varnothing$$

$$(\langle A_1, A_2, \ldots, A_k \rangle)^{\sharp\mathsf{INT}} =$$
$$= \begin{cases} \langle A_1 \rangle\,'\,(\langle A_2, \ldots, A_k \rangle)^{\sharp\mathsf{INT}} & \text{if } A_1 = \sim B, \text{ or } A_1 = B \supset C, \text{ or } A_1 \in VAR, \\ (\langle A_2, \ldots, A_k \rangle)^{\sharp\mathsf{INT}} & \text{otherwise.} \end{cases}$$

In summary, calculus $\mathbb{G}^{\mathsf{INT}}$ is the set of rules:

$$\{L^{\mathbb{G}}_{\sim}, R^{\mathbb{G}}_{\sim}, L^{\mathbb{G}i}_{\wedge}, R^{\mathbb{G}i}_{\vee}, R^{\mathbb{G}}_{\supset}, R^{\mathbb{G}i}_{\wedge}, L^{\mathbb{G}i}_{\vee}, L^{\mathbb{G}}_{\supset}, Ax + W\}.$$

The notion of proof in calculus $\mathbb{G}^{\mathsf{INT}}$ is defined in a standard manner:

Definition 7.15 Let A be a formula of $\mathcal{L}_{\mathsf{INT}}$. A *proof of the sequent* $\vdash A$ *in calculus* $\mathbb{G}^{\mathsf{INT}}$ is a finite tree labeled with sequents of the language $\mathcal{L}^{imp}_{\vdash}$, regulated by the rules of calculus $\mathbb{G}^{\mathsf{INT}}$, with the root labeled with the sequent $\vdash A$, and the leaves labeled with the axioms of the calculus.

Let us recall a Socratic proof of $\vdash (p \supset {\sim\sim}\, p)^0$ in $\mathbb{E}^{\mathsf{INT}}$ (to the left), and compare it with a proof of $\vdash p \supset {\sim\sim}\, p$ in $\mathbb{G}^{\mathsf{INT}}$ (to the right). A strict definition of correspondence between the two structures is our last task in this chapter.

$$\cfrac{\cfrac{\cfrac{\cfrac{?(\vdash (p \supset {\sim\sim}\, p)^0)}{?(p^{0,1} \vdash ({\sim\sim}\, p)^{0,1})}\, R_{\supset}}{?(p^{0,1}, ({\sim}\, p)^{0,1,2} \vdash)}\, R_{\sim}}{?(p^{0,1}, ({\sim}\, p)^{0,1,2} \vdash p^2)}\, L_{\sim}}{?(p^{0,1}, p^2, ({\sim}\, p)^{0,1,2} \vdash p^2)}\, R_H \qquad \cfrac{\cfrac{\cfrac{p, {\sim}\, p \vdash p}{p, {\sim}\, p \vdash}\, L^{\mathbb{G}}_{\sim}}{p \vdash {\sim\sim}\, p}\, R^{\mathbb{G}}_{\sim}}{\vdash p \supset {\sim\sim}\, p}\, R^{\mathbb{G}}_{\supset}$$

Regular Socratic transformations in calculus $\mathbb{E}^{\mathsf{INT}}$ are defined according to the general scheme of Algorithm 7.3, with Procedure CLASSICAL supplemented with Procedure SETTLE-INT, and Procedure MODAL supplemented with Procedure INT. The new procedures, in turn, will be variants of procedures used for the modal logic $\mathsf{S4}$. No surprises.

In order to define Procedure INT we need to introduce additional rules that will introduce an order in the way we deal with negations and implications.

Algorithm 7.10: Operations $\sharp^I_j$, $\flat^I_j$ for INT

Data: pair $\langle \Gamma, \Delta \rangle$ of finite sequences of indexed formulas, where $\Gamma = \langle (A_1)^{\varsigma_1}, \ldots, (A_m)^{\varsigma_m} \rangle$
set $R(\phi)$ of pairs of numerals
Result: pair of finite sequences of indexed formulas $\langle \Gamma\sharp^I_j, \Delta\flat^I_j \rangle$
$\Gamma\sharp_j \leftarrow \varnothing$
$\Delta\flat^I_j \leftarrow \Delta$
for $k = 1$ **to** m **do**
  if *k-th term of* Γ *is of the form* $(p_l)^{\varsigma/i}$ *where* $\langle i, j \rangle \in R(\phi)^{tr}$ **then**
    $\Gamma\sharp^I_j \leftarrow \Gamma\sharp^I_j\, '\, \langle (p_l)^{\varsigma/i}, (p_l)^j \rangle$
  else
    if *k-th term of* Γ *is of the form* $({\sim}\, C)^{\varsigma/i}$ *where* $\langle i, j \rangle \in R(\phi)^{tr}$ **then**
      $\Gamma\sharp^I_j \leftarrow \Gamma\sharp^I_j\, '\, \langle ({\sim}\, C)^{\varsigma/i} \rangle$
      $\Delta\flat^I_j \leftarrow \Delta\flat^I_j\, '\, \langle C^j \rangle$
    else
      $\Gamma\sharp^I_j \leftarrow \Gamma\sharp^I_j\, '\, \langle (A_k)^{\varsigma_k} \rangle$
    end
  end
end
return $\langle \Gamma\sharp^I_j, \Delta\flat^I_j \rangle$

$$\frac{?(\,\Phi\,(\,\Gamma_1\, '\, \Gamma_2 \vdash \Delta_1\, '\, ({\sim}\, A)^{\varsigma/i}\, '\, \Delta_2\,)\,)}{?(\,\Phi\,(\,(\Gamma_1\, '\, (A)^{\varsigma/i,j}\, '\, \Gamma_2)\sharp^I_j \vdash (\Delta_1\, '\, \Delta_2)\flat^I_j\,)\,)}\; R^{\sim}_H$$

$$\frac{?(\ \Phi\ (\ \Gamma_1\ '\ \Gamma_2 \vdash \Delta_1\ '\ (A \supset B)^{\varsigma/i}\ '\ \Delta_2\ )\ )}{?(\ \Phi\ (\ (\Gamma_1\ '\ (A)^{\varsigma/i,j}\ '\ \Gamma_2)\sharp^I_j \vdash (\Delta_1\ '(B)^{\varsigma/i,j}\ '\ \Delta_2)\flat^I_j\ )\ )}\ R^{\supset}_H$$

The first step is relatively easy. Rules $R^{\sim}_H$ and $R^{\sim}_H$ gather an application of $R_{\sim}$ (respectively, $R_{\supset}$) followed by a sequence of applications of R_H and $L_{\sim}$ introducing 'settlers' to the new world created by the application of $R_{\sim}$ (respectively, $R_{\supset}$). It is easy to see that:

Corollary 7.12 *Rules $R^{\sim}_H$ and $L^{\supset}_H$ are derivable in $\mathbb{E}^{\mathsf{INT}}$.*

The work is not done, though, since we need to apply $L_{\supset}$ in a similar manner before we allow for a creation of another accessible world. Since $L_{\supset}$ is a branching rule, we shall deal with it with a trick. The information about rule $L_{\supset}$ applied to a sequent with respect to '$(B \supset C)^{\varrho/i}$' will be carried in the form of an expression 'F B^j OR C^j', and only after finishing the search of all settlers, the expression will be used to trigger branching.

Procedure I-CONT(sequent, numeral, ϕ)

Data: sequent $(A_1)^{\varsigma_1}, \ldots, (A_k)^{\varsigma_k} \vdash \Delta$, numeral i

$\Gamma \leftarrow \varnothing$
for $i = 1$ **to** k **do**
  if $(A_i)^{\varsigma_i}$ *is of the form* $(B \supset C)^{\varrho/i}$ **then**
    $\Gamma \leftarrow \Gamma\ '\ \langle (B \supset C)^{\varrho/i}, \mathrm{F} B^j\ \mathrm{OR}\ C^j \rangle$
  else
    $\Gamma \leftarrow \Gamma\ '\ \langle (A_i)^{\varsigma_i} \rangle$
  end
end
$\phi \leftarrow \Gamma \vdash \Delta$

Once the additional expressions are introduced, we need to remove them from a question. This will be done according to the following transformation.

$$\frac{?(\ \Phi\ (\ \Gamma_1\ '\ \mathrm{F}\ B^j\ \mathrm{OR}\ C^j\ '\ \Gamma_2 \vdash \Delta\ )\ )}{?(\ \Phi\ (\ \Gamma_1\ '\ \Gamma_2 \vdash \Delta\ '\ B^j\ ;\ \Gamma_1\ '\ C^j\ '\ \Gamma_2 \vdash \Delta \rangle\ )\ )}\ R_{clean}$$

Let us observe that R_{clean} is not a rule of an erotetic calculus. It is a transformation defined for the use of Procedure SETTLE-INT.

By $L^{ref}_{\sim}$ and $L^{ref}_{\supset}$ we shall mean subsets of rules $L_{\sim}$ and $L_{\supset}$ such that the numerals i, j displayed in the schemes of these rules are equal. Let us recall the rules for ease of reference.

$$\frac{?(\ \Phi\ (\ \Gamma\ (\ (\sim A)^{\varsigma/i}\ ) \vdash \Delta\ (\_)\ )\ )}{?(\ \Phi\ (\ \Gamma\ (\ (\sim A)^{\varsigma/i}\ ) \vdash \Delta\ (\ A^i\ )\ )\ )}\ L^{ref}_{\sim}$$

Procedure SETTLE-INT(question,transformation,Q, s)

Data: question $= ?(\Phi(\phi))$, `transformation`, $\Gamma = \langle (A_1)^{\varsigma_1}, \ldots, (A_m)^{\varsigma_m} \rangle$ is the sequence of all indexed formulas occurring in the antecedent of ϕ

s $\leftarrow$ `transformation`
$Q \leftarrow ?(\Phi(\phi))$
$j \leftarrow 1$
while $j \leqslant m$ **do**
    if $(A_j)^{\varsigma/i}$ *is a formula-premise of* $\mathbf{r} \in \{R_H^{\sim}, R_H^{\supset}\}$ **then**
        $\psi \leftarrow$ sequent-conclusion of **r**
        I-CONT(ψ, i, ψ)
        $s \leftarrow s\,;\ \langle ?(\Phi(\psi)) \rangle$
        $j \leftarrow m+1$; `/* ends the execution of the while-loop */`
    else
        $j \leftarrow j+1$
    end
end

$$\frac{?(\,\Phi\,(\,\Gamma\,(\,(A \supset B)^{\varsigma/i}\,)(\_) \vdash \Delta\,(\_)\,)\,)}{?(\,\Phi\,(\,\Gamma\,(\,(A \supset B)^{\varsigma/i}\,)(\_) \vdash \Delta\,(\,A^i\,)\,;\ \Gamma\,(\,(A \supset B)^{\varsigma/i}\,)(\,B^i\,) \vdash \Delta\,(\_)\,)\,)}\ L_{\supset}^{ref}$$

As we can see, $L_{\sim}^{ref} \subset L_{\sim}$ and $L_{\supset}^{ref} \subset L_{\supset}$.

In the description of Procedure SETTLE-T we have used the bottom labels; their role in transitive and reflexive logics is to avoid redundant repetitions of applications of rules. Procedure SETTLE-INT is defined, in general, as Procedure SETTLE-T, but with a number of exceptions:

- line 13 takes the form
  if $(A_j)^{\varsigma/j}$ is a formula-premise of a classical branching rule **r**, or a subject to transformation by $\mathbf{r} = R_{clean}$ **then**
- line 21 takes the form
  if *j-th term* $(A)^{\varsigma/i}$ *of* Γ *is of the form* $(\sim B)^{\varsigma/i}$ *and it is unlabeled* **then**
- line 22 takes the form (here we add the bottom label)
  $A \leftarrow A_{ref}$;
- line 23 takes the form
  $\Gamma \leftarrow \langle (A_1)^{\varsigma_1}, \ldots, (A_{j-1})^{\varsigma_{j-1}}, (A)^{\varsigma/i}, (A_{j+1})^{\varsigma_{j+1}}, \ldots, (A_m)^{\varsigma_m}, (B)^i \rangle$
- line 24 takes the form
  $\psi \leftarrow$ sequent-conclusion of $L_{\sim}^{ref}$
- line 29 is replaced with another **if-then-else** condition dealing with the applications of rule $L_{\supset}^{ref}$

if *j-th term* $(A)^{\varsigma/i}$ *of* Γ *is of the form* $(B \supset C)^{\varsigma/i}$ *and it is unlabeled* **then**
  $A \leftarrow A_{ref}$
  $\Gamma \leftarrow \langle (A_1)^{\varsigma_1}, \ldots, (A_{j-1})^{\varsigma_{j-1}}, (A)^{\varsigma/i}, (A_{j+1})^{\varsigma_{j+1}}, \ldots, (A_m)^{\varsigma_m}, (B)^i \rangle$
  $\psi_1, \psi_2 \leftarrow$ the left and right sequent-conclusion of $L^{ref}_{\supset}$
  $Q \leftarrow ?(\Phi(\psi_1, \psi_2))$
  $\mathbf{s} \leftarrow \mathbf{s}; \langle Q \rangle$
  $j \leftarrow j + 1$
else
  $j \leftarrow j + 1$
end

Regular Socratic transformations obtained by the Algorithm 7.3, modified as described above, are further subject to the manipulations applied successfully in the case of modal logics: a regular Socratic transformation is turned into a Socratic tree, and this is pruned with Procedure PRUNE. At the next step we assume that we deal with trees obtained from Socratic *proofs*.

Definition 7.16 Let $i \in \mathbb{N}$ and let ϕ be a sequent of $\mathcal{L}^?_{\vdash\mathsf{INT}}$. We define function τ_{INT} as follows:

$$\tau_{\mathsf{INT}}(i, \varnothing) = \varnothing$$

$$\tau_{\mathsf{INT}}(i, \langle A_1^{\varsigma_1}, \ldots, A_k^{\varsigma_k} \rangle) = $$

$$= \begin{cases} \langle A_1 \rangle \, ' \, \tau_{\mathsf{INT}}(i, \langle A_2^{\varsigma_2}, \ldots, A_k^{\varsigma_k} \rangle) & \text{if } \varsigma_1 \text{ ends with } i, \textbf{ or } \varsigma_1 \text{ ends with } k \\ & \text{such that } \langle k, i \rangle \in R^{tr}(\phi), \text{ and} \\ & A_1 = \sim B \text{ or } A_1 = B \supset C \\ \tau_{\mathsf{INT}L}(i, \langle A_2^{\varsigma_2}, \ldots, A_k^{\varsigma_k} \rangle) & \text{otherwise} \end{cases}$$

The final relabelling function, $\tau_{\mathsf{INT}\mathbf{s}}$, is defined analogously to function $\tau_{\mathbf{s}}$ (Definition 7.13) but with τ_{INT} playing the role of τ.

Definition 7.17 Let **s** be a regular Socratic transformation of a question of the form '$?(\vdash (A)^0)$' in $\mathbb{E}^{\mathsf{INT}} \cup \{R^{\sim}_{H}, R^{\supset}_{H}\}$, and let **Tr**(**s**) be the Socratic tree determined by **s**. Then $\tau_{\mathsf{INT}\mathbf{s}}$ is a relabelling function defined recursively as follows:

1. If ϕ labels the root of **Tr**(**s**), then $\tau_{\mathsf{INT}\mathbf{s}}(\phi) = \tau_{\mathsf{INT}}(0, \phi)$.
2. If ϕ labels a node, ψ labels the immediate $R_{\mathbf{s}}$-predecessor of the node, ϕ results from ψ in **s** by a rule for $\wedge$ or $\vee$, and i is the numeral such that $\tau_{\mathsf{INT}\mathbf{s}}(\psi) = \tau_{\mathsf{INT}}(i, \psi)$, then $\tau_{\mathsf{INT}\mathbf{s}}(\phi) = \tau_{\mathsf{INT}}(i, \phi)$.
3. If ϕ is a label of a node that results in **s** from its immediate $R_{\mathbf{s}}$-predecessor by a rule for $\sim$ or $\supset$, then $\tau_{\mathsf{INT}\mathbf{s}}(\phi) = \tau_{\mathsf{INT}}(j, \phi)$, where j is the principal numeral in ϕ.

We finish this section with the formulation of counterparts of Lemma 7.1 and Theorem 7.1.

Lemma 7.7 *Let* **s** *stand for a regular Socratic proof of the sequent '$\vdash (A)^0$' in calculus* $\mathbb{E}^{\mathsf{INT}} \cup \{R_H^{\sim}, R_H^{\supset}\}$, *and let* $\mathcal{T} = \langle X, R, \tau_{\mathsf{INTs}}|_X\rangle$ *be the tree obtained from* **Tr**(**s**) *by procedure PRUNE.*

1. *If a sequent* ψ *labels a node of* $\mathcal{T}$ *that immediately precedes exactly one node labeled by* ϕ, *and these two sequents have the same principal numeral, then the following:*
$$\frac{\tau_{\mathsf{INTs}}(\phi)}{\tau_{\mathsf{INTs}}(\psi)}$$
is either an instance of a classical non-branching rule of $\mathbb{G}^{\mathsf{INT}}$ *or an instance of one of* $L_{\sim}^{ref}$, $L_{\supset}^{ref}$.
2. *as in Lemma 7.1.*
3. *If a sequent* ψ *labels a node of* $\mathcal{T}$ *that immediately precedes exactly one node labeled by* ϕ, *and these two sequents have different principal numerals, then the following:*
$$\frac{\tau_{\mathsf{INTs}}(\phi)}{\tau_{\mathsf{INTs}}(\psi)}$$
is an instance of one of the rules: $L_H^{\sim}$, $L_H^{\supset}$.

Proof As in the previous cases. □

Theorem 7.7 *Let* **s** *be a regular Socratic proof of the sequent '$\vdash (A)^0$' in calculus* $\mathbb{E}^{\mathsf{INT}\tau} \cup \{L_H^{\sim}, L_H^{\supset}\}$. *If* $\langle X, R\rangle$ *is a tree obtained from* **s** *by the Procedure PRUNE, then* $\langle X, R, \tau_{\mathsf{INTs}}|_X\rangle$ *is a proof of '$\vdash A$' in calculus* $\mathbb{G}^{\mathsf{INT}\tau}$.

References

1. Avron, A. (1993). Gentzen-type systems, resolution and tableaux. *Journal of Automated Reasoning, 10*, 265–281.
2. Cormen, T. H., Leiserson, C. E., Rivest, R. L., & Stein, C. (2009). *Introduction to algorithms* (3rd ed.). The MIT Press.
3. Fitting, M. (1983). *Proof methods for modal and intuitionistic logics*. Berlin: Springer.
4. Goré, R. (1999). Tableau methods for modal and temporal logics. In M. D'Agostino, D. M. Gabbay, R. Hähnle, & J. Posegga (Eds.), *Handbook of tableau methods* (pp. 297–396). Netherlands: Springer.
5. Hacking, I. (1994). What is logic? [First published in the Journal of Philosophy, Vol. 76, No. 6, pp. 285–319 (1979).]. In D. M. Gabbay (Ed.), *What is a logical system?* (pp. 1–33). Clarendon Press.
6. Hintikka, J. (1955). Form and content in quantification theory. *Acta Philosophica Fennica, 8*, 3–55.
7. Kripke, S. A. (1963). Semantical analysis of modal logic I. Normal modal propositional calculi. *Zeitschrift f ü r Mathematische Logik und Grundlagen der Mathematik, 9*, 67–96.
8. Leszczyńska, D. (2007). *The method of Socratic proofs for normal modal propositional logics*. Adam Mickiewicz University Press.

9. Poggiolesi, F. (2011). *Gentzen calculi for modal propositional logic*. Berlin: Springer.
10. Rautenberg, W. (1983). Modal tableau calculi and interpolation. *Journal of Philosophical Logic, 12*(4), 403–423.
11. Wansing, H. (2002). Sequent systems for modal logics. In D. M. Gabbay & F. Guenthner (Eds.), *Handbook of philosophical logic* (2nd ed., Vol. 8, pp. 61–145). Kluwer Academic Publishers.

Appendix A: Trees

The notion of tree is commonly used in proof theory to capture the structure lying behind occurrences of formulas which constitute the body of the proof. In most textbooks, however, it is left undefined, as it is assumed that an illustration in the form of diagrams will be sufficient (and usually it is). If defined, a tree is often treated as a graph. The reader can find a full exposition of this account in [4, Chap. 11], or in [1, pp. 1168–1182]. Some other interesting accounts are possible, like in [2, p. 5], where a tree is defined as a set of binary sequences. But in this book we find it useful to use the notion in its set-theoretical account, as in [6, pp. 7–8], [5, pp. 3–4], also [3] tacitly assumes this account.

A binary relation R defined on X, that is such that $R \subseteq X \times X$, is called a *partial order* iff R is reflexive, transitive and antisymmetric in X, that is,

$$\forall x, y \in X(xRy \wedge yRx \rightarrow x = y).$$

A *linear order* is a partial order which makes every two elements of X comparable, that is, such that xRy or yRx for each $x, y \in X$. (See also Table 3.2 on Sect. 3.1 for the definitions of the other properties.) The symbol '$\succeq$' is commonly used for partial orders.

A *tree* is a partially ordered set $\langle X, \succeq\rangle$ with the least element, called its *root*, and a set $\{y \in X : y \succeq x\}$ linearly ordered for each $x \in X$. We will use '$\succ$' for the *strict order corresponding to* $\succeq$. Formally, $x \succ y$ iff $x \succeq y$ and $x \neq y$. The elements of X are called the *nodes* of $\langle X, \succeq\rangle$. A *branch of* $\langle X, \succeq\rangle$ is a linearly ordered subset of X which is maximal with respect to set inclusion. (That is, a branch cannot be extended further.) The maximal elements of a tree, if there are any, are called its *leaves*. Trees are usually displayed 'upside down', like on Fig..1, with the root (labeled with '$\bullet$') at the top and the leaves (labeled with '$\simeq$') at the bottom.

Relations between nodes of a tree are usually described with the use of a 'genealogical' terminology. If $x \succ y$, then x is called *ancestor of* y and y is called *descendant of* x. If $x \succ y$, and there is no 'intermediate relative' z such that $x \succ z \succ y$, then x

D. Leszczyńska-Jasion, *The Method of Socratic Proofs*, Trends in Logic 64,
https://doi.org/10.1007/978-3-031-82451-7

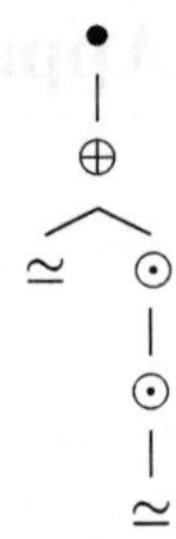

Fig. .1 We plant trees upside down

is called *parent of* y and y is a *child of* x. In the example presented on Fig..1 the root • is the parent of the node with $\oplus$, which is the only child of the root. Each node is the root's descendant except itself, and so on. The more neutral terminology is, respectively, *predecessor* and *successor* and *immediate predecessor* and *immediate successor*. Nodes that share the same parent node are called *siblings*.

We will deal with binary trees, that is, such that every node has at most two children. It will occur very useful to be able to differentiate between the two children (if there are two). It is understandable if we think of rules which have two premises, and we are used to calling them 'the left' and 'the right' premise, respectively.[1] This additional ordering is not essential from the point of view of such crucial properties of the rules like semantic correctness (soundness), but it is handy when it comes to the formulation of algorithms. Formally, it means that a tree $\langle X, \succeq \rangle$ is given together with a function f which may be defined as follows. First, let:

$$Families(X, \succeq) = \{\langle x, y \rangle : x \succ y \wedge \neg \exists z \in X (x \succ z \succ y)\}$$

that is, $Families(X, \succeq)$ is the set of all pairs of the form $\langle parent, child \rangle$, where *parent* is the *child*'s parent in $\langle X, \succeq \rangle$. Then, second, we may set f as a function:

$$f : Families(X, \succeq) \longrightarrow \{0, 1, 2\}$$

such that if *child* is the only child of the node *parent*, then $f(parent, child) = 0$; in this situation *parent* is called *non-branching point*. If there are two children of *parent*: *childA*, *childB*, then f assigns numerals from $\{1, 2\}$ to the respective pairs in an injective way; in this situation *parent* is called *branching point*. *E.g.* in the above example the node labeled with $\oplus$ is a branching point (the only one in the tree). If $f(parent, childA) = 1$ and $f(parent, childB) = 2$, then we call *childA* the *left child*, and *childB* the *right child*.[2] This left-right terminology is extended to whole branches in a natural way. A branch going through the left child is called *left branch*, and the other is called *right branch*. In the sequel we will use the left-right terminology without indicating explicitly function assigning 'the sides'.

---

[1] In the erotetic calculi presented in this book the rules may have two sequent-conclusions, not two sequent-premises. However, this is not a standard proof-theoretic approach.

[2] In [2] the reader will find similar, yet different account of the left-right distinction.

We use the standard definition of *level of a node in a tree*. If $\langle X, R\rangle$ is a tree, then the root of the tree is at level 0. Further, if a node $x \in X$ is at level n, then each of its immediate successors (if there are any) is at level $n + 1$. By *length of a finite branch* of a tree we mean the level of the leaf of the branch. By *height of a finite tree* we mean the length of the maximal branch (*i.e.* the one with maximal length).

Another additional device that often goes with trees is a labelling function. In the sequel, this will be indicated explicitly. The need to use such a function is illustrated well by the example on Fig..1, where we can see clearly that the symbols '$\simeq$', '$\odot$' do not represent the very nodes of the tree, since these cannot occur twice in two different positions (recall that a tree is just a set). Therefore we assume that the nodes 'beneath' are some different objects, they are just labeled with the same label. Proofs such as tableaux are labeled trees in this sense. In most cases, the nature of the nodes which lie beneath is absolutely irrelevant, only the labels matter.[3] However, it becomes relevant when it comes to formulation of an algorithm.

References

1. Cormen, T. H., Leiserson, C. E., Rivest, R. L., & Stein, C. (2009). *Introduction to algorithms* (3rd ed.). The MIT Press.
2. Kaye, R. (2008). *The mathematics of logic*. Cambridge University Press.
3. Priest, G. (2008). *An introduction to non-classical logic: From if to is*. Cambridge University Press.
4. Rosen, K. H. (2012). *Discrete mathematics and its applications* (7th ed.). The McGraw-Hill Companies.
5. Smullyan, R. M. (1968). *First-order logic*. Springer-Verlag.
6. Troelstra, A. S., & Schwichtenberg, H. (2000). *Basic proof theory* (2nd ed.). Cambridge University Press.

[3] If we wish to be more specific, then we should say that two proof trees are regarded as essentially the same, if the trees without labels are isomorphic, and the nodes corresponding under the isomorphism are assigned the same labels. See also [6, p. 32].

Appendix B: Translations from $\mathbb{E}^{\mathsf{L}\tau}$ to $\mathbb{G}^{\mathsf{L}\tau}$: A Summary

Numeration of columns:

1. Erotetic calculus (name)
2. Primary rules of the erotetic calculus
3. Rules derived in the erotetic calculus and used in the translation process
4. Procedures used in Algorithm 7.3 (Sect. 7.1) for generating regular Socratic transformations in the erotetic calculus
5. Translation function and function labeling ('relabeling') the final tree (obtained by Procedure PRUNE)

D. Leszczyńska-Jasion, *The Method of Socratic Proofs*, Trends in Logic 64,
https://doi.org/10.1007/978-3-031-82451-7

In the Table below, 'classical' stands for: $L^{\mathbb{E}}_{\neg}$, $R^{\mathbb{E}}_{\neg}$, $L^{\mathbb{E}}_{\wedge}$, $R^{\mathbb{E}}_{\vee}$, $R^{\mathbb{E}}_{\rightarrow}$, $R^{\mathbb{E}}_{\wedge}$, $L^{\mathbb{E}}_{\vee}$, $L^{\mathbb{E}}_{\rightarrow}$ (see Tables 7.1 and 7.2 in Sect. 7.1).

1	2	3	4	5
$\mathbb{E}^{\mathsf{K}\tau}$	Classical, $L^{\mathbb{E}}_{\Diamond}$, $R^{\mathbb{E}}_{\Box}$ and $L^{\mathbb{E}}_{\Box}$, $R^{\mathbb{E}}_{\Diamond}$ with proviso for K (Tables 7.1, 7.2 and 7.3)	$L^{\mathbb{E}}_{\Diamond\Box}$, $R^{\mathbb{E}}_{\Box\Diamond}$ (Table 7.6)	CLASSICAL (Sect. 7.1), MODAL (Sect. 7.1)	τ_s (Definition 7.6, in Sect. 7.1)
$\mathbb{E}^{\mathsf{D}\tau}$	All above and $L^{\mathbb{E}}_{\mathsf{D}}$ (Sect. 7.2), $R^{\mathbb{E}}_{\mathsf{D}}$ (Sect. 7.2)	All above and $R^{\mathbb{E}}_{\mathsf{D}g}$ (Sect. 7.2)	CLASSICAL (Sect. 7.1), MODAL-D (Sect. 7.2)	τ_s (Definition 7.6, Sect. 7.1)
$\mathbb{E}^{\mathsf{T}\tau}$	As in $\mathbb{E}^{\mathsf{K}\tau}$, (also $L^{\mathbb{E}}_{\Box}$, $R^{\mathbb{E}}_{\Diamond}$ with proviso for K), $L^{\mathbb{E}}_{ref}$, $R^{\mathbb{E}}_{ref}$ (Sect. 7.2)	$L^{\mathbb{E}}_{\Diamond\Box}$, $R^{\mathbb{E}}_{\Box\Diamond}$ (Table 7.6)	SETTLE-T (Sect. 7.2), MODAL (Sect. 7.1)	$\tau_{\mathsf{T}s}$ (Definition 7.10, Sect. 7.2) defined by τ_{T} (Definition 7.9, Sect. 7.2)
$\mathbb{E}^{\mathsf{K4}\tau}$	Classical, $L^{\mathbb{E}}_{\Diamond}$, $R^{\mathbb{E}}_{\Box}$ and $L^{\mathbb{E}}_{\Box}$, $R^{\mathbb{E}}_{\Diamond}$ with proviso for K4 (Sect. 7.1)	$L^{\mathbb{E}}_{tr}$, $R^{\mathbb{E}}_{tr}$ (Sect. 7.2)	CLASSICAL (Sect. 7.1), MODAL-K4 (Sect. 7.2)	τ_{4s} (Definition 7.13, Sect. 7.2) defined by τ_4 (Definition 7.11, 7.12, Sect. 7.2)
$\mathbb{E}^{\mathsf{D4}\tau}$	Classical, $L^{\mathbb{E}}_{\Diamond}$, $R^{\mathbb{E}}_{\Box}$, $L^{\mathbb{E}}_{\mathsf{D}}$, $R^{\mathbb{E}}_{\mathsf{D}}$, and $L^{\mathbb{E}}_{\Box}$, $R^{\mathbb{E}}_{\Diamond}$ with proviso for K4 (Sects. 7.1 and 7.2)	$L^{\mathbb{E}}_{tr}$, $R^{\mathbb{E}}_{tr}$ (Sect. 7.2), $L^{\mathbb{E}}_{\mathsf{D}g}$, $R^{\mathbb{E}}_{\mathsf{D}g2}$ (Sect. 7.2)	CLASSICAL (Sect. 7.1), MODAL-D4 (Sect. 7.2)	as in $\mathbb{E}^{\mathsf{K4}\tau}$ above
$\mathbb{E}^{\mathsf{S4}\tau}$	Classical, $L^{\mathbb{E}}_{\Diamond}$, $R^{\mathbb{E}}_{\Box}$ and $L^{\mathbb{E}}_{\Box}$, $R^{\mathbb{E}}_{\Diamond}$ with proviso for K4 (Sect. 7.1), $L^{\mathbb{E}}_{ref}$, $R^{\mathbb{E}}_{ref}$ (Sect. 7.2)	$L^{\mathbb{E}}_{tr}$, $R^{\mathbb{E}}_{tr}$ (Sect. 7.2)	SETTLE-T (Sect. 7.2), MODAL-K4 (Sect. 7.2)	$\tau_{\mathsf{S4}s}$ (Definition 7.14) defined with the use of τ_4 (Definition 7.11, 7.12)
$\mathbb{E}^{\mathsf{INT}\tau}$	$L_{\wedge}$, $L_{\vee}$, $R_{\wedge}$, $R_{\vee}$, R_H, $L_{\sim}$, $R_{\sim}$, $L_{\supset}$, $R_{\supset}$ (Sect. 4.1)	$R^{\sim}_{H}$, $R^{\supset}_{H}$ (Sect. 7.3), R_{clean} (Sect. 7.3)	SETTLE-INT (Sect. 7.3), SETTLE-INT (Sect. 7.3)	$\tau_{\mathsf{INT}s}$ (Definition 7.17) defined with the use of τ_{INT} (Definition 7.16)

Subject Index

D. Leszczyńska-Jasion, *The Method of Socratic Proofs*, Trends in Logic 64,
https://doi.org/10.1007/978-3-031-82451-7

Index of Names

D. Leszczyńska-Jasion, *The Method of Socratic Proofs*, Trends in Logic 64,
https://doi.org/10.1007/978-3-031-82451-7

MIX
Papier aus verantwortungsvollen Quellen
Paper from responsible sources
FSC® C105338

If you have any concerns about our products,
you can contact us on
ProductSafety@springernature.com

In case Publisher is established outside the EU,
the EU authorized representative is:
Springer Nature Customer Service Center GmbH
Europaplatz 3, 69115 Heidelberg, Germany

Printed by Libri Plureos GmbH
in Hamburg, Germany